Graduate Texts in Mathematics   252

T0202573

Graduate Texts in Mathematics

(continued after index)

Gerd Grubb

Distributions and Operators

 Springer

Gerd Grubb
University of Copenhagen
Department of Mathematical Sciences
Universitetsparken 5
2100 Koebenhavn
Denmark
grubb@math.ku.dk

ISBN: 978-1-4419-2743-9   e-ISBN: 978-0-387-84895-2

Mathematics Subject Classification (2000): 47Bxx:47B06

Preface

This textbook gives an introduction to distribution theory with emphasis on applications using functional analysis. In more advanced parts of the book, pseudodifferential methods are introduced.

Distribution theory has been developed primarily to deal with partial (and ordinary) differential equations in general situations. Functional analysis in, say, Hilbert spaces has powerful tools to establish operators with good mapping properties and invertibility properties. A combination of the two allows showing solvability of suitable concrete partial differential equations (PDE).

When partial differential operators are realized as operators in $L_2(\Omega)$ for an open subset Ω of $\mathbb{R}^n$, they come out as unbounded operators. Basic courses in functional analysis are often limited to the study of bounded operators, but we here meet the necessity of treating suitable types of unbounded operators; primarily those that are densely defined and closed. Moreover, the emphasis in functional analysis is often placed on selfadjoint or normal operators, for which beautiful results can be obtained by means of spectral theory, but the cases of interest in PDE include many nonselfadjoint operators, where diagonalization by spectral theory is not very useful. We include in this book a chapter on unbounded operators in Hilbert space (Chapter 12), where classes of convenient operators are set up, in particular the variational operators, including selfadjoint semibounded cases (e.g., the Friedrichs extension of a symmetric operator), but with a much wider scope.

Whereas the functional analysis definition of the operators is relatively clean and simple, the interpretation to PDE is more messy and complicated. It is here that distribution theory comes in useful. Some textbooks on PDE are limited to weak definitions, taking place e.g. in $L_2(\Omega)$. In our experience this is not quite satisfactory; one needs for example the Sobolev spaces with negative exponents to fully understand the structure. Also, Sobolev spaces with noninteger exponents are important, in studies of boundary conditions. Such spaces can be very nicely explained in terms of Fourier transformation of (temperate) distributions, which is also useful for many further aspects of the treatment of PDE.

In addition to the direct application of distribution theory to interpret partial differential operators by functional analysis, we have included some more advanced material, which allows a further interpretation of the solution operators, namely, the theory of pseudodifferential operators, and its extension to pseudodifferential boundary operators.

The basic part of the book is Part I (Chapters 1–3), Chapter 12, and Part II (Chapters 4–6). Here the theory of distributions over open sets is introduced in an unhurried way, their rules of calculus are established by duality, further properties are developed, and some immediate applications are worked out. For a correct deduction of distribution theory, one needs a certain knowledge of Fréchet spaces and their inductive limits. We have tried to keep the technicalities at a minimum by relegating the explanation of such spaces to an appendix (Appendix B), from which one can simply draw on the statements, or go more in depth with the proofs if preferred. The functional analysis needed for the applications is explained in Chapter 12. The Fourier transformation plays an important role in Part II, from Chapter 5 on.

The auxiliary tools from functional analysis, primarily in Hilbert spaces, are collected in Part V. Besides Chapter 12 introducing unbounded operators, there is Chapter 13 on extension theory and Chapter 14 on semigroups.

Part III is written in a more compact style. We here extend the PDE theory by the introduction of x-dependent pseudodifferential operators (ψdo's), over open sets (Chapter 7) as well as over compact C^∞ manifolds (Chapter 8). This is an important application of distribution theory and leads to a very useful "algebra" of operators including partial differential operators and the solution operators for the elliptic ones. Fredholm theory is explained and used to establish the existence of an index of elliptic operators.

Pseudodifferential operators are by many people regarded as a very sophisticated tool, and indeed it is so, perhaps most of all because of the imprecisions in the theory: There are asymptotic series that are not supposed to converge, the calculus works "modulo operators of order $-\infty$", etc. We have tried to sum up the most important points in a straightforward way.

Part IV deals with boundary value problems. Homogeneous boundary conditions for some basic cases were considered in Chapter 4 (with reference to the variational theory in Chapter 12); this was supplied with the general Gårding inequality at the end of Chapter 7. Now we present an instructive example in Chapter 9, where explicit solution operators for nonhomogeneous Dirichlet and Neumann problems are found, and the role of half-order Sobolev spaces over the boundary (also of negative order) is demonstrated. Moreover, we here discuss some other Neumann-type conditions (that are not always elliptic), and interpret the abstract characterization of extensions of operators in Hilbert space presented in Chapter 13, in terms of boundary conditions.

Whereas Chapter 9 is "elementary", in the sense that it can be read directly in succession to Parts I and II, the next two chapters, based on Part III, contain more heavy material. It is our point of view that a modern treatment of boundary value problems with smooth coefficients goes most efficiently by a pseudodifferential technique related to the one on closed manifolds. In Chapter 10 we give an introduction to the calculus of pseudodifferential boundary operators (ψdbo's) initiated by Boutet de Monvel, with full details of the explanation in local coordinates. In Chapter 11 we introduce the Calderón projector for an elliptic differential operator on a manifold with boundary, and show some of its applications.

The contents of the book have been used frequently at the University of Copenhagen for a one-semester first-year graduate course, covering Chapters 1–3, 12 and 4–6 (in that order) with some omissions. Chapters 7–9 plus summaries of Chapters 10, 11 and 13 were used for a subsequent graduate course. Moreover, Chapters 12–14, together with excursions into Chapters 4 and 5 and supplements on parabolic equations, have been used for a graduate course.

The bibliography exposes the many sources that were consulted while the material was collected. It is not meant to be a complete literature list of the available works in this area.

It is my hope that the text is relatively free of errors, but I will be interested to be informed if readers find something to correct; then I may possibly set up a homepage with the information.

Besides drawing on many books and papers, as referred to in the text, I have benefited from the work of colleagues in Denmark, reading their earlier notes for related courses, and getting their advice on my courses. My thanks go especially to Esben Kehlet, and to Jon Johnsen, Henrik Schlichtkrull, Mogens Flensted-Jensen and Christian Berg. I also thank all those who have helped me improve the text while participating in the courses as graduate students through the years. Moreover, my thanks go to Mads Haar and Jan Caesar for creating the figures, and to Jan Caesar for his invaluable help in adapting the manuscript to Springer's style.

Copenhagen, June 2008   Gerd Grubb

Contents

Part I
Distributions and derivatives

Chapter 1
Motivation and overview

1.1 Introduction

In the study of ordinary differential equations one can get very far by using just the classical concept of differentiability, working with spaces of continuously differentiable functions on an interval $I \subset \mathbb{R}$:

$$C^m(I) = \{ u : I \to \mathbb{C} \mid \tfrac{d^j}{dx^j}u \text{ exists and is continuous on } I \text{ for } 0 \le j \le m \}. \tag{1.1}$$

The need for more general concepts comes up for example in the study of eigenvalue problems for second-order operators on an interval $[a, b]$ with boundary conditions at the endpoints a, b, by Hilbert space methods. But here it usually suffices to extend the notions to *absolutely continuous functions*, i.e., functions $u(x)$ of the form

$$u(x) = \int_{x_0}^{x} v(y)\, dy + c, \quad v \text{ locally integrable on } I. \tag{1.2}$$

Here c denotes a constant, and "locally integrable" means integrable on compact subsets of I. The function v is regarded as the derivative $\tfrac{d}{dx}u$ of u, and the fundamental formula

$$u(x) = u(x_0) + \int_{x_0}^{x} \tfrac{d}{dy}u(y)\, dy \tag{1.3}$$

still holds.

But for partial differential equations one finds when using methods from functional analysis that the spaces C^m are inadequate, and there is no good concept of absolute continuity in the case of functions of several real variables. One can get some ways by using the concept of *weak derivatives*: When u and v are locally integrable on an open subset Ω of $\mathbb{R}^n$, we say that $v = \tfrac{\partial}{\partial x_j}u$ in the weak sense, when

$$-\int_\Omega u \frac{\partial}{\partial x_j}\varphi \, dx = \int_\Omega v\varphi \, dx, \text{ for all } \varphi \in C_0^\infty(\Omega); \qquad (1.4)$$

here $C_0^\infty(\Omega)$ denotes the space of C^∞ functions on Ω with compact support in Ω. (The support supp f of a function f is the complement of the largest open set where the function is zero.) This criterion is modeled after the fact that the formula (1.4) holds when $u \in C^1(\Omega)$, with $v = \frac{\partial}{\partial x_j}u$.

Sometimes even the concept of weak derivatives is not sufficient, and the need arises to define derivatives that are not functions, but are more general objects. Some measures and derivatives of measures will enter. For example, there is the Dirac measure δ_0 that assigns 1 to every Lebesgue measurable set in $\mathbb{R}^n$ containing $\{0\}$, and 0 to any Lebesgue measurable set not containing $\{0\}$. For $n = 1$, δ_0 is the derivative of the Heaviside function defined in (1.8) below. In the book of Laurent Schwartz [S61] there is also a description of the *derivative of* δ_0 (on $\mathbb{R}$) — which is not even a measure — as a "dipole", with some kind of physical explanation.

For the purpose of setting up the rules for a general theory of differention where classical differentiability fails, Schwartz brought forward around 1950 the concept of *distributions*: a class of objects containing the locally integrable functions and allowing differentiations of any order.

This book gives an introduction to distribution theory, based on the work of Schwartz and of many other people. Our aim is also to show how the theory is combined with the study of operators in Hilbert space by methods of functional analysis, with applications to ordinary and partial differential equations. In some chapters of a more advanced character, we show how the distribution theory is used to define pseudodifferential operators and how they are applied in the discussion of solvability of PDE, with or without boundary conditions. A bibliography of relevant books and papers is collected at the end.

Plan

Part I gives an introduction to distributions.

In the rest of Chapter 1 we begin the discussion of taking derivatives in the distribution sense, motivating the study of function spaces in the following chapter.

Notation and prerequisites are collected in Appendix A.

Chapter 2 studies the spaces of C^∞-functions (and C^k-functions) needed in the theory, and their relations to L_p-spaces.

The relevant topological considerations are collected in Appendix B.

In Chapter 3 we introduce distributions in full generality and show the most prominent rules of calculus for them.

Part II connects the distribution concept with differential equations and Fourier transformation.

Chapter 4 is aimed at linking distribution theory to the treatment of partial differential equations (PDE) by Hilbert space methods. Here we introduce Sobolev spaces and realizations of differential operators, both in the (relatively simple) one-dimensional case and in n-space, and study some applications.

Here we use some of the basic results on unbounded operators in Hilbert space that are collected in Chapter 12.

In Chapter 5, we study the Fourier transformation in the framework of temperate distributions.

Chapter 6 gives a further development of Sobolev spaces as well as applications to PDE by use of Fourier theory, and shows a fundamental result on the structure of distributions.

Part III introduces a more advanced tool, namely, pseudodifferential operators (ψdo's), a generalization of partial differential operators containing also the solution operators for elliptic problems.

Chapter 7 gives the basic ingredients of the local calculus of pseudodifferential operators. Applications include a proof of the Gårding inequality.

Chapter 8 shows how to define ψdo's on manifolds, and how they in the elliptic case define Fredholm operators, with solvability properties modulo finite-dimensional spaces. (An introduction to Fredholm operators is included.)

Part IV treats boundary value problems.

Chapter 9 (independent of Chapter 7 and 8) takes up the study of boundary value problems by use of Fourier transformation. The main effort is spent on an important constant-coefficient case which, as an example, shows how Sobolev spaces of noninteger and negative order can enter. Also, a connection is made to the abstract theory of Chapter 13. This chapter can be read directly after Parts I and II.

In Chapter 10 we present the basic ingredients in a pseudodifferential theory of boundary value problems introduced originally by L. Boutet de Monvel; this builds on the methods of Chapters 7 and 8 and the example in Chapter 9, introducing new operator types.

Chapter 11 shows how the theory of Chapter 10 can be used to discuss solvability of elliptic boundary value problems, by use of the Calderón projector, that we construct in detail. As a special example, regularity of solutions of the Dirichlet problem is shown. Some other boundary value problems are taken up in the exercises.

Part V gives the supplementing topics needed from Hilbert space theory.

Chapter 12, departing from the knowledge of bounded linear operators in Hilbert spaces, shows some basic results for unbounded operators, and develops the theory of variational operators.

Chapter 13 gives a systematic presentation of closed extensions of adjoint pairs, with consequences for symmetric and semibounded operators; this is of interest for the study of boundary value problems for elliptic PDE and their positivity properties. We moreover include a recent development concerning resolvents, their M-functions and Kreĭn formulas.

Chapter 14 establishes some basic results on semigroups of operators, relevant for parabolic PDE (problems with a time parameter), and appealing to positivity and variationality properties discussed in earlier chapters.

Finally, there are three appendices. In Appendix A, we recall some basic rules of calculus and set up the notation.

Appendix B gives some elements of the theory of topological vector spaces, that can be invoked when one wants the correct topological formulation of the properties of distributions.

Appendix C introduces some function spaces, as a continuation of Chapter 2, but needed only later in the text.

1.2 On the definition of distributions

The definition of a weak derivative $\partial_j u$ was mentioned in (1.4) above. Here both u and its weak derivative v are locally integrable functions on Ω. Observe that the right-hand side is a linear functional on $C_0^\infty(\Omega)$, i.e., a linear mapping Λ_v of $C_0^\infty(\Omega)$ into $\mathbb{C}$, here defined by

$$\Lambda_v : \varphi \mapsto \Lambda_v(\varphi) = \int_\Omega v\varphi \, dx. \tag{1.5}$$

The idea of Distribution Theory is to allow much more general functionals than this one. In fact, when Λ is any linear functional on $C_0^\infty(\Omega)$ such that

$$-\int_\Omega u\partial_j\varphi \, dx = \Lambda(\varphi) \text{ for all } \varphi \in C_0^\infty(\Omega), \tag{1.6}$$

we shall say that

$$\partial_j u = \Lambda \text{ in the distribution sense,} \tag{1.7}$$

even if there is no function v (locally integrable) such that Λ can be defined from it as in (1.5).

Example 1.1. Here is the most famous example in the theory: Let $\Omega = \mathbb{R}$ and consider the Heaviside function $H(x)$; it is defined by

$$H(x) = \begin{cases} 1 & \text{for } x > 0, \\ 0 & \text{for } x \leq 0. \end{cases} \tag{1.8}$$

It is locally integrable on $\mathbb{R}$. But there is no locally integrable function v such that (1.4) holds with $u = H$:

$$-\int_{\mathbb{R}} H \tfrac{d}{dx}\varphi\, dx = \int v\varphi\, dx, \quad \text{for all } \varphi \in C_0^\infty(\mathbb{R}). \tag{1.9}$$

For, assume that v were such a function, and let $\varphi \in C_0^\infty(\mathbb{R})$ with $\varphi(0) = 1$ and set $\varphi_N(x) = \varphi(Nx)$. Note that $\max|\varphi(x)| = \max|\varphi_N(x)|$ for all N, and that when φ is supported in $[-R, R]$, φ_N is supported in $[-R/N, R/N]$. Thus by the theorem of Lebesgue,

$$\int_{\mathbb{R}} v\varphi_N\, dx \to 0 \text{ for } N \to \infty, \tag{1.10}$$

but on the other hand,

$$-\int_{\mathbb{R}} H \tfrac{d}{dx}\varphi_N\, dx = -\int_0^\infty N\varphi'(Nx)\, dx = -\int_0^\infty \varphi'(y)\, dy = \varphi(0) = 1. \tag{1.11}$$

So (1.9) cannot hold for this sequence of functions φ_N, and we conclude that a locally integrable function v for which (1.9) holds for all $\varphi \in C_0^\infty(\mathbb{R})$ cannot exist.

A linear functional that does match H in a formula (1.6) is the following one:

$$\Lambda : \varphi \mapsto \varphi(0) \tag{1.12}$$

(as seen by a calculation as in (1.11)). This is the famous delta-distribution, usually denoted δ_0. (It identifies with the Dirac measure mentioned earlier.)

There are some technical things that have to be cleared up before we can define distributions in a proper way.

For one thing, we have to look more carefully at the elements of $C_0^\infty(\Omega)$. We must demonstrate that such functions really do exist, and we need to show that there are elements with convenient properties (such as having the support in a prescribed set and being 1 on a smaller prescribed set).

Moreover, we have to describe what is meant by convergence in $C_0^\infty(\Omega)$, in terms of a suitable topology. There are also some other spaces of C^∞ or C^k functions with suitable support or integrability properties that we need to introduce.

These preparatory steps will take some time, before we begin to introduce distributions in full generality. (The theories that go into giving $C_0^\infty(\Omega)$ a good topology are quite advanced, and will partly be relegated to Appendix B. In fact, the urge to do this in all details has been something of an obstacle to making the tool of distributions available to everybody working with PDE — so we shall here take the point of view of giving full details of how one *operates with* distributions, but tone down the topological discussion to some statements one can use without necessarily checking all proofs.)

The reader is urged to consult Appendix A (with notation and prerequisites) before starting to read the next chapters.

Chapter 2
Function spaces and approximation

2.1 The space of test functions

Notation and prerequisites are collected in Appendix A.

Let Ω be an open subset of $\mathbb{R}^n$. The space $C_0^\infty(\Omega)$, consisting of the C^∞-functions on Ω with compact support in Ω, is called the space of *test functions* (on Ω). The support supp u of a function $u \in L_{1,\text{loc}}(\Omega)$ is defined as the complement of the largest open set where u vanishes; we can write it as

$$\text{supp}\, u = \Omega \setminus \left(\bigcup \{\omega \text{ open in } \Omega \mid u|_\omega = 0 \} \right). \tag{2.1}$$

We show first of all that there *exist* test functions:

Lemma 2.1. 1° *Let $R > r > 0$. There is a function $\chi_{r,R}(x) \in C_0^\infty(\mathbb{R}^n)$ with the properties: $\chi_{r,R}(x) = 1$ for $|x| \leq r$, $\chi_{r,R}(x) \in [0,1]$ for $r \leq |x| \leq R$, $\chi_{r,R}(x) = 0$ for $|x| \geq R$.*
2° *There is a function $h \in C_0^\infty(\mathbb{R}^n)$ satisfying:*

$$\text{supp}\, h = \underline{B}(0,1), \quad h(x) > 0 \text{ for } |x| < 1, \quad \int h(x)\, dx = 1. \tag{2.2}$$

Proof. 1°. The function

$$f(t) = \begin{cases} e^{-1/t} & \text{for } t > 0, \\ 0 & \text{for } t \leq 0, \end{cases}$$

is a C^∞-function on $\mathbb{R}$. For $t \neq 0$ this is obvious. At the point $t = 0$ we have that $f(t) \to 0$ for $t \searrow 0$, and that the derivatives of $f(t)$ for $t \neq 0$ are of the form

$$\partial_t^k f(t) = \begin{cases} p_k(1/t) e^{-1/t} & \text{for } t > 0, \\ 0 & \text{for } t < 0, \end{cases}$$

for certain polynomials p_k, $k \in \mathbb{N}_0$. Since any polynomial p satisfies $p(1/t)e^{-1/t} \to 0$ for $t \searrow 0$, f and its derivatives are differentiable at 0.

From f we construct the functions (see the figure)

$$f_1(t) = f(t-r)f(R-t), \quad f_2(t) = \int_t^\infty f_1(s)\,ds.$$

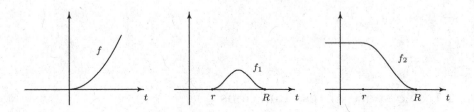

Here we see that $f_2(x) \geq 0$ for all x, equals 0 for $t \geq R$ and equals

$$C = \int_r^R f_1(s)\,ds > 0$$

for $t \leq r$. The function

$$\chi_{r,R}(x) = \frac{1}{C}f_2(|x|), \quad x \in \mathbb{R}^n,$$

then has the desired properties.

$2°$. Here one can for example take

$$h(x) = \frac{\chi_{\frac{1}{2},1}(x)}{\int \chi_{\frac{1}{2},1}(x)\,dx}.$$

□

Note that *analytic functions* (functions defined by a converging Taylor expansion) cannot be in $C_0^\infty(\mathbb{R})$ without being identically zero! So we have to go outside the elementary functions (such as $\cos t$, e^t, e^{-t^2}, etc.) to find nontrivial C^∞-functions. The construction in Lemma 2.1 can be viewed from a "plumber's point of view": We want a C^∞-function that is 0 on a certain interval and takes a certain positive value on another; we can get it by twisting the graph suitably. But analyticity is lost then.

For later reference we shall from now on denote by χ a function in $C_0^\infty(\mathbb{R}^n)$ satisfying

$$\chi(x) \begin{cases} = 1 & \text{for } |x| \le 1 \,, \\ \in [0,1] & \text{for } 1 \le |x| \le 2 \,, \\ = 0 & \text{for } |x| \ge 2 \,, \end{cases} \tag{2.3}$$

one can for example take $\chi_{1,2}$ constructed in Lemma 2.1. A C_0^∞-function that is 1 on a given set and vanishes outside some larger given set is often called a *cut-off function*. Of course we get some other cut-off functions by translating the functions $\chi_{r,R}$ around. More refined examples will be constructed later by convolution, see e.g. Theorem 2.13. These functions are all examples of test functions, when their support is compact.

We use throughout the following convention (of "extension by zero") for test functions: If $\varphi \in C_0^\infty(\Omega)$, Ω open $\subset \mathbb{R}^n$, we also denote the function obtained by extending by zero on $\mathbb{R}^n \setminus \Omega$ by φ; it is in $C_0^\infty(\mathbb{R}^n)$. When $\varphi \in C_0^\infty(\mathbb{R}^n)$ and its support is contained in Ω, we can regard it as an element of $C_0^\infty(\Omega)$ and again denote it φ. Similarly, we can view a C^∞-function φ with compact support in $\Omega \cap \Omega'$ (Ω and Ω' open) as an element of $C_0^\infty(\Omega)$ or $C_0^\infty(\Omega')$, whatever is convenient.

Before we describe the topology of the space $C_0^\infty(\Omega)$ we recall how some other useful spaces are topologized. The reader can find the necessary information on topological vector spaces in Appendix B and its problem session.

When we consider an open subset Ω of $\mathbb{R}^n$, the compact subsets play an important role.

Lemma 2.2. *Let Ω be a nonempty open subset of $\mathbb{R}^n$. There exists a sequence of compact subsets $(K_j)_{j \in \mathbb{N}}$ such that*

$$K_1 \subset K_2^\circ \subset K_2 \subset \cdots \subset K_j^\circ \subset K_j \subset \dots, \quad \bigcup_{j \in \mathbb{N}} K_j^\circ = \Omega \,. \tag{2.4}$$

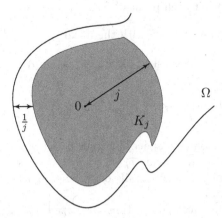

Proof. We can for example take

$$K_j = \big\{\, x \in \Omega \mid |x| \le j \ \text{ and } \ \operatorname{dist}(x, \complement\Omega) \ge \tfrac{1}{j} \,\big\}; \tag{2.5}$$

the interior of this set is defined by the formula with $\leq$ and $\geq$ replaced by $<$ and $>$. (If $\complement\Omega = \emptyset$, the condition dist $(x, \complement\Omega) \geq \frac{1}{j}$ is left out.) If necessary, we can omit the first, at most finitely many, sets with $K_j^\circ = \emptyset$ and modify the indexation. $\qquad\qquad\qquad\qquad\qquad\qquad\qquad\qquad\qquad\qquad\qquad\qquad\qquad\qquad$ $\square$

When K is a compact subset of Ω, it is covered by the system of open sets $\{K_j^\circ\}_{j\in\mathbb{N}}$ and hence by a finite subsystem, say with $j \leq j_0$. Then $K \subset K_j$ for all $j \geq j_0$.

Recall that when $[a, b]$ is a compact interval of $\mathbb{R}$, $C^k([a, b])$ (in one of the versions $C^k([a, b], \mathbb{C})$ or $C^k([a, b], \mathbb{R})$) is defined as the Banach space of complex resp. real functions having continuous derivatives up to order k, provided with a norm

$$\|f\|'_{C^k} = \sum_{0\leq j\leq k} \sup_x |f^{(j)}(x)|, \text{ or the equivalent norm}$$

$$\|f\|_{C^k} = \sup\{\, |f^{(j)}(x)| \mid x \in [a, b], 0 \leq j \leq k \,\}. \qquad\qquad (2.6)$$

In the proof that these normed spaces are complete one uses the well-known theorem that when f_l is a sequence of C^1-functions such that f_l and f_l' converge uniformly to f resp. g for $l \to \infty$, then f is C^1 with derivative $f' = g$. There is a similar result for functions of several variables:

Lemma 2.3. *Let $J = [a_1, b_1] \times \cdots \times [a_n, b_n]$ be a closed box in $\mathbb{R}^n$ and let f_l be a sequence of functions in $C^1(J)$ such that $f_l \to f$ and $\partial_j f_l \to g_j$ uniformly on J for $j = 1, \ldots, n$. Then $f \in C^1(J)$ with $\partial_j f = g_j$ for each j.*

Proof. For each j we use the above-mentioned theorem in situations where all but one coordinate are fixed. This shows that f has continuous partial derivatives $\partial_j f = g_j$ at each point of J. $\qquad\qquad\qquad\qquad\qquad\qquad$ $\square$

So $C^k(J)$ is a Banach space with the norm

$$\|u\|_{C^k(J)} = \sup\{\, |\partial^\alpha u(x)| \mid x \in J, |\alpha| \leq k \,\}. \qquad\qquad (2.7)$$

We define

$$C^\infty(J) = \bigcap_{k\in\mathbb{N}_0} C^k(J). \qquad\qquad (2.8)$$

This is no longer a Banach space, but can be shown to be a Fréchet space with the family of (semi)norms $p_k(f) = \|f\|_{C^k(J)}$, $k \in \mathbb{N}_0$, by arguments as in Lemma 2.4 below. (For details on Fréchet spaces, see Appendix B, in particular Theorem B.9.)

For spaces of differentiable functions over *open* sets, the full sup-norms are unsatisfactory since the functions and their derivatives need not be bounded. We here use sup-norms over compact subsets to define a Fréchet topology. Let Ω be open and let K_j be an increasing sequence of compact subsets as in Lemma 2.2. Define the system of seminorms

$$p_{k,j}(f) = \sup\{\,|\partial^\alpha f(x)|\mid |\alpha| \le k\,,\ x \in K_j\,\}, \quad \text{for } j \in \mathbb{N}. \tag{2.9}$$

Lemma 2.4. $1°$ *For each $k \in \mathbb{N}_0$, $C^k(\Omega)$ is a Fréchet space when provided with the family of seminorms $\{p_{k,j}\}_{j\in\mathbb{N}}$.*

$2°$ *The space $C^\infty(\Omega) = \bigcap_{k\in\mathbb{N}_0} C^k(\Omega)$ is a Fréchet space when provided with the family of seminorms $\{p_{k,j}\}_{k\in\mathbb{N}_0,j\in\mathbb{N}}$.*

Proof. $1°$. The family $\{p_{k,j}\}_{j\in\mathbb{N}}$ is separating, for when $f \in C^k(\Omega)$ is $\ne 0$, then there is a point x_0 where $f(x_0) \ne 0$, and $x_0 \in K_j$ for j sufficiently large; for such j, $p_{k,j}(f) > 0$. The seminorms then define a translation invariant metric by Theorem B.9. We just have to show that the space is complete under this metric. Here we use Lemma 2.3: Let $(f_l)_{l\in\mathbb{N}}$ be a Cauchy sequence in $C^k(\Omega)$. Let $x_0 = \{x_{01},\dots,x_{0n}\} \in \Omega$ and consider a box $J = J_{x_0,\delta} = \{\,x \mid |x_m - x_{0m}| \le \delta,\ m = 1,\dots,n\,\}$ around x_0, with δ taken so small that $J \subset \Omega$. Since $J \subset K_{j_0}$ for a certain j_0, the Cauchy sequence property implies that f_l defines a Cauchy sequence in $C^k(J)$, i.e., f_l and its derivatives up to order k are Cauchy sequences with respect to uniform convergence on J. So there is a limit f_J in $C^k(J)$. We use similar arguments for other boxes J' in Ω and find that the limits $f_{J'}$ and f_J are the same on the overlap of J and J'. In this way we can define a C^k-function f that is the limit of the sequence in $C^k(J)$ for all boxes $J \subset \Omega$. Finally, $p_{j,k}(f_l - f) \to 0$ for all j, since each K_j can be covered by a finite number of box-interiors $J°$. Then f_l has the limit f in the Fréchet topology of $C^k(\Omega)$.

$2°$. The proof in this case is a variant of the preceding proof, where we now investigate $p_{j,k}$ for all k also. $\qquad\square$

The family (2.9) has the max-property (see Remark B.6), so the sets

$$V(p_{k,j},\varepsilon) = \{\,f \in C^\infty(\Omega)\mid |\partial^\alpha f(x)| < \varepsilon \text{ for } |\alpha| \le k,\ x \in K_j\,\} \tag{2.10}$$

constitute a local basis for the system of neighborhoods at 0. One could in fact make do with the sequence of seminorms $\{p_{k,k}\}_{k\in\mathbb{N}}$, which increase with k.

For any compact subset K of Ω we define

$$C_K^\infty(\Omega) = \{\,u \in C^\infty(\Omega)\mid \operatorname{supp} u \subset K\,\}, \tag{2.11}$$

the space of C^∞-functions with support in K (cf. (2.1)); this space is provided with the topology inherited from $C^\infty(\Omega)$.

The space $C_K^\infty(\Omega)$ is a closed subspace of $C^\infty(\Omega)$ (so it is a Fréchet space). The topology is for example defined by the family of seminorms $\{p_{k,j_0}\}_{k\in\mathbb{N}_0}$ (cf. (2.9)) with j_0 taken so large that $K \subset K_{j_0}$. This family has the max-property.

In the theory of distributions we need not only the Fréchet spaces $C^\infty(\Omega)$ and $C_K^\infty(\Omega)$ but also the space

$$C_0^\infty(\Omega) = \{\,\varphi \in C^\infty(\Omega)\mid \operatorname{supp}\varphi \text{ is compact in } \Omega\,\}. \tag{2.12}$$

As already mentioned, it is called the space of *test functions*, and it is also denoted $\mathscr{D}(\Omega)$.

If we provide this space with the topology inherited from $C^\infty(\Omega)$, we get an incomplete metric space. For example, if Ω is the interval $I = \,]0,3[$ and $\varphi(x)$ is a C^∞-function on I with $\operatorname{supp}\varphi = [1,2]$, then $\varphi_l(x) = \varphi(x - 1 + \frac{1}{l})$, $l \in \mathbb{N}$, is a sequence of functions in $C_0^\infty(I)$ which converges in $C^\infty(I)$ to the function $\varphi(x-1) \in C^\infty(I) \setminus C_0^\infty(I)$.

We prefer to provide $C_0^\infty(\Omega)$ with a stronger and somewhat more complicated vector space topology that makes it a *sequentially complete* (but not metric) space. More precisely, we regard $C_0^\infty(\Omega)$ as

$$C_0^\infty(\Omega) = \bigcup_{j=1}^\infty C_{K_j}^\infty(\Omega)\,, \qquad (2.13)$$

where K_j is an increasing sequence of compact subsets as in (2.4) and the topology is the *inductive limit topology*, cf. Theorem B.17 (also called the $\mathcal{LF}$-topology). The spaces $C_{K_j}^\infty(\Omega)$ are provided with Fréchet space topologies by families of seminorms (2.9).

The properties of this space that we shall need are summed up in the following theorem, which just specifies the general properties given in Appendix B (Theorem B.18 and Corollary B.19):

Theorem 2.5. *The topology on $C_0^\infty(\Omega)$ has the following properties:*

(a) *A sequence $(\varphi_l)_{l\in\mathbb{N}}$ of test functions converges to φ_0 in $C_0^\infty(\Omega)$ if and only if there is a $j \in \mathbb{N}$ such that $\operatorname{supp}\varphi_l \subset K_j$ for all $l \in \mathbb{N}_0$, and $\varphi_l \to \varphi_0$ in $C_{K_j}^\infty(\Omega)$:*

$$\sup_{x\in K_j} |\partial^\alpha \varphi_l(x) - \partial^\alpha \varphi_0(x)| \to 0 \quad \text{for } l \to \infty, \qquad (2.14)$$

for all $\alpha \in \mathbb{N}_0^n$.

(b) *A set $E \subset C_0^\infty(\Omega)$ is bounded if and only if there exists a $j \in \mathbb{N}$ such that E is a bounded subset of $C_{K_j}^\infty(\Omega)$. In particular, if $(\varphi_l)_{l\in\mathbb{N}}$ is a Cauchy sequence in $C_0^\infty(\Omega)$, then there is a j such that $\operatorname{supp}\varphi_l \subset K_j$ for all l, and φ_l is convergent in $C_{K_j}^\infty(\Omega)$ (and then also in $C_0^\infty(\Omega)$).*

(c) *Let Y be a locally convex topological vector space. A mapping T from $C_0^\infty(\Omega)$ to Y is continuous if and only if $T : C_{K_j}^\infty(\Omega) \to Y$ is continuous for each $j \in \mathbb{N}$.*

(d) *A linear functional $\Lambda : C_0^\infty(\Omega) \to \mathbb{C}$ is continuous if and only if there is an $N_j \in \mathbb{N}_0$ and a $c_j > 0$ for any $j \in \mathbb{N}$, such that*

$$|\Lambda(\varphi)| \le c_j \sup\{|\partial^\alpha \varphi(x)| \mid x \in K_j,\, |\alpha| \le N_j\} \qquad (2.15)$$

for all $\varphi \in C_{K_j}^\infty(\Omega)$.

Note that (a) is a very *strong* assumption on the sequence φ_l. Convergence in $C_0^\infty(\Omega)$ implies convergence in practically all the other spaces we shall meet. On the other hand, (d) is a very *mild* assumption on the functional Λ; practically all the functionals that we shall meet will have this property.

We underline that a sequence can only be a Cauchy sequence when there is a j such that all functions in the sequence have support in K_j and the sequence is Cauchy in $C_{K_j}^\infty(\Omega)$. Then the sequence converges because of the completeness of the Fréchet space $C_{K_j}^\infty(\Omega)$. In the example mentioned above, the sequence $\varphi_l(x) = \varphi(x - 1 + \frac{1}{l})$ in $C_0^\infty(\,]0,3\,[)$ is clearly not a Cauchy sequence with respect to this topology on $C_0^\infty(\,]0,3\,[)$.

It is not hard to show that when $(K_j')_{j \in \mathbb{N}}$ is another sequence of compact subsets as in (2.4), the topology on $C_0^\infty(\Omega)$ defined by use of this sequence is the same as that based on the first one (Exercise 2.2).

We now consider two important operators on these spaces. One is *differentiation*, the other is *multiplication* (by a C^∞-function f); both will be shown to be continuous. The operators are denoted ∂^α (with $D^\alpha = (-i)^{|\alpha|}\partial^\alpha$) resp. M_f. We also write $M_f\varphi$ as $f\varphi$ — and the same notation will be used later for generalizations of these operators.

Theorem 2.6. $1°$ *The mapping $\partial^\alpha : \varphi \mapsto \partial^\alpha\varphi$ is a continuous linear operator in $C_0^\infty(\Omega)$. The same holds for D^α.*

$2°$ *For any $f \in C^\infty(\Omega)$, the mapping $M_f : \varphi \mapsto f\varphi$ is a continuous linear operator in $C_0^\infty(\Omega)$.*

Proof. Clearly, ∂^α and M_f are linear operators from $C_0^\infty(\Omega)$ to itself. As for the continuity it suffices, according to 2.5 (c), to show that ∂^α resp. M_f is continuous from $C_{K_j}^\infty(\Omega)$ to $C_0^\infty(\Omega)$ for each j. Since the operators satisfy

$$\operatorname{supp}\partial^\alpha\varphi \subset \operatorname{supp}\varphi\,, \quad \operatorname{supp}M_f\varphi \subset \operatorname{supp}\varphi, \tag{2.16}$$

for all φ, the range space can for each j be replaced by $C_{K_j}^\infty(\Omega)$. Here we have for each k:

$$\begin{aligned}
p_{k,j}(\partial^\alpha\varphi) &= \sup\{\,|\partial^\beta\partial^\alpha\varphi|\mid |\beta| \le k\,, x \in K_j\,\} \\
&\le \sup\{\,|\partial^\gamma\varphi|\mid |\gamma| \le k + |\alpha|\,, x \in K_j\,\} = p_{k+|\alpha|,j}(\varphi)\,,
\end{aligned} \tag{2.17}$$

which shows the continuity of ∂^α. The result extends immediately to D^α. By the Leibniz rule (A.7) we have for each k:

$$\begin{aligned}
p_{k,j}(f\varphi) &= \sup\{\,|\partial^\alpha(f\varphi)|\mid |\alpha| \le k\,, x \in K_j\,\} \\
&\le \sup\{\,\textstyle\sum_{\beta \le \alpha}|c_{\alpha,\beta}\partial^\beta f\partial^{\alpha-\beta}\varphi|\mid |\alpha| \le k\,, x \in K_j\,\} \\
&\le C_k p_{k,j}(f)p_{k,j}(\varphi),
\end{aligned} \tag{2.18}$$

for a suitably large constant C_k; this shows the continuity of M_f. $\square$

2.2 Some other function spaces

$C_0^\infty(\Omega)$ is contained in practically every other space defined in connection with Ω, that we shall meet. For example,

$$C_0^\infty(\Omega) \subset L_p(\Omega) \quad \text{for } p \in [1, \infty] \,,$$
$$C_0^\infty(\Omega) \subset C^\infty(\Omega) \,, \tag{2.19}$$

and these injections are continuous: According to Theorem 2.5 (c) it suffices to show that the corresponding injections of $C_{K_j}^\infty(\Omega)$ are continuous, for each j (with K_j as in (2.4)). For $\varphi \in C_{K_j}^\infty(\Omega)$ we have when $p < \infty$:

$$\|\varphi\|_{L_p} \leq \sup_{x \in K_j} |\varphi(x)| \, \mathrm{vol}\,(K_j)^{1/p}, \tag{2.20}$$

which shows that the injection $J : C_{K_j}^\infty(\Omega) \to L_p(\Omega)$ maps the basic neighborhood

$$V(p_{0,j}, \varepsilon) = \{\, \varphi \mid \sup_x |\varphi(x)| < \varepsilon \,\}$$

into the ball $B(0, r)$ in $L_p(\Omega)$ with $r = \varepsilon \, \mathrm{vol}\,(K_j)^{1/p}$. The continuity of the other injection in (2.19) follows from the fact that $C_{K_j}^\infty(\Omega)$ had the inherited topology as a subspace of $C^\infty(\Omega)$.

We now introduce some further spaces of functions.

It is typical for the space $C^\infty(\Omega)$ that it gives no restriction on the *global* behavior of the elements (their behavior on Ω as a whole): from the knowledge of the behavior of a function on the compact subsets of Ω one can determine whether it belongs to $C^\infty(\Omega)$. This does not hold for $L_p(\Omega)$ where a certain globally defined number (the norm) must be finite in order for the function to belong to $L_p(\Omega)$. Sometimes one needs the following "locally" defined variant of $L_p(\Omega)$:

$$L_{p,\mathrm{loc}}(\Omega) = \{\, u \text{ measurable} \mid u|_K \in L_p(K) \text{ when } K \text{ compact } \subset \Omega \,\}, \tag{2.21}$$

with the usual identification of functions that are equal almost everywhere.

$L_{p,\mathrm{loc}}(\Omega)$ is provided with the Fréchet space topology defined from the family of seminorms

$$p_j(u) = \|1_{K_j} u\|_{L_p(\Omega)}, \; j = 1, 2, \ldots,$$

where K_j is as in (2.4). For K compact $\subset \Omega$, we can identify $L_p(K)$ with

$$L_{p,K}(\Omega) = \{\, u \in L_p(\Omega) \mid \mathrm{supp}\, u \subset K \,\}, \tag{2.22}$$

by extension by zero in $\Omega \setminus K$; it is a closed subspace of $L_p(\Omega)$. The completeness of the spaces $L_{p,\text{loc}}(\Omega)$ can be deduced from the completeness of the spaces $L_{p,K_j}(\Omega)$.

In analogy with the subspace $C_0^\infty(\Omega)$ of $C^\infty(\Omega)$ (with a stronger topology) we define the subspace $L_{p,\text{comp}}(\Omega)$ of $L_{p,\text{loc}}(\Omega)$ (and of $L_p(\Omega)$) by

$$L_{p,\text{comp}}(\Omega) = \{\, u \in L_p(\Omega) \mid \operatorname{supp} u \text{ compact } \subset \Omega \,\}. \qquad (2.23)$$

It is provided with the inductive limit topology, when written as

$$L_{p,\text{comp}}(\Omega) = \bigcup_{j=1}^{\infty} L_{p,K_j}(\Omega). \qquad (2.24)$$

In this way, $L_{p,\text{comp}}(\Omega)$ is an $\mathcal{LF}$-space (cf. Appendix B), with a stronger topology than the one inherited from $L_p(\Omega)$.

Remark 2.7. The above choices of topology assure for example that $L_{2,\text{loc}}(\Omega)$ and $L_{2,\text{comp}}(\Omega)$ may be identified with the dual space of one another, in such a way that the duality is a generalization of the integral $\int_\Omega u\bar{v}\,dx$ (Exercises 2.4 and 2.8).

It is not hard to show (cf. (A.21), (A.31)) that

$$\begin{aligned}
C^\infty(\Omega) &\subset L_{p,\text{loc}}(\Omega) \subset L_{q,\text{loc}}(\Omega)\,, \\
C_0^\infty(\Omega) &\subset L_{p,\text{comp}}(\Omega) \subset L_{q,\text{comp}}(\Omega)\,, \qquad \text{for } p > q\,,
\end{aligned} \qquad (2.25)$$

with continuous injections.

More function spaces are defined in Appendix C. The reader can bypass them until needed in the text. There is a large number of spaces that one can define for various purposes, and rather than learning all these spaces by heart, the reader should strive to be able to introduce the appropriate space with the appropriate topology ("do-it-yourself") when needed.

2.3 Approximation theorems

From the test functions constructed in Lemma 2.1 one can construct a wealth of other test functions by convolution. Recall that when f and g are measurable functions on $\mathbb{R}^n$ and the product $f(y)g(x-y)$ is an integrable function of y for a fixed x, then the convolution product $(f * g)(x)$ is defined by

$$(f * g)(x) = \int_{\mathbb{R}^n} f(y)g(x-y)\,dy\,. \qquad (2.26)$$

Note that $(f * g)(x) = \int_{\mathbb{R}^n} f(x-y)g(y)\,dy = (g * f)(x)$. (2.26) is for example defined when $f \in L_{1,\text{loc}}$ and g is bounded and one of them has compact support (or both have support in a conical set such as for example $\{\, x \mid x_1 \geq 0, \ldots, x_n \geq 0 \,\}$). (2.26) is also well-defined when $f \in L_p$ and $g \in L_q$ with $1/p + 1/q = 1$.

It will be convenient to use the following general convergence principles that follow from the theorem of Lebesgue:

Lemma 2.8. *Let $M \subset \mathbb{R}^n$ be measurable and let I be an interval of $\mathbb{R}$. Let $f(x,a)$ be a family of functions of $x \in M$ depending on the parameter $a \in I$, such that for each $a \in I$, $f(x,a) \in L_1(M)$. Consider the function F on I defined by*

$$F(a) = \int_M f(x,a)\, dx \text{ for } a \in I. \tag{2.27}$$

1° Assume that for each $x \in M$, $f(x,a)$ is a continuous function of a at the point $a_0 \in I$, and that there is a function $g(x) \in L^1(M)$ such that $|f(x,a)| \leq g(x)$ for all $(x,a) \in M \times I$. Then $F(a)$ is continuous at the point a_0.

A similar statement holds when the parameter a runs in a ball $B(a_0, r)$ in $\mathbb{R}^k$.

2° Assume that $\frac{\partial}{\partial a} f(x,a)$ exists for all $(x,a) \in M \times I$ and that there is a function $g(x) \in L^1(M)$ such that

$$|\frac{\partial}{\partial a} f(x,a)| \leq g(x) \text{ for all } (x,a) \in M \times I. \tag{2.28}$$

Then $F(a)$ is a differentiable function of $a \in I$, and

$$\frac{d}{da} F(a) = \int_M \frac{\partial}{\partial a} f(x,a)\, dx. \tag{2.29}$$

Let $h(x)$ be a function with the properties

$$h \in C_0^\infty(\mathbb{R}^n),\ h \geq 0,\ \int_{\mathbb{R}^n} h(x)\, dx = 1,\ \operatorname{supp} h \subset \underline{B}(0,1). \tag{2.30}$$

Such functions exist according to Lemma 2.1. For $j \in \mathbb{N}$ we set

$$h_j(x) = j^n h(jx); \tag{2.31}$$

then we have for each j,

$$h_j \in C_0^\infty(\mathbb{R}^n),\ h_j \geq 0,\ \int h_j(x)\, dx = 1,\ \operatorname{supp} h_j \subset \underline{B}(0,\tfrac{1}{j}). \tag{2.32}$$

The sequence $(h_j)_{j \in \mathbb{N}}$ is often called an *approximate unit*. This refers to the approximation property shown in Theorem 2.10 below, generalized to distributions in Chapter 3.

We shall study convolutions with the functions h_j. Let $u \in L_{1,\text{loc}}(\mathbb{R}^n)$ and consider $h_j * u$,

$$(h_j * u)(x) = \int_{B(0,\frac{1}{j})} h_j(y)u(x-y)\,dy = \int_{B(x,\frac{1}{j})} h_j(x-y)u(y)\,dy \,. \quad (2.33)$$

Concerning supports of these functions it is clear that if dist $(x, \mathrm{supp}\, u) > \frac{1}{j}$, then $(h_j * u)(x) = 0$. Thus $\mathrm{supp}(h_j * u)$ is contained in the closed set

$$\mathrm{supp}(h_j * u) \subset \mathrm{supp}\, u + \underline{B}(0, \tfrac{1}{j}) \,. \quad (2.34)$$

In particular, if u has compact support, then $h_j * u$ has a slightly larger compact support.

Lemma 2.9. *When* $u \in L_{1,\mathrm{loc}}(\mathbb{R}^n)$, *then* $h_j * u \in C^\infty(\mathbb{R}^n)$, *and*

$$\partial^\alpha(h_j * u) = (\partial^\alpha h_j) * u \quad \text{for all } \alpha \in \mathbb{N}_0^n \,. \quad (2.35)$$

Proof. Let x_0 be an arbitrary point of $\mathbb{R}^n$; we shall show that $h_j * u$ is C^∞ on a neighborhood of the point and satisfies (2.35) there. When $x \in B(x_0, 1)$, then $h_j(x - y)$ vanishes for $y \notin B(x_0, 2)$, so we may write

$$(h_j * u)(x) = \int_{B(x_0,2)} h_j(x-y)u(y)\,dy \,, \quad (2.36)$$

for such x. Note that

$$\partial_x^\alpha h_j(x) = j^{n+|\alpha|}\partial_y^\alpha h(y)|_{y=jx}, \text{ so } \sup_x |\partial_x^\alpha h_j(x)| = j^{n+|\alpha|}\sup_x |\partial_x^\alpha h(x)|, \quad (2.37)$$

and hence the x-dependent family of functions $h_j(x - y)u(y)$ and its x-derivatives are bounded by multiples of $|u(y)|$:

$$|h_j(x-y)u(y)| \leq C_j|u(y)|, \ |\partial_x^\alpha h_j(x-y)u(y)| \leq C_{\alpha,j}|u(y)|. \quad (2.38)$$

Since u is integrable on $B(x_0, 2)$, we can first use Lemma 2.8 1° (with $k = n$) to see that $(h_j * u)(x)$ is continuous at the points $x \in B(x_0, 1)$. Next, we can use Lemma 2.8 2° for each of the partial derivatives $\frac{\partial}{\partial x_k}$, $k = 1, \ldots, n$, where $x \in B(x_0, 1)$, keeping all but one coordinate fixed. This gives that $\frac{\partial}{\partial x_k}(h_j * u)(x)$ exists and equals the continuous function

$$\frac{\partial}{\partial x_k}(h_j * u)(x) = \int_{B(x_0,2)} \frac{\partial}{\partial x_k}h_j(x-y)u(y)\,dy, \ k = 1, \ldots, n,$$

for $x \in B(x_0, 1)$. Here is a new formula where we can apply the argument again, showing that $\frac{\partial}{\partial x_l}\frac{\partial}{\partial x_k}(h_j * u)(x)$ exists and equals $((\frac{\partial}{\partial x_l}\frac{\partial}{\partial x_k}h_j) * u)(x)$. By induction we include all derivatives and obtain (2.35). $\quad\square$

One can place the differentiations on u, to the extent that u has well-defined partial derivatives, by a variant of the above arguments (Exercise 2.6).

Theorem 2.10. 1° *When v is continuous and has compact support in $\mathbb{R}^n$, i.e., $v \in C_0^0(\mathbb{R}^n)$ (cf. (C.8)), then $h_j * v \to v$ for $j \to \infty$ uniformly, hence also in $L_p(\mathbb{R}^n)$ for any $p \in [1, \infty]$ (in particular in $C_{L_\infty}^0(\mathbb{R}^n)$, cf. (C.12)).*
   2° *For any $p \in [1, \infty]$ one has that*

$$\|h_j * u\|_{L_p} \le \|u\|_{L_p} \text{ for } u \in L_p(\mathbb{R}^n) . \tag{2.39}$$

   3° *When $p \in [1, \infty[$ and $u \in L_p(\mathbb{R}^n)$, then $h_j * u \to u$ in $L_p(\mathbb{R}^n)$ for $j \to \infty$. Moreover, $C_0^\infty(\mathbb{R}^n)$ is dense in $L_p(\mathbb{R}^n)$.*

Proof. 1°. When v is continuous with compact support, then v is *uniformly continuous* and one has for $x \in \mathbb{R}^n$:

$$|(h_j * v)(x) - v(x)| = | \int_{B(0,\frac{1}{j})} v(x - y)h_j(y)dy - \int_{B(0,\frac{1}{j})} v(x)h_j(y)dy |$$

$$\le \sup_{y \in B(0,\frac{1}{j})} |v(x - y) - v(x)| \le \varepsilon_j , \tag{2.40}$$

where $\varepsilon_j \to 0$ for $j \to \infty$, *independently of x.* It follows immediately that $h_j * v \to v$ pointwise and in sup-norm, and one finds by integration over the compact set $\operatorname{supp} v + \underline{B}(0, 1)$ that $h_j * v \to v$ in L_p for $p \in [1, \infty]$.
   2°. The inequality is for $1 < p < \infty$ a consequence of Hölder's inequality (A.24), where we set $f(y) = h_j(x - y)^{1/p} u(y)$ and $g(y) = h_j(x - y)^{1/p'}$:

$$\|h_j * u\|_{L_p}^p = \int | \int h_j(x - y)u(y)dy |^p dx$$

$$\le \int (\int h_j(x - y)|u(y)|^p dy)(\int h_j(x - y)dy)^{p/p'} dx \tag{2.41}$$

$$= \iint h_j(x - y)|u(y)|^p dy \, dx = \|u\|_{L_p}^p,$$

using (2.32) and the Fubini theorem. In the cases $p = 1$ and $p = \infty$ one uses suitable variants of this argument.
   3°. We here use the result known from measure theory that when $p < \infty$, the functions in $L_p(\mathbb{R}^n)$ may be approximated in L_p norm by continuous functions with compact support. Let $u \in L_p(\mathbb{R}^n)$, let $\varepsilon > 0$ and let $v \in C_0^0(\mathbb{R}^n)$ with $\|u - v\|_{L_p} \le \varepsilon/3$. By 1°, j_0 can be chosen so large that

$$\|h_j * v - v\|_{L_p} \le \varepsilon/3 \text{ for } j \ge j_0 .$$

Then by (2.39),

$$\|h_j * u - u\|_{L_p} \le \|h_j * (u - v)\|_{L_p} + \|h_j * v - v\|_{L_p} + \|v - u\|_{L_p}$$

$$\le 2\|v - u\| + \varepsilon/3 \le \varepsilon , \quad \text{for } j \ge j_0 ,$$

which shows that $h_j * u \to u$ in L_p for $j \to \infty$. The last assertion is seen from the fact that in this construction, $h_j * v$ approximates u. $\square$

The theorem shows how sequences of smooth functions $h_j * u$ approximate u in a number of different spaces. We can extend this to still other spaces.

Lemma 2.11. *For every $p \in [1, \infty[$, $C_0^\infty(\mathbb{R}^n)$ is dense in $L_{p,\text{loc}}(\mathbb{R}^n)$.*

Proof. Note first that $\chi(x/N)u \to u$ in $L_{p,\text{loc}}$ for $N \to \infty$ (cf. (2.3)). Namely, $\chi(x/N) = 1$ for $|x| \leq N$, and hence for any j,

$$p_j(\chi(x/N)u - u) \equiv \int_{B(0,j)} |\chi(x/N)u - u|^p \, dx = 0 \quad \text{for } N \geq j, \qquad (2.42)$$

so $p_j(\chi(x/N)u - u) \to 0$ for $N \to \infty$, any j, whereby $\chi(x/N)u - u \to 0$ in $L_{p,\text{loc}}(\mathbb{R}^n)$. The convergence of course holds in any metric defining the topology. Now $\chi(x/N)u \in L_p(\mathbb{R}^n)$, and $h_l * (\chi(x/N)u) \to \chi(x/N)u$ in $L_p(\mathbb{R}^n)$ by Theorem 2.10, with $h_l * (\chi(x/N)u) \in C_0^\infty(\mathbb{R}^n)$ supported in $\underline{B}(0, 2N + \frac{1}{l})$; the convergence also holds in $L_{p,\text{loc}}(\mathbb{R}^n)$. We conclude that the functions in $L_{p,\text{loc}}(\mathbb{R}^n)$ may be approximated by test functions, with respect to the topology of $L_{p,\text{loc}}(\mathbb{R}^n)$. $\qquad \square$

When we consider $u \in L_{p,\text{loc}}(\Omega)$ for an open subset Ω of $\mathbb{R}^n$, the expression $(h_j * u)(x)$ is usually not well-defined for x close to the boundary. But one does have the following result:

Lemma 2.12. *Let $u \in L_{p,\text{loc}}(\Omega)$ for some $p \in [1, \infty[$ and let $\varepsilon > 0$. When $j > 1/\varepsilon$, then*

$$v_j(x) = (h_j * u)(x) = \int_{B(0,\frac{1}{j})} h_j(y)u(x - y)dy \qquad (2.43)$$

is defined for x in the set

$$\Omega_\varepsilon = \{\, x \in \Omega \mid \text{dist}\,(x, \complement\Omega) > \varepsilon \,\}, \qquad (2.44)$$

and one has for any $R > 0$

$$\left(\int_{\Omega_\varepsilon \cap B(0,R)} |u(x) - v_j(x)|^p dx \right)^{1/p} \to 0 \quad \text{for } j \to \infty. \qquad (2.45)$$

Proof. Let $j > 1/\varepsilon$, then $v_j(x)$ is defined for $x \in \Omega_\varepsilon$. In the calculation of the integral (2.45), when $j > 2/\varepsilon$ one only uses the values of u on $K_{\varepsilon,R} = \overline{\Omega}_\varepsilon \cap \underline{B}(0, R) + \underline{B}(0, \varepsilon/2)$, which is a compact subset of Ω. We can then replace u by

$$u_1(x) = \begin{cases} u(x) & \text{for } x \in K_{\varepsilon,R}, \\ 0 & \text{otherwise.} \end{cases} \qquad (2.46)$$

Here $u_1 \in L_p(\mathbb{R}^n)$, whereby $v_j = h_j * u_1$ on $\Omega_\varepsilon \cap B(0, R)$ and the result follows from Theorem 2.10. $\qquad \square$

Other types of approximation results in $L_{p,\mathrm{loc}}(\Omega)$ can be obtained by use of more refined cut-off functions than those in Lemma 2.1.

Theorem 2.13. *Let M be a subset of $\mathbb{R}^n$, let $\varepsilon > 0$, and set $M_{k\varepsilon} = \overline{M} + \underline{B}(0, k\varepsilon)$ for $k > 0$. There exists a function $\eta \in C^\infty(\mathbb{R}^n)$ which is 1 on M_ε and is supported in $M_{3\varepsilon}$, and which satisfies $0 \le \eta(x) \le 1$ for all $x \in \mathbb{R}^n$.*

Proof. The function

$$\psi(x) = \begin{cases} 1 & \text{on } M_{2\varepsilon}, \\ 0 & \text{on } \mathbb{R}^n \setminus M_{2\varepsilon}, \end{cases} \tag{2.47}$$

is in $L_{1,\mathrm{loc}}(\mathbb{R}^n)$, and for $j \ge 1/\varepsilon$, the function $h_j * \psi$ is nonnegative and C^∞ with support in

$$M_{2\varepsilon} + \underline{B}(0, \tfrac{1}{j}) \subset M_{3\varepsilon}.$$

When $x \in M_\varepsilon$, we have that $\psi = 1$ on the ball $\underline{B}(x, \varepsilon)$, and hence $(h_j * \psi)(x) = \int_{B(0,\frac{1}{j})} h_j(y)\psi(x-y)dy = 1$ when $j \ge 1/\varepsilon$. The function takes values in $[0,1]$ elsewhere. Thus, as the function η we can use $h_j * \psi$ for $j \ge 1/\varepsilon$. $\qquad \square$

Observe that η in Theorem 2.13 has compact support when $\overline{M}$ is compact. One often needs the following special cases:

Corollary 2.14. $1°$ *Let Ω be open and let K be compact $\subset \Omega$. There is a function $\eta \in C_0^\infty(\Omega)$ taking values in $[0,1]$ such that $\eta = 1$ on a neighborhood of K.*

$2°$ *Let K_j, $j \in \mathbb{N}$, be a sequence of compact sets as in (2.4). There is a sequence of functions $\eta_j \in C_0^\infty(\Omega)$ taking values in $[0,1]$ such that $\eta_j = 1$ on a neighborhood of K_j and $\operatorname{supp}\eta_j \subset K_{j+1}^\circ$.*

Proof. We use Theorem 2.13, noting that $\operatorname{dist}(K, \complement\Omega) > 0$ and that for all j, $\operatorname{dist}(K_j, \complement K_{j+1}) > 0$. $\qquad \square$

Using these functions we can moreover show:

Theorem 2.15. *Let Ω be open $\subset \mathbb{R}^n$.*

$1°$ *$C_0^\infty(\Omega)$ is dense in $C^\infty(\Omega)$.*
$2°$ *$C_0^\infty(\Omega)$ is dense in $L_{p,\mathrm{loc}}(\Omega)$ for all $p \in [1, \infty[$.*
$3°$ *$C_0^\infty(\Omega)$ is dense in $L_p(\Omega)$ for all $p \in [1, \infty[$.*

Proof. $1°$. Let $u \in C^\infty(\Omega)$. Choosing η_j as in Corollary 2.14 one has that $\eta_j u \in C_0^\infty(\Omega)$ and $\eta_j u \to u$ in $C^\infty(\Omega)$ for $j \to \infty$ (since $\eta_l u = u$ on K_j for $l \ge j$).

$2°$. Let $u \in L_{p,\mathrm{loc}}(\Omega)$. Now $\eta_l u \in L_p(\Omega)$ with support in K_{l+1}, and $\eta_l u \to u$ in $L_{p,\mathrm{loc}}(\Omega)$ for $l \to \infty$, since $\eta_l u = u$ on K_j for $l \ge j$. Next, $h_k * \eta_l u \to \eta_l u$ in $L_p(\mathbb{R}^n)$ for $k \to \infty$ by Theorem 2.10. Since $\operatorname{supp}(h_k * \eta_l u) \subset K_{l+2}$ for k sufficiently large, this is also a convergence in $L_{p,\mathrm{loc}}(\Omega)$.

$3°$. Let $u \in L_p(\Omega)$. Again, $\eta_l u \in L_p(\Omega)$, and now $\eta_l u \to u$ in $L_p(\Omega)$ by the theorem of Lebesgue (since $0 \le \eta_l \le 1$). The proof is completed as under $2°$. $\qquad \square$

2.4 Partitions of unity

The special test functions from Corollary 2.14 are also used in the construction of a so-called "partition of unity", i.e., a system of smooth functions with supports in given sets and sum 1 on a given set.

We shall show two versions, in Theorems 2.16 and 2.17; it is the latter that is most often used. In this text we shall not need it until Lemma 3.11.

For Theorem 2.16 we fill out an open set Ω with a countable family of bounded open subsets V_j that is *locally finite* in Ω, i.e., each compact subset of Ω has nonempty intersection with only a finite number of the V_j's. Moreover, we require that the V_j can be shrunk slightly to open sets V_j' with $\overline{V_j'} \subset V_j$ such that the union of the V_j' still covers Ω. As an example of this situation, take the sets

$$V_0 = K_4^{\circ}, \quad V_j = K_{j+4}^{\circ} \setminus K_j \text{ for } j \in \mathbb{N},$$
$$V_0' = K_3^{\circ}, \quad V_j' = K_{j+3}^{\circ} \setminus K_{j+1} \text{ for } j \in \mathbb{N},$$
(2.48)

where the K_j are as in (2.4). This system is locally finite since every compact subset of Ω is contained in some K_{j_0}, hence does not meet the V_j with $j \geq j_0$.

Theorem 2.16. *Let the open set Ω be a union of bounded open sets V_j with $\overline{V_j} \subset \Omega$, $j \in \mathbb{N}_0$, for which there are open subsets V_j' such that $\overline{V_j'} \subset V_j$ and we still have $\bigcup_{j \in \mathbb{N}_0} V_j' = \Omega$. Assume, moreover, that the cover $\{V_j\}_{j \in \mathbb{N}_0}$ is locally finite in Ω. Then there is a family of functions $\psi_j \in C_0^{\infty}(V_j)$ taking values in $[0, 1]$ such that*

$$\sum_{j \in \mathbb{N}_0} \psi_j(x) = 1 \text{ for all } x \in \Omega.$$
(2.49)

Proof. Since $\overline{V_j'}$ is a compact subset of V_j, we can for each j choose a function $\zeta_j \in C_0^{\infty}(V_j)$ that is 1 on V_j' and takes values in $[0, 1]$, by Corollary 2.14. Now

$$\Psi(x) = \sum_{j \in \mathbb{N}_0} \zeta_j(x)$$

is a well-defined C^{∞}-function on Ω, since any point $x \in \Omega$ has a compact neighborhood in Ω where only finitely many of the functions ζ_j are nonzero. Moreover, $\Psi(x) \geq 1$ at all $x \in \Omega$, since each x is in V_j' for some j. Then let

$$\psi_j(x) = \frac{\zeta_j(x)}{\Psi(x)} \text{ for } x \in \Omega, \ j \in \mathbb{N}_0.$$

The system $\{\psi_j\}_{j \in \mathbb{N}_0}$ has the desired properties. □

We say that $\{\psi_j\}_{j \in \mathbb{N}_0}$ is a *partition of unity* (cf. (2.49)) for Ω subordinate to the cover $\{V_j\}_{j \in \mathbb{N}_0}$.

The other partition of unity version we need is as follows:

Theorem 2.17. *Let K be a compact subset of $\mathbb{R}^n$, and let $\{V_j\}_{j=0}^N$ be a bounded open cover of K (i.e., the V_j are bounded and open in $\mathbb{R}^n$, and $K \subset \bigcup_{j=0}^N V_j$). There exists a family of functions $\psi_j \in C_0^\infty(V_j)$ taking values in $[0,1]$ such that*

$$\sum_{j=0}^N \psi_j(x) = 1 \text{ for } x \in K. \tag{2.50}$$

Proof. Let us first show that there exist open sets $V_j' \subset V_j$, still forming a cover $\{V_j'\}_{j=0}^N$ of K, such that $\overline{V_j'}$ is a compact subset of V_j for each j. For this, let $V_{jl} = \{x \in V_j \mid \operatorname{dist}(x, \partial V_j) > \frac{1}{l}\}$, then the family of sets $\{V_{jl}\}_{j=0,\dots,N; l \in \mathbb{N}}$ forms an open cover of K. Since K is compact, there is a finite subfamily that still covers K; here since $V_{jl} \subset V_{jl'}$ for $l < l'$, we can reduce to a system where there is at most one l for each j. Use these V_{jl} as V_j', and supplement by $V_j' = V_{j1}$ for each of those values of j that are not represented in the system.

Now use Corollary 2.14 for each j to choose $\zeta_j \in C_0^\infty(V_j)$, equal to 1 on V_j' and taking values in $[0,1]$. Then

$$\Psi(x) = \sum_{j=0}^N \zeta_j(x) \geq 1 \text{ for } x \in \bigcup_{j=0}^N V_j' \supset K. \tag{2.51}$$

Since $\bigcup_{j=0}^N V_j'$ is an open set containing K, we can use Corollary 2.14 once more to find a function $\varphi \in C_0^\infty(\bigcup_{j=0}^N V_j')$ that is 1 on K and takes values in $[0,1]$. Now set

$$\psi_j(x) = \begin{cases} \zeta_j(x)\dfrac{\varphi(x)}{\Psi(x)} & \text{on } \bigcup_{j=0}^N V_j', \\ 0 & \text{elsewhere;} \end{cases} \tag{2.52}$$

it is a well-defined C^∞-function supported in a compact subset of V_j and taking values in $[0,1]$, and the family of functions ψ_j clearly satisfies (2.50). $\square$

In this case we say that $\{\psi_j\}_{i=0}^N$ is a partition of unity for K subordinate to the cover $\{V_j\}_{i=0}^N$.

Exercises for Chapter 2

2.1. Let $\varphi(x)$ be analytic on an open interval I of $\mathbb{R}$. Show that if $\varphi \in C_0^\infty(I)$, then $\varphi \equiv 0$.

2.2. Show that the topology on $C_0^\infty(\Omega)$ is independent of which system of compact subsets satisfying (2.4) is used. (This amounts to a comparison of the corresponding systems of seminorms.)

2.3. Show that convergence of a sequence in $C_0^\infty(\Omega)$ implies convergence of the sequence in $C^\infty(\Omega)$, in $L_p(\Omega)$ and in $L_{p,\text{loc}}(\Omega)$ (for $p \in [1, \infty]$).

2.4. Show that $L_{2,\text{comp}}(\Omega)$ can be identified with the dual space $(L_{2,\text{loc}}(\Omega))^*$ of $L_{2,\text{loc}}(\Omega)$ (the space of continuous linear functionals on $L_{2,\text{loc}}(\Omega)$) in such a way that the element $v \in L_{2,\text{comp}}(\Omega)$ corresponds to the functional

$$u \mapsto \int u(x)\overline{v}(x)\, dx\ , \quad u \in L_{2,\text{loc}}(\Omega)\ .$$

One can use Lemma B.7. (You are just asked to establish the identification for each element.)

2.5. Show that when $\varphi \in C_0^\infty(\mathbb{R}^n)$ and $\text{supp}\,\varphi \subset \underline{B}(0, R)$, then

$$\sup |\varphi(x)| \le 2R \sup |\partial_{x_1}\varphi(x)|\ .$$

(*Hint.* Express φ as an integral of $\partial_{x_1}\varphi$.)

2.6. Show that when $u \in C^1(\mathbb{R}^n)$, then

$$\partial_k(h_j * u) = h_j * \partial_k u\ .$$

(Even if u is not assumed to have compact support, it is only the behavior of u on a compact set that is used when one investigates the derivative at a point.)

2.7. (a) Show that $C_0^\infty(\Omega)$ is dense in $C^k(\Omega)$ for each $k \in \mathbb{N}_0$.
(b) Find out whether $C_0^\infty(\Omega)$ is dense in $C^k(\overline{\Omega})$, in the case $\Omega =]0, 1[\subset \mathbb{R}$, for $k \in \mathbb{N}_0$.

2.8. Show that $L_{2,\text{loc}}(\Omega)$ can be identified with $(L_{2,\text{comp}}(\Omega))^*$ in such a way that the element $v \in L_{2,\text{loc}}(\Omega)$ corresponds to the functional

$$u \mapsto \int u(x)\overline{v}(x)\, dx\ , \quad u \in L_{2,\text{comp}}(\Omega)\ .$$

2.9. Verify (2.34) in detail.

Chapter 3
Distributions. Examples and rules of calculus

3.1 Distributions

The space $C_0^\infty(\Omega)$ is often denoted $\mathscr{D}(\Omega)$ in the literature. The distributions are simply the elements of the dual space:

Definition 3.1. A distribution on Ω is a continuous linear functional on $C_0^\infty(\Omega)$. The vector space of distributions on Ω is denoted $\mathscr{D}'(\Omega)$. When $\Lambda \in \mathscr{D}'(\Omega)$, we denote the value of Λ on $\varphi \in C_0^\infty(\Omega)$ by $\Lambda(\varphi)$ or $\langle \Lambda, \varphi \rangle$.

The tradition is here to take linear (rather than conjugate linear) functionals. But it is easy to change to conjugate linear functionals if needed, for $\varphi \mapsto \Lambda(\varphi)$ is a linear functional on $C_0^\infty(\Omega)$ if and only if $\varphi \mapsto \Lambda(\overline{\varphi})$ is a conjugate linear functional.

See Theorem 2.5 (d) for how the continuity of a functional on $C_0^\infty(\Omega)$ is checked.

The space $\mathscr{D}'(\Omega)$ itself is provided with the weak* topology, i.e., the topology defined by the system of seminorms p_φ on $\mathscr{D}'(\Omega)$:

$$p_\varphi : u \mapsto |\langle u, \varphi \rangle|, \tag{3.1}$$

where φ runs through $C_0^\infty(\Omega)$. We here use Theorem B.5, noting that the family of seminorms is separating (since $u \neq 0$ in $\mathscr{D}'(\Omega)$ means that $\langle u, \varphi \rangle \neq 0$ for some φ).

Let us consider some *examples*. When $f \in L_{1,\mathrm{loc}}(\Omega)$, then the map

$$\Lambda_f : \varphi \mapsto \int_\Omega f(x)\varphi(x)\, dx \tag{3.2}$$

is a distribution. Indeed, we have on every K_j (cf. (2.4)), when $\varphi \in C_{K_j}^\infty(\Omega)$,

$$|\Lambda_f(\varphi)| = \left| \int_{K_j} f(x)\varphi(x)dx \right| \leq \sup |\varphi(x)| \int_{K_j} |f(x)|dx, \tag{3.3}$$

so (2.15) is satisfied with $N_j = 0$ and $c_j = \|f\|_{L_1(K_j)}$. Here one can in fact *identify* Λ_f *with* f, in view of the following fact:

Lemma 3.2. *When* $f \in L_{1,\mathrm{loc}}(\Omega)$ *with* $\int f(x)\varphi(x)dx = 0$ *for all* $\varphi \in C_0^\infty(\Omega)$, *then* $f = 0$.

Proof. Let $\varepsilon > 0$ and consider $v_j(x) = (h_j * f)(x)$ for $j > 1/\varepsilon$ as in Lemma 2.12. When $x \in \Omega_\varepsilon$, then $h_j(x - y) \in C_0^\infty(\Omega)$, so that $v_j(x) = 0$ in Ω_ε. From (2.45) we conclude that $f = 0$ in $\Omega_\varepsilon \cap B(0,R)$. Since ε and R can take all values in $\mathbb{R}_+$, it follows that $f = 0$ in Ω. $\square$

The lemma (and variants of it) is sometimes called "the fundamental lemma of the calculus of variations" or "Du Bois-Reymond's lemma".

The lemma implies that when the *distribution* Λ_f defined from $f \in L_{1,\mathrm{loc}}(\Omega)$ by (3.2) gives 0 on all test functions, then the *function* f is equal to 0 as an element of $L_{1,\mathrm{loc}}(\Omega)$. Then the map $f \mapsto \Lambda_f$ is *injective* from $L_{1,\mathrm{loc}}(\Omega)$ to $\mathscr{D}'(\Omega)$, so that we may identify f with Λ_f and write

$$L_{1,\mathrm{loc}}(\Omega) \subset \mathscr{D}'(\Omega). \tag{3.4}$$

The element 0 of $\mathscr{D}'(\Omega)$ will from now on be identified with the function 0 (where we as usual take the continuous representative).

Since $L_{p,\mathrm{loc}}(\Omega) \subset L_{1,\mathrm{loc}}(\Omega)$ for $p > 1$, these spaces are also naturally injected in $\mathscr{D}'(\Omega)$.

Remark 3.3. Let us also mention how Radon measures fit in here. The space $C_0^0(\Omega)$ of continuous functions with compact support in Ω is defined in (C.8). In topological measure theory it is shown how the vector space $\mathcal{M}(\Omega)$ of complex *Radon measures* μ on Ω can be identified with the space of continuous linear functionals Λ_μ on $C_0^0(\Omega)$ in such a way that

$$\Lambda_\mu(\varphi) = \int_{\mathrm{supp}\,\varphi} \varphi\,d\mu \quad \text{for} \quad \varphi \in C_0^0(\Omega).$$

Since one has that

$$|\Lambda_\mu(\varphi)| \leq |\mu|(\mathrm{supp}\,\varphi) \cdot \sup|\varphi(x)|, \tag{3.5}$$

Λ_μ is continuous on $C_0^\infty(\Omega)$, hence defines a distribution $\Lambda_\mu' \in \mathscr{D}'(\Omega)$. Since $C_0^\infty(\Omega)$ *is dense in* $C_0^0(\Omega)$ (cf. Theorem 2.15 1°), the map $\Lambda_\mu \mapsto \Lambda_\mu'$ is *injective*. Then the space of complex Radon measures identifies with a subset of $\mathscr{D}'(\Omega)$:

$$\mathcal{M}(\Omega) \subset \mathscr{D}'(\Omega). \tag{3.6}$$

The inclusions (3.4) and (3.6) place $L_{1,\mathrm{loc}}(\Omega)$ and $\mathcal{M}(\Omega)$ as subspaces of $\mathscr{D}'(\Omega)$. They are consistent with the usual injection of $L_{1,\mathrm{loc}}(\Omega)$ in $\mathcal{M}(\Omega)$, where a function $f \in L_{1,\mathrm{loc}}(\Omega)$ defines the Radon measure μ_f by the formula

$$\mu_f(K) = \int_K f\,dx \ \text{ for } K \text{ compact } \subset \Omega. \tag{3.7}$$

Indeed, it is known from measure theory that

$$\int f\varphi\,dx = \int \varphi\,d\mu_f \quad \text{for all } \varphi \in C_0^0(\Omega) \tag{3.8}$$

(hence in particular for $\varphi \in C_0^\infty(\Omega)$), so the distributions Λ_f and Λ_{μ_f} coincide. When $f \in L_{1,\text{loc}}(\Omega)$, we shall usually write f instead of Λ_f; then we also write

$$\Lambda_f(\varphi) = \langle \Lambda_f, \varphi \rangle = \langle f, \varphi \rangle = \int_\Omega f(x)\varphi(x)dx. \tag{3.9}$$

Moreover, one often writes μ instead of Λ_μ when $\mu \in \mathcal{M}(\Omega)$. In the following we shall even use the notation f or u (resembling a function) to indicate an arbitrary distribution!

In the systematical theory we will in particular be concerned with the inclusions

$$C_0^\infty(\Omega) \subset L_2(\Omega) \subset \mathscr{D}'(\Omega) \tag{3.10}$$

(and other L_2-inclusions of importance in Hilbert space theory). We shall show how the large gaps between $C_0^\infty(\Omega)$ and $L_2(\Omega)$, and between $L_2(\Omega)$ and $\mathscr{D}'(\Omega)$, are filled out by Sobolev spaces.

Here is another important *example*.
Let x_0 be a point in Ω. The map

$$\delta_{x_0} : \varphi \mapsto \varphi(x_0) \tag{3.11}$$

sending a test function into its *value at x_0* is a distribution, for it is clearly a linear map from $C_0^\infty(\Omega)$ to $\mathbb{C}$, and one has for any j, when $\text{supp}\,\varphi \subset K_j$ (where K_j is as in (2.4)),

$$|\langle \delta_{x_0}, \varphi \rangle| = |\varphi(x_0)| \le \sup\{\,|\varphi(x)| \mid x \in K_j\,\} \tag{3.12}$$

(note that $\varphi(x_0) = 0$ when $x_0 \notin K_j$). Here (2.15) is satisfied with $c_j = 1$, $N_j = 0$, for all j. In a similar way one finds that the maps

$$\Lambda_\alpha : \varphi \mapsto (D^\alpha \varphi)(x_0) \tag{3.13}$$

are distributions, now with $c_j = 1$ and $N_j = |\alpha|$ for each j. The distribution (3.11) is the famous "Dirac's δ-function" or "δ-measure". The notation *measure* is correct, for we can write

$$\langle \delta_{x_0}, \varphi \rangle = \int \varphi\,d\mu_{x_0}, \tag{3.14}$$

where μ_{x_0} is the point measure that has the value 1 on the set $\{x_0\}$ and the value 0 on compact sets disjoint from x_0. The notation δ-*function* is a wild "abuse of notation" (see also (3.22) ff. later). Maybe it has survived because it is so bad that the motivation for introducing the concept of distributions becomes clear.

The distribution δ_0 is often just denoted δ.

Still other distributions are obtained in the following way: Let $f \in L_{1,loc}(\Omega)$ and let $\alpha \in \mathbb{N}_0^n$. Then the map

$$\Lambda_{f,\alpha} : \varphi \mapsto \int f(x)(D^\alpha \varphi)(x)dx \ , \ \varphi \in C_0^\infty(\Omega), \tag{3.15}$$

is a distribution, since we have for any $\varphi \in C_{K_j}^\infty(\Omega)$:

$$|\langle \Lambda_{f,\alpha}, \varphi \rangle| = \left| \int_{K_j} f \, D^\alpha \varphi \, dx \right| \leq \int_{K_j} |f(x)| dx \cdot \sup_{x \in K_j} |D^\alpha \varphi(x)|; \tag{3.16}$$

here (2.15) is satisfied with $c_j = \|f\|_{L_1(K_j)}$ and $N_j = |\alpha|$ for each j.

One can show that the most general distributions are not much worse than this last example. One has in fact that when Λ is an arbitrary distribution, then for any fixed compact set $K \subset \Omega$ there is an N (depending on K) and a system of functions $f_\alpha \in C^0(\Omega)$ for $|\alpha| \leq N$ such that

$$\langle \Lambda, \varphi \rangle = \sum_{|\alpha| \leq N} \langle f_\alpha, D^\alpha \varphi \rangle \ \text{ for } \ \varphi \in C_K^\infty(\Omega) \tag{3.17}$$

(the *Structure theorem*). We shall show this later in connection with the theorem of Sobolev in Chapter 6.

In the fulfillment of (2.15) one cannot always find an N that works for *all* $K_j \subset \Omega$ (only one N_j for each K_j); another way of expressing this is to say that a distribution does not necessarily have a finite *order*, where the concept of order is defined as follows:

Definition 3.4. We say that $\Lambda \in \mathscr{D}'(\Omega)$ is **of order** $N \in \mathbb{N}_0$ when the inequalities (2.15) hold for Λ with $N_j \leq N$ for all j (but the constants c_j may very well depend on j). Λ is said to be of **infinite order** if it is not of order N for any N; otherwise it is said to be of finite order. **The order** of Λ is the smallest N that can be used, resp. ∞.

In all the examples we have given, the order is finite. Namely, $L_{1,loc}(\Omega)$ and $\mathcal{M}(\Omega)$ define distributions of order 0 (cf. (3.3), (3.5) and (3.12)), whereas Λ_α and $\Lambda_{f,\alpha}$ in (3.13) and (3.15) are of order $|\alpha|$. To see an example of a distribution of infinte order we consider the distribution $\Lambda \in \mathscr{D}'(\mathbb{R})$ defined by

$$\langle \Lambda, \varphi \rangle = \sum_{N=1}^\infty \langle 1_{[N,2N]}, \varphi^{(N)}(x) \rangle, \tag{3.18}$$

cf. (A.27). (As soon as we have defined the notion of *support* of a distribution it will be clear that when a distribution has *compact* support in Ω, its order is finite, cf. Theorem 3.12 below.)

The theory of distributions was introduced systematically by L. Schwartz; his monograph [S50] is still a principal reference in the literature on distributions.

3.2 Rules of calculus for distributions

When T is a continuous linear operator in $C_0^\infty(\Omega)$, and $\Lambda \in \mathscr{D}'(\Omega)$, then the composition defines another element $\Lambda T \in \mathscr{D}'(\Omega)$, namely, the functional

$$(\Lambda T)(\varphi) = \langle \Lambda, T\varphi \rangle.$$

The map $T^\times : \Lambda \mapsto \Lambda T$ in $\mathscr{D}'(\Omega)$ is simply the *adjoint map* of the map $\varphi \mapsto T\varphi$. (We write $T^\times$ to avoid conflict with the notation for taking adjoints of operators in complex Hilbert spaces, where a certain conjugate linearity has to be taken into account. The notation T' may also be used, but the prime could be misunderstood as differentiation.)

As shown in Theorem 2.6, the following simple maps are continuous in $C_0^\infty(\Omega)$:

$$M_f : \varphi \mapsto f\varphi, \quad \text{when } f \in C^\infty(\Omega),$$
$$D^\alpha : \varphi \mapsto D^\alpha \varphi.$$

They induce two maps in $\mathscr{D}'(\Omega)$ that we shall temporarily denote $M_f^\times$ and $(D^\alpha)^\times$:

$$\langle M_f^\times \Lambda, \varphi \rangle = \langle \Lambda, f\varphi \rangle,$$
$$\langle (D^\alpha)^\times \Lambda, \varphi \rangle = \langle \Lambda, D^\alpha \varphi \rangle,$$

for $\Lambda \in \mathscr{D}'(\Omega)$ and $\varphi \in C_0^\infty(\Omega)$.

How do these new maps look when Λ itself is a function? If $\Lambda = v \in L_{1,\text{loc}}(\Omega)$, then

$$\langle M_f^\times v, \varphi \rangle = \langle v, f\varphi \rangle = \int v(x)f(x)\varphi(x)dx = \langle fv, \varphi \rangle;$$

hence

$$M_f^\times v = fv, \quad \text{when } v \in L_{1,\text{loc}}(\Omega). \tag{3.19}$$

When $v \in C^\infty(\Omega)$,

$$\langle (D^\alpha)^\times v, \varphi \rangle = \langle v, D^\alpha \varphi \rangle = \int v(x)(D^\alpha \varphi)(x)dx$$

$$= (-1)^{|\alpha|} \int (D^\alpha v)(x)\varphi(x)dx = \langle (-1)^{|\alpha|} D^\alpha v, \varphi \rangle,$$

so that

$$(-1)^{|\alpha|} (D^\alpha)^\times v = D^\alpha v , \quad \text{when } v \in C^\infty(\Omega). \tag{3.20}$$

These formulas motivate the following definition.

Definition 3.5. 1° When $f \in C^\infty(\Omega)$, we define the multiplication operator M_f in $\mathscr{D}'(\Omega)$ by

$$\langle M_f \Lambda, \varphi \rangle = \langle \Lambda, f\varphi \rangle \quad \text{for } \varphi \in C_0^\infty(\Omega).$$

Instead of M_f we often just write f.

2° For any $\alpha \in \mathbb{N}_0^n$, the differentiation operator D^α in $\mathscr{D}'(\Omega)$ is defined by

$$\langle D^\alpha \Lambda, \varphi \rangle = \langle \Lambda, (-1)^{|\alpha|} D^\alpha \varphi \rangle \quad \text{for } \varphi \in C_0^\infty(\Omega).$$

Similarly, we define the operator ∂^α in $\mathscr{D}'(\Omega)$ by

$$\langle \partial^\alpha \Lambda, \varphi \rangle = \langle \Lambda, (-1)^{|\alpha|} \partial^\alpha \varphi \rangle \quad \text{for } \varphi \in C_0^\infty(\Omega).$$

In particular, these extensions still satisfy: $D^\alpha \Lambda = (-i)^{|\alpha|} \partial^\alpha \Lambda$.

The definition really just says that we denote the adjoint of $M_f : \mathscr{D}(\Omega) \to \mathscr{D}(\Omega)$ by M_f again (usually abbreviated to f), and that we denote the adjoint of $(-1)^{|\alpha|} D^\alpha : \mathscr{D}(\Omega) \to \mathscr{D}(\Omega)$ by D^α; the motivation for this "abuse of notation" lies in the consistency with classical formulas shown in (3.19) and (3.20). As a matter of fact, the abuse is not very grave, since one can show that $C^\infty(\Omega)$ is a *dense* subset of $\mathscr{D}'(\Omega)$, when the latter is provided with the weak* topology, cf. Theorem 3.18 below, so that the extension of the operators f and D^α from elements $v \in C^\infty(\Omega)$ to $\Lambda \in \mathscr{D}'(\Omega)$ is *uniquely* determined.

Observe also that when $v \in C^k(\Omega)$, the distribution derivatives $D^\alpha v$ coincide with the usual partial derivatives for $|\alpha| \le k$, because of the usual formulas for integration by parts. We may write $(-1)^{|\alpha|} D^\alpha$ as $(-D)^\alpha$.

The exciting aspect of Definition 3.5 is that we can now define *derivatives of distributions* — hence, in particular, derivatives of functions in $L_{1,\text{loc}}$ which were not differentiable in the original sense.

Note that Λ_α and $\Lambda_{f,\alpha}$ defined in (3.13) and (3.15) satisfy

$$\langle \Lambda_\alpha, \varphi \rangle = \langle (-D)^\alpha \delta_{x_0}, \varphi \rangle , \quad \langle \Lambda_{f,\alpha}, \varphi \rangle = \langle (-D)^\alpha f, \varphi \rangle, \tag{3.21}$$

for $\varphi \in C_0^\infty(\Omega)$. Let us consider an important example (already mentioned in Chapter 1):

By $H(x)$ we denote the function on $\mathbb{R}$ defined by

$$H(x) = 1_{\{x>0\}} \qquad (3.22)$$

(cf. (A.27)); it is called the Heaviside function. Since $H \in L_{1,\mathrm{loc}}(\mathbb{R})$, we have that $H \in \mathscr{D}'(\mathbb{R})$. The derivative in $\mathscr{D}'(\mathbb{R})$ is found as follows:

$$\langle \frac{d}{dx}H, \varphi \rangle = \langle H, -\frac{d}{dx}\varphi \rangle = -\int_0^\infty \varphi'(x)dx$$
$$= \varphi(0) = \langle \delta_0, \varphi \rangle \quad \text{for } \varphi \in C_0^\infty(\mathbb{R}).$$

We see that

$$\frac{d}{dx}H = \delta_0, \qquad (3.23)$$

the delta-measure at 0! H and $\frac{d}{dx}H$ are distributions of order 0, while the higher derivatives $\frac{d^k}{dx^k}H$ are of order $k - 1$. As shown already in Example 1.1, there is no $L_{1,\mathrm{loc}}(\mathbb{R})$-function that identifies with δ_0.

There is a similar calculation in higher dimensions, based on the Gauss formula (A.18). Let Ω be an open subset of $\mathbb{R}^n$ with C^1-boundary. The function 1_Ω (cf. (A.27)) has distribution derivatives described as follows: For $\varphi \in C_0^\infty(\mathbb{R}^n)$,

$$\langle \partial_j 1_\Omega, \varphi \rangle \equiv -\int_\Omega \partial_j\varphi \, dx = \int_{\partial\Omega} \nu_j(x)\varphi(x) \, d\sigma. \qquad (3.24)$$

Since

$$\left| \int_{\partial\Omega} \nu_j(x)\varphi(x) \, d\sigma \right| \le \int_{\partial\Omega\cap K} 1 \, d\sigma \cdot \sup_{x \in \partial\Omega\cap K} |\varphi(x)|,$$

when K is a compact set containing supp φ, $\partial_j 1_\Omega$ is a distribution in $\mathscr{D}'(\mathbb{R}^n)$ of order 0; (3.24) shows precisely how it acts.

Another important aspect is that the distributions theory allows us to define derivatives of functions which only to a mild degree lack classical derivatives. Recall that the classical concept of differentiation for functions of several variables only works really well when the partial derivatives are *continuous*, for then we can exchange the order of differentiation. More precisely, $\partial_1\partial_2 u = \partial_2\partial_1 u$ holds when u is C^2, whereas the rule often fails for more general functions (e.g., for $u(x_1, x_2) = |x_1|$, where $\partial_1\partial_2 u$ but not $\partial_2\partial_1 u$ has a classical meaning on $\mathbb{R}^2$).

The new concept of derivative is *insensitive* to the order of differentiation. In fact, $\partial_1\partial_2$ and $\partial_2\partial_1$ define the same operator in $\mathscr{D}'$, since they are carried over to C_0^∞ where they have the same effect:

$$\langle \partial_1\partial_2 u, \varphi \rangle = \langle u, (-\partial_2)(-\partial_1)\varphi \rangle = \langle u, \partial_2\partial_1\varphi \rangle = \langle u, \partial_1\partial_2\varphi \rangle = \langle \partial_2\partial_1 u, \varphi \rangle.$$

In the next lemma, we consider a useful special case of how the distribution definition works for a function that lacks classical derivatives on part of the domain.

Lemma 3.6. *Let $R > 0$, and let $\Omega = B(0, R)$ in $\mathbb{R}^n$; define also $\Omega_\pm = \Omega \cap \mathbb{R}^n_\pm$. Let $k > 0$, and let $u \in C^{k-1}(\overline{\Omega})$ with k-th derivatives defined in Ω_+ and Ω_- in such a way that they extend to continuous functions on $\overline{\Omega}_+$ resp. $\overline{\Omega}_-$ (so u is piecewise C^k). For $|\alpha| = k$, the α-th derivative in the distribution sense is then equal to the function $v \in L_1(\Omega)$ defined by*

$$v = \begin{cases} \partial^\alpha u & on \ \Omega_+, \\ \partial^\alpha u & on \ \Omega_-. \end{cases} \tag{3.25}$$

Proof. Let $|\alpha| = k$, and write $\partial^\alpha = \partial_j \partial^\beta$, where $|\beta| = k - 1$. When $\varphi \in C_0^\infty(\Omega)$, we have if $j = n$ (using the notation $x' = (x_1, \ldots, x_{n-1})$):

$$\langle \partial^\alpha u, \varphi \rangle = -\langle \partial^\beta u, \partial_n \varphi \rangle = -\int_{\Omega_-} \partial^\beta u \partial_n \varphi \, dx - \int_{\Omega_+} \partial^\beta u \partial_n \varphi \, dx$$

$$= \int_{\Omega_-} [(\partial_n \partial^\beta u)\varphi - \partial_n(\partial^\beta u \varphi)] dx + \int_{\Omega_+} [(\partial_n \partial^\beta u)\varphi - \partial_n(\partial^\beta u \varphi)] dx$$

$$= \int_{\Omega_-} \partial^\alpha u \varphi \, dx - \int_{|x'| < R} (\lim_{x_n \to 0-} \partial^\beta u \varphi - \lim_{x_n \to 0+} \partial^\beta u \varphi) dx' + \int_{\Omega_+} \partial^\alpha u \varphi \, dx$$

$$= \int_\Omega v \varphi \, dx;$$

we use here that the two contributions from $\{x_n = 0\}$ cancel each other since $\partial^\beta u$ is continuous on Ω. If $j < n$, we get more simply that

$$-\int_{\Omega_\pm} \partial^\beta u \partial_j \varphi \, dx = \int_{\Omega_\pm} \partial^\alpha u \varphi \, dx,$$

using that integration by parts in the x_j-direction gives no boundary contributions since $\mathrm{supp}\,\varphi \subset \Omega$. It follows that the distribution $\partial^\alpha u$ equals v. □

We note, as a special case of the lemma, that the derivative of the function $|x|$ on the interval $]-1, 1[$ is what it should be, namely, the discontinuous (but integrable) function

$$\mathrm{sign}\, x = \begin{cases} +1 & \text{for } x > 0, \\ -1 & \text{for } x < 0. \end{cases} \tag{3.26}$$

The operations multiplication by a smooth function and differentiation are combined in the following rule of calculus:

Lemma 3.7 (THE LEIBNIZ FORMULA). *When $u \in \mathscr{D}'(\Omega)$, $f \in C^\infty(\Omega)$ and $\alpha \in \mathbb{N}_0^n$, then*

$$\partial^\alpha(f\,u) = \sum_{\beta \le \alpha} \binom{\alpha}{\beta} \partial^\beta f \partial^{\alpha-\beta} u,$$

$$D^\alpha(f\,u) = \sum_{\beta \le \alpha} \binom{\alpha}{\beta} D^\beta f D^{\alpha-\beta} u. \tag{3.27}$$

Proof. When f and u are C^∞-functions, the first formula is obtained by induction from the simplest case

$$\partial_j(f\,u) = (\partial_j f)u + f\partial_j u. \tag{3.28}$$

The same induction works in the distribution case, if we can only show (3.28) in that case. This is done by use of the definitions: For $\varphi \in C_0^\infty(\Omega)$,

$$\langle \partial_j(fu), \varphi \rangle = \langle fu, -\partial_j\varphi \rangle = \langle u, -f\partial_j\varphi \rangle = \langle u, -\partial_j(f\varphi) + (\partial_j f)\varphi \rangle$$
$$= \langle \partial_j u, f\varphi \rangle + \langle (\partial_j f)u, \varphi \rangle = \langle f\partial_j u + (\partial_j f)u, \varphi \rangle.$$

The second formula is an immediate consequence. □

Recall that the space $\mathscr{D}'(\Omega)$ is provided with the weak* topology, i.e., the topology defined by the system of seminorms (3.1)ff.

Theorem 3.8. *Let T be a continuous linear operator in $\mathscr{D}(\Omega)$. Then the adjoint operator in $\mathscr{D}'(\Omega)$, defined by*

$$\langle T^\times u, \varphi \rangle = \langle u, T\varphi \rangle \ \text{for } u \in \mathscr{D}'(\Omega), \ \varphi \in \mathscr{D}(\Omega), \tag{3.29}$$

is a continuous linear operator in $\mathscr{D}'(\Omega)$.

In particular, when $f \in C^\infty(\Omega)$ and $\alpha \in \mathbb{N}_0^n$, then the operators M_f and D^α introduced in Definition 3.5 are continuous in $\mathscr{D}'(\Omega)$.

Proof. Let W be a neighborhood of 0 in $\mathscr{D}'(\Omega)$. Then W contains a neighborhood W_0 of 0 of the form

$$W_0 = W(\varphi_1, \dots, \varphi_N, \varepsilon)$$
$$\equiv \{v \in \mathscr{D}'(\Omega) \mid |\langle v, \varphi_1 \rangle| < \varepsilon, \dots, |\langle v, \varphi_N \rangle| < \varepsilon\}, \tag{3.30}$$

where $\varphi_1, \dots, \varphi_N \in C_0^\infty(\Omega)$. Since $T\varphi_1, \dots, T\varphi_N$ belong to $C_0^\infty(\Omega)$, we can define the neighborhood

$$V = W(T\varphi_1, \dots, T\varphi_N, \varepsilon).$$

Since $\langle T^\times u, \varphi_j \rangle = \langle u, T\varphi_j \rangle$ for each φ_j, we see that $T^\times$ sends V into W_0. This shows the continuity of $T^\times$, and it follows for the operators M_f and D^α in $\mathscr{D}'$, since they are defined as adjoints of continuous operators in $\mathscr{D}(\Omega)$. □

Further discussions of the topology of $\mathscr{D}'$ are found in Section 3.5 below.

The topology in $L_{1,\mathrm{loc}}(\Omega)$ is clearly stronger than the topology induced from $\mathscr{D}'(\Omega)$. One has in general that convergence in $C_0^\infty(\Omega)$, $L_p(\Omega)$ or $L_{p,\mathrm{loc}}(\Omega)$ ($p \in [1,\infty]$) implies convergence in $\mathscr{D}'(\Omega)$.

By use of the Banach-Steinhaus theorem (as applied in Appendix B) one obtains the following fundamental property of $\mathscr{D}'(\Omega)$:

Theorem 3.9 (THE LIMIT THEOREM). *A sequence of distributions $u_k \in \mathscr{D}'(\Omega)$ ($k \in \mathbb{N}$) is convergent in $\mathscr{D}'(\Omega)$ for $k \to \infty$ if and only if the sequence $\langle u_k, \varphi \rangle$ is convergent in $\mathbb{C}$ for all $\varphi \in C_0^\infty(\Omega)$. The limit of u_k in $\mathscr{D}'(\Omega)$ is then the functional u defined by*

$$\langle u, \varphi \rangle = \lim_{k \to \infty} \langle u_k, \varphi \rangle \, , \quad for \ \varphi \in C_0^\infty(\Omega). \tag{3.31}$$

Then also $fD^\alpha u_k \to fD^\alpha u$ in $\mathscr{D}'(\Omega)$ for all $f \in C^\infty(\Omega)$, all $\alpha \in \mathbb{N}_0^n$.

Proof. When the topology is defined by the seminorms (3.1) (cf. Theorem B.5), then $u_k \to v$ in $\mathscr{D}'(\Omega)$ if and only if

$$\langle u_k - v, \varphi \rangle \to 0 \ for \ k \to \infty$$

holds for all $\varphi \in C_0^\infty(\Omega)$.

We will show that when we just know that the sequences $\langle u_k, \varphi \rangle$ converge, then there is a distribution $u \in \mathscr{D}'(\Omega)$ so that $\langle u_k - u, \varphi \rangle \to 0$ for all φ. Here we use Corollary B.14 and Theorem 2.5. Define the functional Λ by

$$\Lambda(\varphi) = \lim_{k \to \infty} \langle u_k, \varphi \rangle \quad for \ \varphi \in C_0^\infty(\Omega).$$

According to Theorem 2.5 (c), Λ is continuous from $C_0^\infty(\Omega)$ to $\mathbb{C}$ if and only if Λ defines continuous maps from $C_{K_j}^\infty(\Omega)$ to $\mathbb{C}$ for each K_j. Since $C_{K_j}^\infty(\Omega)$ is a Fréchet space, we can apply Corollary B.14 to the map $\Lambda : C_{K_j}^\infty(\Omega) \to \mathbb{C}$, as the limit for $k \to \infty$ of the functionals $u_k : C_{K_j}^\infty(\Omega) \to \mathbb{C}$; this gives the desired continuity. The last assertion now follows immediately from Theorem 3.8. $\square$

One has for example that $h_j \to \delta$ in $\mathscr{D}'(\mathbb{R}^n)$ for $j \to \infty$. (The reader is encouraged to verify this.) Also more general convergence concepts (for nets) can be allowed, by use of Theorem B.13.

3.3 Distributions with compact support

In the following we often use a convention of "extension by zero" as mentioned for test functions in Section 2.1, namely, that a function f defined on a subset ω of Ω is identified with the function on Ω that equals f on ω and equals 0 on $\Omega \setminus \omega$.

Definition 3.10. Let $u \in \mathscr{D}'(\Omega)$.

$1°$ We say that u is 0 on the open subset $\omega \subset \Omega$ when

$$\langle u, \varphi \rangle = 0 \quad \text{for all} \quad \varphi \in C_0^\infty(\omega). \tag{3.32}$$

$2°$ The **support of** u is defined as the set

$$\operatorname{supp} u = \Omega \setminus \left(\bigcup \{ \omega \mid \omega \text{ open } \subset \Omega, \ u \text{ is 0 on } \omega \} \right). \tag{3.33}$$

Observe for example that the support of the nontrivial distribution $\partial_j 1_\Omega$ defined in (3.24) is contained in $\partial\Omega$ (a deeper analysis will show that $\operatorname{supp} \partial_j 1_\Omega = \partial\Omega$). Since the support of $\partial_j 1_\Omega$ is a null-set in $\mathbb{R}^n$, and 0 is the only $L_{1,\mathrm{loc}}$-function with support in a null-set, $\partial_j 1_\Omega$ cannot be a function in $L_{1,\mathrm{loc}}(\mathbb{R}^n)$ (see also the discussion after Lemma 3.2).

Lemma 3.11. *Let* $(\omega_\lambda)_{\lambda \in \Lambda}$ *be a family of open subsets of* Ω. *If* $u \in \mathscr{D}'(\Omega)$ *is 0 on* ω_λ *for each* $\lambda \in \Lambda$, *then* u *is 0 on the union* $\bigcup_{\lambda \in \Lambda} \omega_\lambda$.

Proof. Let $\varphi \in C_0^\infty(\Omega)$ with support $K \subset \bigcup_{\lambda \in \Lambda} \omega_\lambda$; we must show that $\langle u, \varphi \rangle = 0$. The compact set K is covered by a finite system of the ω_λ's, say $\omega_1, \ldots, \omega_N$. According to Theorem 2.17, there exist $\psi_1, \ldots, \psi_N \in C_0^\infty(\Omega)$ with $\psi_1 + \cdots + \psi_N = 1$ on K and $\operatorname{supp}\psi_j \subset \omega_j$ for each j. Now let $\varphi_j = \psi_j\varphi$, then $\varphi = \sum_{j=1}^N \varphi_j$, and $\langle u, \varphi \rangle = \sum_{j=1}^N \langle u, \varphi_j \rangle = 0$ by assumption. $\square$

Because of this lemma, we can also describe the support as *the complement of the largest open set where* u *is 0.*

An interesting subset of $\mathscr{D}'(\Omega)$ is the set of *distributions with compact support in* Ω. It is usually denoted $\mathscr{E}'(\Omega)$,

$$\mathscr{E}'(\Omega) = \{ u \in \mathscr{D}'(\Omega) \mid \operatorname{supp} u \text{ is compact } \subset \Omega \}. \tag{3.34}$$

When $u \in \mathscr{E}'(\Omega)$, there is a j such that $\operatorname{supp} u \subset K_{j-1} \subset K_j^\circ$ (cf. (2.4)). Since $u \in \mathscr{D}'(\Omega)$, there exist c_j and N_j so that

$$|\langle u, \psi \rangle| \leq c_j \sup \{ |D^\alpha \psi(x)| \mid x \in K_j, \ |\alpha| \leq N_j \},$$

for all ψ with support in K_j. Choose a function $\eta \in C_0^\infty(\Omega)$ which is 1 on a neighborhood of K_{j-1} and has support in K_j° (cf. Corollary 2.14). An *arbitrary test function* $\varphi \in C_0^\infty(\Omega)$ can then be written as

$$\varphi = \eta\varphi + (1 - \eta)\varphi,$$

where $\operatorname{supp}\eta\varphi \subset K_j^\circ$ and $\operatorname{supp}(1 - \eta)\varphi \subset \complement K_{j-1}$. Since u is 0 on $\complement K_{j-1}$, $\langle u, (1 - \eta)\varphi \rangle = 0$, so that

$$\begin{aligned}
|\langle u, \varphi \rangle| = |\langle u, \eta\varphi \rangle| &\leq c_j \sup \{ |D^\alpha(\eta(x)\varphi(x))| \mid x \in K_j, \ |\alpha| \leq N_j \} \\
&\leq c' \sup \{ |D^\alpha \varphi(x)| \mid x \in \operatorname{supp}\varphi, \ |\alpha| \leq N_j \},
\end{aligned} \tag{3.35}$$

where c' depends on the derivatives of η up to order N_j (by the Leibniz formula, cf. also (2.18)). Since φ was arbitrary, this shows that u *has order* N_j (it shows even more: that we can use the same constant c' on all compact sets $K_m \subset \Omega$). We have shown:

Theorem 3.12. *When $u \in \mathscr{E}'(\Omega)$, there is an $N \in \mathbb{N}_0$ so that u has order N.*

Let us also observe that when $u \in \mathscr{D}'(\Omega)$ has compact support, then $\langle u, \varphi \rangle$ can be given a sense also for $\varphi \in C^\infty(\Omega)$ (since it is only the behavior of φ on a neighborhood of the support of u that in reality enters in the expression). The space $\mathscr{E}'(\Omega)$ may in fact be *identified with* the space of continuous functionals on $C^\infty(\Omega)$ (which is sometimes denoted $\mathscr{E}(\Omega)$; this explains the terminology $\mathscr{E}'(\Omega)$ for the dual space). See Exercise 3.11.

Remark 3.13. When Ω' is an open subset of Ω with $\overline{\Omega'}$ compact $\subset \Omega$, and K is compact with $\overline{\Omega'} \subset K^\circ \subset K \subset \Omega$, then an arbitrary distribution $u \in \mathscr{D}'(\Omega)$ can be written as the sum of a distribution supported in K and a distribution which is 0 on Ω':

$$u = \zeta u + (1 - \zeta)u, \tag{3.36}$$

where $\zeta \in C_0^\infty(K^\circ)$ is chosen to be 1 on $\overline{\Omega'}$ (such functions exist according to Corollary 2.14). The distribution ζu has support in K since $\zeta\varphi = 0$ for $\operatorname{supp}\varphi \subset \Omega \setminus K$; and $(1 - \zeta)u$ is 0 on Ω' since $(1 - \zeta)\varphi = 0$ for $\operatorname{supp}\varphi \subset \Omega'$.

In this connection we shall also consider *restrictions* of distributions, and describe how distributions are *glued together*.

When $u \in \mathscr{D}'(\Omega)$ and Ω' is an open subset of Ω, we define the *restriction* of u to Ω' as the element $u|_{\Omega'} \in \mathscr{D}'(\Omega')$ defined by

$$\langle u|_{\Omega'}, \varphi \rangle_{\Omega'} = \langle u, \varphi \rangle_\Omega \quad \text{for } \varphi \in C_0^\infty(\Omega'). \tag{3.37}$$

(For the sake of precision, we here indicate the duality between $\mathscr{D}'(\omega)$ and $C_0^\infty(\omega)$ by $\langle \ , \ \rangle_\omega$, when ω is an open set.)

When $u_1 \in \mathscr{D}'(\Omega_1)$ and $u_2 \in \mathscr{D}'(\Omega_2)$, and ω is an open subset of $\Omega_1 \cap \Omega_2$, we say that $u_1 = u_2$ *on ω, when*

$$u_1|_\omega - u_2|_\omega = 0 \quad \text{as an element of } \mathscr{D}'(\omega). \tag{3.38}$$

The following theorem is well-known for continuous functions and for $L_{1,\text{loc}}$-functions.

Theorem 3.14 (GLUING DISTRIBUTIONS TOGETHER). *Let $(\omega_\lambda)_{\lambda \in \Lambda}$ be an arbitrary system of open sets in $\mathbb{R}^n$ and let $\Omega = \bigcup_{\lambda \in \Lambda} \omega_\lambda$. Assume that there is given a system of distributions $u_\lambda \in \mathscr{D}'(\omega_\lambda)$ with the property that u_λ equals u_μ on $\omega_\lambda \cap \omega_\mu$, for each pair of indices $\lambda, \mu \in \Lambda$. Then there exists one and only one distribution $u \in \mathscr{D}'(\Omega)$ such that $u|_{\omega_\lambda} = u_\lambda$ for all $\lambda \in \Lambda$.*

Proof. Observe to begin with that there is at most one solution u. Namely, if u and v are solutions, then $(u - v)|_{\omega_\lambda} = 0$ for all λ. This implies that $u - v = 0$, by Lemma 3.11.

We construct u as follows: Let $(K_l)_{l \in \mathbb{N}}$ be a sequence of compact sets as in (2.4) and consider a fixed l. Since K_l is compact, it is covered by a finite subfamily $(\Omega_j)_{j=1,\dots,N}$ of the sets $(\omega_\lambda)_{\lambda \in \Lambda}$; we denote u_j the associated distributions given in $\mathscr{D}'(\Omega_j)$, respectively. By Theorem 2.17 there is a partition of unity $\psi_1, \dots, \psi_N$ consisting of functions $\psi_j \in C_0^\infty(\Omega_j)$ satisfying $\psi_1 + \cdots + \psi_N = 1$ on K_l. For $\varphi \in C_{K_l}^\infty(\Omega)$ we set

$$\langle u, \varphi \rangle_\Omega = \langle u, \sum_{j=1}^N \psi_j \varphi \rangle_\Omega = \sum_{j=1}^N \langle u_j, \psi_j \varphi \rangle_{\Omega_j}. \tag{3.39}$$

In this way, we have given $\langle u, \varphi \rangle$ a value which apparently depends on a lot of choices (of l, of the subfamily $(\Omega_j)_{j=1,\dots,N}$ and of the partition of unity $\{\psi_j\}$). But if $(\Omega_k')_{k=1,\dots,M}$ is another subfamily covering K_l, and $\psi_1', \dots, \psi_M'$ is an associated partition of unity, we have, with u_k' denoting the distribution given on Ω_k':

$$\sum_{j=1}^N \langle u_j, \psi_j \varphi \rangle_{\Omega_j} = \sum_{j=1}^N \sum_{k=1}^M \langle u_j, \psi_k' \psi_j \varphi \rangle_{\Omega_j} = \sum_{j=1}^N \sum_{k=1}^M \langle u_j, \psi_k' \psi_j \varphi \rangle_{\Omega_j \cap \Omega_k'}$$

$$= \sum_{j=1}^N \sum_{k=1}^M \langle u_k', \psi_k' \psi_j \varphi \rangle_{\Omega_k'} = \sum_{k=1}^M \langle u_k', \psi_k' \varphi \rangle_{\Omega_k'},$$

since $u_j = u_k'$ on $\Omega_j \cap \Omega_k'$. This shows that u has been defined for $\varphi \in C_{K_l}^\infty(\Omega)$ independently of the choice of finite subcovering of K_l and associated partition of unity. If we use such a definition for each K_l, $l = 1, 2, \dots$, we find moreover that these definitions are consistent with each other. Indeed, for both K_l and K_{l+1} one can use one cover and partition of unity chosen for K_{l+1}. (In a similar way one finds that u does not depend on the choice of the sequence $(K_l)_{l \in \mathbb{N}}$.) This defines u as an element of $\mathscr{D}'(\Omega)$.

Now we check the consistency of u with each u_λ as follows: Let $\lambda \in \Lambda$. For each $\varphi \in C_0^\infty(\omega_\lambda)$ there is an l such that $\varphi \in C_{K_l}^\infty(\Omega)$. Then $\langle u, \varphi \rangle$ can be defined by (3.39). Here

$$\langle u, \varphi \rangle_\Omega = \langle u, \sum_{j=1}^N \psi_j \varphi \rangle_\Omega = \sum_{j=1}^N \langle u_j, \psi_j \varphi \rangle_{\Omega_j}$$

$$= \sum_{j=1}^N \langle u_j, \psi_j \varphi \rangle_{\Omega_j \cap \omega_\lambda} = \sum_{j=1}^N \langle u_\lambda, \psi_j \varphi \rangle_{\Omega_j \cap \omega_\lambda} = \langle u_\lambda, \varphi \rangle_{\omega_\lambda},$$

which shows that $u|_{\omega_\lambda} = u_\lambda$. $\qquad\square$

In the French literature the procedure is called "recollement des morceaux" (gluing the pieces together).

A very simple example is the case where $u \in \mathscr{E}'(\Omega)$ is glued together with the 0-distribution on a neighborhood of $\mathbb{R}^n \setminus \Omega$. In other words, u is "extended by 0" to a distribution in $\mathscr{E}'(\mathbb{R}^n)$. Such an extension is often tacitly understood.

3.4 Convolutions and coordinate changes

We here give two other useful applications of Theorem 3.8, namely, an extension to $\mathscr{D}'(\mathbb{R}^n)$ of the definition of *convolutions with* φ, and a generalization of *coordinate changes*. First we consider convolutions:

When φ and ψ are in $C_0^\infty(\mathbb{R}^n)$, then $\varphi * \psi$ (recall (2.26)) is in $C_0^\infty(\mathbb{R}^n)$ and satisfies $\partial^\alpha(\varphi * \psi) = \varphi * \partial^\alpha \psi$ for each α. Note here that $\varphi * \psi(x)$ is 0 except if $x - y \in \operatorname{supp}\varphi$ for some $y \in \operatorname{supp}\psi$; the latter means that $x \in \operatorname{supp}\varphi + y$ for some $y \in \operatorname{supp}\psi$, i.e., $x \in \operatorname{supp}\varphi + \operatorname{supp}\psi$. Thus

$$\operatorname{supp}\varphi * \psi \subset \operatorname{supp}\varphi + \operatorname{supp}\psi. \tag{3.40}$$

The map $\psi \mapsto \varphi * \psi$ is continuous, for if K is an arbitrary subset of Ω, then the application of $\varphi *$ to $C_K^\infty(\mathbb{R}^n)$ gives a continuous map into $C_{K + \operatorname{supp}\varphi}^\infty(\mathbb{R}^n)$, since one has for $k \in \mathbb{N}_0$:

$$\sup\{|\partial^\alpha(\varphi * \psi)(x)| \mid x \in \mathbb{R}^n, |\alpha| \le k\}$$
$$= \sup\{|\varphi * \partial^\alpha \psi(x)| \mid x \in \mathbb{R}^n, |\alpha| \le k\}$$
$$\le \|\varphi\|_{L_1} \cdot \sup\{|\partial^\alpha \psi(x)| \mid x \in K, |\alpha| \le k\} \quad \text{for } \psi \text{ in } C_K^\infty(\mathbb{R}^n).$$

Denote $\varphi(-x)$ by $\check{\varphi}(x)$ (the operator $S : \varphi(x) \mapsto \varphi(-x)$ can be called the antipodal operator). One has for φ and χ in $C_0^\infty(\mathbb{R}^n)$, $u \in L_{1,\mathrm{loc}}(\mathbb{R}^n)$ that

$$\langle \varphi * u, \chi \rangle = \int_{\mathbb{R}^n} (\varphi * u)(y)\chi(y)\,dy = \int_{\mathbb{R}^n} \int_{\mathbb{R}^n} \varphi(x)u(y - x)\chi(y)\,dx\,dy$$
$$= \int_{\mathbb{R}^n} u(x)(\check{\varphi} * \chi)(x)\,dx = \langle u, \check{\varphi} * \chi \rangle,$$

by the Fubini theorem. So we see that the adjoint $T^\times$ of $T = \check{\varphi} * : \mathscr{D}(\mathbb{R}^n) \to \mathscr{D}(\mathbb{R}^n)$ acts like $\varphi *$ on functions in $L_{1,\mathrm{loc}}(\mathbb{R}^n)$. Therefore we define the operator $\varphi *$ on distributions as the adjoint of the operator $\check{\varphi} *$ on test functions:

$$\langle \varphi * u, \chi \rangle = \langle u, \check{\varphi} * \chi \rangle, \ u \in \mathscr{D}'(\mathbb{R}^n), \varphi, \chi \in C_0^\infty(\mathbb{R}^n); \tag{3.41}$$

this makes $u \mapsto \varphi * u$ a continuous operator on $\mathscr{D}'(\mathbb{R}^n)$ by Theorem 3.8. The rule

$$\partial^\alpha(\varphi * u) = (\partial^\alpha\varphi) * u = \varphi * (\partial^\alpha u), \quad \text{for } \varphi \in C_0^\infty(\mathbb{R}^n), \, u \in \mathscr{D}'(\mathbb{R}^n), \quad (3.42)$$

follows by use of the defining formulas and calculations on test functions:

$$\langle \partial^\alpha(\varphi * u), \chi \rangle = \langle \varphi * u, (-\partial)^\alpha \chi \rangle = \langle u, \check{\varphi} * (-\partial)^\alpha \chi \rangle$$
$$= \langle u, (-\partial)^\alpha(\check{\varphi} * \chi) \rangle = \langle \partial^\alpha u, \check{\varphi} * \chi \rangle = \langle \varphi * \partial^\alpha u, \chi \rangle, \text{ also}$$
$$= \langle u, (-\partial)^\alpha \check{\varphi} * \chi \rangle = \langle u, (\partial^\alpha \varphi)^\vee * \chi \rangle = \langle \partial^\alpha \varphi * u, \chi \rangle.$$

In a similar way one verifies the rule

$$(\varphi * \psi) * u = \varphi * (\psi * u), \quad \text{for } \varphi, \psi \in C_0^\infty(\mathbb{R}^n), \, u \in \mathscr{D}'(\mathbb{R}^n). \quad (3.43)$$

We have then obtained:

Theorem 3.15. *When $\varphi \in C_0^\infty(\mathbb{R}^n)$, the convolution map $u \mapsto \varphi * u$ defined by (3.41) is continuous in $\mathscr{D}'(\mathbb{R}^n)$; it satisfies (3.42) and (3.43) there.*

One can define the convolution in higher generality, with more general objects in the place of φ, for example a distribution $v \in \mathscr{E}'(\mathbb{R}^n)$. The procedure does not extend to completely general $v \in \mathscr{D}'(\mathbb{R}^n)$ without any support conditions or growth conditions. But if for example u and v are distributions with support in $[0, \infty[^n$, then $v * u$ can be given a sense. (More about convolutions in [S50] and [H83].)

When $u \in L_{1,\mathrm{loc}}(\mathbb{R}^n)$, then we have as in Lemma 2.9 that $\varphi * u$ is a C^∞-function. Note moreover that for each $x \in \mathbb{R}^n$,

$$\varphi * u(x) = \int \varphi(x - y)u(y) \, dy = \langle u, \varphi(x - \cdot) \rangle. \quad (3.44)$$

We shall show that this formula extends to general distributions and defines a C^∞-function even then:

Theorem 3.16. *When $u \in \mathscr{D}'(\mathbb{R}^n)$ and $\varphi \in \mathscr{D}(\mathbb{R}^n)$, then $\varphi * u$ equals the function of $x \in \mathbb{R}^n$ defined by $\langle u, \varphi(x - \cdot) \rangle$, it is in $C^\infty(\mathbb{R}^n)$.*

Proof. Note first that $x \mapsto \varphi(x - \cdot)$ is continuous from $\mathbb{R}^n$ to $\mathscr{D}(\mathbb{R}^n)$. Then the map $x \mapsto \langle u, \varphi(x - \cdot) \rangle$ is continuous from $\mathbb{R}^n$ to $\mathbb{C}$ (you are asked to think about such situations in Exercise 3.14); let us denote this continuous function $v(x) = \langle u, \varphi(x - \cdot) \rangle$. To see that v is differentiable, one can use the mean value theorem to verify that $\frac{1}{h}[\varphi(x + he_i - \cdot) - \varphi(x - \cdot)]$ converges to $\partial_i\varphi(x - \cdot)$ in $\mathscr{D}(\mathbb{R}^n)$ for $h \to 0$; then (b) in Exercise 3.14 applies. Higher derivatives are included by iteration of the argument.

We have to show that

$$\langle v, \psi \rangle = \langle \varphi * u, \psi \rangle \quad \text{for all } \psi \in \mathscr{D}.$$

To do this, denote $\operatorname{supp} \psi = K$ and write $\langle v, \psi \rangle$ as a limit of Riemann sums:

$$\langle v, \psi \rangle = \int v(x)\psi(x)\, dx = \lim_{h \to 0+} \sum_{z \in \mathbb{Z}^n,\, hz \in K} h^n v(hz)\psi(hz)$$

$$= \lim_{h \to 0+} \sum_{z \in \mathbb{Z}^n,\, hz \in K} h^n \langle u, \varphi(hz - \cdot) \rangle \psi(hz)$$

$$= \lim_{h \to 0+} \langle u, \sum_{z \in \mathbb{Z}^n,\, hz \in K} h^n \varphi(hz - \cdot)\psi(hz) \rangle.$$

Here we observe that $\sum_{z \in \mathbb{Z}^n,\, hz \in K} h^n \varphi(hz - y)\psi(hz)$ is a Riemann sum for $(\check{\varphi} * \psi)(y)$, so it converges to $(\check{\varphi} * \psi)(y)$ for $h \to 0+$, each y. The reader can check that this holds not only pointwise, but uniformly in y; uniform convergence can also be shown for the y-derivatives, and the support (with respect to y) is contained in the compact set $\operatorname{supp}\check{\varphi} + \operatorname{supp}\psi$ for all h. Thus

$$\sum_{z \in \mathbb{Z}^n,\, hz \in K} h^n \varphi(hz - \cdot)\psi(hz) \to \check{\varphi} * \psi \text{ in } \mathscr{D}(\mathbb{R}^n), \quad \text{for } h \to 0+. \tag{3.45}$$

Applying this to the preceding calculation, we find that

$$\langle v, \psi \rangle = \langle u, \check{\varphi} * \psi \rangle = \langle \varphi * u, \psi \rangle,$$

as was to be shown.   □

Convolution is often used for approximation techniques:

Lemma 3.17. *Let $(h_j)_{j \in \mathbb{N}}$ be a sequence as in (2.32). Then (for $j \to \infty$) $h_j * \varphi \to \varphi$ in $C_0^\infty(\mathbb{R}^n)$ when φ in $C_0^\infty(\mathbb{R}^n)$, and $h_j * u \to u$ in $\mathscr{D}'(\mathbb{R}^n)$ when u in $\mathscr{D}'(\mathbb{R}^n)$.*

Proof. For any α, $\partial^\alpha(h_j * \varphi) = h_j * \partial^\alpha \varphi \to \partial^\alpha \varphi$ uniformly (cf. (2.40)). Then $h_j * \varphi \to \varphi$ in $C_0^\infty(\mathbb{R}^n)$. Moreover, $(\check{h}_j)_{j \in \mathbb{N}}$ has the properties (2.32), so

$$\langle h_j * u, \varphi \rangle = \langle u, \check{h}_j * \varphi \rangle \to \langle u, \varphi \rangle, \text{ when } u \in \mathscr{D}'(\mathbb{R}^n), \varphi \in C_0^\infty(\mathbb{R}^n).$$

This shows that $h_j * u \to u$ in $\mathscr{D}'(\mathbb{R}^n)$.   □

Because of these convergence properties we call a sequence $\{h_j\}$ as in (2.35) an *approximate unit in $C_0^\infty(\mathbb{R}^n)$* (the name was mentioned already in Chapter 2). Note that the approximating sequence $h_j * u$ consists of C^∞-functions, by Theorem 3.16.

The idea can be modified to show that any distribution in $\mathscr{D}'(\Omega)$ is a limit of functions in $C_0^\infty(\Omega)$:

Theorem 3.18. *Let Ω be open $\subset \mathbb{R}^n$. For any $u \in \mathscr{D}'(\Omega)$ there exists a sequence of functions $u_j \in C_0^\infty(\Omega)$ so that $u_j \to u$ in $\mathscr{D}'(\Omega)$ for $j \to \infty$.*

Proof. Choose K_j and η_j as in Corollary 2.14 $2°$; then $\eta_j u \to u$ for $j \to \infty$, and each $\eta_j u$ identifies with a distribution in $\mathscr{D}'(\mathbb{R}^n)$. For each j, choose

$k_j \geq j$ so large that $\operatorname{supp} \eta_j + \underline{B}(0, \frac{1}{k_j}) \subset K^{\circ}_{j+1}$; then $u_j = h_{k_j} * (\eta_j u)$ is well-defined and belongs to $C_0^{\infty}(\Omega)$ (by Theorem 3.16). When $\varphi \in C_0^{\infty}(\Omega)$, write

$$\langle u_j, \varphi \rangle = \langle h_{k_j} * (\eta_j u), \varphi \rangle = \langle \eta_j u, \check{h}_{k_j} * \varphi \rangle = I_j.$$

Since φ has compact support, there is a j_0 such that for $j \geq j_0$, $\check{h}_{k_j} * \varphi$ is supported in K_{j_0}, hence in K_j for $j \geq j_0$. For such j, we can continue the calculation as follows:

$$I_j = \langle u, \eta_j \cdot (\check{h}_{k_j} * \varphi) \rangle = \langle u, \check{h}_{k_j} * \varphi \rangle \to \langle u, \varphi \rangle, \text{ for } j \to \infty.$$

In the last step we used that $\check{h}_k$ has similar properties as h_k, so that Lemma 3.17 applies to $\check{h}_k * \varphi$. $\qquad \square$

Hence $C_0^{\infty}(\Omega)$ is dense in $\mathscr{D}'(\Omega)$. Thanks to this theorem, we can carry many rules of calculus over from C_0^{∞} to $\mathscr{D}'$ by approximation instead of via adjoints. For example, the Leibniz formula (Lemma 3.7) can be deduced from the C_0^{∞} case as follows: We know that (3.27) holds if $f \in C^{\infty}(\Omega)$ and $u \in C_0^{\infty}(\Omega)$. If $u \in \mathscr{D}'(\Omega)$, let $u_j \to u$ in $\mathscr{D}'(\Omega)$, $u_j \in C_0^{\infty}(\Omega)$. We have that

$$\partial^{\alpha}(f \, u_j) = \sum_{\beta \leq \alpha} \binom{\alpha}{\beta} \partial^{\beta} f \partial^{\alpha - \beta} u_j$$

holds for each j. By Theorem 3.8, each side converges to the corresponding expression with u_j replaced by u, so the rule for u follows.

Finally, we consider coordinate changes. A C^{∞}-coordinate change (a diffeomorphism) carries C^{∞}-functions resp. $L_{1,\text{loc}}$-functions into C^{∞}-functions resp. $L_{1,\text{loc}}$-functions. We sometimes need a similar concept for distributions. As usual, we base the concept on analogy with functions.

Let Ω and Ξ be open sets in $\mathbb{R}^n$, and let κ be a diffeomorphism of Ω onto Ξ. More precisely, κ is a bijective map

$$\kappa : x = (x_1, \ldots, x_n) \mapsto (\kappa_1(x_1, \ldots, x_n), \ldots, \kappa_n(x_1, \ldots, x_n)), \qquad (3.46)$$

where each κ_j is a C^{∞}-function from Ω to $\mathbb{R}$, and the modulus of the functional determinant

$$J(x) = \left| \det \begin{pmatrix} \frac{\partial \kappa_1}{\partial x_1} & \cdots & \frac{\partial \kappa_1}{\partial x_n} \\ \vdots & & \vdots \\ \frac{\partial \kappa_n}{\partial x_1} & \cdots & \frac{\partial \kappa_n}{\partial x_n} \end{pmatrix} \right| \qquad (3.47)$$

is > 0 for all $x \in \Omega$ (so that $J(x)$ and $1/J(x)$ are C^{∞}-functions). A function $f(x)$ on Ω is carried over to a function $(Tf)(y)$ on Ξ by the definition

$$(Tf)(y) = f(\kappa^{-1}(y)). \qquad (3.48)$$

The usual rules for coordinate changes show that T is a linear operator from $C_0^\infty(\Omega)$ to $C_0^\infty(\Xi)$, from $C^\infty(\Omega)$ to $C^\infty(\Xi)$ and from $L_{1,\mathrm{loc}}(\Omega)$ to $L_{1,\mathrm{loc}}(\Xi)$. Concerning integration, we have when $f \in L_{1,\mathrm{loc}}(\Omega)$ and $\psi \in C_0^\infty(\Xi)$,

$$
\begin{aligned}
\langle Tf, \psi(y) \rangle_\Xi &= \int_\Xi f(\kappa^{-1}(y))\psi(y)dy = \int_\Omega f(x)\psi(\kappa(x))J(x)dx \\
&= \langle f, J(x)\psi(\kappa(x)) \rangle_\Omega = \langle f, JT^{-1}\psi \rangle_\Omega.
\end{aligned}
\tag{3.49}
$$

We carry this over to distributions by analogy:

Definition 3.19. When $\kappa = (\kappa_1, \ldots, \kappa_n)$ is a diffeomorphism of Ω onto Ξ and $J(x) = |\det(\frac{\partial \kappa_i}{\partial x_j}(x))_{i,j=1,\ldots,n}|$, we define the coordinate change map $T : \mathscr{D}'(\Omega) \to \mathscr{D}'(\Xi)$ by

$$
\langle Tu, \psi(y) \rangle_\Xi = \langle u, J(x)\psi(\kappa(x)) \rangle_\Omega,
\tag{3.50}
$$

for $\psi \in C_0^\infty(\Xi)$.

Clearly, Tu is a linear functional on $C_0^\infty(\Xi)$. The continuity of this functional follows from the fact that one has for $\psi \in C_0^\infty(\Xi)$ supported in K:

$$
|D_x^\alpha(J(x)\psi(\kappa(x)))| \le c_K \sup\{ |D_y^\beta \psi(y)| \mid y \in K, \, \beta \le \alpha \}
\tag{3.51}
$$

by the Leibniz formula and the chain rule for differentiation of composed functions. In this way, the map T has been defined such that it is consistent with (3.48) when u is a locally integrable function. (There is a peculiar asymmetry in the transformation rule for f and for ψ in (3.49). In some texts this is removed by introduction of a definition where one views the distributions as a generalization of measures with the functional determinant built in, in a suitable way; so-called *densities*. More on this in Section 8.2, with reference to [H83, Section 6.3].)

Since $T : \mathscr{D}'(\Omega) \to \mathscr{D}'(\Xi)$ is defined as the adjoint of the map $J \circ T^{-1}$ from $\mathscr{D}(\Xi)$ to $\mathscr{D}(\Omega)$, T is *continuous* from $\mathscr{D}'(\Omega)$ to $\mathscr{D}'(\Xi)$ by Theorem 3.8 (generalized to the case of a map from $\mathscr{D}(\Omega)$ to $\mathscr{D}(\Xi)$ with two different open sets Ω and Ξ).

Definition 3.19 is useful for example when we consider smooth open subsets of $\mathbb{R}^n$, where we use a coordinate change to "straighten out" the boundary; cf. Appendix C. It can also be used to extend Lemma 3.6 to functions with discontinuities along curved surfaces:

Theorem 3.20. *Let Ω be a smooth open bounded subset of $\mathbb{R}^n$. Let $k \in \mathbb{N}$. If $u \in C^{k-1}(\mathbb{R}^n)$ is such that its k-th derivatives in Ω and in $\mathbb{R}^n \setminus \overline{\Omega}$ exist and can be extended to continuous functions on $\overline{\Omega}$ resp. $\mathbb{R}^n \setminus \Omega$, then the distribution derivatives of u of order k are in $L_{1,\mathrm{loc}}(\mathbb{R}^n)$ and coincide with the usual derivatives in Ω and in $\mathbb{R}^n \setminus \overline{\Omega}$ (this determines the derivatives).*

Proof. For each boundary point x we have an open set U_x and a diffeomorphism $\kappa_x : U_x \to B(0,1)$ according to Definition C.1; let $U_x' = \kappa_x^{-1}(B(0,\frac{1}{2}))$.

Since $\overline{\Omega}$ is compact, the covering of $\partial\Omega$ with the sets U'_x can be reduced to a finite covering system $(\Omega_i)_{i=1,\dots,N}$; the associated diffeomorphisms from Ω_i onto $B(0, \frac{1}{2})$ will be denoted $\kappa_{(i)}$. By the diffeomorphism $\kappa_{(i)}$, $u|_{\Omega_i}$ is carried over to a function v on $B(0, \frac{1}{2})$ satisfying the hypotheses of Lemma 3.6. Thus the k-th derivatives of v in the distribution sense are functions, defined by the usual rules for differentiation inside the two parts of $B(0, \frac{1}{2})$. Since the effect of the diffeomorphism on distributions is consistent with the effect on functions, we see that $u|_{\Omega_i}$ has k-th derivatives that are functions, coinciding with the functions defined by the usual rules of differentiation in $\Omega_i \cap \Omega$ resp. $\Omega_i \cap (\mathbb{R}^n \setminus \overline{\Omega})$. Finally, since u is C^k in the open sets Ω and $\mathbb{R}^n \setminus \overline{\Omega}$, we get the final result by use of the fact that u equals the distribution (function) obtained by gluing the distributions $u|_{\Omega_i} (i = 1, \dots, N)$, $u|_\Omega$ and $u|_{\mathbb{R}^n \setminus \overline{\Omega}}$ together (cf. Theorem 3.14). $\qquad\square$

We shall also use the coordinate changes in Chapter 4 (where for example translation plays a role in some proofs) and in Chapter 5, where the Fourier transforms of some particular functions are determined by use of their invariance properties under certain coordinate changes. Moreover, one needs to know what happens under coordinate changes when one wants to consider differential operators on *manifolds*; this will be taken up in Chapter 8.

Example 3.21. Some simple coordinate changes in $\mathbb{R}^n$ that are often used are *translation*

$$\tau_a(x) = x - a \quad (\text{where } a \in \mathbb{R}^n), \tag{3.52}$$

and *dilation*

$$\mu_\lambda(x) = \lambda x \quad (\text{where } \lambda \in \mathbb{R} \setminus \{0\}). \tag{3.53}$$

They lead to the coordinate change maps $T(\tau_a)$ and $T(\mu_\lambda)$, which look as follows for functions on $\mathbb{R}^n$:

$$(T(\tau_a)u)(y) = u(\tau_a^{-1}y) = u(y + a) = u(x), \quad \text{where } y = x - a, \tag{3.54}$$

$$(T(\mu_\lambda)u)(y) = u(\mu_\lambda^{-1}y) = u(y/\lambda) = u(x), \quad \text{where } y = \lambda x, \tag{3.55}$$

and therefore look as follows for distributions:

$$\langle T(\tau_a)u, \psi(y)\rangle_{\mathbb{R}^n_y} = \langle u, \psi(x - a)\rangle_{\mathbb{R}^n_x} = \langle u, T(\tau_{-a})\psi\rangle, \tag{3.56}$$

$$\langle T(\mu_\lambda)u, \psi(y)\rangle_{\mathbb{R}^n_y} = \langle u, |\lambda^n|\psi(\lambda x)\rangle_{\mathbb{R}^n_x} = \langle u, |\lambda^n|T(\mu_{1/\lambda})\psi\rangle, \tag{3.57}$$

since the functional determinants are 1 resp. λ^n.

Another example is an *orthogonal transformation* O (a unitary operator in the real Hilbert space $\mathbb{R}^n$), where the coordinate change for functions on $\mathbb{R}^n$ is described by the formula

$$(T(O)u)(y) = u(O^{-1}y) = u(x), \quad \text{where } y = Ox, \tag{3.58}$$

and hence for distributions must take the form

$$\langle T(O)u, \psi(y)\rangle_{\mathbb{R}^n_y} = \langle u, \psi(Ox)\rangle_{\mathbb{R}^n_x} = \langle u, T(O^{-1})\psi\rangle, \qquad (3.59)$$

since the modulus of the functional determinant is 1.

We shall *write* the coordinate changes as in (3.54), (3.55), (3.58) also when they are applied to distributions; the precise interpretation is then (3.56), (3.57), resp. (3.59).

The chain rule for coordinate changes is easily carried over to distributions by use of Theorem 3.18: When $u \in C_0^\infty(\Omega)$, differentiation of $Tu = u \circ \kappa^{-1} \in C_0^\infty(\Xi)$ is governed by the rule

$$\partial_i(u \circ \kappa^{-1})(y) = \sum_{l=1}^{n} \frac{\partial u}{\partial x_l}(\kappa^{-1}(y))\frac{\partial \kappa^{-1}_l}{\partial y_i}(y), \qquad (3.60)$$

that may also be written

$$\partial_i(Tu) = \sum_{l=1}^{n} \frac{\partial \kappa^{-1}_l}{\partial y_i} T(\partial_l u), \qquad (3.61)$$

by definition of T. For a general distribution u, choose a sequence u_j in $C_0^\infty(\Omega)$ that converges to u in $\mathscr{D}'(\Omega)$. Since (3.61) holds with u replaced by u_j, and T is continuous from $\mathscr{D}'(\Omega)$ to $\mathscr{D}'(\Xi)$, the validity for u follows by convergence from the validity for the u_j, in view of Theorem 3.8.

3.5 The calculation rules and the weak* topology on $\mathscr{D}'$

For completeness, we also include a more formal and fast deduction of the rules given above, obtained by a direct appeal to general results for topological vector spaces. (Thanks are due to Esben Kehlet for providing this supplement to an earlier version of the text.)

Let E be a locally convex Hausdorff topological vector space over $\mathbb{C}$. Let E' denote the dual space consisting of the continuous linear maps of E into $\mathbb{C}$. The topology $\sigma(E, E')$ on E defined by the family $(e \mapsto |\eta(e)|)_{\eta \in E'}$ of seminorms is called the weak topology on E, and E provided with the weak topology is a locally convex Hausdorff topological vector space with the dual space E'.

The topology $\sigma(E', E)$ on E' defined by the family $(\eta \mapsto |\eta(e)|)_{e \in E}$ of seminorms is called the weak* topology on E', and E' provided with the weak* topology is a locally convex Hausdorff topological vector space with dual space E.

Let also F denote a locally convex Hausdorff topological vector space, and let T be a linear map of E into F.

If T is continuous, then $\varphi \circ T$ is in E' for φ in F', and $\varphi \mapsto \varphi \circ T$ defines a linear map $T^\times$ of F' into E'. This adjoint map $T^\times$ is weak*-weak* continuous.

Indeed, for each e in E, $\varphi \mapsto (T^\times \varphi)(e) = \varphi(Te)$, $\varphi \in F'$, is weak* continuous on F'. The situation is symmetrical: If S is a weak*-weak* continuous linear map of F' into E', then $S^\times$, defined by $\varphi(S^\times e) = (S\varphi)(e)$ for e in E and φ in F', is a weak-weak continuous linear map of E into F.

If T is continuous, it is also weak-weak continuous, since $e \mapsto \varphi(Te) = (T^\times \varphi)(e)$, $e \in E$, is weakly continuous for each φ in F'. (When E and F are Fréchet spaces, the converse also holds, since a weak-weak continuous linear map has a closed graph.)

Lemma 3.22. *Let M be a subspace of E'. If*

$$\{ e \in E \mid \forall \eta \in M : \eta(e) = 0 \} = \{0\}, \tag{3.62}$$

then M is weak* dense in E'.

Proof. Assume that there is an η_0 in E' which does not lie in the weak* closure of M. Let U be an open convex neighborhood of η_0 disjoint from M. According to a Hahn–Banach theorem there exists e_0 in E and t in $\mathbb{R}$ such that $\mathrm{Re}\,\psi(e_0) \le t$ for ψ in M and $\mathrm{Re}\,\eta_0(e_0) > t$. Since $0 \in M$, $0 \le t$. For ψ in M and an arbitrary scalar λ in $\mathbb{C}$, one has that $\mathrm{Re}[\lambda\psi(e_0)] \le t$; thus $\psi(e_0) = 0$ for ψ in M. By hypothesis, e_0 must be 0, but this contradicts the fact that $\mathrm{Re}\,\eta_0(e_0) > t \ge 0$. □

Let Ω be a given open set in a Euclidean space $\mathbb{R}^a$, $a \in \mathbb{N}$. We consider the space $C_0^\infty(\Omega)$ of test functions on Ω provided with the (locally convex) topology as an inductive limit of the Fréchet spaces $C_K^\infty(\Omega)$, K compact $\subset \Omega$, and the dual space $\mathscr{D}'(\Omega)$ of distributions on Ω provided with the weak* topology.

For each f in $L_{1,\mathrm{loc}}(\Omega)$, the map $\varphi \mapsto \int_\Omega f\varphi\,dx$, $\varphi \in C_0^\infty(\Omega)$ is a distribution Λ_f on Ω. The map $f \mapsto \Lambda_f$ is a continuous injective linear map of $L_{1,\mathrm{loc}}(\Omega)$ into $\mathscr{D}'(\Omega)$ (in view of the Du Bois-Reymond lemma, Lemma 3.2).

Theorem 3.23. *The subspace $\{\Lambda_\varphi \mid \varphi \in C_0^\infty(\Omega)\}$ is weak* dense in $\mathscr{D}'(\Omega)$.*

Proof. It suffices to show that 0 is the only function ψ in $C_0^\infty(\Omega)$ for which $0 = \Lambda_\varphi(\psi) = \int_\Omega \varphi\psi\,dx$ for every function φ in $C_0^\infty(\Omega)$; this follows from the Du Bois-Reymond lemma. □

Theorem 3.24. *Let there be given open sets Ω in $\mathbb{R}^a$ and Ξ in $\mathbb{R}^b$, $a,b \in \mathbb{N}$ together with a weak-weak continuous linear map A of $C_0^\infty(\Omega)$ into $C_0^\infty(\Xi)$. There is at most one weak*-weak* continuous linear map $\tilde{A}$ of $\mathscr{D}'(\Omega)$ into $\mathscr{D}'(\Xi)$ with $\tilde{A}\Lambda_\varphi = \Lambda_{A\varphi}$ for all φ in $C_0^\infty(\Omega)$. Such a map $\tilde{A}$ exists if and only if there is a weak-weak continuous linear map B of $C_0^\infty(\Xi)$ into $C_0^\infty(\Omega)$ so that*

$$\int_\Xi (A\varphi)\psi\,dy = \int_\Omega \varphi(B\psi)dx \quad \text{for} \quad \varphi \in C_0^\infty(\Omega)\,, \ \psi \in C_0^\infty(\Xi). \tag{3.63}$$

In the affirmative case, $B = \tilde{A}^\times$, $\tilde{A} = B^\times$, and $\Lambda_{B\psi} = A^\times(\Lambda_\psi)$, $\psi \in C_0^\infty(\Xi)$.

Remark 3.25. The symbol Λ is here used both for the map of $C_0^\infty(\Omega)$ into $\mathscr{D}'(\Omega)$ and for the corresponding map of $C_0^\infty(\Xi)$ into $\mathscr{D}'(\Xi)$.

Proof. The uniqueness is an immediate consequence of Theorem 3.23.

In the rest of the proof, we set $E = C_0^\infty(\Omega)$, $F = C_0^\infty(\Xi)$.

Assume that $\tilde{A}$ exists as desired, then $\tilde{A}^\times$ is a weak-weak continuous map of F into E with

$$\int_\Omega \varphi(\tilde{A}^\times \psi)dx = \Lambda_\varphi(\tilde{A}^\times \psi) = \tilde{A}(\Lambda_\varphi)(\psi) = \Lambda_{A\varphi}(\psi) = \int_\Xi (A\varphi)\psi \, dy,$$

for $\varphi \in E$, $\psi \in F$, so we can use $\tilde{A}^\times$ as B.

Assume instead that B exists; then $B^\times$ is a weak*-weak* continuous linear map of $\mathscr{D}'(\Omega)$ into $\mathscr{D}'(\Xi)$, and

$$(B^\times \Lambda_\varphi)(\psi) = \Lambda_\varphi(B\psi) = \int_\Omega \varphi(B\psi)dx = \int_\Xi (A\varphi)\psi \, dy = \Lambda_{A\varphi}(\psi),$$

for $\varphi \in E$, $\psi \in F$, so that $B^\times \Lambda_\varphi = \Lambda_{A\varphi}$, $\varphi \in E$; hence we can use $B^\times$ as $\tilde{A}$. Moreover we observe that

$$\Lambda_{B\psi}(\varphi) = \int_\Omega \varphi(B\psi)dx = \int_\Xi (A\varphi)\psi \, dy = \Lambda_\psi(A\varphi) = A^\times(\Lambda_\psi)(\varphi),$$

for $\varphi \in E$, $\psi \in F$, so that $\Lambda_{B\psi} = A^\times(\Lambda_\psi)$, $\psi \in F$.   □

Remark 3.26. If a weak-weak continuous linear map

$$A : C_0^\infty(\Omega) \to C_0^\infty(\Xi)$$

has the property that for each compact subset K of Ω there exists a compact subset L of Ξ so that $A(C_K^\infty(\Omega)) \subseteq C_L^\infty(\Xi)$, then A is continuous, since $A|_{C_K^\infty(\Omega)}$ is closed for each K. Actually, all the operators we shall consider are continuous.

PROGRAM: When you meet a continuous linear map $A : C_0^\infty(\Omega) \to C_0^\infty(\Xi)$, you should look for a corresponding map B. When B has been found, drop the tildas ("lägg bort tildarna"[1]) and define $(Au)(\psi) = u(B\psi)$, $u \in \mathscr{D}'(\Omega)$, $\psi \in C_0^\infty(\Xi)$. It often happens that A is the restriction to $C_0^\infty(\Omega)$ of an operator defined on a larger space of functions. One should therefore think about which functions f in $L_{1,\text{loc}}(\Omega)$ have the property $A\Lambda_f = \Lambda_{Af}$.

The program looks as follows for the operators discussed above in Sections 3.2 and 3.4.

Example 3.27 (MULTIPLICATION). Let f be a function in $C^\infty(\Omega)$. The multiplication by f defines a continuous operator $M_f : \varphi \mapsto f\varphi$ on $C_0^\infty(\Omega)$. Since

---

[1] In Swedish, "lägg bort titlarna" means "put away titles" — go over to using first names.

$$\int_\Omega (f\varphi)\psi\,dx = \int_\Omega \varphi(f\psi)dx\,, \ \varphi,\psi \in C_0^\infty(\Omega)\,,$$

we define $M_f u = fu$ by

$$(fu)(\varphi) = u(f\varphi)\,, \ u \in \mathscr{D}'(\Omega)\,, \ \varphi \in C_0^\infty(\Omega)\,;$$

M_f is a continuous operator on $\mathscr{D}'(\Omega)$.

For g in $L_{1,\mathrm{loc}}(\Omega)$,

$$(f\Lambda_g)(\varphi) = \int_\Omega gf\varphi\,dx = \Lambda_{fg}(\varphi)\,, \ \varphi \in C_0^\infty(\Omega)\,.$$

Example 3.28 (DIFFERENTIATION). For $\alpha \in \mathbb{N}_0^a$, ∂^α is a continuous operator on $C_0^\infty(\Omega)$. For φ and ψ in $C_0^\infty(\Omega)$,

$$\int_\Omega (\partial^\alpha \varphi)\psi\,dx = (-1)^{|\alpha|}\int_\Omega \varphi(\partial^\alpha \psi)dx\,.$$

We therefore define a continuous operator ∂^α on $\mathscr{D}'(\Omega)$ by

$$(\partial^\alpha u)(\varphi) = (-1)^{|\alpha|}u(\partial^\alpha \varphi)\,, \ u \in \mathscr{D}'(\Omega)\,, \ \varphi \in C_0^\infty(\Omega)\,.$$

If we identify f with Λ_f for f in $L_{1,\mathrm{loc}}(\Omega)$, we have given $\partial^\alpha f$ a sense for any function f in $L_{1,\mathrm{loc}}(\Omega)$.

When f is so smooth that we can use the formula for integration by parts, e.g., for f in $C^{|\alpha|}(\Omega)$,

$$(\partial^\alpha \Lambda_f)(\varphi) = (-1)^{|\alpha|}\int_\Omega f\partial^\alpha \varphi\,dx = \int_\Omega (\partial^\alpha f)\varphi\,dx = \Lambda_{\partial^\alpha f}(\varphi),$$

for $\varphi \in C_0^\infty(\Omega)$. The Leibniz formula now follows directly from the smooth case by extension by continuity in view of Theorem 3.23.

Example 3.29 (CONVOLUTION). When φ and ψ are in $C_0^\infty(\mathbb{R}^n)$, then, as noted earlier, $\varphi * \psi$ is in $C_0^\infty(\mathbb{R}^n)$ and satisfies $\partial^\alpha(\varphi * \psi) = \varphi * \partial^\alpha \psi$ for each α, and the map $\psi \mapsto \varphi * \psi$ is continuous.

For φ, ψ and χ in $C_0^\infty(\mathbb{R}^n)$ we have, denoting $\varphi(-x)$ by $\check{\varphi}(x)$, that

$$\int_{\mathbb{R}^n} \varphi * \psi(y)\chi(y)dy = \int_{\mathbb{R}^n}\int_{\mathbb{R}^n} \psi(x)\varphi(y-x)\chi(y)dxdy = \int_{\mathbb{R}^n} \psi(x)\chi * \check{\varphi}(x)dx\,;$$

therefore we define

$$(\varphi * u)(\chi) = u(\check{\varphi} * \chi)\,, \ u \in \mathscr{D}'(\mathbb{R}^n)\,, \ \varphi,\chi \in C_0^\infty(\mathbb{R}^n)\,;$$

this makes $u \mapsto \varphi * u$ a continuous operator on $\mathscr{D}'(\mathbb{R}^n)$.

For f in $L_{1,\mathrm{loc}}(\mathbb{R}^n)$,

$$(\varphi * \Lambda_f)(\psi) = \int_{\mathbb{R}^n} \int_{\mathbb{R}^n} f(y)\varphi(x-y)\psi(x)dxdy = \int_{\mathbb{R}^n} \varphi * f(x)\psi(x)dx$$

by the Fubini theorem, so $\varphi * \Lambda_f = \Lambda_{\varphi * f}$ for $\varphi \in C_0^\infty(\mathbb{R}^n)$, $f \in L_{1,\text{loc}}(\mathbb{R}^n)$. For $\varphi \in C_0^\infty(\mathbb{R}^n)$ and $u \in \mathscr{D}'(\mathbb{R}^n)$, the property

$$\partial^\alpha(\varphi * u) = (\partial^\alpha \varphi) * u = \varphi * (\partial^\alpha u),$$

now follows simply by extension by continuity.

Example 3.30 (CHANGE OF COORDINATES). Coordinate changes can also be handled in this way. Let κ be a C^∞ diffeomorphism of Ω onto Ξ with the modulus of the functional determinant equal to J. Define $T(\kappa) : C_0^\infty(\Omega) \to C_0^\infty(\Xi)$ by

$$(T(\kappa)\varphi)(y) = \varphi(\kappa^{-1}(y)), \; \varphi \in C_0^\infty(\Omega), \; y \in \Xi.$$

The map $T(\kappa)$ is continuous according to the chain rule and the Leibniz formula and we have that

$$\int_\Xi T(\kappa)\varphi \cdot \psi \, dy = \int_\Omega \varphi \cdot \psi \circ \kappa \cdot J \, dx \quad \text{for } \varphi \in C_0^\infty(\Omega), \; \psi \in C_0^\infty(\Xi).$$

Then

$$(T(\kappa)u)(\psi) = u(\psi \circ \kappa \cdot J), \; \psi \in C_0^\infty(\Xi), \; u \in \mathscr{D}'(\Omega),$$

defines a continuous linear map $T(\kappa)$ of $\mathscr{D}'(\Omega)$ into $\mathscr{D}'(\Xi)$.

It is easily seen that $T(\kappa)\Lambda_f = \Lambda_{f \circ \kappa^{-1}}$ for $f \in L_{1,\text{loc}}(\Omega)$.

Exercises for Chapter 3

3.1. Show that convergence of a sequence in $C_0^\infty(\Omega)$, $C^\infty(\Omega)$, $L_p(\Omega)$ or $L_{p,\text{loc}}(\Omega)$ $(p \in [1, \infty])$ implies convergence in $\mathscr{D}'(\Omega)$.

3.2. (a) With $f_n(x)$ defined by

$$f_n(x) = \begin{cases} n & \text{for } x \in \left[-\frac{1}{2n}, \frac{1}{2n}\right], \\ 0 & \text{for } x \in \mathbb{R} \setminus \left[-\frac{1}{2n}, \frac{1}{2n}\right], \end{cases}$$

show that $f_n \to \delta$ in $\mathscr{D}'(\mathbb{R})$ for $n \to \infty$.
(b) With

$$g_n(x) = \frac{1}{\pi} \frac{\sin nx}{x},$$

show that $g_n \to \delta$ in $\mathscr{D}'(\mathbb{R})$, for $n \to \infty$.
(One can use the Riemann-Lebesgue lemma from Fourier theory.)

3.3. Let $f(x)$ be a function on $\mathbb{R}$ such that f is C^∞ on each of the intervals $]-\infty, x_0[$ and $]x_0, +\infty[$, and such that the limits $\lim_{x\to x_0+} f^{(k)}(x)$ and $\lim_{x\to x_0-} f^{(k)}(x)$ exist for all $k \in \mathbb{N}_0$. Denote by $f_k(x)$ the *function* that equals $f^{(k)}(x)$ for $x \neq x_0$. Show that the *distribution* $f \in \mathscr{D}'(\mathbb{R})$ is such that its derivative ∂f identifies with the sum of the function f_1 (considered as a distribution) and the distribution $c\delta_{x_0}$, where $c = \lim_{x\to x_0+} f(x) - \lim_{x\to x_0-} f(x)$; briefly expressed:

$$\partial f = f_1 + c\delta_{x_0} \quad \text{in } \mathscr{D}'(\mathbb{R}).$$

Find similar expressions for $\partial^k f$, for all $k \in \mathbb{N}$.

3.4. Consider the series $\sum_{k\in\mathbb{Z}} e^{ikx}$ for $x \in I =]-\pi, \pi[$ (this series is in the usual sense divergent at all points $x \in I$).

(a) Show that the sequences $\sum_{0<k\leq M} e^{ikx}$ and $\sum_{-M\leq k<0} e^{ikx}$ converge to distributions Λ_+ resp. Λ_- in $\mathscr{D}'(I)$ for $M \to \infty$, and find $\Lambda = \Lambda_+ + \Lambda_-$. (We say that the series $\sum_{k\in\mathbb{Z}} e^{ikx}$ converges to Λ in $\mathscr{D}'(I)$.)

(b) Show that for any $N \in \mathbb{N}$, the series $\sum_{k\in\mathbb{Z}} k^N e^{ikx}$ converges to a distribution Λ_N in $\mathscr{D}'(I)$, and show that $\Lambda_N = D^N \Lambda$.

3.5. For $a \in \mathbb{R}_+$, let

$$f_a(x) = \frac{a}{\pi} \frac{1}{x^2 + a^2} \quad \text{for } x \in \mathbb{R}.$$

Show that $f_a \to \delta$ in $\mathscr{D}'(\mathbb{R})$ for $a \to 0+$.

3.6 (DISTRIBUTIONS SUPPORTED IN A POINT). Let u be a distribution on $\mathbb{R}^n$ with support $= \{0\}$. Then there exists an N so that u has order N. Denote $\chi(x/r) = \zeta_r(x)$ for $r \in]0,1]$.

(a) *The case $N = 0$* Show that there is a constant c_1 so that

$$|\langle u, \varphi\rangle| \leq c_1 |\varphi(0)| \quad \text{for all } \varphi \in C_0^\infty(\mathbb{R}^n).$$

(Apply the distribution to $\varphi = \zeta_r\varphi + (1-\zeta_r)\varphi$ and let $r \to 0$.) Show that there is a constant a so that

$$u = a\delta.$$

(*Hint.* One can show that $\langle u, \varphi\rangle = \langle u, \zeta_1\varphi\rangle + 0 = \langle u, \zeta_1\rangle\varphi(0)$.)

(b) *The case $N > 0$.* Show that the function ζ_r satisfies

$$|\partial^\alpha \zeta_r(x)| \leq c_\alpha r^{-|\alpha|} \quad \text{for each } \alpha \in \mathbb{N}_0^n,$$

when $r \in]0,1]$. Let $V = \{\psi \in C_0^\infty(\mathbb{R}^n) \mid \partial^\alpha\psi(0) = 0 \text{ for all } |\alpha| \leq N\}$, and show that there are inequalities for each $\psi \in V$:

$$|\psi(x)| \leq c|x|^{N+1} \quad \text{for } x \in \mathbb{R}^n;$$
$$|\partial^\alpha(\zeta_r(x)\psi(x))| \leq c' r^{N+1-|\alpha|} \quad \text{for } x \in \mathbb{R}^n, \; r \in]0,1] \text{ and } |\alpha| \leq N;$$

$$|\langle u, \zeta_r \psi \rangle| \le c'r \quad \text{for all} \quad r \in]0,1];$$

and hence $\langle u, \psi \rangle = 0$ when $\psi \in V$. Show that there are constants a_α so that

$$u = \sum_{|\alpha| \le N} a_\alpha \partial^\alpha \delta.$$

(*Hint.* One may use that $\langle u, \varphi \rangle = \langle u, \zeta_1 \varphi \rangle = \langle u, \zeta_1 \sum_{|\alpha| \le N} \frac{\partial^\alpha \varphi(0)}{\alpha!} x^\alpha + \psi(x) \rangle = \sum_{|\alpha| \le N} \langle u, \zeta_1 \frac{(-x)^\alpha}{\alpha!} \rangle (-\partial)^\alpha \varphi(0)$.)

3.7.   We consider $\mathscr{D}'(\mathbb{R}^n)$ for $n \ge 2$.

(a) Show that the function $f(x) = \dfrac{x_1}{|x|}$ is bounded and belongs to $L_{1,\text{loc}}(\mathbb{R}^n)$.

(b) Show that the first-order classical derivatives of f, defined for $x \ne 0$, are functions in $L_{1,\text{loc}}(\mathbb{R}^n)$.

(c) Show that the first-order derivatives of f defined in the distribution sense on $\mathbb{R}^n$ equal the functions defined under (b).
(*Hint.* It is sufficient to consider f and $\partial_{x_j} f$ on $B(0,1)$. One can here calculate $\langle \partial_{x_j} f, \varphi \rangle = -\langle f, \partial_{x_j} \varphi \rangle$ for $\varphi \in C_0^\infty(B(0,1))$ as an integral over $B(0,1) = [B(0,1) \setminus B(0,\varepsilon)] \cup B(0,\varepsilon)$, using formula (A.20) and letting $\varepsilon \to 0$.)

3.8.   (a) Let $\varphi \in C_0^\infty(\mathbb{R}^n)$. Show that $\langle \delta, \varphi \rangle = 0 \implies \varphi \delta = 0$.

(b) Consider $u \in \mathscr{D}'(\mathbb{R}^n)$ and $\varphi \in C_0^\infty(\mathbb{R}^n)$. Find out whether one of the following implications holds for arbitrary u and φ:

$$\langle u, \varphi \rangle = 0 \implies \varphi u = 0 \ ?$$
$$\varphi u = 0 \implies \langle u, \varphi \rangle = 0 \ ?$$

3.9.   (a) Let $\Omega = \mathbb{R}^n$. Show that *the order* of the distribution $D^\alpha \delta$ equals $|\alpha|$. Show that when M is an interval $[a,b]$ of $\mathbb{R}$ ($a < b$), then the order of $D^j 1_M \in \mathscr{D}'(\mathbb{R})$ equals $j - 1$.

(b) Let $\Omega = \mathbb{R}$. Show that the functional Λ_1 defined by

$$\langle \Lambda_1, \varphi \rangle = \sum_{N=1}^\infty \varphi^{(N)}(N) \quad \text{for} \quad \varphi \in C_0^\infty(\mathbb{R}),$$

is a distribution on $\mathbb{R}$ whose order equals ∞. Show that the functional Λ defined by (3.18) is a distribution whose order equals ∞.

3.10.   Let Ω be a smooth open subset of $\mathbb{R}^n$, or let $\Omega = \mathbb{R}^n_+$.

(a) Show that $\operatorname{supp} \partial_j 1_\Omega \subset \partial \Omega$.

(b) Show that the distribution $(-\Delta) 1_\Omega$ on $\mathbb{R}^n$ satisfies

$$\langle (-\Delta) 1_\Omega, \varphi \rangle = \int_{\partial \Omega} \frac{\partial \varphi}{\partial \nu} \, d\sigma \quad \text{for} \quad \varphi \in C_0^\infty(\mathbb{R}^n)$$

(cf. (A.20)), and determine the order and support of the distribution in the case $\Omega = \mathbb{R}^n_+$.

3.11. Show that the space $C^\infty(\Omega)'$ of continuous linear functionals on $C^\infty(\Omega)$ can be identified with the subspace $\mathscr{E}'(\Omega)$ of $\mathscr{D}'(\Omega)$, in such a way that when $\Lambda \in C^\infty(\Omega)'$ is identified with $\Lambda_1 \in \mathscr{D}'(\Omega)$, then

$$\Lambda(\varphi) = \langle \Lambda_1, \varphi \rangle \quad \text{for} \ \varphi \in C_0^\infty(\Omega).$$

3.12. One often meets the notation $\delta(x)$ for the distribution δ_0. Moreover it is customary (e.g., in physics texts) to write $\delta(x - a)$ for the distribution δ_a, $a \in \mathbb{R}$; this is motivated by the heuristic calculation

$$\int_{-\infty}^\infty \delta(x - a)\varphi(x)\,dx = \int_{-\infty}^\infty \delta(y)\varphi(y + a)\,dy = \varphi(a), \ \text{for} \ \varphi \in C_0^\infty(\mathbb{R}).$$

(a) Motivate by a similar calculation the formula

$$\delta(ax) = \frac{1}{|a|}\delta(x), \quad \text{for} \ a \in \mathbb{R} \setminus \{0\}.$$

(b) Motivate the following formula:

$$\delta(x^2 - a^2) = \frac{1}{2a}\big(\delta(x - a) + \delta(x + a)\big), \quad \text{for} \ a > 0.$$

(*Hint.* One can calculate the integral $\int_{-\infty}^\infty \delta(x^2 - a^2)\varphi(x)\,dx$ heuristically by decomposing it into integrals over $]-\infty, 0[$ and $]0, \infty[$ and use of the change of variables $x = \pm\sqrt{y}$. A precise account of how to compose distributions and functions — in the present case δ composed with $f(x) = x^2 - a^2$ — can be found in [H83, Chapter 3.1].)

3.13. Denote by e_j the j-th coordinate vector in $\mathbb{R}^n$ and define the difference quotient $\Delta_{j,h}u$ of an arbitrary distribution $u \in \mathscr{D}'(\mathbb{R}^n)$ by

$$\Delta_{j,h}u = \frac{1}{h}(T(\tau_{he_j})u - u), \quad \text{for} \ h \in \mathbb{R} \setminus \{0\},$$

cf. Example 3.21. Show that $\Delta_{j,h}u \to \partial_j u$ in $\mathscr{D}'(\mathbb{R}^n)$ for $h \to 0$.

3.14. For an open subset Ω of $\mathbb{R}^n$ and an open interval I of $\mathbb{R}$ we consider a parametrized family of functions $\varphi(x, t)$ belonging to $C_0^\infty(\Omega)$ as functions of x for each value of the parameter t.

(a) Show that when the map $t \mapsto \varphi(x, t)$ is continuous from I to $C_0^\infty(\Omega)$, then the function $t \mapsto f(t) = \langle u, \varphi(x, t) \rangle$ is continuous from I to $\mathbb{C}$, for any distribution $u \in \mathscr{D}'(\Omega)$.

(b) We say that the map $t \mapsto \varphi(x,t)$ is differentiable from I til $C_0^\infty(\Omega)$, when we have for each $t \in I$ that $[\varphi(x,t+h) - \varphi(x,t)]/h$ (defined for h so small that $t + h \in I$) converges in $C_0^\infty(\Omega)$ to a function $\psi(x,t)$ for $h \to 0$; observe that ψ in that case is the usual partial derivative $\partial_t \varphi$. If this function $\partial_t \varphi(x,t)$ moreover is continuous from I to $C_0^\infty(\Omega)$, we say that $\varphi(x,t)$ is C^1 from I to $C_0^\infty(\Omega)$. C^k-maps are similarly defined.

Show that when $t \mapsto \varphi(x,t)$ is differentiable (resp. C^k for some $k \geq 1$) from I to $C_0^\infty(\Omega)$, then the function $t \mapsto f(t) = \langle u, \varphi(x,t) \rangle$ is differentiable (resp. C^k) for any distribution $u \in \mathscr{D}'(\Omega)$, and one has at each $t \in I$:

$$\partial_t f(t) = \langle u, \partial_t \varphi(x,t) \rangle; \quad \text{resp.} \quad \partial_t^k f(t) = \langle u, \partial_t^k \varphi(x,t) \rangle.$$

3.15. Let $\Omega = \Omega' \times \mathbb{R}$, where Ω' is an open subset of $\mathbb{R}^{n-1}$ (the points in Ω, Ω' resp. $\mathbb{R}$ are denoted x, x' resp. x_n).

(a) Show that if $u \in \mathscr{D}'(\Omega)$ satisfies $\partial_{x_n} u = 0$, then u is *invariant under x_n-translation*, i.e., $T_h u = u$ for all $h \in \mathbb{R}$, where T_h is the translation coordinate change (denoted $T(\tau_{he_n})$ in Example 3.21) defined by

$$\langle T_h u, \varphi(x', x_n) \rangle = \langle u, \varphi(x', x_n + h) \rangle \quad \text{for } \varphi \in C_0^\infty(\Omega).$$

(*Hint.* Introduce the function $f(h) = \langle T_h u, \varphi \rangle$ and apply the Taylor formula (A.8) and Exercise 3.14 to this function.)

(b) Show that if u is a continuous function on Ω satisfying $\partial_{x_n} u = 0$ in the distribution sense, then $u(x', x_n) = u(x', 0)$ for all $x' \in \Omega'$.

3.16. (a) Let $\Omega = \Omega' \times \mathbb{R}$, as in Exercise 3.15. Show that if u and $u_1 = \partial_{x_n} u$ are continuous functions on Ω (where $\partial_{x_n} u$ is defined in the distribution sense), then u is differentiable in the original sense with respect to x_n at every point $x \in \Omega$, with the derivative $u_1(x)$.

(*Hint.* Let v be the function defined by $v(x', x_n) = \int_0^{x_n} u_1(x', t)\, dt$; show that $\partial_{x_n}(u - v) = 0$ in $\mathscr{D}'(\Omega)$, and apply Exercise 3.15.)

(b) Show that the conclusion in (a) also holds when Ω is replaced by an arbitrary open set in $\mathbb{R}^n$.

3.17. The distribution $\frac{d^k}{dx^k}\delta$ is often denoted $\delta^{(k)}$; for $k = 1, 2, 3$, the notation δ', δ'', δ''' (respectively) is also used. Let $f \in C^\infty(\mathbb{R})$.

(a) Show that there are constants c_0 and c_1 such that one has the identity:

$$f\delta' = c_0 \delta + c_1 \delta';$$

find these.

(b) For a general $k \in \mathbb{N}_0$, show that for suitable constants c_{kj} (to be determined), one has an identity:

$$f\delta^{(k)} = \sum_{j=0}^{k} c_{kj} \delta^{(j)}.$$

Part II
Extensions and applications

Chapter 4
Realizations and Sobolev spaces

4.1 Realizations of differential operators

From now on, a familiarity with general results for unbounded operators in Hilbert space is assumed. The relevant material is found in Chapter 12.

There are various general methods to associate operators in Hilbert spaces to differential operators. We consider two types: the so-called "strong definitions" and the so-called "weak definitions". These definitions can be formulated without having distribution theory available, but in fact the weak definition is closely related to the ideas of distribution theory.

First we observe that nontrivial differential operators cannot be bounded in L_2-spaces. Just take the simplest example of $\frac{d}{dx}$ acting on functions on the interval $J = [0,1]$. Let $f_n(x) = \frac{1}{n} \sin nx$, $n \in \mathbb{N}$. Then clearly $f_n(x) \to 0$ in $L_2(J)$ for $n \to \infty$, whereas the sequence $\frac{d}{dx} f_n(x) = \cos nx$ is not convergent in $L_2(J)$. Another example is $g_n(x) = x^n$, which goes to 0 in $L_2(J)$, whereas the L_2-norms of $g'_n(x) = nx^{n-1}$ go to ∞ for $n \to \infty$. So, at best, the differential operators can be viewed as suitable unbounded operators in L_2-spaces.

Let A be a differential operator of order m with C^∞ coefficients a_α on an open set $\Omega \subset \mathbb{R}^n$:

$$Au = \sum_{|\alpha| \leq m} a_\alpha(x) D^\alpha u . \qquad (4.1)$$

When $u \in C^m(\Omega)$ and $\varphi \in C_0^\infty(\Omega)$, we find by integration by parts (using the notation (u,v) for $\int_\Omega u\bar{v}\, dx$ when $u\bar{v}$ is integrable):

$$(Au, \varphi)_{L_2(\Omega)} = \int_\Omega \sum_{|\alpha| \leq m} (a_\alpha D^\alpha u)\, \bar{\varphi}\, dx$$

$$= \int_\Omega u \sum_{|\alpha| \leq m} \overline{D^\alpha(\bar{a}_\alpha \varphi)}\, dx = (u, A'\varphi)_{L_2(\Omega)}, \qquad (4.2)$$

where we have defined A' by $A'v = \sum_{|\alpha| \leq m} D^\alpha(\bar{a}_\alpha v)$. In (4.2), we used the second formula in (A.20) to carry all differentiations over to φ, using that the integral can be replaced by an integral over a bounded set Ω_1 with smooth boundary chosen such that $\operatorname{supp}\varphi \subset \Omega_1 \subset \bar{\Omega}_1 \subset \Omega$; then all contributions from the boundary are 0.

The operator A' is called the formal adjoint of A; it satisfies

$$A'v = \sum_{|\alpha| \leq m} D^\alpha(\bar{a}_\alpha v) = \sum_{|\alpha| \leq m} a'_\alpha(x)D^\alpha v, \qquad (4.3)$$

for suitable functions $a'_\alpha(x)$ determined by the Leibniz formula (cf. (A.7)); note in particular that

$$a'_\alpha = \bar{a}_\alpha \quad \text{for } |\alpha| = m. \qquad (4.4)$$

When $a_\alpha = a'_\alpha$ for all $|\alpha| \leq m$, i.e., $A = A'$, we say that A is formally selfadjoint.

The formula $A = \sum_{|\alpha| \leq m} a_\alpha D^\alpha$ is regarded as a formal expression. In the following, we write $A|_M$ for an operator acting like A and defined on a set M of functions u for which Au has a meaning (that we specify in each case). As an elementary example, Au is defined classically as a function in $C_0^\infty(\Omega)$ when $u \in C_0^\infty(\Omega)$. We can then consider $A|_{C_0^\infty(\Omega)}$ as a densely defined operator in $L_2(\Omega)$. Similar operators can of course be defined for A'.

Now we introduce some concrete versions of A acting in $L_2(\Omega)$. We take the opportunity to present first the weak definition of differential operators in L_2 (which does not require distribution theory in its formulation), since it identifies with the definition of the maximal realization. Afterwards, the connection with distribution theory will be made.

Definition 4.1. 1° When u and $f \in L_2(\Omega)$, we say that $Au = f$ weakly in $L_2(\Omega)$, when

$$(f, \varphi) = (u, A'\varphi) \quad \text{for all } \varphi \in C_0^\infty(\Omega). \qquad (4.5)$$

2° The maximal realization $A_{\max}$ associated with A in $L_2(\Omega)$ is defined by

$$D(A_{\max}) = \{u \in L_2(\Omega) \mid \exists f \in L_2(\Omega) \text{ such that } Au = f \text{ weakly}\}, \qquad (4.6)$$
$$A_{\max}u = f.$$

There is at most one f for each u in (4.6), since $C_0^\infty(\Omega)$ is dense in $L_2(\Omega)$. Observe that in view of (4.5), the operator $A_{\max}$ is the adjoint of the densely defined operator $A'|_{C_0^\infty(\Omega)}$ in $L_2(\Omega)$:

$$A_{\max} = (A'|_{C_0^\infty(\Omega)})^*. \qquad (4.7)$$

Note in particular that $A_{\max}$ is closed (cf. Lemma 12.4). We see from (4.2) that

$$A|_{C_0^\infty(\Omega)} \subset A_{\max}; \qquad (4.8)$$

so it follows that $A|_{C_0^\infty(\Omega)}$ is closable as an operator in $L_2(\Omega)$.

Definition 4.2. The minimal realization $A_{\min}$ associated with A is defined by

$$A_{\min} = \text{the closure of } A|_{C_0^\infty(\Omega)} \text{ as an operator in } L_2(\Omega). \qquad (4.9)$$

The inclusion (4.8) extends to the closure:

$$A_{\min} \subset A_{\max}. \qquad (4.10)$$

We have moreover:

Lemma 4.3. *The operators $A_{\max}$ and $A'_{\min}$ are the adjoints of one another, as operators in $L_2(\Omega)$.*

Proof. Since $A'_{\min}$ is the closure of $A'|_{C_0^\infty(\Omega)}$, and $A_{\max}$ is the Hilbert space adjoint of the latter operator, we see from the rule $T^* = (\overline{T})^*$ that $A_{\max} = (A'_{\min})^*$, and from the rule $T^{**} = \overline{T}$ that $(A_{\max})^* = A'_{\min}$ (cf. Corollary 12.6). $\qquad\square$

The above definitions were formulated without reference to distribution theory, but we now observe that there is a close link:

$$D(A_{\max}) = \{u \in L_2(\Omega) \mid Au \in \mathscr{D}'(\Omega) \text{ satisfies } Au \in L_2(\Omega)\}, \qquad (4.11)$$

$A_{\max}$ acts like A in the distribution sense.

Indeed, $Au \in \mathscr{D}'(\Omega)$ is the distribution f such that

$$\langle f, \overline{\varphi} \rangle = \langle u, \sum_{|\alpha| \le m} (-D)^\alpha (a_\alpha \overline{\varphi}) \rangle, \text{ for all } \varphi \in C_0^\infty(\Omega). \qquad (4.12)$$

Here $\sum_{|\alpha| \le m} (-D)^\alpha (a_\alpha \overline{\varphi}) = \overline{A'\varphi}$. Then when $f \in L_2(\Omega)$, (4.12) can be written as (4.5), so $Au = f$ weakly in $L_2(\Omega)$. Conversely, if $Au = f$ weakly in $L_2(\Omega)$, (4.12) holds with $f \in L_2(\Omega)$.

Whereas the definition of $A_{\max}$ is called *weak* since it is based on duality, the definition of $A_{\min}$ is called a *strong* definition since the operator is obtained by closure of a classically defined operator.

Observe that $A_{\max}$ is the largest possible operator in $L_2(\Omega)$ associated to A by distribution theory. We have in particular that $D(A_{\max})$ is closed with respect to the graph norm $(\|u\|_{L_2}^2 + \|Au\|_{L_2}^2)^{\frac{1}{2}}$, and that $D(A_{\min})$ is the closure of $C_0^\infty(\Omega)$ in $D(A_{\max})$ with respect to the graph norm.

Whereas $A_{\max}$ is the largest operator in $L_2(\Omega)$ associated with A, $A_{\min}$ is the smallest closed restriction of $A_{\max}$ whose domain contains $C_0^\infty(\Omega)$. The operators $\widetilde{A}$ satisfying

$$A_{\min} \subset \widetilde{A} \subset A_{\max} \qquad (4.13)$$

are called the *realizations of A*, here $A_{\min}$ and $A_{\max}$ are in themselves examples of realizations.

Note that if $A = A'$ (i.e., A is formally selfadjoint) *and* $D(A_{\max}) = D(A_{\min})$, then $A_{\max}$ is selfadjoint as an unbounded operator in $L_2(\Omega)$. This will often be the case when $\Omega = \mathbb{R}^n$.

In the treatment of differential equations involving A, one of the available tools is to apply results from functional analysis to suitable realizations of A. On the one hand, it is then important to find out when a realization can be constructed such that the functional analysis results apply, for example whether the operator is selfadjoint, or lower bounded, or variational (see Chapter 12). On the other hand, one should at the same time keep track of how the realization corresponds to a concrete problem, for example whether it represents a specific boundary condition. This is an interesting interface between abstract functional analysis and concrete problems for differential operators.

Remark 4.4. One can also define other relevant *strong* realizations than $A_{\min}$. (The reader may skip this remark in a first reading.) For example, let

$$M_1 = \{\, u \in C^m(\Omega) \mid u \text{ and } Au \in L_2(\Omega) \,\},$$
$$M_2 = C_{L_2}^m(\Omega) = \{\, u \in C^m(\Omega) \mid D^\alpha u \in L_2(\Omega) \text{ for all } |\alpha| \le m \,\}, \qquad (4.14)$$
$$M_3 = C_{L_2}^m(\overline{\Omega}) = \{\, u \in C^m(\overline{\Omega}) \mid D^\alpha u \in L_2(\Omega) \text{ for all } |\alpha| \le m \,\}.$$

Then

$$A|_{C_0^\infty(\Omega)} \subset A|_{M_1} \subset A_{\max},$$

and, when the coefficient functions a_α are bounded,

$$A|_{C_0^\infty(\Omega)} \subset A|_{M_3} \subset A|_{M_2} \subset A|_{M_1} \subset A_{\max}.$$

Defining

$$A_{s_i} = \overline{A|_{M_i}}, \text{ for } i = 1, 2, 3,$$

we then have in general

$$A_{\min} \subset A_{s_1} \subset A_{\max}. \qquad (4.15)$$

Furthermore,

$$A_{\min} \subset A_{s_3} \subset A_{s_2} \subset A_{s_1} \subset A_{\max} \qquad (4.16)$$

holds when the a_α's are bounded. For certain types of operators A and domains Ω one can show that some of these A_{s_i}'s coincide with each other or coincide with $A_{\max}$. This makes it possible to show properties of realizations by approximation from properties of classically defined operators.

In the one-dimensional case where Ω is an interval I, one will often find that $D(A_{\max})$ differs from $D(A_{\min})$ by a finite dimensional space (and that the three realizations A_{s_i} in the above remark coincide with $A_{\max}$). Then the various realizations represent various concrete boundary conditions which can be completely analyzed. (More about this in Section 4.3.) In the one-dimensional case, the weak definition of $\frac{d}{dx}$ is related to absolute continuity, as mentioned in Chapter 1, and the large effort to introduce distributions is not strictly necessary.

But in higher dimensional cases ($n \geq 2$) where Ω is different from $\mathbb{R}^n$, there will usually be an infinite dimensional difference between $D(A_{\max})$ and $D(A_{\min})$, so there is room for a lot of different realizations. A new difficult phenomenon here is that for a function $u \in D(A_{\max})$, the "intermediate" derivatives $D^\alpha u$ (with $|\alpha| \leq m$) need not exist as functions on Ω, even though Au does so.

For example, the function f on $\Omega =]0, 1[\times]0, 1[\subset \mathbb{R}^2$ defined by

$$f(x, y) = \begin{cases} 1 & \text{for } x > y, \\ 0 & \text{for } x \leq y, \end{cases} \tag{4.17}$$

is in $L_2(\Omega)$ and may be shown to belong to $D(A_{\max})$ for the second-order operator $A = \partial_x^2 - \partial_y^2$ considered on Ω (Exercise 4.3). But $\partial_x f$ and $\partial_y f$ do not have a good sense as L_2-functions (f does not belong to the domains of the maximal realizations of ∂_x or ∂_y, Exercise 4.4).

Also for the Laplace operator Δ one can give examples where $u \in D(A_{\max})$ but the first derivatives are not in $L_2(\Omega)$ (see Exercise 4.5).

It is here that we find great help in distribution theory, which gives a precise explanation of which sense we can give to these derivatives.

4.2 Sobolev spaces

The domains of realizations are often described by the help of various *Sobolev spaces* that we shall now define.

Definition 4.5. Let Ω be an open subset of $\mathbb{R}^n$. Let $m \in \mathbb{N}_0$.
1° The Sobolev space $H^m(\Omega)$ is defined by

$$H^m(\Omega) = \{\, u \in L_2(\Omega) \mid D^\alpha u \in L_2(\Omega) \text{ for } |\alpha| \leq m \,\}, \tag{4.18}$$

where D^α is applied in the distribution sense. (Equivalently, $H^m(\Omega)$ consists of the $u \in L_2(\Omega)$ for which $D^\alpha u$ exists weakly in $L_2(\Omega)$ for $|\alpha| \leq m$.)
$H^m(\Omega)$ is provided with the scalar product and norm (the m-norm)

$$(u, v)_m = \sum_{|\alpha| \leq m} (D^\alpha u, D^\alpha v)_{L_2(\Omega)}, \quad \|u\|_m = (u, u)_m^{\frac{1}{2}}. \tag{4.19}$$

2° The Sobolev space $H_0^m(\Omega)$ is defined as the closure of $C_0^\infty(\Omega)$ in $H^m(\Omega)$.

It is clear that $(u, v)_m$ is a scalar product with associated norm $\|u\|_m$ (since $\|u\|_m \geq \|u\|_0 \equiv \|u\|_{L_2(\Omega)}$), so that $H^m(\Omega)$ is a pre-Hilbert space. That the space is complete is easily obtained from distribution theory: Let $(u_k)_{k\in\mathbb{N}}$ be a Cauchy sequence in $H^m(\Omega)$. Then (u_k) is in particular a Cauchy sequence in $L_2(\Omega)$, hence has a limit u in $L_2(\Omega)$. The sequences $(D^\alpha u_k)_{k\in\mathbb{N}}$ with $|\alpha| \leq m$ are also likewise Cauchy sequences in $L_2(\Omega)$ with limits u_α. Since $u_k \to u$ in $L_2(\Omega)$, we also have that $u_k \to u$ in $\mathscr{D}'(\Omega)$, so that $D^\alpha u_k \to D^\alpha u$ in $\mathscr{D}'(\Omega)$ (cf. Theorem 3.8 or 3.9). When we compare this with the fact that $D^\alpha u_k \to u_\alpha$ in $L_2(\Omega)$, we see that $u_\alpha = D^\alpha u$ for any $|\alpha| \leq m$, and hence $u \in H^m(\Omega)$ and $u_k \to u$ in $H^m(\Omega)$. (When using the weak definition, one can instead appeal to the closedness of each $D^\alpha_{\max}$.) The subspace $H_0^m(\Omega)$ is now also a Hilbert space, with the induced norm. We have shown:

Lemma 4.6. $H^m(\Omega)$ and $H_0^m(\Omega)$ are Hilbert spaces.

Note that we have continuous injections

$$C_0^\infty(\Omega) \subset H_0^m(\Omega) \subset H^m(\Omega) \subset L_2(\Omega) \subset \mathscr{D}'(\Omega) , \qquad (4.20)$$

in particular, convergence in $H^m(\Omega)$ implies convergence in $\mathscr{D}'(\Omega)$. Note also that when A is an m-th order differential operator with bounded C^∞-coefficients, then

$$H_0^m(\Omega) \subset D(A_{\min}) \subset D(A_{\max}) , \qquad (4.21)$$

since convergence in H^m implies convergence in the graph-norm. For *elliptic* operators of order m, the first inclusion in (4.21) can be shown to be an identity (cf. Theorem 6.29 for operators with constant coefficients in the principal part), while the second inclusion is not usually so, when $\Omega \neq \mathbb{R}^n$ (the case $\Omega \subset \mathbb{R}$ is treated later in this chapter). $D(A_{\max})$ is, for $n > 1$, usually strictly larger than $H^m(\Omega)$, cf. Exercises 4.3–4.5.

Remark 4.7. One can also define similar Sobolev spaces associated with L_p spaces for general $1 \leq p \leq \infty$; here one uses the norms written in (C.11) with $\partial^\alpha u$ taken in the distribution sense. These spaces are Banach spaces; they are often denoted $W_p^m(\Omega)$ (or $W^{m,p}(\Omega)$; the notation $H_p^m(\Omega)$ may also be used). They are useful for example in nonlinear problems where it may be advantageous to use several values of p at the same time. (For example, if the nonlinearity involves a power of u, one can use that $u \in L_p(\mathbb{R}^n)$ implies $u^a \in L_{p/a}(\mathbb{R}^n)$.)

The above definition of $H^m(\Omega)$ is a *weak* definition, in the sense that the derivatives are defined by the help of duality. For the sake of our applications, we shall compare this with various strong definitions.

First we show that $h_j * u$ is a good approximation to u in $H^m(\mathbb{R}^n)$:

Lemma 4.8. Let $u \in H^m(\mathbb{R}^n)$. Then $h_j * u \in C^\infty \cap H^m(\mathbb{R}^n)$ with

$$\begin{aligned} D^\alpha(h_j * u) &= h_j * D^\alpha u \quad \text{for } |\alpha| \leq m; \\ h_j * u &\to u \quad \text{in } H^m(\mathbb{R}^n) \text{ for } j \to \infty. \end{aligned} \qquad (4.22)$$

Proof. When $u \in H^m(\mathbb{R}^n)$, then $h_j * D^\alpha u \in L_2(\mathbb{R}^n) \cap C^\infty(\mathbb{R}^n)$ for each $|\alpha| \leq m$, and $h_j * D^\alpha u \to D^\alpha u$ in $L_2(\mathbb{R}^n)$, by Theorem 2.10. Here

$$h_j * D^\alpha u = D^\alpha(h_j * u), \qquad (4.23)$$

according to Theorem 3.15, so it follows that $h_j * v \to v$ in $H^m(\mathbb{R}^n)$ for $j \to \infty$. $\qquad \square$

For general sets this can be used to prove:

Theorem 4.9. *Let m be integer ≥ 0 and let Ω be any open set in $\mathbb{R}^n$. Then $C^\infty(\Omega) \cap H^m(\Omega)$ is dense in $H^m(\Omega)$.*

Proof. We can assume that Ω is covered by a locally finite sequence of open sets V_j, $j \in \mathbb{N}_0$, with an associated partition of unity ψ_j as in Theorem 2.16 (cf. (2.4), (2.48)).

Let $u \in H^m(\Omega)$; we have to show that it can be approximated arbitrarily well in m-norm by functions in $C^\infty(\Omega) \cap H^m(\Omega)$. First we set $u_j = \psi_j u$, so that we can write $u = \sum_{j \in \mathbb{N}_0} u_j$, Here u_j has compact support in V_j. Let $\varepsilon > 0$ be given. Let $u_j' = h_{k_j} * u_j$, where k_j is taken so large that $\operatorname{supp} u_j' \subset V_j$ and $\|u_j' - u_j\|_{H^m(\Omega)} \leq \varepsilon 2^{-j}$. The first property of k_j can be obtained since $\operatorname{supp} u_j' \subset \operatorname{supp} u_j + \underline{B}(0, \frac{1}{k_j})$ and $\operatorname{supp} u_j$ has positive distance from ∂V_j (recall (2.34)); the second property can be obtained from Lemma 4.8 since the $H^m(\Omega)$ norms of u_j, u_j' and $u_j - u_j'$ are the same as the respective $H^m(\mathbb{R}^n)$ norms, where we identify the functions with their extensions by 0 outside the support.

Let $v = \sum_{j \in \mathbb{N}_0} u_j'$; it exists as a function on Ω since the cover $\{V_j\}_{j \in \mathbb{N}_0}$ is locally finite, and v is C^∞ on Ω. Let K be a compact subset of Ω, it meets a finite number of the sets V_j, say, for $j \leq j_0$. Then

$$\begin{aligned} \|u - v\|_{H^m(K^\circ)} = \| \sum_{j \leq j_0} (u_j - u_j')\|_{H^m(K^\circ)} \\ \leq \sum_{j \leq j_0} \|u_j - u_j'\|_{H^m(\Omega)} \leq \varepsilon \sum_{j \leq j_0} 2^{-j} \leq 2\varepsilon. \end{aligned} \qquad (4.24)$$

Since this can be shown for any $K \subset \Omega$, it follows that $v \in H^m(\Omega)$ and $\|u - v\|_{H^m(\Omega)} \leq 2\varepsilon$. $\qquad \square$

The procedure of approximating a function u by smooth functions by convolution by h_j is in part of the literature called "mollifying", and the operator $h_j *$ called "the Friedrichs mollifier" after K. O. Friedrichs, who introduced it in an important application.

Sometimes the above result is not informative enough for our purposes, since there is no control of how the approximating C^∞ functions behave near the boundary. When the boundary of Ω is sufficiently nice, we can show that C^∞ functions with a controlled behavior are dense.

We recall for the following theorem that for a smooth open set Ω (cf. Definition C.1) we set

$$C_{(0)}^{\infty}(\overline{\Omega}) = \{\, u \in C^{\infty}(\overline{\Omega}) \mid \operatorname{supp} u \text{ compact } \subset \overline{\Omega}\,\}. \tag{4.25}$$

When $\Omega = \mathbb{R}^n$ (and only then), $C_0^{\infty}(\mathbb{R}^n)$ and $C_{(0)}^{\infty}(\mathbb{R}^n)$ coincide.

Theorem 4.10. *Let $\Omega = \mathbb{R}^n$ or $\mathbb{R}_+^n$, or let Ω be a smooth open bounded set. Then $C_{(0)}^{\infty}(\overline{\Omega})$ is dense in $H^m(\Omega)$.*

Proof. 1°. *The case $\Omega = \mathbb{R}^n$.* We already know from Lemma 4.8 that $C^{\infty}(\mathbb{R}^n) \cap H^m(\mathbb{R}^n)$ is dense in $H^m(\mathbb{R}^n)$. Now let $u \in C^{\infty}(\mathbb{R}^n) \cap H^m(\mathbb{R}^n)$, it must be approximated by C_0^{∞} functions. Here we apply the technique of "truncation". With $\chi(x)$ defined in (2.3), we clearly have that $\chi(x/N)u \to u$ in $L_2(\mathbb{R}^n)$ for $N \to \infty$. For the derivatives we have by the Leibniz formula:

$$D^{\alpha}(\chi(x/N)u) = \sum_{\beta \le \alpha} \binom{\alpha}{\beta} D^{\beta}(\chi(x/N))D^{\alpha-\beta}u \tag{4.26}$$

$$= \chi(x/N)D^{\alpha}u + \sum_{\substack{\beta \le \alpha \\ \beta \ne 0}} \binom{\alpha}{\beta} D^{\beta}(\chi(x/N))D^{\alpha-\beta}u\,.$$

The first term converges to $D^{\alpha}u$ in $L_2(\mathbb{R}^n)$ for $N \to \infty$. For the other terms we use that $\sup_{x \in \mathbb{R}^n} |D^{\beta}(\chi(x/N))|$ is $O(N^{-1})$ for $N \to \infty$, for each $\beta \ne 0$. Then the contribution from the sum over $\beta \ne 0$ goes to 0 in $L_2(\mathbb{R}^n)$ for $N \to \infty$. It follows that $\chi(x/N)u \to u$ in $H^m(\mathbb{R}^n)$; here $\chi(x/N)u \in C_0^{\infty}(\mathbb{R}^n)$. This proves 1°.

2°. *The case $\Omega = \mathbb{R}_+^n$.* We now combine the preceding methods (mollification and truncation) with a third technique: "translation". It is used here in a way where the truncated function is pushed a little *outwards*, across the boundary of the domain, before mollification, so that we are only using the mollified functions on a set where the convergence requirements can be verified.

First note that the truncation argument given under 1° works equally well, when $\mathbb{R}^n$ is replaced by an arbitrary open set Ω, so we can replace $u \in H^m(\mathbb{R}_+^n)$ by $v_N = \chi_N u$ having bounded support. (Using Theorem 4.9, we could even assume $u \in C^{\infty}(\mathbb{R}_+^n) \cap H^m(\mathbb{R}_+^n)$, but it has an interest to show a proof that departs from $u \in H^m(\mathbb{R}_+^n)$.)

Defining the translation operator τ_h by

$$\tau_h u(x) = u(x_1, \ldots, x_n - h) \quad \text{for } h \in \mathbb{R}\,, \tag{4.27}$$

we have for $u \in L_2(\mathbb{R}^n)$ that

$$\int_M |u - \tau_h u|^2 dx \to 0 \quad \text{for } h \to 0\,, \tag{4.28}$$

when M is a measurable subset of $\mathbb{R}^n$. Indeed (as used in Theorem 2.10), for any ε there is a $u' \in C_0^0(\mathbb{R}^n)$ with $\|u - u'\|_{L^2(\mathbb{R}^n)} \leq \varepsilon$, and this satisfies: $\|u' - \tau_h u'\|_{L_2(\mathbb{R}^n)} \to 0$ in view of the uniform continuity and compact support. Taking h_0 so that $\|u' - \tau_h u'\|_{L_2(\mathbb{R}^n)} \leq \varepsilon$ for $|h| \leq h_0$, we have that

$$\|u - \tau_h u\|_{L^2(M)} \leq \|u - u'\|_{L^2(M)} + \|u' - \tau_h u'\|_{L^2(M)} + \|\tau_h u' - \tau_h u\|_{L^2(M)}$$
$$\leq 2\|u - u'\|_{L^2(\mathbb{R}^n)} + \|u' - \tau_h u'\|_{L^2(\mathbb{R}^n)} \leq 3\varepsilon$$

for $|h| \leq h_0$.

Note that when $h \geq 0$, then τ_{-h} carries $H^m(\mathbb{R}^n_+)$ over to $H^m(\Omega_{-h})$, where $\Omega_{-h} = \{x \in \mathbb{R}^n \mid x_n > -h\} \supset \mathbb{R}^n_+$. When $u \in L_2(\mathbb{R}^n_+)$, we denote its extension by 0 for $x_n < 0$ by $e^+ u$. In particular, for $u \in H^m(\mathbb{R}^n_+)$, one can consider the functions $e^+ D^\alpha u$ in $L_2(\mathbb{R}^n)$. Here $\tau_{-h} e^+ D^\alpha u$ equals $\tau_{-h} D^\alpha u$ on Ω_{-h}, in particular on $\mathbb{R}^n_+$. Then the above consideration shows that

$$\int_{\mathbb{R}^n_+} |D^\alpha u - \tau_{-h} D^\alpha u|^2 dx \to 0 \quad \text{for } h \to 0+. \tag{4.29}$$

We have moreover, for $\varphi \in C_0^\infty(\mathbb{R}^n_+)$ and $\Omega_h = \{x \mid x_n > h\}$ (still assuming $h \geq 0$),

$$\langle \tau_{-h} D^\alpha u, \varphi \rangle_{\mathbb{R}^n_+} = \langle D^\alpha u, \tau_h \varphi \rangle_{\Omega_h} = \langle u, (-D)^\alpha \tau_h \varphi \rangle_{\Omega_h}$$
$$= \langle u, \tau_h (-D)^\alpha \varphi \rangle_{\Omega_h} = \langle D^\alpha \tau_{-h} u, \varphi \rangle_{\mathbb{R}^n_+},$$

so that D^α and τ_{-h} may be interchanged here. Then $(D^\alpha \tau_{-h} u)|_{\mathbb{R}^n_+}$ is in $L_2(\mathbb{R}^n_+)$ (for $|\alpha| \leq m$), and $(\tau_{-h} u)|_{\mathbb{R}^n_+}$ converges to u in $H^m(\mathbb{R}^n_+)$ for $h \to 0+$.

Returning to our $v_N = \chi_N u$, we thus have that $w_{N,h} = (\tau_{-h} v_N)|_{\mathbb{R}^n_+}$ approximates v_N in $H^m(\mathbb{R}^n_+)$ for $h \to 0+$. We shall end the proof of this case by approximating $w_{N,h}$ in $H^m(\mathbb{R}^n_+)$ by $C_{(0)}^\infty(\overline{\mathbb{R}}^n_+)$-functions. This can be done because $\tau_{-h} v_N$ is in fact in $H^m(\Omega_{-h})$. Recall from Lemma 2.12 that for $j > \frac{1}{\varepsilon}$, $(h_j * \tau_{-h} v_N)(x)$ is a well-defined C^∞-function on the set of x with distance $> \varepsilon$ from the boundary, and there is L_2-convergence to $\tau_{-h} v_N$ on this set, for $j \to \infty$. Let us take $\varepsilon = \frac{h}{2}$; then the set equals $\Omega_{-\frac{h}{2}}$ and contains $\mathbb{R}^n_+$. Clearly also $(h_j * D^\alpha \tau_{-h} v_N)(x)$ converges to $D^\alpha \tau_{-h} v_N$ in L_2 on $\Omega_{-\frac{h}{2}}$, hence on $\mathbb{R}^n_+$, when $|\alpha| \leq m$. Now we show that for $j > \frac{2}{h}$, $D^\alpha(h_j * \tau_{-h} v_N) = h_j * D^\alpha(\tau_{-h} v_N)$ on $\mathbb{R}^n_+$ (it even holds on $\Omega_{-\frac{h}{2}}$): For any $\varphi \in \mathscr{D}(\mathbb{R}^n_+)$,

$$\langle D^\alpha(h_j * \tau_{-h} v_N), \varphi \rangle_{\mathbb{R}^n_+} = \langle h_j * \tau_{-h} v_N, (-D)^\alpha \varphi \rangle_{\mathbb{R}^n_+}$$
$$= \int_{x \in \mathbb{R}^n_+} \int_{y \in B(0, \frac{1}{j})} h_j(x - y)(\tau_{-h} v_N)(y)(-D)^\alpha \varphi(x) \, dy dx$$
$$= \int_{y \in \mathbb{R}^n_+ + B(0, \frac{1}{j})} (\tau_{-h} v_N)(y)(\check{h}_j * (-D)^\alpha \varphi)(y) \, dy$$

$$= \int_{y \in \mathbb{R}^n_+ + B(0, \frac{1}{j})} (\tau_{-h} v_N)(y)(-D)^\alpha (\check{h}_j * \varphi)(y) \, dy$$

$$= \int_{y \in \mathbb{R}^n_+ + B(0, \frac{1}{j})} D^\alpha (\tau_{-h} v_N)(y)(\check{h}_j * \varphi)(y) \, dy = \langle h_j * D^\alpha (\tau_{-h} v_N), \varphi \rangle_{\mathbb{R}^n_+}.$$

Thus also $D^\alpha(h_j * \tau_{-h} v_N)$ converges to $D^\alpha \tau_{-h} v_N$ in $L_2(\mathbb{R}^n_+)$ for each $|\alpha| \le m$; so $(h_j * \tau_{-h} v_N)|_{\mathbb{R}^n_+}$ converges to $(\tau_{-h} v_N)|_{\mathbb{R}^n_+} = w_{N,h}$ in $H^m(\mathbb{R}^n_+)$ for $j \to \infty$. Since $(h_j * \tau_{-h} v_N)|_{\mathbb{R}^n_+}$ is in $C^\infty_{(0)}(\overline{\mathbb{R}}^n_+)$, this ends the proof of $2°$.

3°. *The case where Ω is smooth, open and bounded.* Here we moreover include the technique of "localization." We cover $\overline{\Omega}$ by open sets $\Omega_0, \Omega_1, \ldots, \Omega_N$, where $\Omega_1, \ldots, \Omega_N$ are of the type U described in Definition C.1, and $\overline{\Omega}_0 \subset \Omega$. Let $\psi_0, \ldots, \psi_N$ be a partition of unity with $\psi_l \in C^\infty_0(\Omega_l)$ and $\psi_0 + \cdots + \psi_N = 1$ on $\overline{\Omega}$ (cf. Theorem 2.17). The function $\psi_0 u$ has compact support in Ω and gives by extension by 0 outside of Ω a function $\widetilde{\psi_0 u}$ in $H^m(\mathbb{R}^n)$ (since $D^\alpha(\psi_0 u) = \sum_{\beta \le \alpha} \binom{\alpha}{\beta} D^\beta \psi_0 D^{\alpha - \beta} u$ has support in supp ψ_0, so that $(D^\alpha(\psi_0 u))^\sim$ equals $D^\alpha(\widetilde{\psi_0 u})$). Now $\widetilde{\psi_0 u}$ can be approximated according to 1°, and restriction to Ω then gives the desired approximation of $\psi_0 u$. The functions $\psi_l u, l = 1, \ldots, N$, are by the diffeomorphisms associated with each Ω_l carried into functions v_l in $H^m(\mathbb{R}^n_+)$ (with support in $B(0,1))$,[1] which is approximated in $H^m(\mathbb{R}^n_+)$ according to 2°. (Since supp v_l is compact $\subset B(0,1)$, the translated function stays supported in the ball for sufficiently small h.) Transforming this back to Ω_l, we get an approximation of $\psi_l u$, for each l. The sum of the approximations of the $\psi_l u, l = 0, \ldots, N$, approximates u. $\square$

The result in 3° holds also under weaker regularity requirements on Ω, where the idea of translation across the boundary can still be used.

Corollary 4.11. 1° $H^m(\mathbb{R}^n) = H^m_0(\mathbb{R}^n)$ *for all $m \in \mathbb{N}_0$; i.e., $C^\infty_0(\mathbb{R}^n)$ is dense in $H^m(\mathbb{R}^n)$ for all $m \in \mathbb{N}_0$.*

2° *For Ω smooth open and bounded, $C^\infty(\overline{\Omega})$ is dense in $H^m(\Omega)$, for all $m \in \mathbb{N}_0$.*

Proof. 1° follows from Theorem 4.10 and the fact that $C^\infty_{(0)}(\mathbb{R}^n) = C^\infty_0(\mathbb{R}^n)$, cf. Definition 4.5 2°. 2° follows from Theorem 4.10 and the fact that for Ω smooth, open and bounded, $C^\infty_{(0)}(\overline{\Omega}) = C^\infty(\overline{\Omega})$. $\square$

Of course we always have that $H^0(\Omega) = L_2(\Omega) = H^0_0(\Omega)$ for arbitrary Ω (since $C^\infty_0(\Omega)$ is dense in $L_2(\Omega)$). But for $m > 0$, $H^m(\Omega) \ne H^m_0(\Omega)$ when $\mathbb{R}^n \setminus \overline{\Omega} \ne \emptyset$.

We can now also show an extension theorem.

---

[1] Here we use the chain rule for distributions, cf. (3.43). Since κ, κ^{-1} and their derivatives are C^∞ functions, the property $u \in H^m$ is invariant under diffeomorphisms κ where κ, κ^{-1} and their derivatives are bounded.

Theorem 4.12. *Let $m \in \mathbb{N}$, and let Ω be smooth open and bounded, or equal to $\mathbb{R}^n_+$. There is a continuous linear operator $E : H^m(\Omega) \to H^m(\mathbb{R}^n)$ so that $u = (Eu)|_\Omega$ for $u \in H^m(\Omega)$.*

Proof. 1°. *The case $\Omega = \mathbb{R}^n_+$.* Choose in an arbitrary way a set of $m + 1$ different positive numbers $\lambda_0, \ldots, \lambda_m$. Let $\{\alpha_0, \ldots, \alpha_m\}$ be the solution of the system of equations

$$
\begin{aligned}
\textstyle\sum_{k=0}^m \alpha_k &= 1 \,, \\
\textstyle\sum_{k=0}^m \lambda_k \alpha_k &= -1 \,, \\
&\;\;\vdots \\
\textstyle\sum_{k=0}^m \lambda_k^m \alpha_k &= (-1)^m \,.
\end{aligned}
\tag{4.30}
$$

The solution exists and is uniquely determined, since the determinant of the system is the Vandermonde determinant,

$$
\det \begin{pmatrix} 1 & \cdots & 1 \\ \lambda_0 & \cdots & \lambda_m \\ \vdots & & \vdots \\ \lambda_0^m & \cdots & \lambda_m^m \end{pmatrix} = \prod_{0 \le i < j \le m} (\lambda_j - \lambda_i) \ne 0 \,.
\tag{4.31}
$$

Now when $u \in C^\infty_{(0)}(\overline{\mathbb{R}}^n_+)$, we define Eu by

$$
(Eu)(x) = \begin{cases} u(x) & \text{for } x_n \ge 0 \,, \\ \sum_{k=0}^m \alpha_k u(x', -\lambda_k x_n) & \text{for } x_n < 0 \,. \end{cases}
\tag{4.32}
$$

Because of (4.30), we have that $D^\alpha Eu$ is continuous on $\mathbb{R}^n$ for all $|\alpha| \le m$, so that $Eu \in H^m(\mathbb{R}^n)$; it is easy to verify by use of (4.32) that

$$
\|Eu\|_{H^m(\mathbb{R}^n)} \le c\|u\|_{H^m(\mathbb{R}^n_+)}
\tag{4.33}
$$

for some constant c. Since the operator $E : u \mapsto Eu$ is defined linearly on the dense subset $C^\infty_{(0)}(\overline{\mathbb{R}}^n_+)$ of $H^m(\mathbb{R}^n_+)$ and by (4.33) is continuous in m-norm, it extends by continuity to a continuous map of $H^m(\mathbb{R}^n_+)$ into $H^m(\mathbb{R}^n)$ with the desired properties.

2°. *The case where Ω is smooth, open and bounded.* This is reduced to an application of the preceding case by use of a covering $\bigcup_{l=0}^N \Omega_l$ of $\overline{\Omega}$ as in the proof of Theorem 4.10 3° (with $\overline{\Omega}_0 \subset \Omega$) and associated diffeomorphisms $\kappa_{(l)} : \Omega_l \to B(0,1)$ for $l > 0$, together with a partition of unity $\psi_0, \ldots, \psi_n$ with $\psi_l \in C^\infty_0(\Omega_l)$ and $\psi_0 + \cdots + \psi_N = 1$ on $\overline{\Omega}$. Let $u \in C^\infty(\overline{\Omega})$ $(= C^\infty_{(0)}(\overline{\Omega}))$. For $\psi_0 u$, we take $E(\psi_0 u) = \widetilde{\psi_0 u}$ (extension by 0 outside Ω). For each $l > 0$, $\psi_l u$ is by the diffeomorphism $\kappa_{(l)}$ carried over to a function $v_l \in C^\infty_{(0)}(\overline{\mathbb{R}}^n_+)$ supported in $B(0,1) \cap \overline{\mathbb{R}}^n_+$. Here we use the extension operator $E_{\mathbb{R}^n_+}$ constructed in 1°, choosing λ_k to be > 1 for each k (e.g., $\lambda_k = k + 2$), so that the support of $E_{\mathbb{R}^n_+} v_l$ is a compact subset of $B(0,1)$. With the notation

$$(T_l v)(x) = v(\kappa_{(l)}(x)), \quad l = 1, \ldots, N ,$$

we set

$$E_\Omega u = \widetilde{\psi_0 u} + \sum_{l=1}^n T_l(E_{\mathbb{R}^n_+} v_l) .$$

Defining E_Ω in this way for smooth functions, we get by extension by continuity a map E_Ω having the desired properties.   □

Part $1°$ of the proof could actually have been shown using only m numbers $\lambda_0, \ldots, \lambda_{m-1}$ and m equations (4.30) (with m replaced by $m-1$); for we could then appeal to Theorem 3.20 in the proof that the m-th order derivatives of Eu were in L_2.

There exist constructions where the extension operator does not at all depend on m, and where the boundary is allowed to be considerably less smooth, cf. e.g. [EE87, Th. V 4.11–12].

The analysis of Sobolev spaces will be continued in connection with studies of boundary value problems and in connection with the Fourier transformation. A standard reference for the use of Sobolev spaces in the treatment of boundary value problems is J.-L. Lions and E. Magenes [LM68].

Theorems 4.9 and 4.10 show the equivalence of weak and strong definitions of differential operators in the particular case where we consider the whole family of differential operators $\{D^\alpha\}_{|\alpha| \le m}$ taken together. For a general operator $A = \sum_{|\alpha| \le m} a_\alpha D^\alpha$, such properties are harder to show (and need not hold when there is a boundary). For example, in an application of Friedrichs' mollifier, one will have to treat $h_j * (\sum_\alpha a_\alpha D^\alpha u) - \sum_\alpha a_\alpha D^\alpha(h_j * u)$, which requires further techniques when the a_α depend on x.

For operators in one variable one can often take recourse to absolute continuity (explained below). For some operators on $\mathbb{R}^n$ (for example the Laplace operator), the Fourier transformation is extremely useful, as we shall see in Chapters 5 and 6.

4.3 The one-dimensional case

In the special case where Ω is an interval of $\mathbb{R}$, we use the notation $u', u'', u^{(k)}$ along with $\partial u, \partial^2 u, \partial^k u$, also for distribution derivatives. Here the derivative defined on Sobolev spaces is closely related to the derivative of absolutely continuous functions.

Traditionally, a function f is said to be absolutely continuous on $\overline{I}$, $I =]\alpha, \beta[$, when f has the property that for any $\varepsilon > 0$ there exists a $\delta > 0$ such that for every set $J_1, \ldots, J_N$ of disjoint subintervals $J_k = [\alpha_k, \beta_k]$ of $\overline{I}$ with total length $\leq \delta$ (i.e., with $\sum_{k=1,\ldots,N}(\beta_k - \alpha_k) \leq \delta$),

$$\sum_{k=1,\ldots,N} |f(\beta_k) - f(\alpha_k)| \leq \varepsilon.$$

It is known from measure theory that this property is equivalent with the existence of an integrable function g on I and a number k such that

$$f(x) = \int_\alpha^x g(s)ds + k, \text{ for } x \in \overline{I}. \tag{4.34}$$

We shall show that $H^1(I)$ consists precisely of such functions with $g \in L_2(I)$.

Theorem 4.13. *Let* $I =]\alpha, \beta[$ *for some* $\alpha < \beta$. *Then*

$$C^0(\overline{I}) \supset H^1(I) \supset C^1(\overline{I}); \text{ with} \tag{4.35}$$
$$\|u\|_{L_\infty(I)} \leq c_1 \|u\|_{H^1(I)} \leq c_2(\|u\|_{L_\infty(I)} + \|\partial u\|_{L_\infty(I)}), \tag{4.36}$$

for some constants c_1 *and* $c_2 > 0$. *Moreover,*

$$H^1(I) = \{ f \mid f(x) = \int_\alpha^x g(s)ds + k, \ g \in L_2(I), \ k \in \mathbb{C} \}. \tag{4.37}$$

Proof. It is obvious that $C^1(\overline{I}) \subset H^1(I)$ (the second inclusion in (4.35)), with the inequality

$$\|u\|_{H^1(I)}^2 = \int_\alpha^\beta (|u(x)|^2 + |u'(x)|^2) \, dx \leq (\beta - \alpha)(\|u\|_{L_\infty(I)}^2 + \|u'\|_{L_\infty(I)}^2);$$

this implies the second inequality in (4.36).

Now let $u \in C^1(\overline{I})$ and set $v(x) = \int_\alpha^x u'(s) \, ds$, so that

$$u(x) = u(\alpha) + \int_\alpha^x u'(s) \, ds = u(\alpha) + v(x). \tag{4.38}$$

Clearly,

$$\int_\alpha^\beta |v(x)|^2 \, dx \leq (\beta - \alpha)\|v\|_{L_\infty(I)}^2.$$

Moreover, by the Cauchy-Schwarz inequality,

$$|v(x)|^2 = |\int_\alpha^x u'(s) \, ds|^2 \leq \int_\alpha^x 1 \, ds \int_\alpha^x |u'(s)|^2 \, ds \leq (\beta - \alpha) \int_\alpha^\beta |u'(s)|^2 \, ds;$$

we collect the inequalities in

$$(\beta - \alpha)^{-1}\|v\|_{L_2(I)}^2 \leq \|v\|_{L_\infty(I)}^2 \leq (\beta - \alpha)\|u'\|_{L_2(I)}^2. \tag{4.39}$$

It follows (cf. (4.38)) that

$$|u(\alpha)| = \left((\beta - \alpha)^{-1} \int_\alpha^\beta |u(\alpha)|^2 \, dx\right)^{\frac{1}{2}} = (\beta - \alpha)^{-\frac{1}{2}} \|u - v\|_{L_2(I)}$$
$$\leq (\beta - \alpha)^{-\frac{1}{2}} \|u\|_{L_2(I)} + (\beta - \alpha)^{-\frac{1}{2}} \|v\|_{L_2(I)}$$
$$\leq (\beta - \alpha)^{-\frac{1}{2}} \|u\|_{L_2(I)} + (\beta - \alpha)^{\frac{1}{2}} \|u'\|_{L_2(I)};$$

and then furthermore:

$$|u(x)| = |v(x) + u(\alpha)| \leq \|v\|_{L_\infty(I)} + |u(\alpha)|$$
$$\leq (\beta - \alpha)^{-\frac{1}{2}} \|u\|_{L_2(I)} + 2(\beta - \alpha)^{\frac{1}{2}} \|u'\|_{L_2(I)}, \text{ for all } x \in \overline{I}.$$

This shows the first inequality in (4.36) for $u \in C^1(\overline{I})$. For a general $u \in H^1(I)$, let $u_n \in C^\infty(\overline{I})$ be a sequence converging to u in $H^1(I)$ for $n \to \infty$. Then u_n is a Cauchy sequence in $L_\infty(I)$, and hence since the u_n are continuous, has a continuous limit u_0 in sup-norm. Since $u_n \to u_0$ also in $L_2(I)$, u_0 is a continuous representative of u; we use this representative in the following, denoting it u again. We have hereby shown the first inclusion in (4.35), and the first inequality in (4.36) extends to $u \in H^1(I)$.

Now let us show (4.37). Let $u \in H^1(I)$ and let $u_n \in C^\infty(\overline{I})$, $u_n \to u$ in $H^1(I)$. Then $u_n' \to u'$ in $L_2(I)$, and $u_n(x) \to u(x)$ uniformly for $x \in \overline{I}$, as we have just shown. Then

$$u_n(x) = \int_\alpha^x u_n'(s) \, ds + u_n(\alpha) \text{ implies } u(x) = \int_\alpha^x u'(s) \, ds + u(\alpha),$$

by passage to the limit, so u belongs to the right-hand side of (4.37).

Conversely, let f satisfy

$$f(x) = \int_\alpha^x g(s) \, ds + k,$$

for some $g \in L_2(I)$, $k \in \mathbb{C}$. Clearly, $f \in C^0(\overline{I})$ with $f(\alpha) = k$. We shall show that ∂f, taken in the distribution sense, equals g; this will imply $f \in H^1(I)$. We have for any $\varphi \in C_0^\infty(I)$, using the Fubini theorem and the fact that $\varphi(\alpha) = \varphi(\beta) = 0$ (see the figure):

$$\langle \partial f, \varphi \rangle = -\langle f, \partial \varphi \rangle = -\int_\alpha^\beta f(x)\varphi'(x) \, dx$$
$$= -\int_\alpha^\beta \int_\alpha^x g(s)\varphi'(x) \, ds dx - \int_\alpha^\beta k\varphi'(x) \, dx$$
$$= -\int_\alpha^\beta \int_s^\beta g(s)\varphi'(x) \, dx ds - k(\varphi(\beta) - \varphi(\alpha))$$

$$= -\int_\alpha^\beta g(s)(\varphi(\beta) - \varphi(s))\, ds = \int_\alpha^\beta g(s)\varphi(s)\, ds = \langle g, \varphi \rangle.$$

Thus $\partial f = g$ in $\mathscr{D}'(I)$, and the proof is complete. $\qquad\square$

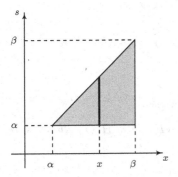
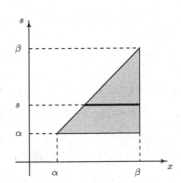

We have furthermore:

Theorem 4.14. *When u and $v \in H^1(I)$, then also $uv \in H^1(I)$, and*

$$\partial(uv) \overset{.}{=} (\partial u)v + u(\partial v); \qquad with \qquad (4.40)$$

$$\|uv\|_{H^1(I)} \le 5^{\frac{1}{2}} c_1 \|u\|_{H^1(I)} \|v\|_{H^1(I)}. \qquad (4.41)$$

Moreover,

$$(\partial u, v)_{L_2(I)} + (u, \partial v)_{L_2(I)} = u(\beta)\bar{v}(\beta) - u(\alpha)\bar{v}(\alpha), \qquad (4.42)$$

$$(Du, v)_{L_2(I)} - (u, Dv)_{L_2(I)} = -iu(\beta)\bar{v}(\beta) + iu(\alpha)\bar{v}(\alpha). \qquad (4.43)$$

Proof. Let $u, v \in H^1(I)$ and let $u_n, v_n \in C^1(\overline{I})$ with $u_n \to u$ and $v_n \to v$ in $H^1(I)$ for $n \to \infty$. By Theorem 4.13, the convergences hold in $C^0(\overline{I})$, so $u_n v_n \to uv$ in $C^0(\overline{I})$, hence also in $\mathscr{D}'(I)$. Moreover, $(\partial u_n)v_n \to (\partial u)v$ and $u_n(\partial v_n) \to u(\partial v)$ in $L_2(I)$, hence in $\mathscr{D}'(I)$, so the formula

$$\partial(u_n v_n) = (\partial u_n)v_n + u_n(\partial v_n),$$

valid for each n, implies

$$\partial(uv) = (\partial u)v + u(\partial v) \text{ in } \mathscr{D}'(I),$$

by Theorem 3.9. This shows (4.40), and (4.41) follows since

$$\|uv\|_{L_2(I)} \le \|u\|_{L_\infty(I)}\|v\|_{L_2(I)} \le c_1 \|u\|_{H^1(I)}\|v\|_{H^1(I)},$$

72 4 Realizations and Sobolev spaces

$$\|\partial(uv)\|_{L_2(I)} = \|(\partial u)v + u(\partial v)\|_{L_2(I)}$$
$$\leq \|\partial u\|_{L_2(I)}\|v\|_{L_\infty(I)} + \|u\|_{L_\infty(I)}\|\partial v\|_{L_2(I)}$$
$$\leq 2c_1\|u\|_{H^1(I)}\|v\|_{H^1(I)},$$

by (4.36). The formula (4.42) is shown by an application of (4.37) to $u\bar{v}$:

$$u(\beta)\bar{v}(\beta) - u(\alpha)\bar{v}(\alpha) = \int_\alpha^\beta \partial(u(s)\bar{v}(s))\,ds$$
$$= \int_\alpha^\beta (\partial u(s)\,\bar{v}(s) + u(s)\partial\bar{v}(s))\,ds,$$

and (4.43) follows by multiplication by $-i$. □

The subspace $H_0^1(I)$ is characterized as follows:

Theorem 4.15. *Let* $I =]\alpha, \beta[$. *The subspace* $H_0^1(I)$ *of* $H^1(I)$ *(the closure of* $C_0^\infty(I)$ *in* $H^1(I)$*) satisfies*

$$H_0^1(I) = \{\, u \in H^1(I) \mid u(\alpha) = u(\beta) = 0 \,\}$$
$$= \{\, u(x) = \int_\alpha^x g(s)\,ds \mid g \in L_2(I), (g,1)_{L_2(I)} = 0\}. \quad (4.44)$$

Proof. When $u = \int_\alpha^x g(s)\,ds + k$, then $u(\alpha) = 0$ if and only if $k = 0$, and when this holds, $u(\beta) = 0$ if and only if $\int_\alpha^\beta g(s)\,ds = 0$. In view of (4.37), this proves the second equality in (4.44).

Let $u \in H_0^1(I)$, then it is the limit in $H^1(I)$ of a sequence of functions $u_n \in C_0^\infty(I)$. Since $u_n(\alpha) = u_n(\beta) = 0$ and the convergence holds in $C^0(\bar{I})$ (cf. (4.35) and (4.36)), $u(\alpha) = u(\beta) = 0$.

Conversely, let $u \in H^1(I)$ with $u(\alpha) = u(\beta) = 0$. Then $\tilde{u}$, the extension by 0 outside $[\alpha, \beta]$, is in $H^1(\mathbb{R})$, since $\tilde{u}(x) = \int_\alpha^x \tilde{u}'\,ds$ for any $x \in \mathbb{R}$, where $\tilde{u}'$ is the extension of u' by 0 outside $[\alpha, \beta]$. Let us "shrink" $\tilde{u}$ by defining, for $0 < \delta \leq \frac{1}{2}$,

$$\tilde{v}_\delta = \tilde{u}(\tfrac{1}{1-\delta}(x - \tfrac{\alpha+\beta}{2}));$$

since $\tilde{u}$ vanishes for $|x - \frac{\alpha+\beta}{2}| \geq \frac{\beta-\alpha}{2}$, $\tilde{v}_\delta$ vanishes for $|x - \frac{\alpha+\beta}{2}| \geq (1-\delta)\frac{\beta-\alpha}{2}$, i.e., is supported in the interval

$$[\alpha + \delta\tfrac{\beta-\alpha}{2}, \beta - \delta\tfrac{\beta-\alpha}{2}] \subset \,]\alpha, \beta[.$$

Clearly, $\tilde{v}_\delta \to \tilde{u}$ in $H^1(\mathbb{R})$ for $\delta \to 0$. Mollifying $\tilde{v}_\delta$ to $h_j * \tilde{v}_\delta$ for $\frac{1}{j} < \delta\frac{\beta-\alpha}{2}$, we get a C^∞-function with compact support in I, such that $h_j * \tilde{v}_\delta \to \tilde{v}_\delta$ in $H^1(I)$ for $j \to \infty$. Thus we can find a $C_0^\infty(I)$-function $(h_j * \tilde{v}_\delta)|_I$ arbitrarily close to u in $H^1(I)$ by choosing first δ small enough and then j large enough. This completes the proof of the first equality in (4.44). □

Now let us consider realizations of the basic differential operator $A = D = \frac{1}{i}\partial$. By definition (see Definition 4.1 ff.),

$$D(A_{\max}) = H^1(I),$$
$$D(A_{\min}) = H_0^1(I). \tag{4.45}$$

Equation (4.43) implies

$$(A_{\min}u, v) - (u, A_{\min}v) = 0,$$

so $A_{\min}$ is symmetric, and $A_{\max} = A_{\min}{}^*$, cf. (4.7). Here $A_{\min}$ is too small and $A_{\max}$ is too large to be selfadjoint; in fact we shall show that the realization $A_{\#}$ defined by

$$D(A_{\#}) = \{\, u \in H^1(I) \mid u(\alpha) = u(\beta) \,\} \tag{4.46}$$

(the "perioodic boundary condition") is selfadjoint. To see this, note that $A_{\#}$ is symmetric since the right-hand side of (4.43) vanishes if u and $v \in D(A_{\#})$, so $A_{\#} \subset A_{\#}^*$. Since $A_{\min} \subset A_{\#}$, $A_{\#}^* \subset A_{\max}$. Then we show $A_{\#}^* \subset A_{\#}$ as follows: Let $u \in D(A_{\#}^*)$, then for any $v \in D(A_{\#})$ with $v(\alpha) = v(\beta) = k$,

$$0 = (A_{\#}^* u, v) - (u, A_{\#}v) = (Du, v) - (u, Dv) = -iu(\beta)\bar{k} + iu(\alpha)\bar{k}.$$

Since k can be arbitrary, this implies that $u(\beta) = u(\alpha)$, hence $u \in D(A_{\#})$.

We have shown:

Theorem 4.16. *Consider the realizations $A_{\max}$, $A_{\min}$ and $A_{\#}$ of $A = D$ with domains described in (4.45) and (4.46). Then*

1° $A_{\min}$ *is symmetric,*
2° $A_{\min}$ *and $A_{\max}$ are adjoints of one another,*
3° $A_{\#}$ *is selfadjoint.*

For general m, the characterizations of the Sobolev spaces extend as follows:

Theorem 4.17. *Let $I = \,]\alpha, \beta[\,$.*
1° *$H^m(I)$ consists of the functions $u \in C^{m-1}(\overline{I})$ such that $u^{(m-1)} \in H^1(I)$. The inequality*

$$\sum_{j \leq m-1} \|\partial^j u\|_{L_\infty(I)}^2 \leq C \sum_{k \leq m} \|\partial^k u\|_{L_2(I)}^2 \tag{4.47}$$

holds for all $u \in H^m(I)$, with some constant $C > 0$.
2° *The subspace $H_0^m(I)$ (the closure of $C_0^\infty(I)$ in $H^m(I)$) satisfies*

$$H_0^m(I) = \{\, u \in H^m(I) \mid u^{(j)}(\alpha) = u^{(j)}(\beta) = 0 \text{ for } j = 0, \ldots, m-1 \,\}. \tag{4.48}$$

Proof. We can assume $m > 1$.
1°. It is clear from the definition that a function u is in $H^m(I)$ if and only if $u, u', \ldots, u^{(m-1)}$ (defined in the distribution sense) belong to $H^1(I)$. This holds in particular if $u \in C^{m-1}(\overline{I})$ with $u^{(m-1)} \in H^1(I)$. To show that an arbitrary function $u \in H^m(I)$ is in $C^{m-1}(\overline{I})$, note that by Theorem 4.13,

$$u^{(m-1)} \in C^0(\overline{I}),$$

$$u^{(j)}(x) = \int_\alpha^x u^{(j+1)}(s)\, ds + u^{(j)}(\alpha) \text{ for } j < m;$$

the latter gives by successive application for $j = m - 1, m - 2, \ldots, 0$ that $u \in C^{m-1}(\overline{I})$. The inequality (4.47) follows by applying (4.36) in each step.

2°. When $u \in C_0^\infty(I)$, then all derivatives at α and β are 0, so by passage to the limit in m-norm we find that $u^{(j)}(\alpha) = u^{(j)}(\beta) = 0$ for $j \le m - 1$ in view of (4.47). Conversely, if u belongs to the right-hand side of (4.48), one can check that the extension $\tilde{u}$ by zero outside $[\alpha, \beta]$ is in $H^m(\mathbb{R})$ and proceed as in the proof of Theorem 4.15. □

The "integration by parts" formulas (4.42) and (4.43) have the following generalization to m-th order operators on $I =]\alpha, \beta[$, often called the *Lagrange formula*:

$$(D^m u, v)_{L_2(I)} - (u, D^m v)_{L_2(I)} = \sum_{k=0}^{m-1} \left[(D^{m-k} u, D^k v) - (D^{m-k-1} u, D^{k+1} v) \right]$$

$$= -i \sum_{k=0}^{m-1} \left[D^{m-1-k} u(\beta) \overline{D^k v(\beta)} - D^{m-1-k} u(\alpha) \overline{D^k v(\alpha)} \right],$$

for $u, v \in H^m(I)$. $\qquad\qquad\qquad\qquad\qquad\qquad\qquad$ (4.49)

Unbounded intervals are included in the analysis by the following theorem:

Theorem 4.18. 1° *An inequality* (4.47) *holds also when I is an unbounded interval of $\mathbb{R}$.*

2° *For $I = \mathbb{R}$, $H^m(I) = H_0^m(I)$. For $I =]\alpha, \infty[$,*

$$H_0^m(I) = \left\{ u \in H^m(I) \mid u^{(j)}(\alpha) = 0 \quad \text{for } j = 0, \ldots, m - 1 \right\}. \qquad (4.50)$$

There is a similar characterization of $H_0^m(]-\infty, \beta[)$.

Proof. 1°. We already have (4.47) for bounded intervals I', with constants $C(I')$. Since derivatives commute with translation,

$$\max_{x \in \overline{I'} + a} \| \partial^j u(x - a) \| = \max_{x \in \overline{I'}} \| \partial^j u(x) \|, \quad \| \partial^j u(\cdot - a) \|_{L_2(I' + a)} = \| \partial^j u(\cdot) \|_{L_2(I')};$$

then the constant $C(I')$ can be taken to be invariant under translation, depending only on m and the length of I'. When I is unbounded, every point $x \in \overline{I}$ lies in an interval $\overline{I_x} \subset \overline{I}$ of length 1, and then

$$\sum_{0 \le j \le m-1} |\partial^j u(x)|^2 \le C(1) \sum_{0 \le j \le m} \| \partial^j u \|_{L_2(I_x)}^2 \le C(1) \sum_{0 \le j \le m} \| \partial^j u \|_{L_2(I)}^2,$$

where $C(1)$ is the constant used for intervals of length 1.

2°. The first statement is a special case of Corollary 4.11.

For the second statement, the inclusion '⊂' is an obvious consequence of 1°. For the inclusion '⊃' let $u \in H^m(I)$ and consider the decomposition

$$u = \chi_N u + (1 - \chi_N)u,$$

where $\chi_N(x) = \chi(x/N)$ with N taken $\geq 2|\alpha|$ (cf. (2.3)). We see as in Theorem 4.10 that $\chi_N u \to u$ in $H^m(I)$ for $N \to \infty$. Here, if $u^{(k)}(\alpha) = 0$ for $k = 0, 1, \ldots, m-1$, then $\chi_N u$ lies in $H_0^m(]\alpha, 2N[)$ according to (4.48), hence can be approximated in m-norm by functions in $C_0^\infty(]\alpha, 2N[)$. So, by taking first N sufficiently large and next choosing $\varphi \in C_0^\infty(]\alpha, 2N[)$ close to $\chi_N u$, one can approximate u in m-norm by functions in $C_0^\infty(]\alpha, \infty[)$. $\square$

The functions in $H^m(\mathbb{R})$ satisfy $u^{(j)}(x) \to 0$ for $x \to \pm\infty$, $j < m$ (Exercise 4.21); and the formulas (4.42) and (4.43) take the form

$$(\partial u, v)_{L_2(I)} + (u, \partial v)_{L_2(I)} = -u(\alpha)\bar{v}(\alpha), \tag{4.51}$$

$$(Du, v)_{L_2(I)} - (u, Dv)_{L_2(I)} = iu(\alpha)\bar{v}(\alpha), \tag{4.52}$$

for $u, v \in H^1(I)$, $I =]\alpha, \infty[$. We also have the Lagrange formula

$$(D^m u, v)_{L_2(I)} - (u, D^m v)_{L_2(I)} = i \sum_{k=0}^{m-1} D^{m-1-k}u(\alpha)\overline{D^k v(\alpha)}, \tag{4.53}$$

for $u, v \in H^m(I)$, $I =]\alpha, \infty[$.

To prepare for the analysis of realizations of D^m, we shall show an interesting "uniqueness theorem". It shows that there are no other distribution solutions of the equation $D^k u = 0$ than the classical (polynomial) solutions.

Theorem 4.19. *Let I be an open interval of $\mathbb{R}$, let $k \geq 1$ and let $u \in \mathscr{D}'(I)$. If $Du = 0$, then u equals a constant. If $D^k u = 0$, then u is a polynomial of degree $\leq k - 1$.*

Proof. The theorem is shown by induction in k. For the case $k = 1$, one has when $Du = 0$ that

$$0 = \langle Du, \varphi \rangle = -\langle u, D\varphi \rangle \quad \text{for all} \quad \varphi \in C_0^\infty(I).$$

Choose a function $h \in C_0^\infty(I)$ with $\langle 1, h \rangle = 1$ and define, for $\varphi \in C_0^\infty(I)$,

$$\psi(x) = i \int_{-\infty}^x [\varphi(s) - \langle 1, \varphi \rangle h(s)]ds.$$

Here $D\psi = \varphi - \langle 1, \varphi \rangle h$; and $\psi \in C_0^\infty(I)$ since

$$\int_{-\infty}^x [\varphi(s) - \langle 1, \varphi \rangle h(s)]ds = \langle 1, \varphi \rangle - \langle 1, \varphi \rangle \langle 1, h \rangle = 0$$

when $x > \sup(\operatorname{supp}\varphi \cup \operatorname{supp} h)$. Hence φ can be written as

$$\varphi = \varphi - \langle 1, \varphi\rangle h + \langle 1, \varphi\rangle h = D\psi + \langle 1, \varphi\rangle h , \qquad (4.54)$$

whereby

$$\langle u, \varphi\rangle = \langle u, D\psi\rangle + \langle 1, \varphi\rangle\langle u, h\rangle = \langle\langle u, h\rangle 1, \varphi\rangle = \langle c_u, \varphi\rangle,$$

for all $\varphi \in C_0^\infty(I)$; here c_u is the constant $\langle u, h\rangle$. This shows that $u = c_u$.

The induction step goes as follows: Assume that the theorem has been proved up to the index k. If $D^{k+1}u = 0$, then $D^k Du = 0$ (think over why!), so Du is a polynomial $p(x)$ of degree $\le k - 1$. Let $P(x)$ be an integral of $ip(x)$; then $D(u - P(x)) = 0$ and hence $u = P(x) + c$.   □

Theorem 4.20. *Let I be an open interval of $\mathbb{R}$. Let $m \ge 1$. If $u \in \mathscr{D}'(I)$ and $D^m u \in L_2(I)$, then $u \in H^m(I')$ for each bounded subinterval I' of I.*

Proof. It suffices to show the result for a bounded interval $I =]\alpha, \beta[$. By successive integration of $D^m u$ we obtain a function

$$v(t) = i^m \int_\alpha^t ds_1 \int_\alpha^{s_1} ds_2 \int_\alpha^{s_2} \cdots \int_\alpha^{s_{m-1}} D^m u(s_m) ds_m , \qquad (4.55)$$

which belongs to $H^m(I)$. Now $D^m v = D^m u$, hence $u - v$ is a polynomial p (of degree $\le m - 1$) according to Theorem 4.19, hence a C^∞-function. In particular, $u = v + p \in H^m(I)$.   □

Remark 4.21. For unbounded intervals, the property $D^m u \in L_2(I)$ is not sufficient for the conclusion that $u \in H^m(I)$ *globally*. But here one can show that when *both u and $D^m u$* are in $L_2(I)$, then u is in $H^m(I)$. It is easily shown by use of the Fourier transformation studied later on, so we save the proof till then. (See Exercise 5.10.)

Remark 4.22. Most of the above theorems can easily be generalized to Sobolev spaces $W_p^m(\Omega)$ (cf. Remark 4.7), with certain modifications for the case $p = \infty$. The L_1 case is of particular interest, for example Theorem 4.19 carries over immediately to $W_1^m(I)$.

The differential operator D^m can now be treated in a similar way as D:

Theorem 4.23. *Let $I =]\alpha, \beta[$ and let $m > 0$. Let $A = D^m$. Then*

$$\begin{aligned} D(A_{\max}) &= H^m(I), \\ D(A_{\min}) &= H_0^m(I). \end{aligned} \qquad (4.56)$$

There exists a constant $C_1 > 0$ so that

$$(\|u\|_0^2 + \|D^m u\|_0^2)^{\frac{1}{2}} \le \|u\|_m \le C_1(\|u\|_0^2 + \|D^m u\|_0^2)^{\frac{1}{2}} \qquad (4.57)$$

for $u \in D(A_{\max})$.

Proof. Clearly, $H^m(I) \subset D(A_{\max})$. The opposite inclusion follows from Theorem 4.20. The first inequality in (4.57) follows from the definition of the m-norm; thus the injection of $H^m(I)$ into $D(A_{\max})$ (which is a Hilbert space with respect to the graph-norm since $A_{\max}$ is closed) is continuous. We have just shown that this injection is surjective, then it follows from the open mapping principle (cf. Theorem B.15) that it is a homeomorphism, so also the second inequality holds.

Now consider $A_{\min}$. Since D^m is formally selfadjoint, $A_{\min} = A_{\max}{}^*$ and $A_{\min} \subset A_{\max}$. Since $D(A_{\min})$ is the closure of $C_0^\infty(I)$ in the graph-norm, and the graph-norm on $D(A_{\max})$ is equivalent with the m-norm, $D(A_{\min})$ equals the closure of $C_0^\infty(I)$ in the m-norm, hence by definition equals $H_0^m(I)$.   □

Also here, the minimal operator is too small and the maximal operator too large to be selfadjoint, and one can find intermediate realizations that are selfadjoint. The operators D^m are quite simple since they have constant coefficients; questions for operators with x-dependent coefficients are harder to deal with.

As a concrete example of a differential operator in one variable with variable coefficients, we shall briefly consider a *regular Sturm-Liouville operator*: Let $I =]\alpha, \beta[$, and let L be defined by

$$(Lu)(x) = -\frac{d}{dx}\left(p(x)\frac{du(x)}{dx}\right) + q(x)u(x) ; \tag{4.58}$$

where $p \in C^\infty(I) \cap C^1(\overline{I})$ and $q \in C^\infty(I) \cap C^0(\overline{I})$, and

$$p(x) \geq c > 0, \quad q(x) \geq 0 \quad \text{on } \overline{I} . \tag{4.59}$$

(Sturm-Liouville operators are often given with a positive factor $1/\varrho(x)$ on the whole expression; this is left out here for simplicity. When the factor is present, the realizations of L should be considered in $L_2(I, \varrho(x)dx)$.)

In an analysis laid out in Exercises 4.11–4.13 one shows that $D(L_{\max}) = H^2(I)$ and $D(L_{\min}) = H_0^2(I)$, and that the realizations of L are characterized by boundary conditions (linear conditions on the boundary values $u(\alpha), u'(\alpha), u(\beta), u'(\beta)$); moreover, symmetric, selfadjoint and semibounded realizations are discussed.

4.4 Boundary value problems in higher dimensions

Consider an open subset Ω of $\mathbb{R}^n$, where $n \geq 1$. For $H^1(\Omega)$ and $H_0^1(\Omega)$ defined in Definition 4.5 one may expect that, similarly to the case $n = 1$, $\Omega =]\alpha, \beta[$, the *boundary value* $u|_{\partial\Omega}$ has a good meaning when $u \in H^1(\Omega)$ and that the boundary value *is 0 exactly when* $u \in H_0^1(\Omega)$. But the functions in $H^1(\Omega)$ need not even be continuous when $n > 1$. For the cases $n \geq 3$ we have the example: $x_1/(x_1^2 + \cdots + x_n^2)^{\frac{1}{2}}$, which is in $H^1(B(0,1))$; it is bounded but discontinuous (Exercises 3.7 and 4.24). An example in the case $n = 2$ is the function $\log|\log(x_1^2 + x_2^2)^{\frac{1}{2}}|$, which is in $H^1(B(0, \frac{1}{2}))$ and is unbounded at 0 (Exercise 4.16).

However, the concept of boundary value can here be introduced in a more sophisticated way. We shall now show a result which can be considered an easy generalization of (4.47). (The information about which space the boundary value belongs to may be further sharpened: There is a continuous extension of γ_0 as a mapping from $H^1(\overline{\mathbb{R}}_+^n)$ onto $H^{\frac{1}{2}}(\mathbb{R}^{n-1})$, see Theorem 9.2 later. Sobolev spaces of noninteger order will be introduced in Chapter 6.)

Theorem 4.24. *The trace map* $\gamma_0 : u(x', x_n) \mapsto u(x', 0)$ *that sends* $C_{(0)}^\infty(\overline{\mathbb{R}}_+^n)$ *into* $C_0^\infty(\mathbb{R}^{n-1})$ *extends by continuity to a continuous map (also denoted* γ_0*) from* $H^1(\mathbb{R}_+^n)$ *into* $L_2(\mathbb{R}^{n-1})$.

Proof. As earlier, we denote $(x_1, \ldots, x_{n-1}) = x'$. For $u \in C_{(0)}^\infty(\overline{\mathbb{R}}_+^n)$, we have the inequality (using that $|2\operatorname{Re} ab| \le |a|^2 + |b|^2$ for $a, b \in \mathbb{C}$):

$$
\begin{aligned}
|u(x', 0)|^2 &= -\int_0^\infty \partial_n(u(x', x_n)\bar{u}(x', x_n))\, dx_n \\
&= -\int_0^\infty 2\operatorname{Re}(\partial_n u(x', x_n)\bar{u}(x', x_n))\, dx_n \\
&\le \int_0^\infty (|u(x', x_n)|^2 + |\partial_n u(x', x_n)|^2)\, dx_n.
\end{aligned}
$$

(Inequalities of this kind occur already in (4.47), Theorem 4.18.) Integrating with respect to x', we find that

$$
\|\gamma_0 u\|_{L_2(\mathbb{R}^{n-1})}^2 \le \|u\|_{L_2(\mathbb{R}_+^n)}^2 + \|\partial_n u\|_{L_2(\mathbb{R}_+^n)}^2 \le \|u\|_{H^1(\mathbb{R}_+^n)}^2.
$$

Hence the map γ_0, considered on $C_{(0)}^\infty(\overline{\mathbb{R}}_+^n)$, is bounded with respect to the mentioned norms; and since $C_{(0)}^\infty(\overline{\mathbb{R}}_+^n)$ is dense in $H^1(\mathbb{R}_+^n)$ according to Theorem 4.10, the map γ_0 extends by closure to a continuous map of $H^1(\mathbb{R}_+^n)$ into $L_2(\mathbb{R}^{n-1})$. □

Since $\gamma_0 u$ is 0 for $u \in C_0^\infty(\mathbb{R}_+^n)$, it follows from the continuity that $\gamma_0 u = 0$ for $u \in H_0^1(\mathbb{R}_+^n)$. The converse also holds:

Theorem 4.25. *The following identity holds:*

$$
H_0^1(\mathbb{R}_+^n) = \{\, u \in H^1(\mathbb{R}_+^n) \mid \gamma_0 u = 0 \,\}. \tag{4.60}
$$

Proof. As already noted, the inclusion '⊂' follows immediately from the fact that $C_0^\infty(\mathbb{R}_+^n)$ is dense in $H_0^1(\mathbb{R}_+^n)$.

The converse demands more effort. Let $u \in H^1(\mathbb{R}_+^n)$ be such that $\gamma_0 u = 0$, then we shall show how u may be approximated by functions in $C_0^\infty(\mathbb{R}_+^n)$.

According to the density shown in Theorem 4.10, there is a sequence of functions $v_k \in C_{(0)}^\infty(\overline{\mathbb{R}}_+^n)$ such that $v_k \to u$ in $H^1(\mathbb{R}_+^n)$ for $k \to \infty$. Since $\gamma_0 u = 0$, it follows from Theorem 4.24 that $\gamma_0 v_k \to 0$ in $L_2(\mathbb{R}^{n-1})$ for $k \to \infty$. From the inequality

$$|v_k(x', x_n)| \le |v_k(x', 0)| + \int_0^{x_n} |\partial_n v_k(x', y_n)| \, dy_n, \text{ for } x_n > 0,$$

follows, by the Cauchy-Schwarz inequality,

$$|v_k(x', x_n)|^2 \le 2|v_k(x', 0)|^2 + 2\Big(\int_0^{x_n} |\partial_n v_k(x', y_n)| \, dy_n\Big)^2$$

$$\le 2|v_k(x', 0)|^2 + 2\int_0^{x_n} 1 \, dy_n \int_0^{x_n} |\partial_n v_k(x', y_n)|^2 \, dy_n$$

$$= 2|v_k(x', 0)|^2 + 2x_n \int_0^{x_n} |\partial_n v_k(x', y_n)|^2 \, dy_n.$$

Integration with respect to $x' \in \mathbb{R}^{n-1}$ and $x_n \in \,]0, a[$ gives

$$\int_0^a \int_{\mathbb{R}^{n-1}} |v_k(x', x_n)|^2 \, dx' dx_n$$

$$\le 2a \int_{\mathbb{R}^{n-1}} |v_k(x', 0)|^2 \, dx' + 2\int_0^a x_n \int_0^{x_n} \int_{\mathbb{R}^{n-1}} |\partial_n v_k(x', y_n)|^2 \, dx' dy_n dx_n$$

$$\le 2a \int_{\mathbb{R}^{n-1}} |v_k(x', 0)|^2 \, dx' + 2\int_0^a x_n \, dx_n \int_0^a \int_{\mathbb{R}^{n-1}} |\partial_n v_k(x', y_n)|^2 \, dx' dy_n$$

$$= 2a \int_{\mathbb{R}^{n-1}} |v_k(x', 0)|^2 \, dx' + a^2 \int_0^a \int_{\mathbb{R}^{n-1}} |\partial_n v_k(x', y_n)|^2 \, dx' dy_n.$$

For $k \to \infty$ we then find, since $\gamma_0 v_k \to 0$ in $L_2(\mathbb{R}^{n-1})$,

$$\int_0^a \int_{\mathbb{R}^{n-1}} |u(x', x_n)|^2 \, dx' dx_n \le a^2 \int_0^a \int_{\mathbb{R}^{n-1}} |\partial_n u(x', y_n)|^2 \, dy_n dx', \quad (4.61)$$

for $a > 0$. For $\varepsilon > 0$, consider

$$u_\varepsilon(x', x_n) = (1 - \chi(x_n/\varepsilon)) u(x', x_n),$$

with χ defined in (2.3). Clearly,

$$u_\varepsilon \to u \text{ in } L_2(\mathbb{R}_+^n) \text{ for } \varepsilon \to 0,$$
$$\partial_j u_\varepsilon = (1 - \chi(x_n/\varepsilon)) \partial_j u \to \partial_j u \text{ for } \varepsilon \to 0 \text{ when } j < n,$$
$$\partial_n u_\varepsilon = (1 - \chi(x_n/\varepsilon)) \partial_n u - \tfrac{1}{\varepsilon} \chi'(x_n/\varepsilon) u, \text{ where}$$
$$(1 - \chi(x_n/\varepsilon)) \partial_n u \to \partial_n u \text{ for } \varepsilon \to 0.$$

For the remaining term, we have, denoting $\sup |\chi'| = C_1$ and using (4.61),

$$\|\partial_n u_\varepsilon - (1-\chi(x_n/\varepsilon))\partial_n u\|^2_{L_2(\mathbb{R}^n_+)} = \|\tfrac{1}{\varepsilon}\chi'(x_n/\varepsilon)u\|^2_{L_2(\mathbb{R}^n_+)}$$

$$\le \varepsilon^{-2}C_1^2 \int_0^{2\varepsilon}\int_{\mathbb{R}^{n-1}} |u(x',x_n)|^2\, dx' dx_n$$

$$\le 4C_1^2 \int_0^{2\varepsilon}\int_{\mathbb{R}^{n-1}} |\partial_n u(x',y_n)|^2\, dx' dy_n$$

$$\to 0 \quad \text{for } \varepsilon \to 0, \text{ since } \partial_n u \in L_2(\mathbb{R}^n_+).$$

So we may conclude that $u_\varepsilon \to u$ in $H^1(\mathbb{R}^n_+)$ for $\varepsilon \to 0$. Since the support of u_ε is inside $\mathbb{R}^n_+$, with distance $\ge \varepsilon$ from the boundary $\{x_n = 0\}$, u_ε may by truncation and mollification as in the proofs of Lemma 4.8 and Theorem 4.10 be approximated in 1-norm by functions in $C_0^\infty(\mathbb{R}^n_+)$. □

There are other proofs in the literature. For example, one can approximate u by a sequence of continuous functions in H^1 with boundary value 0, and combine truncation and mollification with a translation of the approximating sequence *into* $\mathbb{R}^n_+$, in order to get an approximating sequence in $C_0^\infty(\mathbb{R}^n_+)$.

Remark 4.26. On the basis of Theorem 4.25 one can also introduce a boundary map γ_0 from $H^1(\Omega)$ to $L_2(\partial\Omega)$ for smooth open bounded sets Ω, by working in local coordinates as in Definition C.1. We omit details here, but will just mention that one has again:

$$H_0^1(\Omega) = \{\, u \in H^1(\Omega) \mid \gamma_0 u = 0 \,\}, \tag{4.62}$$

like for $\Omega = \mathbb{R}^n_+$. A systematic presentation is given e.g. in [LM68].

A very important partial differential operator is the Laplace operator Δ,

$$\Delta u = \partial^2_{x_1} u + \cdots + \partial^2_{x_n} u. \tag{4.63}$$

We shall now consider some realizations of $A = -\Delta$ on an open subset Ω of $\mathbb{R}^n$. Introduce first the auxiliary operator S in $L_2(\Omega)$ with domain $D(S) = C_0^\infty(\Omega)$ and action $Su = -\Delta u$. The minimal operator $A_{\min}$ equals $\overline{S}$. Since Δ is clearly formally selfadjoint, S is symmetric; moreover, it is ≥ 0:

$$(Su, v) = \int_\Omega (-\partial_1^2 u - \cdots - \partial_n^2 u)\bar{v}\, dx \tag{4.64}$$

$$= \int_\Omega (\partial_1 u \overline{\partial_1 v} + \cdots + \partial_n u \overline{\partial_n v})dx = (u, Sv), \text{ for } u, v \in C_0^\infty(\Omega);$$

where the third expression shows that $(Su, u) \ge 0$ for all $u \in D(S)$. It follows that the closure $A_{\min}$ is likewise symmetric and ≥ 0. Then, moreover, $A_{\max} = A_{\min}^*$.

It is clear that $H^2(\Omega) \subset D(A_{\max})$. The inclusion is strict (unless $n = 1$), cf. Exercise 4.5, and Exercise 6.2 later.

For $A_{\min}$ it can be shown that $D(A_{\min}) = H_0^2(\Omega)$, cf. Theorem 6.24 later.

As usual, $A_{\min}$ is too small and $A_{\max}$ is too large to be selfadjoint, but we can use general results for unbounded operators in Hilbert space, as presented in Chapter 12, to construct various selfadjoint realizations.

In fact, Theorem 12.24 shows that S (or $A_{\min}$) has an interesting selfadjoint extension T, namely, the Friedrichs extension. It has the same lower bound as S; $m(T) = m(S)$. Let us find out what the associated space V and sesquilinear form $s(u,v)$ are. Following the notation in Chapter 12, we see that s is an extension of

$$s_0(u,v) = \sum_{k=1}^{n}(\partial_k u, \partial_k v),$$

defined on $D(s_0) = D(S) = C_0^\infty(\Omega)$. Since $m(S) \geq 0$, $s_0(u,v) + \alpha \cdot (u,v)$ is a scalar product on $D(s_0)$ for any $\alpha > 0$. (We shall see later in Theorem 4.29 that $m(S) > 0$ when Ω is bounded, so then it is not necessary to add a multiple of (u,v).) Clearly, the associated norm is *equivalent with the 1-norm on Ω*. But then

$$V = \text{ the completion of } C_0^\infty(\Omega) \text{ in } H^1\text{-norm } = H_0^1(\Omega),$$
$$s(u,v) = \sum_{k=1}^{n}(\partial_k u, \partial_k v) \text{ on } H_0^1(\Omega). \tag{4.65}$$

This also explains how T arises from the Lax-Milgram construction (cf. Theorem 12.18); it is the variational operator associated with $(L_2(\Omega), H_0^1(\Omega), s)$, with s defined in (4.65). We can now formulate:

Theorem 4.27. *The Friedrichs extension T of $S = -\Delta|_{C_0^\infty(\Omega)}$ is a selfadjoint realization of $-\Delta$. Its lower bound $m(T)$ equals $m(S)$. T is the variational operator determined from the triple (H,V,s) with $H = L_2(\Omega)$, $V = H_0^1(\Omega)$, $s(u,v) = \sum_{k=1}^{n}(\partial_k u, \partial_k v)$.*

T is the unique lower bounded selfadjoint realization of $-\Delta$ with domain contained in $H_0^1(\Omega)$. The domain equals

$$D(T) = D(A_{\max}) \cap H_0^1(\Omega). \tag{4.66}$$

In this sense, T represents the boundary condition

$$\gamma_0 u = 0, \tag{4.67}$$

i.e., the Dirichlet condition.

Proof. It remains to account for the second paragraph. The uniqueness follows from Corollary 12.25. Note that since T is a selfadjoint extension of $A_{\min}$, $T \subset A_{\max}$, so it acts like $-\Delta$. In formula (4.66), the inclusion '$\subset$' is clear since $T \subset A_{\max}$. Conversely, if $u \in D(A_{\max}) \cap H_0^1(\Omega)$, then for any $\varphi \in C_0^\infty(\Omega)$,

$$s(u, \varphi) = \sum_{k=1}^{n} (\partial_k u, \partial_k \varphi) = -\langle u, \sum_{k=1}^{n} \overline{\partial_k^2 \varphi} \rangle = (A_{\max} u, \varphi),$$

where the equality of the first and last expressions extends to $v \in H_0^1(\Omega)$ by closure:

$$s(u, v) = (A_{\max} u, v), \text{ for } v \in H_0^1(\Omega). \tag{4.68}$$

Then by definition (cf. (12.45)), u is in the domain of the variational operator defined from (H, V, s), and it acts like $A_{\max}$ on u. □

T is called the Dirichlet realization of $A = -\Delta$.

It is shown further below (Corollary 4.30) that $m(T)$ is positive when Ω has finite width.

We note that this theorem assures that the first derivatives are well-defined on $D(T)$ as L_2-functions. It is a deeper fact that is much harder to show, that when Ω is smooth and bounded, also the second derivatives are well-defined as L_2-functions on $D(T)$ (i.e., $D(T) \subset H^2(\Omega)$). Some special cases are treated in Chapter 9, and the general result is shown at the end of Chapter 11.

The Lax-Milgram lemma can also be used to define other realizations of $-\Delta$ which are not reached by the Friedrichs construction departing from S. Let

$$(H, V_1, s_1(u, v)) = \left(L_2(\Omega), H^1(\Omega), \sum_{k=1}^{n} (\partial_k u, \partial_k v) \right); \tag{4.69}$$

here s_1 is V_1-coercive and ≥ 0, but not V_1-elliptic if Ω is bounded, since for example $u \equiv 1$ gives $s_1(u, u) = 0$. Let T_1 be the associated operator in H defined by Corollary 12.19; it is ≥ 0, and selfadjoint since s_1 is symmetric. Since

$$s_1(\varphi, v) = \sum_{k=1}^{n} (\partial_k \varphi, \partial_k v) = (-\sum_{k=1}^{n} \partial_k^2 \varphi, v) \text{ for } \varphi \in C_0^\infty(\Omega), v \in H^1(\Omega),$$

T_1 is an extension of $S = -\Delta|_{C_0^\infty(\Omega)}$. Since T_1 is then a selfadjoint extension of $A_{\min}$, $T_1 \subset A_{\max}$; hence it acts like $-\Delta$.

Now the domain $D(T_1)$ consists of the $u \in H^1(\Omega) \cap D(A_{\max})$ for which

$$(-\Delta u, v) = s_1(u, v), \text{ for all } v \in H^1(\Omega), \tag{4.70}$$

cf. Definition 12.14. Let us investigate those elements $u \in D(T_1)$ which moreover belong to $C^2(\overline{\Omega})$, and let us assume that Ω is smooth. For functions in $H^1(\Omega) \cap C^2(\overline{\Omega})$, we have using (A.20):

$$(-\Delta u, v) - s_1(u, v) = \int_{\partial\Omega} \frac{\partial u}{\partial \nu} \overline{v} \, d\sigma, \text{ when } v \in C^1(\overline{\Omega}) \cap H^1(\Omega). \tag{4.71}$$

Then if $u \in D(T_1)$, the right-hand side of (4.71) must be 0 for all $v \in C^1(\overline{\Omega}) \cap H^1(\Omega)$, so we conclude that

$$\frac{\partial u}{\partial \nu} = 0 \text{ on } \partial\Omega. \tag{4.72}$$

Conversely, if $u \in C^2(\overline{\Omega}) \cap H^1(\Omega)$ and satisfies (4.72), then (4.71) shows that $u \in D(T_1)$. So we see that

$$D(T_1) \cap C^2(\overline{\Omega}) = \{u \in C^2(\overline{\Omega}) \cap H^1(\Omega) \mid \frac{\partial u}{\partial \nu} = 0 \text{ on } \partial\Omega\}. \tag{4.73}$$

In this sense, T_1 represents the *Neumann condition* (4.72). One can in fact show the validity of (4.72) for general elements of $D(T_1)$ using generalizations of (A.20) to Sobolev spaces. Like for the Dirichlet problem there is a deep result showing that $D(T_1) \subset H^2(\Omega)$; here $-\Delta$, $\gamma_0 u$ and $\frac{\partial u}{\partial \nu}|_{\partial\Omega}$ have a good sense. See Sections 9.3 and 11.3. When Ω is bounded, $u \equiv 1$ is in $D(T_1)$, so $m(T_1) = 0$.

We have then obtained:

Theorem 4.28. *The variational operator T_1 defined from the triple* (4.69) *is a selfadjoint realization of* $-\Delta$, *with* $m(T_1) \geq 0$ *(= 0 if Ω is bounded), and*

$$D(T_1) = \{u \in H^1(\Omega) \cap D(A_{\max}) \mid (-\Delta u, v) = s_1(u, v) \text{ for all } v \in H^1(\Omega)\}.$$

The domain $D(T_1)$ is dense in $H^1(\Omega)$ and satisfies (4.73). *The operator represents the Neumann problem for* $-\Delta$ *(with* (4.72)*) in a generalized sense.*

T_1 is called the Neumann realization of $A = -\Delta$.

One can show a sharpening of the lower bound of S for suitably limited domains. For this, we first show:

Theorem 4.29 (POINCARÉ INEQUALITY). *Let $b > 0$. When φ is in $C_0^\infty(\mathbb{R}^n)$ with* supp φ *contained in a "slab" $\Omega_b = \{x \in \mathbb{R}^n \mid 0 \leq x_j \leq b\}$ for a j between 1 and n, then*

$$\|\varphi\|_0 \leq \frac{b}{\sqrt{2}}\|\partial_j \varphi\|_0 . \tag{4.74}$$

The inequality extends to $u \in H_0^1(\Omega_b)$.

Proof. We can assume that $j = n$. Let $\varphi \in C_0^\infty(\mathbb{R}^n)$ with support in the slab. Since $\varphi(x', 0) = 0$, we find for $x_n \in [0, b]$, using the Cauchy-Schwarz inequality,

$$|\varphi(x', x_n)|^2 = |\int_0^{x_n} \partial_{x_n} \varphi(x', t)\, dt|^2 \leq x_n \int_0^{x_n} |\partial_{x_n} \varphi(x', t)|^2\, dt$$

$$\leq x_n \int_0^b |\partial_n \varphi(x', t)|^2\, dt, \quad \text{and hence}$$

$$\int_0^b |\varphi(x', x_n)|^2 \, dx_n \le \int_0^b x_n \, dx_n \int_0^b |\partial_n \varphi(x', t)|^2 \, dt$$

$$= \tfrac{1}{2} b^2 \int_0^b |\partial_n \varphi(x', t)|^2 \, dt.$$

This implies by integration with respect to x' (since $\varphi(x)$ is 0 for $x_n \notin [0, b]$):

$$\|\varphi(x', x_n)\|_{L_2(\mathbb{R}^n)}^2 \le \tfrac{1}{2} b^2 \int_{\mathbb{R}^{n-1}} \int_0^b |\partial_n \varphi(x', t)|^2 \, dt dx'$$

$$= \tfrac{1}{2} b^2 \|\partial_n \varphi\|_{L_2(\mathbb{R}^n)}^2.$$

The inequality extends to functions in $H_0^1(\Omega_R)$ by approximation in H^1-norm by functions in $C_0^\infty(\Omega_b)$. (This proof has some ingredients in common with the proof of Theorem 4.25.)   □

When φ is as in the lemma, we have

$$(S\varphi, \varphi) = \sum_{k=1}^n \|\partial_k \varphi\|_0^2 \ge \frac{2}{b^2} \|\varphi\|_0^2. \tag{4.75}$$

The resulting inequality extends to $A_{\min}$ by closure. The inequality of course also holds for $\varphi \in C_0^\infty(\Omega)$ when Ω is contained in a translated slab

$$\Omega \subset \{ x \mid a \le x_j \le a + b \}, \tag{4.76}$$

for some $a \in \mathbb{R}$, $b > 0$ and $j \in \{1, \dots, n\}$. It holds in particular when Ω is bounded with diameter $\le b$. The inequality moreover holds when Ω is contained in a slab with thickness b in an arbitrary direction, since this situation by an orthogonal change of coordinates can be carried over into a position as in (4.76), and $-\Delta$ as well as the 1-norm are invariant under orthogonal coordinate changes. So we have, recalling the notation for the lower bound from (12.22):

Corollary 4.30. *When Ω is bounded with diameter $\le b$, or is just contained in a slab with thickness b, then*

$$m(S) = m(A_{\min}) \ge 2b^{-2}. \tag{4.77}$$

It follows that the lower bound of the Friedrichs extension, alias the Dirichlet realization, is positive in this case;

$$m(T) \ge 2b^{-2}. \tag{4.78}$$

It is seen from the analysis in Chapter 13 that there exist many other selfadjoint realizations of $-\Delta$ than the ones we have considered here.

One can do a very similar analysis when $-\Delta$ is replaced by more general operators A of the form

$$Au = - \sum_{j,k=1}^{n} \partial_j(a_{jk}(x)\partial_k u), \quad a_{jk} \in C^\infty(\overline{\Omega}) \cap L_\infty(\Omega), \tag{4.79}$$

assumed to be *strongly elliptic*; this means that

$$\text{Re} \sum_{j,k=1}^{n} a_{jk}(x)\xi_k\xi_j \geq c|\xi|^2 \text{ for } \xi \in \mathbb{R}^n, \ x \in \Omega, \text{ some } c > 0. \tag{4.80}$$

Along with A one considers the sesquilinear form

$$a(u,v) = \sum_{j,k=1}^{n} (a_{jk}\partial_k u, \partial_j v); \tag{4.81}$$

it is $H_0^1(\Omega)$-coercive. (This follows from the general results taken up at the end of Chapter 7; the Gårding inequality.) Then the Lax-Milgram lemma applied to a with $H = L_2(\Omega)$, $V = H_0^1(\Omega)$, leads to a variational realization of A representing the *Dirichlet boundary condition*.

The form $a(u,v)$ is not always $H^1(\Omega)$-coercive, but this is assured if the coefficient functions a_{jk} are real (see Exercise 4.25), or if the stronger inequality

$$\text{Re} \sum_{j,k=1}^{n} a_{jk}(x)\eta_k\bar\eta_j \geq c|\eta|^2 \text{ for } \eta \in \mathbb{C}^n, \ x \in \Omega, \tag{4.82}$$

is satified (Exercise 4.26). (The A satisfying (4.82) are sometimes called "very strongly elliptic".) In both these cases the $H_0^1(\Omega)$-coerciveness comes out as a corollary. Here the application of the Lax-Milgram lemma with $V = H^1(\Omega)$ instead of $H_0^1(\Omega)$ gives a realization of A with a first-order boundary condition (where $\nu = (\nu_1, \ldots, \nu_n)$ is the normal to the boundary, cf. Appendix A):

$$\sum_{j,k=1}^{n} \nu_j a_{jk}\partial_k u = 0 \text{ on } \partial\Omega, \tag{4.83}$$

often called *the oblique Neumann condition* defined from a. (There are many sesquilinear forms associated with the same A; this is discussed e.g. in [G71], [G73].)

Note that these variational operators are not in general selfadjoint; they are usually not normal either (an operator N in H is normal if it is closed, densely defined with $NN^* = N^*N$; then also $D(N^*) = D(N)$).

Exercises for Chapter 4

4.1. Let Ω be open $\subset \mathbb{R}^n$ and let $m \in \mathbb{N}$. Show that $H^m(\Omega)$ has a denumerable orthonormal basis, i.e., is separable.
(*Hint.* One can use that $H^m(\Omega)$ may be identified with a subspace of $\prod_{|\alpha| \leq m} L_2(\Omega)$.)

4.2. Let $A = \sum_{i=1}^n a_i D_i$ be a first-order differential operator on $\mathbb{R}^n$ with constant coefficients. Show that $A_{\max} = A_{\min}$.
(*Hint.* Some ingredients from the proof of Theorem 4.10 may be of use.)

4.3. Let $A = \partial_x^2 - \partial_y^2$ on $\Omega =]0,1[\times]0,1[$ in $\mathbb{R}^2$. Show that $A' = \partial_x^2 - \partial_y^2$. Show that $f(x,y)$ defined by (4.17) belongs to $D(A_{\max})$, with $A_{\max}f = 0$.
(*Hint.* One has to carry out some integrations by part, where it may make the arguments easier to begin by changing to new coordinates $s = x + y$, $t = x - y$.)

4.4. Let $B = \partial_x$ on $\Omega =]0,1[\times]0,1[$ in $\mathbb{R}^2$. Show that $B' = -\partial_x$. Show that f defined by (4.17) does not belong to $D(B_{\max})$.
(*Hint.* If there exists a function $g \in L_2(\Omega)$ such that

$$(f, -\partial_x \varphi)_{L_2(\Omega)} = (g, \varphi)_{L_2(\Omega)} \text{ for all } \varphi \in C_0^\infty(\Omega),$$

then g must be 0 a.e. on $\Omega \setminus \{x = y\}$ (try with $\varphi \in C_0^\infty(\Omega \setminus \{x = y\})$). But then $g = 0$ as an element of $L_2(\Omega)$, and hence $(f, -\partial_x \varphi) = 0$ for all φ. Show by counterexamples that this cannot be true.)

4.5. Let $A = -\Delta$ on $\Omega = B(0,1)$ in $\mathbb{R}^2$. Let $u(x,y)$ be a function given on Ω by

$$u(x,y) = c_0 + \sum_{k \in \mathbb{N}} (c_k (x + iy)^k + c_{-k} (x - iy)^k),$$

i.e., $u(r,\theta) = \sum_{n \in \mathbb{Z}} c_n r^{|n|} e^{in\theta}$ in polar coordinates; assume that $\sup_n |c_n| < \infty$. It is known from elementary PDE (separation of variables methods) that $u \in C^\infty(\Omega)$ with $\Delta u = 0$ on Ω, and that if furthermore $\sum_{n \in \mathbb{Z}} |c_n| < \infty$, then $u \in C^0(\overline{\Omega})$ with boundary value $\sum_{n \in \mathbb{Z}} c_n e^{in\theta} = \varphi(\theta)$.
(a) Show that if merely $\sum_{n \in \mathbb{Z}} |c_n|^2 < \infty$ (corresponding to $\varphi \in L_2([-\pi, \pi])$), then $u \in D(A_{\max})$.
(b) Give an example where $u \in D(A_{\max})$ but $\partial_x u \notin L_2(\Omega)$.
(*Hint.* One can show by integration in polar coordinates that the expressions

$$s_{M,N}(r, \theta) = \sum_{M \leq |n| \leq N} c_n r^{|n|} e^{in\theta}$$

satisfy

$$\|s_{M,N}\|_{L_2(\Omega)}^2 = C \sum_{M \leq |n| \leq N} \frac{|c_n|^2}{|n| + 1},$$

for some constant C.)

4.6. Let $I =]\alpha, \beta[$.

(a) Let $x_0 \in [\alpha, \beta[$ and let $\zeta \in C^\infty(\overline{I})$ with $\zeta(x_0) = 1$ and $\zeta(\beta) = 0$. Show that for $u \in H^1(I)$,

$$|u(x_0)| = |\int_{x_0}^{\beta} \frac{\partial}{\partial x}(\zeta u)\, dx| \leq \|\zeta'\|_{L_2(I)}\|u\|_{L_2(I)} + \|\zeta\|_{L_2(I)}\|u'\|_{L_2(I)}.$$

Deduce (4.47) from this for $m = 1$, by a suitable choice of ζ.

(b) Show that for each $\varepsilon > 0$ there exists a constant $C(\varepsilon)$ so that for $u \in H^1(I)$,

$$|u(x)| \leq \varepsilon\|u\|_1 + C(\varepsilon)\|u\|_0, \quad \text{for all } x \in \overline{I}.$$

(*Hint*: Choose ζ in (a) of the form $\chi_{\delta,2\delta}(x - x_0)$ for a suitably small δ.)

4.7. Are there other selfadjoint realizations $\widetilde{A}$ than $A_\#$, for $A = D$ on $I =]\alpha, \beta[$?

4.8. Find a selfadjoint realization of $A = -\dfrac{d^4}{dx^4}$ on $I =]\alpha, \beta[$, and show which boundary condition it represents.

(*Hint.* One can for example use Theorem 12.11.)

4.9. Let $I =]a, b[$. For $A = D$, show that $A_{\max}A_{\min}$ is the Friedrichs extension of $-\frac{d^2}{dx^2}|_{C_0^\infty(I)}$.

4.10. Let $I =]\alpha, \beta[$ and let $m \in \mathbb{N}$. Let A be the differential operator

$$A = D^m + p_1(x)D^{m-1} + \cdots + p_m(x),$$

with coefficients $p_j(x) \in C^\infty(\overline{I})$. Show that $D(A_{\max}) = H^m(I)$ and that $D(A_{\min}) = H_0^m(I)$.

(*Hint.* For the proof that $D(A_{\max}) = H^m(I)$ one can use that Au may be written in the form $D^m u + D^{m-1}(q_1 u) + \cdots + q_m u$, with coefficients $q_j \in C^\infty(\overline{I})$.)

4.11. Let $I =]\alpha, \beta[$, let L be the regular Sturm-Liouville operator defined in (4.58) and (4.59), and let $c > 0$ and $C > 0$ be chosen such that

$$c \leq p(x) \leq C, \quad |p'(x)| \leq C, \quad 0 \leq q(x) \leq C \quad \text{on } \overline{I}.$$

For $u \in H^2(I)$, ϱu denotes the set of four boundary values

$$\varrho u = \{u(\alpha), u(\beta), u'(\alpha), u'(\beta)\} \in \mathbb{C}^4.$$

(a) Show that L is formally selfadjoint.

(b) Show that

$$\varrho : H^2(I) \to \mathbb{C}^4$$

is a continuous linear map, which is surjective.

(c) Show that $D(L_{\max}) = H^2(I)$ and $D(L_{\min}) = H_0^2(I)$.

(d) Show that the realizations of L, i.e., the operators $\widetilde{L}$ with

$$L_{\min} \subset \widetilde{L} \subset L_{\max},$$

are described by boundary conditions:

$$D(\widetilde{L}) = \{\, u \in H^2(I) \mid \varrho u \in W \,\},$$

where W is a subspace of $\mathbb{C}^4$, such that each subspace W corresponds to exactly one realization $\widetilde{L}$.

4.12. Hypotheses and notation as in Exercise 4.11.

(a) Show that one has for u and $v \in H^2(I)$:

$$(Lu, v)_{L_2(I)} - (u, Lv)_{L_2(I)} = (\mathcal{B}\varrho u, \varrho v)_{\mathbb{C}^4},$$

where

$$\mathcal{B} = \begin{pmatrix} 0 & 0 & p(\alpha) & 0 \\ 0 & 0 & 0 & -p(\beta) \\ -p(\alpha) & 0 & 0 & 0 \\ 0 & p(\beta) & 0 & 0 \end{pmatrix}.$$

(b) Let $\widetilde{L}$ correspond to W as in Exercise 4.11 (d). Show that $\widetilde{L}$ is symmetric if and only if $W \subset (\mathcal{B}W)^\perp$, and that $\widetilde{L}$ then is selfadjoint precisely when, in addition, $\dim W = 2$; in this case, $W = (\mathcal{B}W)^\perp$.

(c) Find out how the realization is in the cases

(1) $W = \{z \in \mathbb{C}^4 \mid az_1 + bz_3 = 0, \ cz_2 + dz_4 = 0\}$, where $(a,b), (c,d) \in \mathbb{R}^2 \setminus \{(0,0)\}$;

(2) $W = \{z \in \mathbb{C}^4 \mid z_1 = z_2, \ az_3 + bz_4 = 0\}$, where $(a,b) \in \mathbb{C}^2 \setminus \{(0,0)\}$;

(3) $W = \{z \in \mathbb{C}^4 \mid \begin{pmatrix} z_3 \\ z_4 \end{pmatrix} = F \begin{pmatrix} z_1 \\ z_2 \end{pmatrix}\}$, where F is a complex 2×2-matrix.

(*Comment.* Since $\mathcal{B}$ is skew-selfadjoint ($\mathcal{B}^* = -\mathcal{B}$) and invertible, the sesquilinear form $b(z, w) = (\mathcal{B}z, w)_{\mathbb{C}^4}$ is what is called a nondegenerate symplectic form. The subspaces W such that $W = (\mathcal{B}W)^\perp$ are maximal with respect to the vanishing of $b(w, w)$ on them; they are called Lagrangian. See e.g. Everitt and Markus [EM99] for a general study of boundary conditions for operators like the Sturm-Liouville operator, using symplectic forms and decompositions as in Exercise 12.19.)

4.13.   Hypotheses and notation as in Exercises 4.11 and 4.12. Let $l(u,v)$ be the sesquilinear form

$$l(u,v) = \int_\alpha^\beta [pu'\bar{v}' + qu\bar{v}]dx$$

defined on $H^1(I)$.

(a) Show that one has for $u \in H^2(I)$, $v \in H^1(I)$:

$$(Lu,v) = -p(\beta)u'(\beta)\bar{v}(\beta) + p(\alpha)u'(\alpha)\bar{v}(\alpha) + l(u,v).$$

(b) Show that the triple $(H, V, l(u,v))$ with $H = L_2(I)$, $V = H_0^1(I)$ and $l_0(u,v) = l(u,v)$ on V, by the Lax-Milgram lemma defines the realization of L determined by the boundary condition $u(\alpha) = u(\beta) = 0$.

(c) Show that when $V = H_0^1(I)$ in (b) is replaced by $V = H^1(I)$, one gets the realization defined by the boundary condition $u'(\alpha) = u'(\beta) = 0$.

(d) Let T_1 be the operator determined by the Lax-Milgram lemma from the triple $(H, V, l_1(u,v))$ with $H = L_2(I)$, $V = H^1(I)$ and

$$l_1(u,v) = l(u,v) + u(\beta)\bar{v}(\beta).$$

Show that T_1 is a realization of L and find the boundary condition it represents.

4.14.   Consider the differential operator $Au = -u'' + \alpha^2 u$ on $\mathbb{R}_+$, where α is a positive constant. (This is a Sturm-Liouville operator with $p = 1$, $q = \alpha^2$ constant > 0.) Verify the following facts:

(a) The maximal operator $A_{\max}$ associated with A in $L_2(\mathbb{R}_+)$ has domain $H^2(\mathbb{R}_+)$.

(b) The minimal operator $A_{\min}$ associated with A in $L_2(\mathbb{R}_+)$ has domain $H_0^2(\mathbb{R}_+)$.

(c) The realization A_γ of A defined by the boundary condition $u(0) = 0$ is variational, associated with the triple (H, V, a), where $H = L_2(\mathbb{R}_+)$, $V = H_0^1(\mathbb{R}_+)$,

$$a(u,v) = (u',v')_{L_2(\mathbb{R}_+)} + \alpha^2(u,v)_{L_2(\mathbb{R}_+)}.$$

Here $m(A_\gamma) \geq \alpha^2$.

4.15.   Let I be an open interval of $\mathbb{R}$ (bounded or unbounded) and let Q_n denote the product set

$$Q_n = \{ (x_1, \ldots, x_n) \in \mathbb{R}^n \mid x_j \in I \text{ for } j = 1, \ldots, n \}.$$

Let $h \in C_0^\infty(I)$ with $\int_I h(t)\,dt = 1$, and let $\tilde{h}(x) = h(x_1) \ldots h(x_n)$; it is a function in $C_0^\infty(Q_n)$ with $\langle 1, \tilde{h} \rangle = 1$.

(a) Show that every function $\varphi \in C_0^\infty(Q_n)$ can be written in the form

$$\varphi = \partial_{x_1}\psi_1 + \cdots + \partial_{x_n}\psi_n + \langle 1, \varphi \rangle \tilde{h} , \qquad (4.84)$$

where $\psi_1, \ldots, \psi_n$ belong to $C_0^\infty(Q_n)$.
(*Hint*. One can for example obtain the formula successively as follows: Let

$$\varphi_1(x_2, \ldots, x_n) = \int_I \varphi(x_1, x_2, \ldots, x_n)\, dx_1 ,$$

and put

$$\zeta_1(x_1, \ldots, x_n) = \int_{-\infty}^{x_1} [\varphi(s, x_2, \ldots, x_n) - h(s)\varphi_1(x_2, \ldots, x_n)]\, ds ;$$

show that $\zeta_1 \in C_0^\infty(Q_n)$ and

$$\varphi = \partial_{x_1}\zeta_1 + h(x_1)\varphi_1(x_2, \ldots, x_n) .$$

Perform the analogous construction for $\varphi_1 \in C_0^\infty(Q_{n-1})$ and insert in the formula for φ; continue until (4.84) has been obtained.)
(b) Show that if $v \in \mathscr{D}'(Q_n)$ satisfies

$$\partial_{x_1} v = \partial_{x_2} v = \cdots = \partial_{x_n} v = 0 ,$$

then v equals a constant (namely, the constant $c = \langle v, \tilde{h} \rangle$).

4.16. (a) Consider the function

$$f(x) = \log|\log(x_1^2 + x_2^2)^{\frac{1}{2}}|$$

on the set $M = \{ x \in \mathbb{R}^2 \mid 0 < |x| < \frac{1}{2} \}$. Show that f, $\partial_1 f$ and $\partial_2 f$ are in $L_2(M)$.
(b) Now consider f as an element of $L_2(B(0, \frac{1}{2}))$. Show that the distribution derivatives of f of order 1 are $L_2(B(0, \frac{1}{2}))$-functions equal to the functions $\partial_1 f$ and $\partial_2 f$ defined above on M.
(c) Show that $f \in H^1(B(0, \frac{1}{2})) \setminus C^0(B(0, \frac{1}{2}))$.

4.17. We denote by $L_2(\mathbb{T})$ the space of L_2-functions on $[-\pi, \pi]$ provided with the scalar product and norm

$$(f, g)_{L_2(\mathbb{T})} = \frac{1}{2\pi} \int_{-\pi}^{\pi} f(\theta)\bar{g}(\theta)\, d\theta, \quad \|f\|_{L_2(\mathbb{T})} = \left(\frac{1}{2\pi} \int_{-\pi}^{\pi} |f(\theta)|^2\, d\theta\right)^{\frac{1}{2}};$$

it identifies with the space of locally square integrable functions on $\mathbb{R}$ with period 2π. (There is the usual convention of identification of functions that are equal almost everywhere.) It is known from the theory of Fourier series that the system of functions $\{e^{in\theta}\}_{n\in\mathbb{Z}}$ is an orthonormal basis of $L_2(\mathbb{T})$ such that when

$$f = \sum_{n\in\mathbb{Z}} c_n e^{in\theta}, \text{ then } \|f\|_{L_2(\mathbb{T})} = \left(\sum_{n\in\mathbb{Z}} |c_n|^2\right)^{\frac{1}{2}}.$$

For m integer ≥ 0, we denote by $C^m(\mathbb{T})$ the space of C^m-functions on $\mathbb{R}$ with period 2π; it identifies with the subspace of $C^m([-\pi, \pi])$ consisting of the functions f with $f^{(j)}(-\pi) = f^{(j)}(\pi)$ for $0 \leq j \leq m$.

(a) Define $H^1(\mathbb{T})$ as the completion of $C^1(\mathbb{T})$ in the norm

$$\|f\|_1 = (\|f\|^2_{L_2(\mathbb{T})} + \|f'\|^2_{L_2(\mathbb{T})})^{\frac{1}{2}}.$$

Show that it identifies with the subspace of $L_2(\mathbb{T})$ consisting of the functions f whose Fourier series satisfy

$$\sum_{n\in\mathbb{Z}} n^2 |c_n|^2 < \infty.$$

(b) Show that

$$H^1(\mathbb{T}) = \{ f(\theta) = \int_{-\pi}^{\theta} g(s)\,ds + k \mid g \in L_2(\mathbb{T}) \text{ with } (g,1)_{L_2(\mathbb{T})} = 0,\ k \in \mathbb{C} \}.$$

(c) Also for higher m, one can define $H^m(\mathbb{T})$ as the completion of $C^m(\mathbb{T})$ in the norm

$$\|f\|_m = \left(\sum_{0\leq j\leq m} \|f^{(j)}\|^2_{L_2(\mathbb{T})} \right)^{\frac{1}{2}},$$

and show that $f \in H^m(\mathbb{T})$ if and only if $\sum_{n\in\mathbb{Z}} n^{2m} |c_n|^2 < \infty$. Moreover, $H^m(\mathbb{T}) \subset C^{m-1}(\mathbb{T})$.

4.18. Let J be the closed interval $[\alpha, \beta]$ and let $\sigma \in {]0, 1[}$.

(a) By $C^\sigma(J)$ we denote the space of Hölder continuous functions of order σ on J, i.e., functions u for which there exists a constant C (depending on u) so that

$$|u(x) - u(y)| \leq C|x - y|^\sigma, \text{ for } x, y \in J.$$

Show that $C^\sigma(J)$ is a Banach space with the norm

$$\|u\|_{C^\sigma} = \sup_{x,y\in J, x\neq y} \frac{|u(x) - u(y)|}{|x - y|^\sigma} + \sup_{x\in J} |u(x)|.$$

(b) By $C^\sigma(\mathbb{T})$ we denote the space of Hölder continuous 2π-periodic functions on $\mathbb{R}$, i.e., functions on $\mathbb{R}$ with period 2π which satisfy the inequality with J replaced by $\mathbb{R}$. Show that it is a Banach space with a norm as in (a) with J replaced by $\mathbb{R}$.

4.19. For $I =]\alpha, \beta[$, show that $H^1(I) \subset C^{\frac{1}{2}}(\overline{I})$. (Notation as in Exercise 4.18.)

4.20. We consider functions u on $\mathbb{R}$ with period 2π, written as trigonometric Fourier series of the type

$$u(x) \sim \sum_{j \in \mathbb{N}} a_j e^{i2^j x}$$

(also called lacunary trigonometric series).

(a) Show that $u \in H^1(\mathbb{T})$ if and only if $\sum_{j \in \mathbb{N}} 2^{2j}|a_j|^2 < \infty$ (cf. Exercise 4.17).

(b) Let $\sigma \in]0,1[$. Show that if $\sum_{j \in \mathbb{N}} 2^{\sigma j}|a_j| < \infty$, then $u \in C^\sigma(\mathbb{T})$ (cf. Exercise 4.18).

(*Hint.* For each term $u_j(x) = a_j e^{i2^j x}$, we have that $|u_j(x)| = |a_j|$ and $|u_j'(x)| = 2^j|a_j|$ for all x, and therefore

$$|u_j(x) - u_j(y)| = |u_j(x) - u_j(y)|^{1-\sigma}|u_j(x) - u_j(y)|^\sigma$$
$$\leq |2a_j|^{1-\sigma}(2^j|a_j|)^\sigma|x - y|^\sigma,$$

by the mean value theorem.)

(c) Show that for each $\sigma \in]0,1[$, $C^\sigma(\mathbb{T}) \setminus H^1(\mathbb{T}) \neq \emptyset$. (*Hint.* Let $a_j = 2^{-j}$.)

4.21. (a) Show that when $u \in H^1(\mathbb{R}_+)$, then $u(x) \to 0$ for $x \to \infty$. (*Hint.* Observe that $|u(x)|^2 \leq \int_x^\infty (|u(y)|^2 + |u'(y)|^2)\, dy$.)

(b) Show (4.51), (4.52) and (4.53).

4.22. Show that the mapping γ_0 defined in Theorem 4.24 sends $H^m(\mathbb{R}^n_+)$ into $H^{m-1}(\mathbb{R}^{n-1})$ for any $m \in \mathbb{N}$, with $D^{\alpha'}\gamma_0 u = \gamma_0 D^{\alpha'} u$ for $u \in H^m(\mathbb{R}^n_+)$ when $\alpha' = \{\alpha_1, \ldots, \alpha_{n-1}\}$ is of length $\leq m - 1$.

4.23. Let $a(u,v)$ be the sesquilinear form on $H^1(\mathbb{R}^n_+)$ defined as the scalar product in $H^1(\mathbb{R}^n_+)$:

$$a(u,v) = (u,v)_{L_2(\mathbb{R}^n_+)} + \sum_{j=1}^n (D_j u, D_j v)_{L_2(\mathbb{R}^n_+)},$$

and let $a_0(u,v)$ denote its restriction to $H^1_0(\mathbb{R}^n_+)$.

(a) With $H = L_2(\mathbb{R}^n_+)$, $V_0 = H^1_0(\mathbb{R}^n_+)$, let A_γ be the variational operator defined from the triple (H, V_0, a_0). Show that A_γ is the realization of $A = I - \Delta$ with domain

$$D(A_\gamma) = D(A_{\max}) \cap H^1_0(\mathbb{R}^n_+).$$

(b) With $V = H^1(\mathbb{R}^n_+)$, let A_ν be the variational operator defined from the triple (H, V, a). Show that A_ν is the realization of $A = I - \Delta$ with domain

$$D(A_\nu) = \{u \in D(A_{\max}) \cap H^1(\mathbb{R}^n_+) \mid (Au, v) = a(u, v) \text{ for all } v \in H^1(\mathbb{R}^n_+)\},$$

and that the $C^2(\overline{\mathbb{R}}^n_+)$-functions in $D(A_\nu)$ satisfy

$$\frac{\partial u}{\partial \nu} = 0 \text{ on } \partial\mathbb{R}^n_+ = \mathbb{R}^{n-1}.$$

(*Comment.* A_γ and A_ν represent the Dirichlet resp. Neumann conditions on A. More on these operators in Sections 9.2 and 9.3, where regularity is shown, and a full justification of the Neumann condition is given.)

4.24. Show that the function $x_1/|x|$ is in $H^1(B(0,1))$ when $n \geq 3$.
(*Hint.* One can use that it is known from Exercise 3.7 that the first distribution derivatives are functions in $L_{1,\text{loc}}(B(0,1))$ that coincide with the derivatives defined outside of 0, when $n \geq 2$.)

4.25. Let

$$a(u, v) = \sum_{j,k=1}^{n} (a_{jk}\partial_k u, \partial_j v)_{L_2(\Omega)},$$

where Ω is open $\subset \mathbb{R}^n$ and the $a_{jk}(x)$ are real bounded functions on Ω such that (4.80) holds.
(a) Show that when u is a real function in $H^1(\Omega)$, then

$$a(u, u) \geq c \sum_{j=1}^{n} \|\partial_j u\|^2_{L_2(\Omega)} = c\|u\|^2_1 - c\|u\|^2_0$$

(Sobolev norms).
(b) Show that when u is a complex function in $H^1(\Omega)$, then

$$\operatorname{Re} a(u, u) \geq c \sum_{j=1}^{n} \|\partial_j u\|^2_{L_2(\Omega)} = c\|u\|^2_1 - c\|u\|^2_0.$$

(Hint. Write $u = v + iw$ with real functions v and w, and apply (a).)
(c) Let

$$a_1(u, v) = a(u, v) + \sum_{j=1}^{n}(b_j\partial_j u, v)_{L_2(\Omega)} + (b_0 u, v)_{L_2(\Omega)},$$

where the $b_j(x)$ are bounded functions on Ω. Show that for some $C > 0$,

$$\operatorname{Re} a_1(u, u) \geq \tfrac{c}{2}\|u\|^2_1 - C\|u\|^2_0, \text{ for } u \in H^1(\Omega).$$

(*Hint.* Recall the inequality $|2ab| \leq \varepsilon^2 |a|^2 + \varepsilon^{-2} |b|^2$, $\varepsilon > 0$.)

4.26.   Let

$$a(u, v) = \sum_{j,k=1}^{n} (a_{jk} \partial_k u, \partial_j v)_{L_2(\Omega)},$$

where Ω is open $\subset \mathbb{R}^n$ and the $a_{jk}(x)$ are complex bounded functions on Ω such that (4.82) holds. Show that for any $u \in H^1(\Omega)$,

$$\operatorname{Re} a(u, u) \geq c \sum_{j=1}^{n} \|\partial_j u\|_{L_2(\Omega)}^2 = c\|u\|_1^2 - c\|u\|_0^2.$$

Chapter 5
Fourier transformation of distributions

5.1 Rapidly decreasing functions

In the following we study an important tool in the treatment of differential operators, the *Fourier transform*. To begin with, it is useful in the study of differential operators on $\mathbb{R}^n$ with constant coefficients, but it can also be used in the more advanced theory to treat operators on subsets Ω and with variable coefficients.

The space $\mathscr{D}'(\mathbb{R}^n)$ is too large to permit a sensible definition of the Fourier transform. We therefore restrict the attention to a somewhat smaller space of distributions, $\mathscr{S}'(\mathbb{R}^n)$, the dual of a space of test functions $\mathscr{S}(\mathbb{R}^n)$ which is slightly larger than $C_0^\infty(\mathbb{R}^n)$. ($\mathscr{S}$ resp. $\mathscr{S}'$ is often called the Schwartz space of test functions resp. distributions, after Laurent Schwartz.)

It will be convenient to introduce the function

$$\langle x \rangle = (1 + |x|^2)^{\frac{1}{2}} , \tag{5.1}$$

and its powers $\langle x \rangle^s$, $s \in \mathbb{R}$. Since $|x|/\langle x \rangle \to 1$ for $|x| \to \infty$, $\langle x \rangle$ is of the same order of magnitude as $|x|$, but has the advantage of being a positive C^∞-function on all of $\mathbb{R}^n$. We note that one has for $m \in \mathbb{N}_0$:

$$\langle x \rangle^{2m} = (1 + x_1^2 + \cdots + x_n^2)^m = \sum_{|\alpha| \le m} C_{m,\alpha} x^{2\alpha} \begin{cases} \le C_m \sum_{|\alpha| \le m} x^{2\alpha}, \\ \ge \sum_{|\alpha| \le m} x^{2\alpha}, \end{cases} \tag{5.2}$$

with positive integers $C_{m,\alpha}$ and C_m. This is shown by multiplying out the expression $(1 + x_1^2 + \cdots + x_n^2)^m$; we recall from (A.9) that $C_{m,\alpha} = \frac{m!}{\alpha!(m-|\alpha|)!}$.

Definition 5.1. The vector space $\mathscr{S}(\mathbb{R}^n)$ (often just denoted $\mathscr{S}$) is defined as the space of C^∞-functions $\varphi(x)$ on $\mathbb{R}^n$ such that $x^\alpha D^\beta \varphi(x)$ is bounded for all multi-indices α and $\beta \in \mathbb{N}_0^n$. $\mathscr{S}$ is provided with the family of seminorms

$$p_M(\varphi) = \sup \left\{ \langle x \rangle^M |D^\alpha \varphi(x)| \,\big|\, x \in \mathbb{R}^n, |\alpha| \le M \right\}, \quad M \in \mathbb{N}_0. \tag{5.3}$$

The functions $\varphi \in \mathscr{S}$ are called **rapidly decreasing functions**.

With this system of seminorms, $\mathscr{S}(\mathbb{R}^n)$ is a Fréchet space. The seminorms are in fact norms and form an increasing family, so in particular, the system has the max-property (cf. Remark B.6). We could also have taken the family of seminorms

$$p_{\alpha,\beta}(\varphi) = \sup\{\, |x^\alpha D^\beta \varphi(x)| \mid x \in \mathbb{R}^n \,\}\,, \tag{5.4}$$

where α and β run through $\mathbb{N}_0^n$; it is seen from the inequalities (5.2) that the family of seminorms (5.4) defines the same topology as the family (5.3).

As a local neighborhood basis at 0 we can take the sets

$$V_{M,\frac{1}{N}} = \{\, \varphi \in \mathscr{S} \mid \langle x \rangle^M |D^\alpha \varphi(x)| < \tfrac{1}{N} \text{ for } |\alpha| \le M \,\}, \tag{5.5}$$

for $M \in \mathbb{N}_0$, $N \in \mathbb{N}$. The topology on $C_0^\infty(\mathbb{R}^n)$ is stronger than the topology induced on this space by $\mathscr{S}(\mathbb{R}^n)$, since the sets $V_{M,\frac{1}{N}} \cap C_0^\infty(\mathbb{R}^n)$ are open, convex balanced neighborhoods of 0 in $C_0^\infty(\mathbb{R}^n) = \bigcup C_{K_j}^\infty(\mathbb{R}^n)$; this follows since their intersections with $C_{K_j}^\infty(\mathbb{R}^n)$ are open neighborhoods of 0 in $C_{K_j}^\infty(\mathbb{R}^n)$. (One may observe that the topology induced from $\mathscr{S}$ is metrizable by Theorem B.9, whereas the usual topology on $C_0^\infty(\mathbb{R}^n)$ is not metrizable.)

Lemma 5.2. 1° *For* $1 \le p \le \infty$, $\mathscr{S}(\mathbb{R}^n)$ *is continuously injected in* $L_p(\mathbb{R}^n)$.
2° *For* $1 \le p \le \infty$, $\mathscr{S}(\mathbb{R}^n) \subset C_{L_p}^\infty(\mathbb{R}^n)$.
3° *For* $1 \le p < \infty$, $\mathscr{S}(\mathbb{R}^n)$ *is dense in* $L_p(\mathbb{R}^n)$.

Proof. 1°. We have for φ in $\mathscr{S}(\mathbb{R}^n)$, M in $\mathbb{N}_0$ and α in $\mathbb{N}_0^n$ with $|\alpha| \le M$ that

$$|D^\alpha \varphi(x)| \le p_M(\varphi) \langle x \rangle^{-M}$$

for all x in $\mathbb{R}^n$. In particular, $\|\varphi\|_{L_\infty} = \sup |\varphi(x)| = p_0(\varphi)$. For $1 \le p < \infty$,

$$\|\varphi\|_{L_p}^p \le p_M(\varphi)^p \int_{\mathbb{R}^n} \langle x \rangle^{-Mp} dx;$$

here we note that

$$\int_{\mathbb{R}^n} \langle x \rangle^{-Mp}\, dx = \int_{\mathbb{R}^n} (1 + |x|^2)^{-Mp/2} dx < \infty \text{ when } M > \tfrac{n}{p}. \tag{5.6}$$

Then $\|\varphi\|_{L_p} \le C_M p_M(\varphi)$ for $M > \tfrac{n}{p}$, completing the proof of 1°.

Now 2° follows, since $D^\alpha f \in \mathscr{S}$ for all α when $f \in \mathscr{S}$. (See (C.10) for the notation.)

3° follows from the fact that the subset $C_0^\infty(\mathbb{R}^n)$ of $\mathscr{S}(\mathbb{R}^n)$ is dense in $L_p(\mathbb{R}^n)$ for $p < \infty$, cf. Theorem 2.10. $\square$

It is obvious that multiplication by a polynomial maps $\mathscr{S}$ into $\mathscr{S}$. There are also other C^∞-functions that define multiplication operators in $\mathscr{S}$. L. Schwartz introduced the following space of functions ("opérateurs de multiplication") containing the polynomials:

Definition 5.3. The vector space $\mathcal{O}_M(\mathbb{R}^n)$ (or just $\mathcal{O}_M$) of **slowly increasing functions** on $\mathbb{R}^n$ consists of the functions $p(x) \in C^\infty(\mathbb{R}^n)$ which satisfy: For any $\alpha \in \mathbb{N}_0^n$ there exists $c > 0$ and $a \in \mathbb{R}$ (depending on p and α) such that

$$|D^\alpha p(x)| \leq c\langle x \rangle^a \quad \text{for all } x \in \mathbb{R}^n .$$

For example, $\langle x \rangle^t \in \mathcal{O}_M$ for any $t \in \mathbb{R}$; this is seen by repeated application of the rule

$$\partial_j \langle x \rangle^s = s\langle x \rangle^{s-2} x_j. \tag{5.7}$$

The elements of $\mathcal{O}_M$ define *multiplication operators* $M_p : f \mapsto pf$ which (by the Leibniz formula) map $\mathcal{S}$ continuously into $\mathcal{S}$. In particular, since $\mathcal{S}(\mathbb{R}^n) \subset \mathcal{O}_M(\mathbb{R}^n)$, we see that $\varphi\psi$ belongs to $\mathcal{S}(\mathbb{R}^n)$ when φ and ψ belong to $\mathcal{S}(\mathbb{R}^n)$.

Clearly, ∂^α and D^α are *continuous operators* in $\mathcal{S}(\mathbb{R}^n)$.

When $f \in L_1(\mathbb{R}^n)$, the Fourier transformed function $(\mathscr{F}f)(\xi)$ is defined by the formula

$$(\mathscr{F}f)(\xi) \equiv \hat{f}(\xi) = \int_{\mathbb{R}^n} e^{-ix\cdot\xi} f(x) \, dx .$$

We now recall some of the more elementary rules for the Fourier transformation of functions (proofs are included for the convenience of the reader):

Theorem 5.4. 1° *The Fourier transform $\mathscr{F}$ is a continuous linear map of $L_1(\mathbb{R}^n)$ into $C_{L_\infty}(\mathbb{R}^n)$, such that when $f \in L_1(\mathbb{R}^n)$, then*

$$\|\hat{f}\|_{L_\infty} \leq \|f\|_{L_1}, \quad \hat{f}(\xi) \to 0 \text{ for } |\xi| \to \infty. \tag{5.8}$$

2° *The Fourier transform is a continuous linear map of $\mathcal{S}(\mathbb{R}^n)$ into $\mathcal{S}(\mathbb{R}^n)$, and one has for $f \in \mathcal{S}(\mathbb{R}^n)$, $\xi \in \mathbb{R}^n$:*

$$\mathscr{F}[x^\alpha D_x^\beta f(x)](\xi) = (-D_\xi)^\alpha (\xi^\beta \, \hat{f}(\xi)), \tag{5.9}$$

for all multiindices α and $\beta \in \mathbb{N}_0^n$.

3° *With the co-Fourier transform $\overline{\mathscr{F}}$ (conjugate Fourier transform) defined by*

$$\overline{\mathscr{F}}f(\xi) = \int_{\mathbb{R}^n} e^{+ix\cdot\xi} f(x) \, dx, \text{ whereby } \overline{\mathscr{F}f} = \overline{\mathscr{F}} \, \bar{f}$$

(it is likewise continuous from $\mathcal{S}$ to $\mathcal{S}$), any $f \in \mathcal{S}(\mathbb{R}^n)$ with $\hat{f} = \mathscr{F}f$ satisfies

$$f(x) = (2\pi)^{-n} \int e^{i\xi \cdot x} \hat{f}(\xi) \, d\xi \quad [\equiv (2\pi)^{-n}\overline{\mathscr{F}}\hat{f}], \tag{5.10}$$

so the operator $\mathscr{F}$ maps $\mathcal{S}(\mathbb{R}^n)$ bijectively onto $\mathcal{S}(\mathbb{R}^n)$ with $\mathscr{F}^{-1} = (2\pi)^{-n}\overline{\mathscr{F}}$.

Proof. 1°. The inequality

$$|\hat{f}(\xi)| = |\int e^{ix\cdot\xi} f(x)|\, dx \leq \int |f(x)|\, dx$$

shows the first statement in (5.8), so $\mathscr{F}$ maps L_1 into L_∞. When $f \in L_1$, the functions $e^{-ix\cdot\xi} f(x)$ have the integrable majorant $|f(x)|$ for all ξ, so the continuity of $\hat{f}(\xi)$ follows from Lemma 2.8 1°. That $\hat{f}(\xi) \to 0$ for $|\xi| \to \infty$ will be shown further below.

2°. When $f \in L_1$ with $x_j f(x) \in L_1$, the functions of x

$$\partial_{\xi_j}(e^{-ix\cdot\xi} f(x)) = -ix_j e^{-ix\cdot\xi} f(x)$$

have the integrable majorant $|x_j f(x)|$, so it follows from Lemma 2.8 2° that $\partial_{\xi_j}\hat{f}(\xi)$ exists and equals $\mathscr{F}(-ix_j f(x))$. Then also $-D_{\xi_j}\hat{f} = \mathscr{F}(x_j f(x))$. When $f \in \mathscr{S}$, we can apply this rule to all derivatives, obtaining the formula $(-D_\xi)^\alpha \hat{f} = \mathscr{F}(x^\alpha f)$.

When $f \in \mathscr{S}$, we find by integration by parts (cf. (A.20)) that

$$\int e^{-ix\cdot\xi} \partial_{x_j} f(x)\, dx = \lim_{R\to\infty} \int_{|x|\leq R} e^{-ix\cdot\xi} \partial_{x_j} f(x)\, dx =$$

$$\lim_{R\to\infty} \left(\int_{|x|\leq R} i\xi_j e^{-ix\cdot\xi} f(x)\, dx + \int_{|x|=R} e^{-ix\cdot\xi} f(x)\tfrac{x_j}{|x|}\, dx \right) = i\xi_j \hat{f}(\xi)$$

(since $R^{n-1} \sup\{|f(x)| \mid |x| = R\} \to 0$ for $R \to \infty$), showing the formula $i\xi_j \hat{f} = \mathscr{F}(\partial_{x_j} f(x))$. Then also $\mathscr{F}(D_{x_j} f) = \xi_j \hat{f}$. Repeated application gives that $\mathscr{F}(D^\beta f) = \xi^\beta \hat{f}$ for all $\beta \in \mathbb{N}_0^n$.

Formula (5.9) follows for $f \in \mathscr{S}$ by combination of the two facts we have shown. Note here that by 1°, the right hand side is bounded and continuous for all α, β; this implies (in view of the Leibniz formula) that $\hat{f} \in \mathscr{S}$.

This shows that $\mathscr{F}$ maps $\mathscr{S}$ into $\mathscr{S}$; the continuity will be shown below. Let us first complete the proof of 1°: Let $f \in L_1$ and let $\varepsilon > 0$. Since $\mathscr{S}$ is dense in L_1 (Lemma 5.2), there is a $g \in \mathscr{S}$ with $\|f - g\|_{L_1} < \varepsilon/2$. Then by 1°, $|\hat{f}(\xi) - \hat{g}(\xi)| < \varepsilon/2$ for all ξ. Since $\hat{g} \in \mathscr{S}$ by 2°, we can find an $R > 0$ such that $|\hat{g}(\xi)| \leq \varepsilon/2$ for $|\xi| \geq R$. Then

$$|\hat{f}(\xi)| \leq |\hat{f}(\xi) - \hat{g}(\xi)| + |\hat{g}(\xi)| < \varepsilon \text{ for } |\xi| \geq R.$$

Now for the continuity: Note that (5.9) implies

$$\begin{aligned}
\mathscr{F}((I - \Delta)f) &= \mathscr{F}((I - \partial_{x_1}^2 - \cdots - \partial_{x_n}^2)f(x)) \\
&= (1 + \xi_1^2 + \cdots + \xi_n^2)\hat{f}(\xi) = \langle\xi\rangle^2 \hat{f}.
\end{aligned} \tag{5.11}$$

Then we have for each $k \in \mathbb{N}_0$, taking $l = \frac{k}{2}$ if k is even, $l = \frac{k+1}{2}$ if k is odd, and recalling (5.6):

$$p_0(\hat{f}) = \sup |\hat{f}(\xi)| \le \|\langle x \rangle^{-n-1} \langle x \rangle^{n+1} f\|_{L_1} \le \|\langle x \rangle^{-n-1}\|_{L_1} p_{n+1}(f),$$

$$p_k(\hat{f}) = \sup_{\xi \in \mathbb{R}^n, |\alpha| \le k} |\langle \xi \rangle^k D_\xi^\alpha \hat{f}(\xi)| \le \sup_{\xi \in \mathbb{R}^n, |\alpha| \le k} |\langle \xi \rangle^{2l} D_\xi^\alpha \hat{f}(\xi)|$$

$$= \sup_{\xi \in \mathbb{R}^n, |\alpha| \le k} |\mathscr{F}[(1-\Delta)^l (x^\alpha f(x))]| \le \sup_{|\alpha| \le k} \|(1-\Delta)^l (x^\alpha f(x))\|_{L_1}$$

$$\le \|\langle x \rangle^{-n-1}\|_{L_1} \sup_{x \in \mathbb{R}^n, |\alpha| \le k} |\langle x \rangle^{n+1} (1-\Delta)^l (x^\alpha f(x))|$$

$$\le C p_{k+n+1}(f). \tag{5.12}$$

This shows that $\mathscr{F}$ is continuous from $\mathscr{S}$ to $\mathscr{S}$.

3°. Observe first that since $\overline{\mathscr{F}} f = \overline{\mathscr{F} \bar{f}}$, the operator $\overline{\mathscr{F}}$ has properties analogous to those of $\mathscr{F}$. To show (5.10), we need to calculate $(2\pi)^{-n} \int e^{i\xi \cdot x} (\int e^{-i\xi \cdot y} f(y) dy) d\xi$ for $f \in \mathscr{S}$. The function $e^{i\xi \cdot (x-y)} f(y)$ is not integrable on $\mathbb{R}^{2n}$, so we cannot simply change the order of integration. Therefore we introduce an integration factor $\psi(\xi) \in \mathscr{S}(\mathbb{R}^n)$, which will be removed later by a passage to the limit. More precisely, we insert a function $\psi(\varepsilon\xi)$ with $\psi \in \mathscr{S}(\mathbb{R}^n)$ and $\varepsilon > 0$. Then we find for each fixed x, by use of the Fubini theorem and the change of variables $(\eta, z) = (\varepsilon\xi, (y-x)/\varepsilon)$:

$$(2\pi)^{-n} \int_{\mathbb{R}^n} e^{i\xi \cdot x} \psi(\varepsilon\xi) \hat{f}(\xi) d\xi = (2\pi)^{-n} \int_{\mathbb{R}^n} e^{i\xi \cdot x} \psi(\varepsilon\xi) \left(\int_{\mathbb{R}^n} e^{-i\xi \cdot y} f(y) dy \right) d\xi$$

$$= (2\pi)^{-n} \int_{\mathbb{R}^{2n}} e^{-i\xi \cdot (y-x)} \psi(\varepsilon\xi) f(y) \, d\xi \, dy$$

$$= (2\pi)^{-n} \int_{\mathbb{R}^{2n}} e^{-i\eta \cdot z} \psi(\eta) f(x + \varepsilon z) \, d\eta \, dz$$

$$= (2\pi)^{-n} \int_{\mathbb{R}^n} \hat{\psi}(z) f(x + \varepsilon z) \, dz,$$

since the functional determinant is 1.

For $\varepsilon \to 0$,

$$e^{i\xi \cdot x} \psi(\varepsilon\xi) \hat{f}(\xi) \to \psi(0) e^{i\xi \cdot x} \hat{f}(\xi) \text{ with } |e^{i\xi \cdot x} \psi(\varepsilon\xi) \hat{f}(\xi)| \le C |\hat{f}(\xi)|,$$

where $C = \sup_\xi |\psi(\xi)|$. Moreover,

$$\hat{\psi}(z) f(x + \varepsilon z) \to \hat{\psi}(z) f(x) \text{ with } |\hat{\psi}(z) f(x + \varepsilon z)| \le C' |\hat{\psi}(z)|,$$

where $C' = \sup_y |f(y)|$. By the theorem of Lebesgue we then find:

$$(2\pi)^{-n} \psi(0) \int e^{i\xi \cdot x} \hat{f}(\xi) \, d\xi = (2\pi)^{-n} f(x) \int \hat{\psi}(z) \, dz.$$

One can in particular use $\psi(\xi) = e^{-\frac{1}{2}|\xi|^2}$. In this case, $\psi(0) = 1$, and $\hat{\psi}(z)$ can be shown to satisfy

$$\mathscr{F}(e^{-\frac{1}{2}|\xi|^2}) = (2\pi)^{\frac{n}{2}} e^{-\frac{1}{2}|z|^2}; \tag{5.13}$$

then $\int \hat{\psi}(z)dz = (2\pi)^{\frac{n}{2}} \int e^{-\frac{1}{2}|z|^2} dz = (2\pi)^n$. This implies formula (5.10).   □

We observe moreover that the map $\mathscr{F} : L_1(\mathbb{R}^n) \to C_{L_\infty}(\mathbb{R}^n)$ is *injective*. Indeed, if $f \in L_1(\mathbb{R}^n)$ is such that $\hat{f}(\xi) = 0$ for all $\xi \in \mathbb{R}^n$, then we have for any $\varphi \in C_0^\infty(\mathbb{R}^n)$, denoting $\mathscr{F}^{-1}\varphi$ by ψ, that

$$\begin{aligned}
0 &= \int_{\mathbb{R}^n} \hat{f}(\xi)\psi(\xi)\, d\xi = \int_{\mathbb{R}^n} \int_{\mathbb{R}^n} e^{-ix\cdot\xi} f(x)\psi(\xi)\, dxd\xi \\
&= \int_{\mathbb{R}^n} f(x)\Big(\int_{\mathbb{R}^n} e^{-ix\cdot\xi}\psi(\xi)\, d\xi \Big)dx = \int_{\mathbb{R}^n} f(x)\varphi(x)\, dx;
\end{aligned}$$

then it follows by the Du Bois-Reymond Lemma 3.2 that $f = 0$ as an element of $L_1(\mathbb{R}^n)$.

There is the following extension to $L_2(\mathbb{R}^n)$:

Theorem 5.5 (PARSEVAL-PLANCHEREL THEOREM). 1° *The Fourier transform $\mathscr{F} : \mathscr{S}(\mathbb{R}^n) \to \mathscr{S}(\mathbb{R}^n)$ extends in a unique way to an isometric isomorphism $\mathscr{F}_2$ of $L_2(\mathbb{R}^n, dx)$ onto $L_2(\mathbb{R}^n, (2\pi)^{-n}dx)$. For $f, g \in L_2(\mathbb{R}^n)$,*

$$\begin{aligned}
\int f(x)\overline{g(x)}\, dx &= (2\pi)^{-n} \int \mathscr{F}_2 f(\xi)\overline{\mathscr{F}_2 g(\xi)}\, d\xi, \\
\int |f(x)|^2 dx &= (2\pi)^{-n} \int |\mathscr{F}_2 f(\xi)|^2 d\xi.
\end{aligned} \tag{5.14}$$

2° *There is the identification*

$$\mathscr{F}_2 f = \mathscr{F} f \text{ for } f \in L_2(\mathbb{R}^n) \cap L_1(\mathbb{R}^n). \tag{5.15}$$

Moreover, when $f \in L_2(\mathbb{R}^n)$ and hence $1_{B(0,N)}f \in L_1(\mathbb{R}^n)$, then the sequence of continuous functions $\mathscr{F}(1_{B(0,N)}f)$ converges in $L_2(\mathbb{R}^n)$ to $\mathscr{F}_2 f$ for $N \to \infty$.

Proof. 1°. We first show (5.14) for $f, g \in \mathscr{S}$. By Theorem 5.4 3°,

$$g(x) = (2\pi)^{-n} \int e^{i\xi\cdot x}\hat{g}(\xi)d\xi,$$

so by the Fubini theorem,

$$\begin{aligned}
\int f(x)\overline{g(x)}\, dx &= (2\pi)^{-n} \int f(x) \int e^{-i\xi\cdot x}\overline{\hat{g}(\xi)}\, d\xi dx \\
&= (2\pi)^{-n} \int \Big(\int f(x)e^{-i\xi\cdot x} dx \Big)\overline{\hat{g}(\xi)}\, d\xi = (2\pi)^{-n} \int \hat{f}(\xi)\overline{\hat{g}(\xi)}\, d\xi.
\end{aligned}$$

This shows the first formula, and the second formula follows by taking $f = g$.

We see from these formulas that $\mathscr{F} : \mathscr{S}(\mathbb{R}^n) \to L_2(\mathbb{R}^n, (2\pi)^{-n}dx)$ is a linear isometry from $\mathscr{S}(\mathbb{R}^n)$ considered as a dense subspace of $L_2(\mathbb{R}^n, dx)$.

Since the target space is complete, $\mathscr{F}$ extends in a unique way to a continuous linear map $\mathscr{F}_2 : L_2(\mathbb{R}^n, dx) \to L_2(\mathbb{R}^n, (2\pi)^{-n}dx)$, which is an isometry. The range $\mathscr{F}_2(L_2(\mathbb{R}^n, dx))$ is then also complete, hence is a closed subspace of $L_2(\mathbb{R}^n, (2\pi)^{-n}dx)$, but since it contains $\mathscr{S}(\mathbb{R}^n)$, which is dense in $L_2(\mathbb{R}^n, (2\pi)^{-n}dx)$, it must equal $L_2(\mathbb{R}^n, (2\pi)^{-n}dx)$. The identities in (5.14) extend by continuity. This shows $1°$.

$2°$. Let $f \in L_2(\mathbb{R}^n)$, then $f_N = 1_{B(0,N)}f$ is clearly in $L_2(\mathbb{R}^n)$, and it is in $L_1(\mathbb{R}^n)$ by (A.25). We first show (5.15) for f_N. Since $\mathscr{S}(\mathbb{R}^n)$ is dense in $L_2(\mathbb{R}^n)$, there is a sequence $\varphi_j \in \mathscr{S}(\mathbb{R}^n)$ with $\varphi_j \to f_N$ in $L_2(\mathbb{R}^n)$ for $j \to \infty$. Then since $1_{\complement B(0,N)}$ is a bounded function, $1_{\complement B(0,N)}\varphi_j \to 1_{\complement B(0,N)}f_N = 0$ in $L_2(\mathbb{R}^n)$. With $\chi(x)$ as in (2.3), let $\psi_j = \chi(x/N)\varphi_j$; then also $\psi_j \to f_N$ in $L_2(\mathbb{R}^n)$ for $j \to \infty$. Since f_N and ψ_j are supported in $B(0, 2N)$, we have by the Cauchy-Schwarz inequality:

$$\|f_N - \psi_j\|_{L_1} \leq \mathrm{vol}(B(0, 2N))^{1/2}\|f_N - \psi_j\|_{L_2},$$

so ψ_j converges to f_N also in $L_1(\mathbb{R}^n)$. Now $1°$ above and Theorem 5.4 $1°$ give that

$$\mathscr{F}\psi_j \to \mathscr{F}f_N \quad \text{in } C_{L_\infty}(\mathbb{R}^n),$$
$$\mathscr{F}\psi_j \to \mathscr{F}_2 f_N \quad \text{in } L_2(\mathbb{R}^n, (2\pi)^{-n}dx).$$

Then the limit in $C_{L_\infty}(\mathbb{R}^n)$ is a continuous representative of the limit in $L_2(\mathbb{R}^n, (2\pi)^{-n}dx)$. This shows (5.15) for f_N.

Using the second formula in (5.14) and the Lebesgue theorem we find:

$$(2\pi)^{-n}\|\mathscr{F}_2 f - \mathscr{F}_2 f_N\|_{L_2}^2 = \|f - f_N\|_{L_2}^2 \to 0 \quad \text{for } N \to \infty,$$

showing the convergence statement in $2°$.

Finally, we obtain (5.15) in general: When $f \in L_1(\mathbb{R}^n) \cap L_2(\mathbb{R}^n)$ and we define f_N by $f_N = 1_{B(0,N)}f$, then $f_N \to f$ in $L_1(\mathbb{R}^n)$ as well as in $L_2(\mathbb{R}^n)$ (by the Lebesgue theorem). Then $\mathscr{F}_2 f_N \to \mathscr{F}_2 f$ in $L_2(\mathbb{R}^n, (2\pi)^{-n}dx)$ and $\mathscr{F}f_N \to \mathscr{F}f$ in $C_{L_\infty}(\mathbb{R}^n)$, so the limits are the same as elements of $L_2(\mathbb{R}^n)$. $\square$

$2°$ shows that the definition of $\mathscr{F}_2$ on L_2 is consistent with the definition of $\mathscr{F}$ on L_1, so we can drop the index 2, writing $\mathscr{F}$ instead of $\mathscr{F}_2$ from now on.

The isometry property can also be expressed in the way that *the operator*

$$F = (2\pi)^{-n/2}\mathscr{F} : L_2(\mathbb{R}^n, dx) \to L_2(\mathbb{R}^n, dx) \tag{5.16}$$

is an isometric isomorphism (i.e., F is a unitary operator in the Hilbert space $L_2(\mathbb{R}^n)$). It is because of this isometry property in connection with (5.9) that the L_2-theory for distributions is particularly useful for the treatment of partial differential equations.

Let $f = u$ and $g = \overline{\mathscr{F}v}$, then we have as a special case of (5.14) (using (5.10)):

$$\int \mathscr{F}u\, v\, dx = \int u\, \mathscr{F}v\, dx, \quad \text{for } u, v \in L_2(\mathbb{R}^n, dx). \tag{5.17}$$

The following rules for *convolution* are known from measure theory: When $f \in L_1(\mathbb{R}^n)$ and $g \in C_{L_\infty}(\mathbb{R}^n)$, then the convolution

$$(f * g)(x) = \int_{\mathbb{R}^n} f(x - y)g(y)dy$$

is defined for all $x \in \mathbb{R}^n$. The function $f * g$ belongs to $C_{L_\infty}(\mathbb{R}^n)$, with

$$\|f * g\|_{L_\infty} \leq \|f\|_{L_1}\|g\|_{L_\infty}. \tag{5.18}$$

For $f, g \in L_1(\mathbb{R}^n)$, the convolution $(f * g)(x)$ is defined for almost all $x \in \mathbb{R}^n$ and gives an element of $L_1(\mathbb{R}^n)$ (also denoted $f * g$), with

$$\|f * g\|_{L_1} \leq \|f\|_{L_1}\|g\|_{L_1}. \tag{5.19}$$

The classical result on Fourier transformation of convolutions is:

Theorem 5.6. *When $f, g \in L_1(\mathbb{R}^n)$, then*

$$\mathscr{F}(f * g) = \mathscr{F}f \cdot \mathscr{F}g. \tag{5.20}$$

Proof. We find by use of the Fubini theorem and a simple change of variables:

$$\begin{aligned}
\mathscr{F}(f * g)(\xi) &= \int_{\mathbb{R}^n} e^{-i\xi \cdot x}\left(\int_{\mathbb{R}^n} f(x - y)g(y)dy\right) dx \\
&= \int_{\mathbb{R}^n} g(y)\left(\int_{\mathbb{R}^n} e^{-i\xi \cdot x} f(x - y)dx\right) dy \\
&= \int_{\mathbb{R}^n} g(y)\left(\int_{\mathbb{R}^n} e^{-i\xi \cdot (x+y)} f(x)dx\right) dy \\
&= \int_{\mathbb{R}^n} e^{-i\xi \cdot x} f(x)dx \int_{\mathbb{R}^n} e^{-i\xi \cdot y} g(y)\, dy = \mathscr{F}f(\xi)\mathscr{F}g(\xi). \quad \square
\end{aligned}$$

Observe furthermore:

Lemma 5.7. *When φ and $\psi \in \mathscr{S}(\mathbb{R}^n)$, then $\varphi * \psi \in \mathscr{S}(\mathbb{R}^n)$, and $\psi \mapsto \varphi * \psi$ is a continuous operator on $\mathscr{S}(\mathbb{R}^n)$.*

Proof. Since φ and ψ belong to $L_1(\mathbb{R}^n) \cap C_{L_\infty}(\mathbb{R}^n)$, the rules (5.18) and (5.19) show that $\varphi * \psi \in L_1(\mathbb{R}^n) \cap C_{L_\infty}(\mathbb{R}^n)$. Since $\mathscr{F}(\varphi * \psi) = (\mathscr{F}\varphi) \cdot (\mathscr{F}\psi)$, $\mathscr{F}(\varphi * \psi)$ belongs to $\mathscr{S}(\mathbb{R}^n)$. Since $\varphi * \psi$ is continuous and $\mathscr{F}$ is injective from $L_1(\mathbb{R}^n)$ to $C_{L_\infty}(\mathbb{R}^n)$, mapping $\mathscr{S}(\mathbb{R}^n)$ onto $\mathscr{S}(\mathbb{R}^n)$, we find that $\varphi * \psi \in \mathscr{S}(\mathbb{R}^n)$. It is seen from the formula $\varphi * \psi = \mathscr{F}^{-1}((\mathscr{F}\varphi) \cdot (\mathscr{F}\psi))$ that the map $\psi \mapsto \varphi * \psi$ is continuous. $\square$

5.2 Temperate distributions

Definition 5.8. $\mathscr{S}'(\mathbb{R}^n)$ (also written as $\mathscr{S}'$) is defined as the vector space of continuous linear functionals on $\mathscr{S}(\mathbb{R}^n)$. The elements of $\mathscr{S}'(\mathbb{R}^n)$ are called **temperate distributions**.[1]

According to Lemma B.7, $\mathscr{S}'$ consists of the linear functionals Λ on $\mathscr{S}$ for which there exists an $M \in \mathbb{N}_0$ and a constant C_M (depending on Λ) such that

$$|\Lambda(\varphi)| \leq C_M p_M(\varphi), \text{ for all } \varphi \in \mathscr{S}. \tag{5.21}$$

$\mathscr{S}'(\mathbb{R}^n)$ is provided with the weak* topology (as around (3.1)); this makes $\mathscr{S}'(\mathbb{R}^n)$ a topological vector space. (Its dual space is $\mathscr{S}(\mathbb{R}^n)$, cf. Section 3.5.)

Note that Definition 5.8 does not require introduction of $\mathcal{LF}$ spaces etc. (Section B.2), but is based solely on the concept of Fréchet spaces. However, it is of interest to set $\mathscr{S}'(\mathbb{R}^n)$ in relation to $\mathscr{D}'(\mathbb{R}^n)$, in particular to justify the use of the word "distribution" in this connection. We first show:

Lemma 5.9. $\mathscr{D}(\mathbb{R}^n) = C_0^\infty(\mathbb{R}^n)$ *is a dense subset of* $\mathscr{S}(\mathbb{R}^n)$, *with a stronger topology.*

Proof. As already noted, $C_0^\infty(\mathbb{R}^n) \subset \mathscr{S}(\mathbb{R}^n)$, and the neighborhood basis (5.5) for $\mathscr{S}$ at zero intersects C_0^∞ with open neighborhoods of 0 there, so that the topology induced on C_0^∞ from $\mathscr{S}$ is weaker than the original topology on C_0^∞.

To show the denseness, let $u \in \mathscr{S}(\mathbb{R}^n)$; then we must show that there is a sequence $u_N \to u$ in $\mathscr{S}(\mathbb{R}^n)$ with $u_N \in C_0^\infty(\mathbb{R}^n)$. For this we take

$$u_N(x) = \chi(x/N)u(x), \tag{5.22}$$

cf. (2.3). Here $u_N \in C_0^\infty(\mathbb{R}^n)$; $u_N(x)$ equals $u(x)$ for $|x| \leq N$ and equals 0 for $|x| \geq 2N$. Note that (as used also in the proof of Theorem 4.10)

$$|D_x^\beta \chi(x/N)| = |N^{-|\beta|} D_y^\beta \chi(y)|_{y=x/N}| \leq C_\beta N^{-|\beta|} \tag{5.23}$$

for each β; here $D_x^\beta \chi(x/N)$ has support in $\{\, x \mid N \leq |x| \leq 2N \,\}$ when $\beta \neq 0$. For each $M \geq 0$, each α, we have:

$$\sup_{x \in \mathbb{R}^n} |\langle x \rangle^M D^\alpha[(1 - \chi(x/N))u(x)]| = \sup_{|x| \geq N} |\langle x \rangle^M D^\alpha[(1 - \chi(x/N))u(x)]|$$

$$\leq \langle N \rangle^{-1} \sup_{|x| \geq N} |\langle x \rangle^{M+1}((\chi(x/N) - 1)D^\alpha u + \sum_{0 \neq \beta \leq \alpha} \tbinom{\alpha}{\beta} D^\beta \chi(x/N) D^{\alpha-\beta} u)|$$

---

[1] The word "temperate" used for the special distributions alludes to the temperate zone (with moderate temperature); the word can also mean "exercising moderation and self-restraint". The word "tempered" is also often used, but it has more to do with temper (mood), or can indicate a modification of a physical condition. The word "temperate" is used in Hörmander's books [H83], [H85].

$$\leq C_{M,\alpha}\langle N\rangle^{-1} \to 0 \text{ for } N \to \infty. \qquad (5.24)$$

It follows that $\chi(x/N)u \to u$ in $\mathscr{S}$ for $N \to \infty$. $\qquad\square$

In particular, a functional $\Lambda \in \mathscr{S}'(\mathbb{R}^n)$ restricts to a continuous functional on $\mathscr{D}(\mathbb{R}^n)$, also documented by the fact that (5.21) implies, when φ is supported in a compact set K:

$$
\begin{aligned}
|\Lambda(\varphi)| &\leq C_M p_M(\varphi) = C_M \sup \left\{ \langle x\rangle^M |D^\alpha\varphi(x)| \,\middle|\, x \in \mathbb{R}^n,\ |\alpha| \leq M \right\} \\
&\leq C_M' \sup \left\{ |D^\alpha\varphi(x)| \,\middle|\, x \in K,\ |\alpha| \leq M \right\},
\end{aligned}
\qquad (5.25)
$$

with $C_M' = C_M \sup_{x \in K}\langle x\rangle^M$.

Theorem 5.10. *The map $J : \Lambda \mapsto \Lambda'$ from $\mathscr{S}'(\mathbb{R}^n)$ to $\mathscr{D}'(\mathbb{R}^n)$ defined by restriction of Λ to $\mathscr{D}(\mathbb{R}^n)$,*

$$\langle \Lambda', \varphi\rangle = \Lambda(\varphi) \text{ for } \varphi \in \mathscr{D}(\mathbb{R}^n),$$

is injective, and hence allows an identification of $J\mathscr{S}'(\mathbb{R}^n)$ with a subspace of $\mathscr{D}'(\mathbb{R}^n)$, also called $\mathscr{S}'(\mathbb{R}^n)$:

$$\mathscr{S}'(\mathbb{R}^n) \subset \mathscr{D}'(\mathbb{R}^n). \qquad (5.26)$$

Proof. The map $J : \Lambda \mapsto \Lambda'$ is injective because of Lemma 5.9, since

$$\langle \Lambda', \varphi\rangle = 0 \quad \text{for all } \varphi \in C_0^\infty(\mathbb{R}^n) \qquad (5.27)$$

implies that Λ is 0 on a dense subset of $\mathscr{S}$, hence equals the 0-functional (since Λ is continuous on $\mathscr{S}$). $\qquad\square$

When $\Lambda \in \mathscr{S}'$, we now write $\Lambda(\varphi)$ as $\langle \Lambda, \varphi\rangle$, also when $\varphi \in \mathscr{S}$. Note that the elements of $\mathscr{D}'(\mathbb{R}^n)$ that lie in $\mathscr{S}'(\mathbb{R}^n)$ are exactly those for which there exist an M and a C_M such that

$$|\Lambda(\varphi)| \leq C_M p_M(\varphi) \text{ for all } \varphi \in \mathscr{D}(\mathbb{R}^n), \qquad (5.28)$$

independently of the support of φ. Namely, they are continuous on $\mathscr{D}$ with respect to the topology of $\mathscr{S}$, hence extend (in view of Lemma 5.9) to continuous functionals on $\mathscr{S}$.

One may observe that $J : \mathscr{S}' \to \mathscr{D}'$ is precisely *the adjoint* of the continuous injection $\iota : \mathscr{D} \to \mathscr{S}$.

Lemma 5.11. *For $1 \leq p \leq \infty$ and f in $L_p(\mathbb{R}^n)$, the map $\varphi \mapsto \int_{\mathbb{R}^n} f\varphi\,dx$, $\varphi \in \mathscr{S}(\mathbb{R}^n)$ defines a temperate distribution. In this way one has for each p a continuous injection of $L_p(\mathbb{R}^n)$ into $\mathscr{S}'(\mathbb{R}^n)$.*

Proof. Denote the map $f \mapsto \int_{\mathbb{R}^n} f\varphi\,dx$ by Λ_f. Let as usual p' be given by $\frac{1}{p} + \frac{1}{p'} = 1$, with $1' = \infty$ and $\infty' = 1$. According to the Hölder inequality,

$$|\Lambda_f(\varphi)| \le \|f\|_{L_p}\|\varphi\|_{L_{p'}},$$

and here we have by Lemma 5.2:

$$\|\varphi\|_{L_{p'}} \begin{cases} \le C_M p_M(\varphi), & \text{if } p' < \infty, \ M > \frac{n}{p'}, \\ = p_0(\varphi), & \text{if } p' = \infty. \end{cases}$$

Hence Λ_f is a continuous functional on $\mathscr{S}$, and therefore belongs to $\mathscr{S}'$. Then Λ_f also defines an element of $\mathscr{D}'$, denoted $J\Lambda_f$ above. Since $J\Lambda_f(\varphi) = \int_{\mathbb{R}^n} f\varphi\,dx$ for $\varphi \in C_0^\infty(\mathbb{R}^n)$, we know from the Du Bois-Reymond lemma that the map $f \mapsto J\Lambda_f$ is injective. Then $f \mapsto \Lambda_f$ must likewise be injective. $\qquad\square$

In particular, $\mathscr{S}(\mathbb{R}^n)$ is continuously injected into $\mathscr{S}'(\mathbb{R}^n)$.

Example 5.12. Here are some examples of elements of $\mathscr{S}'$.

$1°$ Besides the already mentioned functions $u \in L_p(\mathbb{R}^n)$, $p \in [1,\infty]$, all *functions* $v \in L_{1,\text{loc}}(\mathbb{R}^n)$ *with* $|v(x)| \le C\langle x\rangle^N$ for some N are in $\mathscr{S}'$. (Note that these functions include the $\mathscr{O}_M$-functions, in particular the polynomials, but they need not be differentiable.) We see this by observing that for such a function v,

$$|\langle v, \varphi\rangle| = |\int v\varphi\,dx| \le C\int \langle x\rangle^{-n-1}\,dx\ \sup\{\langle x\rangle^{N+n+1}|\varphi(x)| \mid x \in \mathbb{R}^n\}$$
$$\le C' p_{N+n+1}(\varphi), \text{ for } \varphi \in \mathscr{S}.$$

$2°$ The δ-*distribution and its derivatives* $D^\alpha\delta$ are in $\mathscr{S}'$, since

$$|\langle D^\alpha\delta, \varphi\rangle| = |D^\alpha\varphi(0)| \le p_{|\alpha|}(\varphi), \text{ for } \varphi \in \mathscr{S}.$$

$3°$ *Distributions with compact support.* We have shown earlier that a distribution u with compact support satisfies an estimate (3.35); the seminorm in this expression is $\le p_{N_j}$, so the distribution is a continuous linear functional on $\mathscr{S}(\mathbb{R}^n)$. Hence

$$\mathscr{E}'(\mathbb{R}^n) \subset \mathscr{S}'(\mathbb{R}^n) \subset \mathscr{D}'(\mathbb{R}^n). \tag{5.29}$$

There exist distributions in $\mathscr{D}'(\mathbb{R}^n)\backslash\mathscr{S}'(\mathbb{R}^n)$. An example is e^x (for $n = 1$), cf. Exercise 5.1. In fact it grows too fast for $x \to \infty$; this illustrates the use of the word "temperate" in the name for $\mathscr{S}'$, indicating that the elements of $\mathscr{S}'$ grow in a controlled way for $|x| \to \infty$.

5.3 The Fourier transform on $\mathscr{S}'$

The operations D^α and M_p (multiplication by p) for $p \in C^\infty(\mathbb{R}^n)$ are defined on $\mathscr{D}'(\mathbb{R}^n)$ (Definition 3.5). Concerning their action on $\mathscr{S}'$, we have:

Lemma 5.13. $1°$ D^α *maps* $\mathscr{S}'$ *continuously into* $\mathscr{S}'$ *for all* $\alpha \in \mathbb{N}_0^n$.
$2°$ *When* $p \in \mathcal{O}_M$, M_p *maps* $\mathscr{S}'$ *continuously into* $\mathscr{S}'$.

Proof. For $\alpha \in \mathbb{N}_0^n$, $p \in \mathcal{O}_M$, we set

$$\langle D^\alpha u, \varphi \rangle = \langle u, (-D)^\alpha \varphi \rangle,$$
$$\langle pu, \varphi \rangle = \langle u, p\varphi \rangle, \text{ for } \varphi \in \mathscr{S}(\mathbb{R}^n). \tag{5.30}$$

Because of the continuity of the maps in $\mathscr{S}$, it follows that these formulas define distributions in $\mathscr{S}'$, which agree with the original definitions on $\mathscr{D}'(\mathbb{R}^n)$. The continuity of the hereby defined maps in $\mathscr{S}'$ is shown in a similar way as in Theorem 3.8. $\square$

For convolutions we find, similarly to Theorem 3.15:

Lemma 5.14. *For* φ *in* $\mathscr{S}(\mathbb{R}^n)$ *and* u *in* $\mathscr{S}'(\mathbb{R}^n)$, *the prescription*

$$\langle \varphi * u, \psi \rangle = \langle u, \check{\varphi} * \psi \rangle, \quad \psi \in \mathscr{S}(\mathbb{R}^n) \tag{5.31}$$

defines a temperate distribution $\varphi * u$; *and* $u \mapsto \varphi * u$ *is a continuous operator in* $\mathscr{S}'(\mathbb{R}^n)$. *Moreover,*

$$D^\alpha(\varphi * u) = (D^\alpha \varphi) * u = \varphi * D^\alpha u, \text{ for } \varphi \in \mathscr{S}, \ u \in \mathscr{S}', \ \alpha \in \mathbb{N}_0^n,$$
$$(\varphi * \psi) * u = \varphi * (\psi * u), \text{ for } \varphi, \psi \in \mathscr{S}, \ u \in \mathscr{S}'. \tag{5.32}$$

Both in multiplication and in convolution formulas we shall henceforth allow the smooth factor to be written to the right, setting $u \cdot \varphi = \varphi \cdot u$ and $u * \varphi = \varphi * u$.

Remark 5.15. With the notation of Section 3.5, one finds by a bi-annihilator argument (as for $\mathscr{D}$ and $\mathscr{D}'$) that $\mathscr{S}(\mathbb{R}^n)$ is a dense subset of $\mathscr{S}'(\mathbb{R}^n)$. An operator A on $\mathscr{S}(\mathbb{R}^n)$ therefore has at most one extension to a continuous operator on $\mathscr{S}'(\mathbb{R}^n)$. It is then moreover seen that if a continuous operator A on $\mathscr{S}(\mathbb{R}^n)$ has a corresponding continuous operator B on $\mathscr{S}(\mathbb{R}^n)$ such that $\int_{\mathbb{R}^n} (A\varphi)\psi \, dz = \int_{\mathbb{R}^n} \varphi(B\psi) \, dx$ for all φ, ψ in $\mathscr{S}(\mathbb{R}^n)$, then A has a unique extension to a continuous operator on $\mathscr{S}'(\mathbb{R}^n)$, namely, $B^\times$. Moreover, if the restrictions of A and B to $C_0^\infty(\mathbb{R}^n)$ map $C_0^\infty(\mathbb{R}^n)$ continuously into $C_0^\infty(\mathbb{R}^n)$, then the restriction to $\mathscr{S}'(\mathbb{R}^n)$ of the already defined extension of $A|_{C_0^\infty(\mathbb{R}^n)}$ to a $\sigma(\mathscr{D}'(\mathbb{R}^n), C_0^\infty(\mathbb{R}^n))$ continuous operator on $\mathscr{D}'(\mathbb{R}^n)$ is precisely the extension of A to a $\sigma(\mathscr{S}'(\mathbb{R}^n), \mathscr{S}(\mathbb{R}^n))$ continuous operator on $\mathscr{S}'(\mathbb{R}^n)$.

The operators of differentiation, multiplication and convolution introduced on $\mathscr{S}'$ above can be considered from this point of view.

We shall finally introduce the very important *generalization of the Fourier transform* and show how it interacts with the other maps:

Definition 5.16. For $u \in \mathscr{S}'$, the prescription

$$\langle \mathscr{F}u, \varphi \rangle = \langle u, \mathscr{F}\varphi \rangle \quad \text{for all } \varphi \in \mathscr{S} \tag{5.33}$$

defines a temperate distribution $\mathscr{F}u$ (also denoted $\hat{u}$); and $\mathscr{F} : u \mapsto \mathscr{F}u$ is a continuous operator on $\mathscr{S}'(\mathbb{R}^n)$. We also define $F = (2\pi)^{-n/2}\mathscr{F}$.

The definition is of course chosen such that it is consistent with the formula (5.17) for the case where $u \in \mathscr{S}$. It is also consistent with the definition of $\mathscr{F}_2$ on $L_2(\mathbb{R}^n)$, since $\varphi_k \to u$ in $L_2(\mathbb{R}^n)$ implies $\mathscr{F}\varphi_k \to \mathscr{F}_2 u$ in $\mathscr{S}'$. Similarly, the definition is consistent with the definition on $L_1(\mathbb{R}^n)$. That $\mathscr{F}$ is a *continuous* operator on $\mathscr{S}'$ is seen as in Theorem 3.8 or by use of Remark 5.15.

The operator $\overline{\mathscr{F}}$ is similarly extended to $\mathscr{S}'$, on the basis of the identity

$$\langle \overline{\mathscr{F}}u, \varphi \rangle = \langle u, \overline{\mathscr{F}}\varphi \rangle; \tag{5.34}$$

and since

$$(2\pi)^{-n}\overline{\mathscr{F}}\mathscr{F} = (2\pi)^{-n}\mathscr{F}\overline{\mathscr{F}} = I \tag{5.35}$$

on $\mathscr{S}$, this identity is likewise carried over to $\mathscr{S}'$, so we obtain:

Theorem 5.17. $\mathscr{F}$ *is a homeomorphism of* $\mathscr{S}'$ *onto* $\mathscr{S}'$, *with inverse* $\mathscr{F}^{-1} = (2\pi)^{-n}\overline{\mathscr{F}}$.

This extension of $\mathscr{F}$ to an operator on $\mathscr{S}'$ gives an enormous freedom in the use of the Fourier transform. We obtain directly from the theorems for $\mathscr{F}$ on $\mathscr{S}$, Lemma 5.9 and the definitions of the generalized operators:

Theorem 5.18. *For all* $u \in \mathscr{S}'$, *one has when* $\alpha \in \mathbb{N}_0^n$ *and* $\varphi \in \mathscr{S}$:

$$
\begin{aligned}
&\text{(i)} && \mathscr{F}(D^\alpha u) = \xi^\alpha \mathscr{F}u, \\
&\text{(ii)} && \mathscr{F}(x^\alpha u) = (-D_\xi)^\alpha \mathscr{F}u, \\
&\text{(iii)} && \mathscr{F}(\varphi * u) = (\mathscr{F}\varphi) \cdot (\mathscr{F}u), \\
&\text{(iv)} && \mathscr{F}(\varphi \cdot u) = (2\pi)^{-n}(\mathscr{F}\varphi) * (\mathscr{F}u).
\end{aligned}
\tag{5.36}
$$

Let us study some special examples.
For $u = \delta$,

$$\langle \mathscr{F}u, \varphi \rangle = \langle u, \mathscr{F}\varphi \rangle = \hat{\varphi}(0) = \int \varphi(x)\,dx = \langle 1, \varphi \rangle, \text{ for } \varphi \in \mathscr{S},$$

hence

$$\mathscr{F}[\delta] = 1. \tag{5.37}$$

Since clearly also $\overline{\mathscr{F}}[\delta] = 1$ (cf. (5.34)), we get from the inversion formula (5.35) that

$$\mathscr{F}[1] = (2\pi)^n \delta. \tag{5.38}$$

An application of Theorem 5.18 then gives:

$$
\begin{aligned}
\mathscr{F}[D^\alpha \delta] &= \xi^\alpha, \\
\mathscr{F}[(-x)^\alpha] &= (2\pi)^n D_\xi^\alpha \delta.
\end{aligned}
\tag{5.39}
$$

Remark 5.19. We have shown that $\mathscr{F}$ defines a homeomorphism of $\mathscr{S}$ onto $\mathscr{S}$, of L_2 onto L_2 and of $\mathscr{S}'$ onto $\mathscr{S}'$. One can ask for the image by $\mathscr{F}$ of other spaces. For example, $\mathscr{F}(C_0^\infty(\mathbb{R}^n))$ must be a certain subspace of $\mathscr{S}$; but this is *not* contained in $C_0^\infty(\mathbb{R}^n)$. On the contrary, if $\varphi \in C_0^\infty(\mathbb{R}^n)$, then $\hat{\varphi}$ can only have compact support if $\varphi = 0$! For $n = 1$ we can give a quick explanation of this: When $\varphi \in C_0^\infty(\mathbb{R})$, then $\hat{\varphi}(\zeta)$ can be defined for *all* $\zeta \in \mathbb{C}$ by the formula

$$\hat{\varphi}(\zeta) = \int_{\text{supp}\,\varphi} e^{-ix\zeta}\varphi(x)dx \;,$$

and this function $\hat{\varphi}(\zeta)$ is *holomorphic* in $\zeta = \xi + i\eta \in \mathbb{C}$, since $(\partial_\xi + i\partial_\eta)\hat{\varphi}(\xi + i\eta) = 0$ (the Cauchy-Riemann equation), as is seen by differentiation under the integral sign. (One could also appeal to Morera's Theorem.) Now if $\hat{\varphi}(\zeta)$ is identically 0 on an open, nonempty interval of the real axis, then $\hat{\varphi} = 0$ everywhere. The argument can be extended to $n > 1$.

Even for *distributions* u with compact support, $\hat{u}(\zeta)$ is a *function* of ζ which can be defined for all $\zeta \in \mathbb{C}^n$. In fact one can show that $\hat{u}$ coincides with the function

$$\hat{u}(\zeta) = \langle u, \psi(x)e^{-ix\cdot\zeta} \rangle \left[= \langle \underset{\mathscr{E}'}{u}, e^{-ix\cdot\zeta} \underset{\mathscr{E}}{} \rangle \right], \tag{5.40}$$

where $\psi(x)$ is a function $\in C_0^\infty(\mathbb{R}^n)$ which is 1 on a neighborhood of supp u. It is seen as in Exercise 3.14 that this function $\hat{u}(\zeta)$ is C^∞ as a function of $(\xi_1, \eta_1, \ldots, \xi_n, \eta_n) \in \mathbb{R}^{2n}$ $(\zeta_j = \xi_j + i\eta_j)$, with

$$\partial_{\xi_j}\hat{u}(\zeta) = \langle u, \psi(x)\partial_{\xi_j}e^{-ix\cdot\zeta} \rangle,$$

and similarly for ∂_{η_j}. Since $e^{-ix\cdot\zeta}$ satisfies the Cauchy-Riemann equation in each complex variable ζ_j, so does $\hat{u}(\zeta)$, so $\hat{u}(\zeta)$ is a holomorphic function of $\zeta_j \in \mathbb{C}$ for each j. Then it follows also here that the support of $\hat{u}(\zeta)$ cannot be compact unless $u = 0$.

The spaces of holomorphic functions obtained by applying $\mathscr{F}$ to $C_0^\infty(\mathbb{R}^n)$ resp. $\mathscr{E}'(\mathbb{R}^n)$ may be characterized by their growth properties in ζ (the Paley-Wiener Theorem, see e.g. the book of W. Rudin [R74, Theorems 7.22, 7.23], or the book of L. Hörmander [H63, Theorem 1.7.7])

For partial differential operators with constant coefficients, the Fourier transform gives a remarkable simplification. When

$$P(D) = \sum_{|\alpha| \leq m} a_\alpha D^\alpha \tag{5.41}$$

is a differential operator on $\mathbb{R}^n$ with coefficients $a_\alpha \in \mathbb{C}$, the equation

$$P(D)u = f \tag{5.42}$$

(with u and $f \in \mathscr{S}'$) is by Fourier transformation carried over to the multiplication equation

$$p(\xi)\hat{u}(\xi) = \hat{f}(\xi) \tag{5.43}$$

where $p(\xi)$ is the polynomial

$$p(\xi) = \sum_{|\alpha| \leq m} a_\alpha \xi^\alpha \; ; \tag{5.44}$$

it is called *the symbol of* $P(D)$.

The m-th order part of $P(D)$ is called *the principal part* (often denoted $P_m(D)$), and its associated symbol p_m *the principal symbol*, i.e.,

$$P_m(D) = \sum_{|\alpha|=m} a_\alpha D^\alpha , \quad p_m(\xi) = \sum_{|\alpha|=m} a_\alpha \xi^\alpha . \tag{5.45}$$

It is often so that it is the principal part that determines the solvability properties of (5.42). The operator $P(D)$ is in particular called *elliptic* if $p_m(\xi) \neq 0$ for $\xi \neq 0$. Note that $p_m(\xi)$ is a homogeneous polynomial in ξ of degree m.

Example 5.20 ("THE WORLD'S SIMPLEST EXAMPLE"). Consider the operator $P = 1 - \Delta$ on $\mathbb{R}^n$. By Fourier transformation, the equation

$$(1 - \Delta)u = f \quad \text{on} \quad \mathbb{R}^n \tag{5.46}$$

is carried into the equation

$$(1 + |\xi|^2)\hat{u} = \hat{f} \quad \text{on} \quad \mathbb{R}^n , \tag{5.47}$$

and this leads by division with $1 + |\xi|^2 = \langle \xi \rangle^2$ to

$$\hat{u} = \langle \xi \rangle^{-2} \hat{f} .$$

Thus (5.46) has the solution

$$u = \mathscr{F}^{-1}(\langle \xi \rangle^{-2} \mathscr{F} f).$$

We see that for any f given in $\mathscr{S}'$ there is one and only one solution $u \in \mathscr{S}'$, and if f belongs to $\mathscr{S}$, then the solution u belongs to $\mathscr{S}$. When f is given in $L_2(\mathbb{R}^n)$, we see from (5.47) that $(1 + |\xi|^2)\hat{u}(\xi) \in L_2$. This implies not only that $u \in L_2$ (since $\hat{u} \in L_2$), but even that $D_j u$ and $D_i D_j u \in L_2$ for $i, j = 1, \ldots, n$. Indeed, $\xi_j \hat{u}$ and $\xi_i \xi_j \hat{u}$ are in L_2 since $|\xi_j| \leq 1 + |\xi|^2$ and $|\xi_i \xi_j| \leq \frac{1}{2}(|\xi_i|^2 + |\xi_j|^2) \leq |\xi|^2$; here we have used the elementary inequality

$$2ab \leq a^2 + b^2 \quad \text{for} \quad a, b \in \mathbb{R} , \tag{5.48}$$

which follows from $(a - b)^2 \geq 0$.

Thus we obtain:

$$u \in \mathscr{S}'(\mathbb{R}^n) \text{ with } (1-\Delta)u \in L_2(\mathbb{R}^n) \implies u \in H^2(\mathbb{R}^n) \,. \tag{5.49}$$

Conversely, it is clear that $u \in H^2(\mathbb{R}^n) \implies (1-\Delta)u \in L_2(\mathbb{R}^n)$, so that in fact,

$$u \in H^2(\mathbb{R}^n) \iff (1-\Delta)u \in L_2(\mathbb{R}^n) \,. \tag{5.50}$$

In particular, the maximal operator in $L_2(\mathbb{R}^n)$ for $A = 1-\Delta$ has $D(A_{\max}) = H^2(\mathbb{R}^n)$. This resembles to some extent what we found for ordinary differential operators in Section 4.3, and it demonstrates clearly the usefulness of the Fourier transform. Note that our estimates show that the graph-norm on $D(A_{\max})$ is equivalent with the $H^2(\mathbb{R}^n)$-norm. Since $C_0^\infty(\mathbb{R}^n)$ is dense in $H^2(\mathbb{R}^n)$ (Corollary 4.11), we see that $A_{\min} = A_{\max}$ here, and the operator is selfadjoint as an unbounded operator in $L_2(\mathbb{R}^n)$. (More on maximal and minimal operators on $\mathbb{R}^n$ in Theorem 6.3 ff. below.)

It should be noted that $1-\Delta$ is an unusually "nice" operator, since the polynomial $1+|\xi|^2$ is positive everywhere. As soon as there are zeros, the theory becomes more complicated. For example, the wave operator $\partial_t^2 - \Delta_x$ on $\mathbb{R}^{n+1}$ with symbol $-\tau^2 + |\xi|^2$ requires rather different techniques. Even for the Laplace operator Δ, whose symbol $-|\xi|^2$ has just one zero $\xi = 0$, it is less simple to discuss exact solutions.

At any rate, the Laplace operator is elliptic, and it is fairly easy to show qualitative properties of the solutions of the equation $-\Delta u = f$ by use of the Fourier transform. We return to this and a further discussion of differential operators in Chapter 6. First we shall study some properties of the Fourier transform which for example lead to exact results for the equation $-\Delta u = f$.

5.4 Homogeneity

When calculating the Fourier transform of specific functions, one can sometimes profit from symmetry properties. We shall give some useful examples.

The idea is to use the interaction of the Fourier transform with suitable coordinate changes; here we take the *orthogonal transformations* $y = Ox$ and the *dilations* $y = \lambda x(= \mu_\lambda(x))$, described in Example 3.21. As mentioned there, the associated maps are given by

$$\begin{aligned} [T(O)u](y) &= u(O^{-1}y) = u(x), \quad \text{when } y = Ox, \\ [T(\mu_\lambda)u](y) &= u(y/\lambda) = u(x), \quad \text{when } y = \lambda x; \end{aligned} \tag{5.51}$$

they clearly map $\mathscr{S}$ into $\mathscr{S}$ and $\mathscr{S}'$ into $\mathscr{S}'$ (where they are interpreted as in (3.59) and (3.57)). For test functions $\psi \in \mathscr{S}$ we now find (using that $(O^*)^{-1} = O$):

$$\mathscr{F}[T(O)\psi](\xi) = \int e^{-iy\cdot\xi}\psi(O^{-1}y)dy \tag{5.52}$$

$$= \int e^{-iOx\cdot\xi}\psi(x)dx = \int e^{-ix\cdot O^*\xi}\psi(x)dx$$

$$= \mathscr{F}[\psi](O^*\xi) = [T((O^*)^{-1})\hat{\psi}](\xi) = [T(O)\hat{\psi}](\xi) \,;$$

$$\mathscr{F}[T(\mu_\lambda)\psi](\xi) = \int e^{-iy\cdot\xi}\psi(y/\lambda)dy \tag{5.53}$$

$$= \int e^{-i\lambda x\cdot\xi}\psi(x)|\lambda^n|dx = |\lambda^n|\mathscr{F}[\psi](\lambda\xi) = [|\lambda^n|T(\mu_{1/\lambda})\hat{\psi}](\xi) \,.$$

This leads to the general rules for $u \in \mathscr{S}'$:

$$\langle\mathscr{F}[T(O)u],\psi\rangle = \langle T(O)u, \mathscr{F}\psi\rangle = \langle u, T(O^{-1})\mathscr{F}\psi\rangle = \langle u, \mathscr{F}[T(O^*)\psi]\rangle$$

$$= \langle\mathscr{F}u, T(O^*)\psi\rangle = \langle T((O^*)^{-1})\mathscr{F}u,\psi\rangle = \langle T(O)\mathscr{F}u,\psi\rangle \,;$$

$$\langle\mathscr{F}[T(\mu_\lambda)u],\psi\rangle = \langle T(\mu_\lambda)u, \mathscr{F}\psi\rangle = \langle u, |\lambda^n|T(\mu_{1/\lambda})\mathscr{F}\psi\rangle = \langle u, \mathscr{F}[T(\mu_\lambda)\psi]\rangle$$

$$= \langle\mathscr{F}u, T(\mu_\lambda)\psi\rangle = \langle|\lambda^n|T(\mu_{1/\lambda})\mathscr{F}u,\psi\rangle \,.$$

(The rules could also have been obtained from (5.52), (5.53) by extension by continuity, cf. Remark 5.15.) We have shown:

Theorem 5.21. *Let O be an orthogonal transformation in $\mathbb{R}^n$ and let μ_λ be the multiplication by the scalar $\lambda \in \mathbb{R} \setminus \{0\}$. The associated coordinate change maps $T(O)$ and $T(\mu_\lambda)$ in $\mathscr{S}'$ are connected with the Fourier transform in the following way:*

$$\mathscr{F}[T(O)u] = T((O^*)^{-1})\mathscr{F}u = T(O)\mathscr{F}u \,, \tag{5.54}$$

$$\mathscr{F}[T(\mu_\lambda)u] = |\lambda^n|T(\mu_{1/\lambda})\mathscr{F}u \,, \tag{5.55}$$

for $u \in \mathscr{S}'$.

The theorem is used in the treatment of functions with special invariance properties under such coordinate changes.

Those functions $u(x)$ which only depend on the distance $|x|$ to 0 may be characterized as the functions that are *invariant under all orthogonal transformations*, i.e., for which

$$T(O)u = u \quad \text{for all orthogonal transformations } O \tag{5.56}$$

(since the orthogonal transformations are exactly those transformations in $\mathbb{R}^n$ which preserve $|x|$). We shall analogously for $u \in \mathscr{S}'(\mathbb{R}^n)$ say that u depends only on the distance $|x|$ to 0 when (5.56) holds.

A function is *homogeneous of degree r*, when $u(ax) = a^r u(x)$ holds for all $a > 0$ and all $x \in \mathbb{R}^n \setminus \{0\}$, that is,

$$T(\mu_{1/a})u = a^r u \,, \quad \text{for all } a > 0 \,. \tag{5.57}$$

We say analogously that a distribution $u \in \mathscr{S}'(\mathbb{R}^n)$ is homogeneous of degree r when (5.57) holds.

Theorem 5.21 easily implies:

Corollary 5.22. *Let $u \in \mathscr{S}'(\mathbb{R}^n)$, and let $r \in \mathbb{R}$.*

1° *If u only depends on the distance $|x|$ to 0, then the same holds for $\hat{u}$.*

2° *If u is homogeneous of degree r, then $\hat{u}$ is homogeneous of degree $-r-n$.*

Proof. 1°. The identities (5.56) carry over to similar identities for $\hat{u}$ according to (5.54).

2°. When u is homogeneous of degree r, then we have according to (5.55) and (5.57):

$$T(\mu_{1/a})\mathscr{F}u = a^{-n}\mathscr{F}[T(\mu_a)u] = a^{-n}\mathscr{F}[a^{-r}u] = a^{-n-r}\mathscr{F}u \, ,$$

which shows that $\mathscr{F}u$ is homogeneous of degree $-n - r$.   □

Let us apply the theorem to the functions $u(x) = |x|^{-r}$, which have both properties: They are homogeneous of degree $-r$ and depend only on $|x|$.

Let $n/2 < r < n$; then u can be integrated into 0 whereas u^2 can be integrated out to ∞. Then we can write $u = \chi u + (1 - \chi)u$, where $\chi u \in L_1$ and $(1-\chi)u \in L_2$. It follows that $\hat{u} \in C_{L_\infty}(\mathbb{R}^n) + L_2(\mathbb{R}^n) \subset L_{2,\mathrm{loc}}(\mathbb{R}^n) \cap \mathscr{S}'$. We see from Corollary 5.22 that $\hat{u}(\xi)$ is a function which only depends on $|\xi|$ and is homogeneous of degree $r - n$.

To determine $\hat{u}$ more precisely, we shall consider the function $v(\xi)$ defined by

$$v(\xi) = |\xi|^{n-r}\hat{u}(\xi).$$

It is in $L_{2,\mathrm{loc}} \cap \mathscr{S}'$ (since $r < n$), and is *homogeneous of degree 0* and depends only on $|\xi|$ (i.e., it is invariant under dilations and orthogonal transformations).

If v is known to be continuous on $\mathbb{R}^n \setminus \{0\}$, the invariance implies that $v(\xi) = v(\eta)$ for all points ξ and $\eta \in \mathbb{R}^n \setminus \{0\}$, so v equals a constant $c_{n,r}$ on $\mathbb{R}^n \setminus \{0\}$. Since v is a locally integrable function, it identifies with the constant function $c_{n,r}$ on $\mathbb{R}^n$. This gives the formula for $\hat{u}$:

$$\mathscr{F}(|x|^{-r}) = \hat{u}(\xi) = c_{n,r}|\xi|^{-n+r} \, . \tag{5.58}$$

We want to show this formula, but since we only know beforehand that $v \in L_{2,\mathrm{loc}}$, an extra argument is needed. For example, one can reason as sketched in the following:

If a distribution f defined on a product set $Q_n = I^n \subset \mathbb{R}^n$ has $\partial_{x_1} f = \partial_{x_2} f = \cdots = \partial_{x_n} f = 0$ on Q_n, then it equals a constant, by Exercise 4.15. If a locally integrable function $g(x)$ is invariant under translations in the coordinate directions, we see from Exercise 3.13 that its first derivatives in the distribution sense are 0, so by the just-mentioned result, it must equal a constant. We are almost in this situation with $v(\xi)$, except that v is invariant not under translations but under dilations and rotations (orthogonal transformations). But this can be carried over to the rectangular situation by a change of coordinates: Consider v on a conical neighborhood of a point $\xi_0 \neq 0$, say, and express it in terms of spherical coordinates there; then it is invariant under translation in the radial direction as well as in the directions orthogonal to this. Hence it must be constant there, in a neighborhood of every point, hence constant throughout.

Thus v equals a constant function c. Denoting the constant by $c_{n,r}$, we have obtained (5.58).

The constant $c_{n,r}$ is determined by suitable calculations (e.g., integration against $\exp(-|x|^2/2)$), and one finds that

$$c_{n,r} = (4\pi)^{n/2} 2^{-r} \frac{\Gamma(\frac{n-r}{2})}{\Gamma(\frac{r}{2})} \qquad (5.59)$$

for $r \in \,]n/2, n[$ (we have this information from [R74, Exercise 8.6]). It is seen directly that $c_{n,r}$ is real by observing that $\overline{\mathscr{F}}(|x|^{-r}) = \mathscr{F}(|-x|^{-r}) = \mathscr{F}(|x|^{-r})$.

When $r \in \,]0, n/2[$, then $r' = n - r$ lies in $]n/2, n[$, so we find from (5.58) using that $\mathscr{F}^{-1} = (2\pi)^{-n}\mathscr{F}$:

$$|x|^{-n+r} = |x|^{-r'} = (2\pi)^{-n}\overline{\mathscr{F}}(c_{n,r'}|\xi|^{-n+r'}) = (2\pi)^{-n}c_{n,r'}\mathscr{F}(|\xi|^{-r});$$

this shows that (5.58) holds for $r \in \,]0, n/2[$, with

$$c_{n,r} = (2\pi)^n c_{n,n-r}^{-1} = (4\pi)^{n/2} 2^{-r} \frac{\Gamma(\frac{n-r}{2})}{\Gamma(\frac{r}{2})}. \qquad (5.60)$$

So (5.59) is also valid here.

Since $c_{n,r}$ by (5.59) converges to $(2\pi)^{n/2}$ for $r \to n/2$, and $|x|^{-r}$ as well as $|x|^{-n+r}$ converge to $|x|^{-n/2}$ in $L_{1,\text{loc}}(\mathbb{R}^n) \cap \mathscr{S}'$ for $r \to n/2$, formula (5.58) is extended by passage to the limit to hold also for $r = n/2$. We have then obtained:

Theorem 5.23. *When $r \in \,]0, n[$, then*

$$\mathscr{F}(|x|^{-r}) = c_{n,r}|\xi|^{-n+r}, \qquad (5.61)$$

where $c_{n,r}$ satisfies (5.60).

For $u(x) = |x|^{-r}$ with $r \geq n$, it is not evident how to interpret u as a distribution; there are special theories for this (see e.g. the definition of the "principal value" in [S50], and a general theory of homogeneous distributions in Section 3.2 of [H83]). See also Section 5.6.

Important special cases of Theorem 5.23 are the formulas

$$|\xi|^{-2} = \mathscr{F}\left(\frac{1}{4\pi|x|}\right) \text{ for } n = 3 \; ; \quad |\xi|^{-2} = \mathscr{F}\left(\frac{\Gamma(\frac{n}{2}-1)}{4\pi^{\frac{n}{2}}|x|^{n-2}}\right) \text{ for } n \geq 3.$$

$$(5.62)$$

5.5 Application to the Laplace operator

The preceding results make it possible to treat the equation for the Laplace operator

$$-\Delta u = f, \tag{5.63}$$

for a reasonable class of functions.

When u and f are temperate distributions, the equation gives by Fourier transformation:

$$|\xi|^2 \hat{u} = \hat{f} \; .$$

A solution may then be written (provided that we can give it a meaning)

$$\hat{u}(\xi) = |\xi|^{-2} \hat{f}(\xi) \; . \tag{5.64}$$

If f is given in $\mathscr{S}$, we can use (5.36) (iv) and (5.62) for $n \geq 3$, which gives

$$
\begin{aligned}
u(x) &= \mathscr{F}^{-1}(|\xi|^{-2}) * f \\
&= \frac{\Gamma(\frac{n}{2}-1)}{4\pi^{\frac{n}{2}}} \int \frac{f(y)}{|x-y|^{n-2}} \, dy = \frac{\Gamma(\frac{n}{2}-1)}{4\pi^{\frac{n}{2}}} \int \frac{f(x-y)}{|y|^{n-2}} \, dy \;,
\end{aligned}
\tag{5.65}
$$

so this is a solution of (5.63). The solution u is a C^∞-function, since differentiation can be carried under the integral sign. There are many other solutions, namely, all functions $u + w$ where w runs through the solutions of $\Delta w = 0$, the *harmonic functions* (which span an infinitely dimensional vector space; already the harmonic *polynomials* do so).

The function $\frac{\Gamma(\frac{n}{2}-1)}{4\pi^{\frac{n}{2}}}|x|^{-n+2}$ is called the Newton potential. Once we have the formula (5.65), we can try to use it for more general f and thereby extend the applicability. For example, if we insert a continuous function with compact support as f, then we get a function u, which is not always two times differentiable in the classical sense, but still for bounded open sets Ω can be shown to belong to $H^2(\Omega)$ and solve (5.63) in the distribution sense (in fact it solves (5.63) as an H^2-function). See also Remark 6.13 later.

This solution method may in fact be extended to distributions $f \in \mathscr{E}'(\mathbb{R}^n)$, but this requires a generalization of the convolution operator which we refrain from including here.

The operator $-\Delta : u \mapsto f$ is *local*, in the sense that the shape of f in the neighborhood of a point depends only on the shape of u in a neighborhood of the point (this holds for all differential operators). On the other hand, the solution operator $T : f \mapsto u$ defined by (5.65) cannot be expected to be local (we see this explicitly from the expression for T as an integral operator).

Let Ω be a bounded open subset of $\mathbb{R}^n$. By use of T defined above, we define the operator T_Ω as the map that sends $\varphi \in C_0^\infty(\Omega)$ (extended by 0 in $\mathbb{R}^n \setminus \Omega$) into $(T\varphi)|_\Omega$, i.e.,

$$T_\Omega : \varphi \mapsto (T\varphi)|_\Omega \text{ for } \varphi \in C_0^\infty(\Omega) . \tag{5.66}$$

We here have that

$$(-\Delta T_\Omega \varphi)(x) = \varphi(x) \text{ for } x \in \Omega ,$$

because Δ is local. Thus T_Ω is a right inverse of $-\Delta$ on Ω. It is an integral operator

$$T_\Omega \varphi = \int_\Omega G(x,y)\varphi(y) \, dy , \tag{5.67}$$

with the kernel

$$G(x,y) = \frac{\Gamma(\frac{n}{2} - 1)}{4\pi^{\frac{n}{2}}} |x - y|^{-n+2}, \text{ for } x, y \in \Omega .$$

An interesting question concerning this solution operator is whether it is a Hilbert-Schmidt operator (i.e., an integral operator whose kernel is in $L_2(\Omega \times \Omega)$). For this we calculate

$$\int_{\Omega \times \Omega} |G(x,y)|^2 dx \, dy = c \int_{\Omega \times \Omega} |x - y|^{-2n+4} dx \, dy$$
$$\leq c' \int_{|z|,|w| \leq R} |z|^{-2n+4} dz \, dw ,$$

where we used the coordinate change $z = x - y$, $w = x + y$, and chose R so large that $\Omega \times \Omega \subset \{(x,y) \mid |x + y| \leq R, |x - y| \leq R \}$. The integral with respect to z in the last expression (and thereby the full integral) is finite if and only if $-2n + 4 > -n$, i.e., $n < 4$. So T_Ω *is a Hilbert-Schmidt operator in* $L_2(\Omega)$, *when* $n = 3$ (in particular, a compact operator).

One can show more generally that $\overline{T}_\Omega$ for bounded Ω is a compact selfadjoint operator in $L_2(\Omega)$, for which the eigenvalue sequence $(\lambda_j(\overline{T}_\Omega))_{j \in \mathbb{N}}$ is in ℓ^p for $p > n/2$ (i.e, $\overline{T}_\Omega$ belongs to the p-th Schatten class; the Hilbert-Schmidt case is the case $p = 2$).

When Ω is unbounded, $\overline{T}_\Omega$ is in general not a compact operator in $L_2(\Omega)$ (unless Ω is very "thin").

5.6 Distributions associated with nonintegrable functions

We shall investigate some more types of distributions on $\mathbb{R}$ and their Fourier transforms. Theorem 5.21 treated homogeneous functions u of degree a, but the desire to have u and $\hat{u}$ in $L_{1,\text{loc}}$ put essential restrictions on the values of a that could be covered. Now $n = 1$, so the calculations before Theorem 5.23 show how the cases $a \in\,] -1, 0\,[$ may be treated. This neither covers the case $a = 0$ (i.e., functions $u = c_1 H(x) + c_2 H(-x)$) nor the case $a = -1$ (where the functions $u = c_1 \dfrac{H(x)}{x} + c_2 \dfrac{H(-x)}{x}$ are not in $L_{1,\text{loc}}$ in the neighborhood of 0 if $(c_1, c_2) \neq (0,0)$). We shall now consider these cases. One result is a description of the Fourier transform of the Heaviside function (which will allow us to treat the case $a = 0$), another result is that we give sense to a distribution which outside of 0 behaves like $\dfrac{1}{x}$.

When f is a function on $\mathbb{R}$ which is integrable on the intervals $]-\infty, -\varepsilon[$ and $[\varepsilon, \infty[$ for every $\varepsilon > 0$, then we define the *principal value integral* of f over $\mathbb{R}$ by

$$\text{PV} \int_{\mathbb{R}} f(x)dx = \lim_{\varepsilon \to 0} \int_{\mathbb{R}\setminus[-\varepsilon,\varepsilon]} f(x)\,dx \,, \qquad (5.68)$$

when this limit exists. (It is important in the definition that the interval $[-\varepsilon, \varepsilon]$ is *symmetric* around 0; when $f \notin L_{1,\text{loc}}(\mathbb{R})$ there is the risk of getting another limit by cutting out another interval like for example $[-\varepsilon, 2\varepsilon]$.) We now define the distribution $\text{PV} \dfrac{1}{x}$ by

$$\langle \text{PV} \frac{1}{x}, \varphi \rangle = \text{PV} \int_{\mathbb{R}} \frac{\varphi(x)}{x}\,dx \ \text{ for } \varphi \in C_0^\infty(\mathbb{R}) \,. \qquad (5.69)$$

(In some of the literature, this distribution is denoted vp. $\dfrac{1}{x}$, for "valeur principale".) We have to show that the functional in (5.69) is well-defined and continuous on $C_0^\infty(\mathbb{R})$. Here we use that by the Taylor formula,

$$\varphi(x) = \varphi(0) + x \cdot \varphi_1(x) \,, \qquad (5.70)$$

where $\varphi_1(x) = \dfrac{\varphi(x) - \varphi(0)}{x}$ is in $C^\infty(\mathbb{R})$ (the reader should verify this). Moreover, we have for $x \in [-R, R]$, by the mean value theorem,

$$\sup_{|x|\leq R} |\varphi_1(x)| = \sup_{|x|\leq R} \left| \frac{\varphi(x) - \varphi(0)}{x} \right| = \sup_{|x|\leq R} |\varphi'(\theta(x))| \leq \sup_{|x|\leq R} |\varphi'(x)| \,, \quad (5.71)$$

where $\theta(x)$ is a suitable point between 0 and x. This gives for $\mathrm{PV}\,\frac{1}{x}$, when $\operatorname{supp}\varphi \subset [-R,R]$,

$$\langle \mathrm{PV}\,\frac{1}{x},\varphi\rangle = \lim_{\varepsilon\to 0}\int_{|x|>\varepsilon}\frac{\varphi(x)}{x}\,dx \tag{5.72}$$

$$= \lim_{\varepsilon\to 0}\left[\int_{[-R,-\varepsilon]\cup[\varepsilon,R]}\frac{\varphi(0)}{x}\,dx + \int_{[-R,-\varepsilon]\cup[\varepsilon,R]}\varphi_1(x)\,dx\right]$$

$$= \int_{-R}^{R}\varphi_1(x)\,dx\,,$$

since the first integral in the square bracket is 0 because of the symmetry of $\frac{1}{x}$. Thus the functional $\mathrm{PV}\,\frac{1}{x}$ is well-defined, and we see from (5.71) that it is a distribution of order 1:

$$|\langle \mathrm{PV}\,\frac{1}{x},\varphi\rangle| \le 2R \sup_{|x|\le R}|\varphi_1(x)| \le 2R \sup_{|x|\le R}|\varphi'(x)|, \tag{5.73}$$

when $\operatorname{supp}\varphi \subset [-R,R]$.

Remark 5.24. One can also associate distributions with the other functions $\frac{1}{x^m}$, $m\in\mathbb{N}$; here one uses on one hand the principal value concept, on the other hand a modification of $\varphi(x)$ by a Taylor polynomial at 0; cf. Exercise 5.9. The resulting distributions are called $\mathrm{Pf}\,\frac{1}{x^m}$, where Pf stands for pseudo-function; for $m=1$ we have that $\mathrm{Pf}\,\frac{1}{x} = \mathrm{PV}\,\frac{1}{x}$.

Since $\mathrm{PV}\,\frac{1}{x} = \chi\,\mathrm{PV}\,\frac{1}{x} + (1-\chi)\frac{1}{x}$ has its first term in $\mathscr{S}'(\mathbb{R})$ and second term in $L_2(\mathbb{R})$, $v=\mathrm{PV}\,\frac{1}{x}$ belongs to $\mathscr{S}'$, hence has a Fourier transformed $\hat{v}$. We can find it in the following way: Observe that

$$x\cdot\mathrm{PV}\,\frac{1}{x} = 1 \tag{5.74}$$

(using the definitions), so that $\hat{v}$ is a solution of the differential equation in $\mathscr{S}'$ (cf. (5.36) (ii) and (5.38))

$$i\partial_\xi\hat{v}(\xi) = 2\pi\delta\,. \tag{5.75}$$

One solution of this equation is $-2\pi i H(\xi)$ (cf. (3.23)); all other solutions are of the form

$$-2\pi i H(\xi) + c, \tag{5.76}$$

where $c\in\mathbb{C}$, cf. Theorem 4.19. We then just have to determine the constant c. For this we observe that $\frac{1}{x}$ is an odd function and that $v=\mathrm{PV}\,\frac{1}{x}$ is an *odd distribution* (i.e., $\langle v, \varphi(-x) \rangle = -\langle v, \varphi(x) \rangle$ for all φ); then the Fourier transform $\hat{v}$ must likewise be odd.

Let us include a carefully elaborated explanation of what was just said: The antipodal operator

$$S : \varphi(x) \mapsto \varphi(-x), \quad \varphi \in C_0^\infty(\mathbb{R}) \tag{5.77}$$

(for which we have earlier used the notation $\varphi \mapsto \check{\varphi}$) is a special case of a dilation (3.53), namely, with $\lambda = -1$. Hence it carries over to distributions in the usual way (cf. (3.57)):

$$\langle Su, \varphi \rangle = \langle u, S\varphi \rangle \quad \text{for all } \varphi \in C_0^\infty(\mathbb{R}). \tag{5.78}$$

A function u is said to be *even* resp. *odd*, when $Su = u$ resp. $Su = -u$; this notation is now extended to distributions. For the connection with the Fourier transformation we observe that

$$(\mathscr{F}S\varphi)(\xi) = \int e^{-ix\xi} \varphi(-x) dx$$

$$= \int e^{iy\xi} \varphi(y) dy = (\overline{\mathscr{F}}\varphi)(\xi) = (\mathscr{F}\varphi)(-\xi) = (S\mathscr{F}\varphi)(\xi) ,$$

or in short:

$$\mathscr{F}S = \overline{\mathscr{F}} = S\mathscr{F} ; \tag{5.79}$$

these formulas are carried over to distributions by use of (5.78) or (5.54). (The formula (5.35) could be written: $\mathscr{F}^2 = (2\pi)^n S$.) In particular, we see that $Sv = -v$ implies $S\hat{v} = -\hat{v}$.

The only odd function of the form (5.76) is the one with $c = \pi i$, so we finally conclude:

$$\mathscr{F}[\mathrm{PV}\,\frac{1}{x}] = -2\pi i H(\xi) + \pi i = -\pi i \,\mathrm{sign}\,\xi; \tag{5.80}$$

cf. (3.26) for $\mathrm{sign}\,\xi$. (It is possible to find $\hat{v}$ by direct calculations, but then one has to be very careful with the interpretation of convergences of the occurring integrals of functions not in L_1. We have avoided this by building up the Fourier transformation on $\mathscr{S}'$ by duality from the definition on $\mathscr{S}$.)

An application of $\dfrac{1}{2\pi}\overline{\mathscr{F}} = \dfrac{1}{2\pi}\mathscr{F}S$ (cf. (5.79)) to (5.80) gives

$$\mathrm{PV}\,\frac{1}{x} = \frac{1}{2\pi}\mathscr{F}S(-2\pi i H(\xi) + \pi i) = \frac{1}{2\pi}\mathscr{F}(2\pi i H(\xi) - i\pi)$$

$$= i\mathscr{F}H(\xi) - \frac{i}{2}\mathscr{F}[1] = i\mathscr{F}H - \pi i \delta .$$

This also leads to the formula for the Fourier transform of the Heaviside function:

$$\mathscr{F}H(x) = -i\,\mathrm{PV}\,\frac{1}{\xi} + \pi\delta. \tag{5.81}$$

Using this formula, we can find the Fourier transforms of all homogeneous functions of degree 0.

Some further remarks: Corresponding to the decomposition

$$1 = H(x) + H(-x),$$

we now get the following decomposition of δ (which is used in theoretical physics):

$$\delta = (2\pi)^{-1}\mathscr{F}[1] = (2\pi)^{-1}(\mathscr{F}H + \mathscr{F}SH) \tag{5.82}$$

$$= \left(\frac{\delta}{2} + \frac{1}{2\pi i}\,\mathrm{PV}\,\frac{1}{x}\right) + \left(\frac{\delta}{2} - \frac{1}{2\pi i}\,\mathrm{PV}\,\frac{1}{x}\right) = \delta_+ + \delta_- \,,\quad \text{where}$$

$$\delta_\pm = \frac{\delta}{2} \pm \frac{1}{2\pi i}\,\mathrm{PV}\,\frac{1}{x} = \frac{1}{2\pi}\mathscr{F}[H(\pm x)]\,. \tag{5.83}$$

Observe also that since

$$H(x) = \lim_{a\to 0+} H(x)e^{-ax} \quad \text{in } \mathscr{S}',$$

one has that

$$\mathscr{F}H = \lim_{a\to 0+} \frac{1}{a + i\xi} \quad \text{in } \mathscr{S}' \tag{5.84}$$

(cf. Exercise 5.3), and then

$$\delta_+ = \frac{1}{2\pi}\lim_{a\to 0+} \frac{1}{a + ix} \quad \text{in } \mathscr{S}'. \tag{5.85}$$

Remark 5.25. To the nonintegrable function $\dfrac{H(x)}{x}$ we associate the distribution $\mathrm{Pf}\,\dfrac{H(x)}{x}$, defined by

$$\left\langle \mathrm{Pf}\,\frac{H(x)}{x}, \varphi \right\rangle = \lim_{\varepsilon\to 0+}\left[\int_\varepsilon^\infty \frac{\varphi(x)}{x}\,dx + \varphi(0)\log\varepsilon\right], \tag{5.86}$$

cf. [S61, Exercise II-14, p. 114-115]; note that there is a logarithmic correction. In this way, every function on $\mathbb{R}$ which is homogeneous of degree -1 is included in the distribution theory, namely, as a distribution

$$c_1\,\mathrm{Pf}\,\frac{H(x)}{x} + c_2 S\,\mathrm{Pf}\,\frac{H(x)}{x}\,,\quad c_1 \text{ and } c_2 \in \mathbb{C}\,. \tag{5.87}$$

In particular, we define $\mathrm{Pf}\,\dfrac{1}{|x|}$ by

$$\mathrm{Pf}\,\frac{1}{|x|} = \mathrm{Pf}\,\frac{H(x)}{x} + S\,\mathrm{Pf}\,\frac{H(x)}{x}. \tag{5.88}$$

It is shown in Exercise 5.12 that

$$\mathscr{F}(\mathrm{Pf}\,\frac{1}{|x|}) = -2\log|\xi| + C,$$

for some constant C.

Also for distributions on $\mathbb{R}^n$ there appear logarithmic terms, when one wants to include general homogeneous functions and their Fourier transforms in the theory. (A complete discussion of homogeneous distributions is given e.g. in [H83].)

Exercises for Chapter 5

5.1.   Let $n = 1$. Show that $e^x \notin \mathscr{S}'(\mathbb{R})$, whereas $e^x \cos(e^x) \in \mathscr{S}'(\mathbb{R})$. (*Hint.* Find an integral of $e^x \cos(e^x)$.)

5.2.   Show the inequalities (5.2), and show that the systems of seminorms (5.3) and (5.4) define the same topology.

5.3.   Let $a > 0$. With $H(t)$ denoting the Heaviside function, show that

$$\mathscr{F}[H(t)e^{-at}] = \frac{1}{a + i\xi}\,.$$

What is $\mathscr{F}[H(-t)e^{at}]$?

5.4.   (a) Show that for $n = 1$,

$$\mathscr{F}^{-1}\Big[\frac{1}{1 + \xi^2}\Big] = c\,e^{-|x|};$$

determine c. (One can use Exercise 5.3.)

(b) Show that for $n = 3$,

$$\mathscr{F}^{-1}\Big[\frac{1}{1 + |\xi|^2}\Big] = \frac{c}{|x|}e^{-|x|},$$

with $c = \frac{1}{4\pi}$. (One may observe that the function is the unique solution v in $\mathscr{S}'$ of $(1 - \Delta)v = \delta$; or one can apply the rotation invariance directly.)

5.5.   Let M and n be integers with $0 < 2M < n$. Find an integral operator T_M on $\mathscr{S}(\mathbb{R}^n)$ with the following properties:

(i) $\Delta^M T_M f = f$ for $f \in \mathscr{S}(\mathbb{R}^n)$.

(ii)   When Ω is a bounded, open subset of $\mathbb{R}^n$, and $2M > n/2$, then the operator $(T_M)_\Omega$ (defined as in (5.66)) is a Hilbert-Schmidt operator in $L_2(\Omega)$.

5.6.   Show that the differential equation on $\mathbb{R}^3$:

$$\frac{\partial^4 u}{\partial x_1^4} - \frac{\partial^2 u}{\partial x_2^2} - \frac{\partial^2 u}{\partial x_2 \partial x_3} - \frac{\partial^2 u}{\partial x_3^2} + 3u = f$$

has one and only one solution $u \in \mathscr{S}'$ for each $f \in \mathscr{S}'$. Determine the values of $m \in \mathbb{N}_0$ for which u belongs to the Sobolev space $H^m(\mathbb{R}^3)$ when $f \in L_2(\mathbb{R}^3)$.

5.7.   Let $a \in \mathbb{C}$, and show that the distribution $u = e^{-ax} H(x)$ is a solution of the differential equation

$$(\partial_x + a)u = \delta \quad \text{in } \mathscr{D}'(\mathbb{R}) .$$

Can we show this by Fourier transformation?

5.8.   Show that the Cauchy-Riemann equation

$$\left(\frac{\partial}{\partial x} + i\frac{\partial}{\partial y} \right) u(x,y) = f(x,y)$$

on $\mathbb{R}^2$ has a solution for each $f \in \mathscr{S}$; describe such a solution.

5.9.   For $m \in \mathbb{N}$ and $\varphi \in C_0^\infty(\mathbb{R})$ we define the functional Λ_m by

$$\Lambda_m(\varphi) = \mathrm{PV} \int_{-\infty}^{\infty} \{\, x^{-m}\varphi(x) - \sum_{p=0}^{m-2} \frac{x^{p-m}}{p!}\varphi^{(p)}(0) \,\} \, dx .$$

(a) Show that PV ... exists, so that $\Lambda_m(\varphi)$ is well-defined.
(b) Show that $\Lambda_m(\varphi') = m\Lambda_{m+1}(\varphi)$.
(c) Show that Λ_m is a distribution, and that

$$\Lambda_m = (-1)^{m-1}(m-1)!\,\frac{d^m}{dx^m}\log|x|.$$

Λ_m is often called $\mathrm{Pf}\,\dfrac{1}{x^m}$, where Pf stands for pseudo-function.

5.10.   Let $I = \mathbb{R}$ or $I =]a, \infty[$. Show that when u and $D^m u \in L_2(I)$, then $u \in H^m(I)$. (One can show this for $I = \mathbb{R}$ by use of the Fourier transformation. Next, one can show it for $I =]a, \infty[$ by use of a cut-off function. This proves the assertion of Remark 4.21.)

5.11.   Show that

$$\mathscr{F}\,\mathrm{sign}\,x = -2i\,\mathrm{PV}\,\frac{1}{\xi}.$$

5.12.   Consider the locally integrable function $\log |x|$.

(a) Let $u = H(x)\log x$. Show that

$$\log |x| = u + Su.$$

(b) Show that

$$\frac{d}{dx}u = \operatorname{Pf}\frac{H(x)}{x}, \qquad \frac{d}{dx}\log |x| = \operatorname{PV}\frac{1}{x}.$$

(c) Show that $x\operatorname{Pf}\frac{1}{|x|} = \operatorname{sign} x$, and that

$$\partial_\xi \mathscr{F}\left(\operatorname{Pf}\frac{1}{|x|}\right) = -2\operatorname{PV}\frac{1}{\xi};$$

cf. Exercise 5.11.

(d) Show that

$$\mathscr{F}\left(\operatorname{Pf}\frac{1}{|x|}\right) = -2\log|\xi| + C,$$

for some constant C.

(e) Show that

$$\mathscr{F}(\log|x|) = -\pi\operatorname{Pf}\frac{1}{|\xi|} + C_1\delta,$$

for some constant C_1.

(Information on the constant can be found in [S50, p. 258] [S61, Exercise V-10], where the Fourier transformation is normalized in a slightly different way.)

Chapter 6
Applications to differential operators. The Sobolev theorem

6.1 Differential and pseudodifferential operators on $\mathbb{R}^n$

As we saw in (5.41)–(5.44), a differential operator $P(D)$ (with constant coefficients) is by Fourier transformation carried over to a multiplication operator $M_p : f \mapsto pf$, where $p(\xi)$ is a polynomial. One can extend this idea to the more general functions $p(\xi) \in \mathcal{O}_M$, obtaining a class of operators which we call (x-independent) pseudodifferential operators (denoted ψdo's for short).

Definition 6.1. Let $p(\xi) \in \mathcal{O}_M$. The associated pseudodifferential operator $\mathrm{Op}(p(\xi))$, also called $P(D)$, is defined by

$$\mathrm{Op}(p)u \equiv P(D)u = \mathscr{F}^{-1}(p(\xi)\hat{u}(\xi)) ; \tag{6.1}$$

it maps $\mathscr{S}$ into $\mathscr{S}$ and $\mathscr{S}'$ into $\mathscr{S}'$ (continuously). The function $p(\xi)$ is called the **symbol** of $\mathrm{Op}(p)$.

As observed, differential operators with constant coefficients are covered by this definition; but it is interesting that also the solution operator in Example 5.20 is of this type, since it equals $\mathrm{Op}(\langle\xi\rangle^{-2})$.

For these pseudodifferential operators one has the extremely simple rule of calculus:

$$\mathrm{Op}(p)\,\mathrm{Op}(q) = \mathrm{Op}(pq), \tag{6.2}$$

since $\mathrm{Op}(p)\,\mathrm{Op}(q)u = \mathscr{F}^{-1}(p\mathscr{F}\mathscr{F}^{-1}(q\mathscr{F}u)) = \mathscr{F}^{-1}(pq\mathscr{F}u)$. In other words, *composition of operators* corresponds to *multiplication of symbols*. Moreover, if p is a function in $\mathcal{O}_M$ for which $1/p$ belongs to $\mathcal{O}_M$, then the operator $\mathrm{Op}(p)$ has the inverse $\mathrm{Op}(1/p)$:

$$\mathrm{Op}(p)\,\mathrm{Op}(1/p) = \mathrm{Op}(1/p)\,\mathrm{Op}(p) = I. \tag{6.3}$$

For example, $1 - \Delta = \mathrm{Op}(\langle\xi\rangle^2)$ has the inverse $\mathrm{Op}(\langle\xi\rangle^{-2})$, cf. Example 5.20.

Remark 6.2. We here use the notation pseudodifferential operator for all operators that are obtained by Fourier transformation from multiplication operators in $\mathscr{S}$ (and $\mathscr{S}'$). In practical applications, one usually considers restricted classes of symbols with special properties. On the other hand, one allows symbols depending on x also, associating the operator $\mathrm{Op}(p(x, \xi))$ defined by

$$[\mathrm{Op}(p(x, \xi))u](x) = (2\pi)^{-n} \int e^{ix \cdot \xi} p(x, \xi) \hat{u}(\xi) d\xi , \qquad (6.4)$$

to the symbol $p(x, \xi)$. This is consistent with the fact that when P is a differential operator of the form

$$P(x, D)u = \sum_{|\alpha| \le m} a_\alpha(x) D^\alpha u , \qquad (6.5)$$

then $P(x, D) = \mathrm{Op}(p(x, \xi))$, where the *symbol* is

$$p(x, \xi) = \sum_{|\alpha| \le m} a_\alpha(x) \xi^\alpha . \qquad (6.6)$$

Allowing "variable coefficients" makes the theory much more complicated, in particular because the identities (6.2) and (6.3) then no longer hold in an exact way, but in a certain approximative sense, depending on which symbol class one considers. The systematic theory of pseudodifferential operators plays an important role in the modern mathematical literature, as a general framework around differential operators and their solution operators. It is technically more complicated than what we are doing at present, and will be taken up later, in Chapter 7.

Let us consider the L_2-realizations of a pseudodifferential operator $P(D)$. In this "constant-coefficient" case we can appeal to Theorem 12.13 on multiplication operators in L_2.

Theorem 6.3. *Let $p(\xi) \in \mathscr{O}_M$ and let $P(D)$ be the associated pseudodifferential operator $\mathrm{Op}(p)$. The maximal realization $P(D)_{\max}$ of $P(D)$ in $L_2(\mathbb{R}^n)$ with domain*

$$D(P(D)_{\max}) = \{ u \in L_2(\mathbb{R}^n) \mid P(D)u \in L_2(\mathbb{R}^n) \} , \qquad (6.7)$$

is densely defined (with $\mathscr{S} \subset D(P(D)_{\max})$) and closed. Let $P(D)_{\min}$ denote the closure of $P(D)|_{C_0^\infty(\mathbb{R}^n)}$ (the minimal realization); then

$$P(D)_{\max} = P(D)_{\min} . \qquad (6.8)$$

Furthermore, $(P(D)_{\max})^* = P'(D)_{\max}$, where $P'(D) = \mathrm{Op}(\overline{p})$.

Proof. We write P for $P(D)$ and P' for $P'(D)$. It follows immediately from the Parseval-Plancherel theorem (Theorem 5.5) that

$$P_{\max} = \mathscr{F}^{-1} M_p \mathscr{F} \; ; \quad \text{with}$$
$$D(P_{\max}) = \mathscr{F}^{-1} D(M_p) = \mathscr{F}^{-1} \{ f \in L_2(\mathbb{R}^n) \mid pf \in L_2(\mathbb{R}^n) \} \, ,$$

where M_p is the multiplication operator in $L_2(\mathbb{R}^n)$ defined as in Theorem 12.13. In particular, $P_{\max}$ is a closed, densely defined operator, and $\mathscr{S} \subset D(M_p)$ implies $\mathscr{S} \subset D(P(D)_{\max})$. We shall now first show that $P_{\max}$ and $P'_{\min}$ are adjoints of one another. This goes in practically the same way as in Section 4.1: For $u \in \mathscr{S}'$ and $\varphi \in C_0^\infty(\mathbb{R}^n)$ one has:

$$\langle Pu, \overline{\varphi} \rangle = \langle \mathscr{F}^{-1} p \mathscr{F} u, \overline{\varphi} \rangle = \langle p \mathscr{F} u, \mathscr{F}^{-1} \overline{\varphi} \rangle \qquad (6.9)$$
$$= \langle u, \mathscr{F} p \mathscr{F}^{-1} \overline{\varphi} \rangle = \langle u, \overline{\mathscr{F}^{-1} \overline{p} \mathscr{F} \varphi} \rangle = \langle u, \overline{P' \varphi} \rangle \, ,$$

using that $\overline{\mathscr{F}} = (2\pi)^n \mathscr{F}^{-1}$. We see from this on one hand that when $u \in D(P_{\max})$, i.e., u and $Pu \in L_2$, then

$$(Pu, \varphi) = (u, P' \varphi) \quad \text{for all } \varphi \in C_0^\infty \, ,$$

so that

$$P_{\max} \subset (P'|_{C_0^\infty})^* \text{ and } P'|_{C_0^\infty} \subset (P_{\max})^* \, ,$$

and thereby

$$P'_{\min} = \text{ closure of } P'|_{C_0^\infty} \subset (P_{\max})^* \, .$$

On the other hand, we see from (6.9) that when $u \in D((P'|_{C_0^\infty})^*)$, i.e., there exists $v \in L_2$ so that $(u, P' \varphi) = (v, \varphi)$ for all $\varphi \in C_0^\infty$, then v equals Pu, i.e.,

$$(P'|_{C_0^\infty})^* \subset P_{\max} \, .$$

Thus $P_{\max} = (P'|_{C_0^\infty})^* = (P'_{\min})^*$ (cf. Corollary 12.6). So Lemma 4.3 extends to the present situation.

But now we can furthermore use that $(M_p)^* = M_{\overline{p}}$ by Theorem 12.13, which by Fourier transformation is carried over to

$$(P_{\max})^* = P'_{\max} \, .$$

In detail:

$$(P_{\max})^* = (\mathscr{F}^{-1} M_p \mathscr{F})^* = \mathscr{F}^* M_p^* (\mathscr{F}^{-1})^* = \overline{\mathscr{F}} M_{\overline{p}} \overline{\mathscr{F}}^{-1} = \mathscr{F}^{-1} M_{\overline{p}} \mathscr{F} = P'_{\max} \, ,$$

using that $\mathscr{F}^* = \overline{\mathscr{F}} = (2\pi)^n \mathscr{F}^{-1}$.

Since $(P_{\max})^* = P'_{\min}$, it follows that $P'_{\max} = P'_{\min}$, showing that the maximal and the minimal operators coincide, for all these multiplication operators and Fourier transformed multiplication operators. $\qquad \square$

Theorem 6.4. *One has for the operators introduced in Theorem 6.3:*

1° $P(D)_{\max}$ *is a bounded operator in* $L_2(\mathbb{R}^n)$ *if and only if* $p(\xi)$ *is bounded, and the norm satisfies*

$$\|P(D)_{\max}\| = \sup\{\,|p(\xi)| \mid \xi \in \mathbb{R}^n\,\}. \qquad (6.10)$$

2° $P(D)_{\max}$ *is selfadjoint in* $L_2(\mathbb{R}^n)$ *if and only if* p *is real.*

3° $P(D)_{\max}$ *has the lower bound*

$$m(P(D)_{\max}) = \inf\{\,\mathrm{Re}\,p(\xi) \mid \xi \in \mathbb{R}^n\,\} \geq -\infty. \qquad (6.11)$$

Proof. 1°. We have from Theorem 12.13 and the subsequent remarks that M_p is a bounded operator in $L_2(\mathbb{R}^n)$ when p is a bounded function on $\mathbb{R}^n$, and that the norm in that case is precisely $\sup\{|p(\xi)| \mid \xi \in \mathbb{R}^n\}$. If p is unbounded on $\mathbb{R}^n$, one has on the other hand that since p is continuous (hence bounded on compact sets), $C_N = \sup\{|p(\xi)| \mid |\xi| \leq N\} \to \infty$ for $N \to \infty$. Now C_N equals the norm of the operator of multiplication by p on $L_2(B(0,N))$. For every $R > 0$, we can by choosing N so large that $C_N \geq R$ find functions $f \in L_2(B(0,N))$ (thereby in $L_2(\mathbb{R}^n)$ by extension by 0) with norm 1 and $\|M_p f\| \geq R$. Thus M_p is an unbounded operator in $L_2(\mathbb{R}^n)$. This shows that M_p is bounded if and only if p is bounded.

Statement 1° now follows immediately by use of the Parseval-Plancherel theorem, observing that $\|P(D)u\|/\|u\| = \|\mathscr{F}P(D)u\|/\|\mathscr{F}u\| = \|p\hat{u}\|/\|\hat{u}\|$ for $u \neq 0$.

2°. Since $M_p = M_{\bar{p}}$ if and only if $p = \bar{p}$ by Theorem 12.13 ff., the statement follows in view of the Parseval-Plancherel theorem.

3°. Since the lower bound of M_p is $m(M_p) = \inf\{\,\mathrm{Re}\,p(\xi) \mid \xi \in \mathbb{R}^n\,\}$ (cf. Exercise 12.36), it follows from the Parseval-Plancherel theorem that $P(D)_{\max}$ has the lower bound (6.11). Here we use that $(P(D)u,u)/\|u\|^2 = (\mathscr{F}P(D)u, \mathscr{F}u)/\|\mathscr{F}u\|^2 = (p\hat{u}, \hat{u})/\|\hat{u}\|^2$ for $u \in D(P(D)_{\max}) \setminus \{0\}$. $\square$

Note that $P(D)_{\max}$ is the zero operator if and only if p is the zero function.

It follows in particular from this theorem that *for all differential operators with constant coefficients on* $\mathbb{R}^n$, *the maximal realization equals the minimal realization;* we have earlier obtained this for first-order operators (cf. Exercise 4.2, where one could use convolution by h_j and truncation), and for $I - \Delta$ (hence for Δ) at the end of Section 5.3.

Since $|\xi|^2$ is real and has lower bound 0, we get as a special case of Theorem 6.4 the result (which could also be inferred from the considerations in Example 5.20):

Corollary 6.5. *The maximal and minimal realizations of* $-\Delta$ *in* $L_2(\mathbb{R}^n)$ *coincide. It is a selfadjoint operator with lower bound 0.*

6.2 Sobolev spaces of arbitrary real order. The Sobolev theorem

One of the applications of Fourier transformation is that it can be used in the analysis of regularity of solutions of differential equations $P(D)u = f$, even when existence or uniqueness results are not known beforehand. In Example 5.20 we found that any solution $u \in \mathscr{S}'$ of $(1 - \Delta)u = f$ with $f \in L_2$ must belong to $H^2(\mathbb{R}^n)$. We shall now consider the Sobolev spaces in relation to the Fourier transformation.

We first introduce some auxiliary weighted L_p-spaces.

Definition 6.6. For each $s \in \mathbb{R}$ and each $p \in [1, \infty]$, we denote by $L_{p,s}(\mathbb{R}^n)$ (or just $L_{p,s}$) the Banach space

$$L_{p,s}(\mathbb{R}^n) = \{\, u \in L_{1,\mathrm{loc}}(\mathbb{R}^n) \mid \langle x \rangle^s u(x) \in L_p(\mathbb{R}^n) \,\}$$
$$\text{with norm } \|u\|_{L_{p,s}} = \|\langle x \rangle^s u(x)\|_{L_p(\mathbb{R}^n)} \,.$$

For $p = 2$, this is a Hilbert space (namely, $L_2(\mathbb{R}^n, \langle x \rangle^{2s} dx)$) with the scalar product

$$(f, g)_{L_{2,s}} = \int_{\mathbb{R}^n} f(x)\overline{g}(x)\langle x \rangle^{2s} \, dx.$$

Note that multiplication by $\langle x \rangle^t$ defines an isometry of $L_{p,s}$ onto $L_{p,s-t}$ for every $p \in [1, \infty]$ and $s, t \in \mathbb{R}$.

One frequently needs the following inequality.

Lemma 6.7 (THE PEETRE INEQUALITY). *For any $s \in \mathbb{R}$,*

$$\langle x - y \rangle^s \leq c_s \langle x \rangle^s \langle y \rangle^{|s|} \qquad \textit{for } s \in \mathbb{R}, \tag{6.12}$$

with a positive constant c_s.

Proof. First observe that

$$1 + |x - y|^2 \leq 1 + (|x| + |y|)^2 \leq c(1 + |x|^2)(1 + |y|^2);$$

this is easily seen to hold with $c = 2$, and with a little more care one can show it with $c = 4/3$. This implies

$$\langle x - y \rangle^s \leq c^{s/2} \langle x \rangle^s \langle y \rangle^s, \qquad\qquad\qquad \text{when } s \geq 0,$$
$$\langle x - y \rangle^s = \frac{\langle x - y \rangle^{-|s|} \langle x - y + y \rangle^{|s|}}{\langle x \rangle^{|s|}} \leq c^{|s|/2} \langle x \rangle^s \langle y \rangle^{|s|}, \qquad \text{when } s \leq 0.$$

Hence (6.12) holds with

$$c_s = c_1^{|s|/2}, \quad c_1 = 4/3. \tag{6.13}$$

□

In the following we shall use M_f again to denote multiplication by f, with domain adapted to varying needs. Because of the inequalities (5.2) we have:

Lemma 6.8. *For $m \in \mathbb{N}_0$, u belongs to $H^m(\mathbb{R}^n)$ if and only if $\hat{u}$ belongs to $L_{2,m}(\mathbb{R}^n)$. The scalar product*

$$(u, v)_{m,\wedge} = (2\pi)^{-n} \int_{\mathbb{R}^n} \hat{u}(\xi)\overline{\hat{v}(\xi)}\langle \xi \rangle^{2m}\, d\xi = (2\pi)^{-n}(\hat{u}, \hat{v})_{L_{2,m}}$$

defines a norm $\|u\|_{m,\wedge} = (u, u)_{m,\wedge}^{\frac{1}{2}}$ equivalent with the norm introduced in Definition 4.5 (cf. (5.2) for C_m):

$$\begin{aligned}
\|u\|_m \leq \|u\|_{m,\wedge} \leq C_m^{\frac{1}{2}} \|u\|_m, \text{ for } m \geq 0, \\
\|u\|_0 = \|u\|_{0,\wedge}.
\end{aligned} \tag{6.14}$$

Proof. In view of the inequalities (5.2) and the Parseval-Plancherel theorem,

$$\begin{aligned}
u \in H^m(\mathbb{R}^n) &\iff \sum_{|\alpha| \leq m} |\xi^\alpha \hat{u}(\xi)|^2 \in L_1(\mathbb{R}^n) \\
&\iff (1 + |\xi|^2)^m |\hat{u}(\xi)|^2 \in L_1(\mathbb{R}^n) \\
&\iff \hat{u} \in L_{2,m}(\mathbb{R}^n).
\end{aligned}$$

The inequalities between the norms follow straightforwardly. $\qquad \square$

The norm $\| \cdot \|_{m,\wedge}$ is interesting since it is easy to generalize to noninteger or even negative values of m. Consistently with Definition 4.5 we introduce (cf. (5.16)):

Definition 6.9. *For each $s \in \mathbb{R}$, the Sobolev space $H^s(\mathbb{R}^n)$ is defined by*

$$H^s(\mathbb{R}^n) = \{u \in \mathscr{S}'(\mathbb{R}^n) \mid \langle \xi \rangle^s \hat{u}(\xi) \in L_2(\mathbb{R}^n)\} = F^{-1} L_{2,s}(\mathbb{R}^n); \tag{6.15}$$

it is a Hilbert space with the scalar product and norm

$$(u, v)_{s,\wedge} = (2\pi)^{-n} \int_{\mathbb{R}^n} \hat{u}(\xi)\overline{\hat{v}(\xi)}\langle \xi \rangle^{2s}\, d\xi, \quad \|u\|_{s,\wedge} = (2\pi)^{-n/2}\|\langle \xi \rangle^s \hat{u}(\xi)\|_{L_2}. \tag{6.16}$$

The Hilbert space property of $H^s(\mathbb{R}^n)$ follows from the fact that $F = (2\pi)^{-n/2}\mathscr{F}$ by definition gives an *isometry*

$$H^s(\mathbb{R}^n) \xrightarrow{\sim} L_{2,s}(\mathbb{R}^n),$$

cf. (5.16) and Definitions 5.16, 6.6. Since $M_{\langle \xi \rangle^s}$ is an isometry of $L_{2,s}(\mathbb{R}^n)$ onto $L_2(\mathbb{R}^n)$, we have the following *commutative diagram of isometries:*

$$\begin{array}{ccc} H^s(\mathbb{R}^n) & \xrightarrow{\ F\ } & L_{2,s}(\mathbb{R}^n) \\[4pt] \Big\downarrow{\scriptstyle \mathrm{Op}(\langle\xi\rangle^s)} & & \Big\downarrow{\scriptstyle M_{\langle\xi\rangle^s}} \\[4pt] L_2(\mathbb{R}^n) & \xrightarrow[\ F\ ]{} & L_2(\mathbb{R}^n) \end{array} \qquad (6.17)$$

where $M_{\langle\xi\rangle^s}F = F\,\mathrm{Op}(\langle\xi\rangle^s)$.

The operator $\mathrm{Op}(\langle\xi\rangle^s)$ will be denoted Ξ^s, and we clearly have:

$$\Xi^s = \mathrm{Op}(\langle\xi\rangle^s)\,, \quad \Xi^{s+t} = \Xi^s\Xi^t \quad \text{for } s,t \in \mathbb{R}\,. \qquad (6.18)$$

Observe that $\Xi^{2M} = (1-\Delta)^M$ when M is integer ≥ 0, whereas Ξ^s is a pseudodifferential operator for other values of s. Note that Ξ^s is an isometry of $H^t(\mathbb{R}^n)$ onto $H^{t-s}(\mathbb{R}^n)$ for all $t \in \mathbb{R}$, when the norms $\|\cdot\|_{t,\wedge}$ and $\|\cdot\|_{t-s,\wedge}$ are used. We now easily find:

Lemma 6.10. *Let $s \in \mathbb{R}$.*

$1°$ Ξ^s *defines a homeomorphism of $\mathscr{S}$ onto $\mathscr{S}$, and of $\mathscr{S}'$ onto $\mathscr{S}'$, with inverse Ξ^{-s}.*

$2°$ $\mathscr{S}$ *is dense in $L_{2,s}$ and in $H^s(\mathbb{R}^n)$. $C_0^\infty(\mathbb{R}^n)$ is likewise dense in these spaces.*

Proof. As noted earlier, $\mathscr{S}$ is dense in $L_2(\mathbb{R}^n)$, since C_0^∞ is so. Since $\langle\xi\rangle^s \in \mathscr{O}_M$, $M_{\langle\xi\rangle^s}$ maps $\mathscr{S}$ continuously into $\mathscr{S}$, and $\mathscr{S}'$ continuously into $\mathscr{S}'$, for all s; and since $M_{\langle\xi\rangle^{-s}}$ clearly acts as an inverse both for $\mathscr{S}$ and $\mathscr{S}'$, $M_{\langle\xi\rangle^s}$ defines a homeomorphism of $\mathscr{S}$ onto $\mathscr{S}$, and of $\mathscr{S}'$ onto $\mathscr{S}'$. By inverse Fourier transformation it follows that Ξ^s defines a homeomorphism of $\mathscr{S}$ onto $\mathscr{S}$, and of $\mathscr{S}'$ onto $\mathscr{S}'$, with inverse Ξ^{-s}. The denseness of $\mathscr{S}$ in L_2 now implies the denseness of $\mathscr{S}$ in $L_{2,s}$ by use of $M_{\langle\xi\rangle^{-s}}$, and the denseness of $\mathscr{S}$ in H^s by use of Ξ^{-s} (cf. the isometry diagram (6.17)). For the last statement, note that the topology of $\mathscr{S}$ is stronger than that of $L_{2,s}$ resp. H^s, any s. An element $u \in H^s$, say, can be approximated by $\varphi \in \mathscr{S}$ in the metric of H^s, and φ can be approximated by $\psi \in C_0^\infty(\mathbb{R}^n)$ in the metric of $\mathscr{S}$ (cf. Lemma 5.9). $\qquad\square$

The statement $2°$ is for s integer ≥ 0 also covered by Theorem 4.10. Note that we now have established continuous injections

$$\mathscr{S} \subset H^{s'} \subset H^s \subset L_2 \subset H^{-s} \subset H^{-s'} \subset \mathscr{S}'\,, \quad \text{for } s' > s > 0\,, \qquad (6.19)$$

so that the H^s-spaces to some extent "fill in" between $\mathscr{S}$ and L_2, resp. between L_2 and $\mathscr{S}'$. However,

$$\mathscr{S} \subsetneqq \bigcap_{s\in\mathbb{R}} H^s \quad\text{and}\quad \mathscr{S}' \supsetneqq \bigcup_{s\in\mathbb{R}} H^s\,, \qquad (6.20)$$

which follows since we correspondingly have that

$$\mathscr{S} \subsetneqq \bigcap_{s \in \mathbb{R}} L_{2,s}\,, \quad \mathscr{S}' \supsetneqq \bigcup_{s \in \mathbb{R}} L_{2,s}\,, \tag{6.21}$$

where the functions in $\bigcap_{s \in \mathbb{R}} L_{2,s}$ of course need not be differentiable, and the elements in $\mathscr{S}'$ are not all functions. More information on $\bigcap_{s \in \mathbb{R}} H^s$ is given below in (6.26). (There exists another scale of spaces where one combines polynomial growth conditions with differentiability, whose intersection resp. union equals $\mathscr{S}$ resp. $\mathscr{S}'$. Exercise 6.38 treats $\mathscr{S}(\mathbb{R})$.)

We shall now study how the Sobolev spaces are related to spaces of continuously differentiable functions; the main result is the Sobolev theorem.

Theorem 6.11 (THE SOBOLEV THEOREM). *Let m be an integer ≥ 0, and let $s > m + n/2$. Then (cf. (C.10))*

$$H^s(\mathbb{R}^n) \subset C_{L_\infty}^m(\mathbb{R}^n)\,, \tag{6.22}$$

with continuous injection, i.e., there is a constant $C > 0$ such that for $u \in H^s(\mathbb{R}^n)$,

$$\sup\{\,|D^\alpha u(x)| \mid x \in \mathbb{R}^n\,,\ |\alpha| \leq m\,\} \leq C\|u\|_{s,\wedge}\,. \tag{6.23}$$

Proof. For $\varphi \in \mathscr{S}$ one has for $s = m + t$, $t > n/2$ and $|\alpha| \leq m$, cf. (5.2),

$$\sup_{x \in \mathbb{R}^n} |D^\alpha \varphi(x)| = \sup |(2\pi)^{-n} \int_{\mathbb{R}^n} e^{ix\cdot\xi} \xi^\alpha \hat{\varphi}(\xi)\,d\xi|$$

$$\leq (2\pi)^{-n} \int_{\mathbb{R}^n} |\hat{\varphi}(\xi)| \langle\xi\rangle^{m+t} \langle\xi\rangle^{-t}\,d\xi \tag{6.24}$$

$$\leq (2\pi)^{-n} \|\hat{\varphi}\|_{L_{2,s}} \left(\int_{\mathbb{R}^n} \langle\xi\rangle^{-2t}\,d\xi\right)^{\frac{1}{2}} = C\|\varphi\|_{s,\wedge}\,,$$

since the integral of $\langle\xi\rangle^{-2t}$ is finite when $t > n/2$. This shows (6.23) for $\varphi \in \mathscr{S}$. When $u \in H^s$, there exists according to Lemma 6.10 a sequence $\varphi_k \in \mathscr{S}$ so that $\|u - \varphi_k\|_{s,\wedge} \to 0$ for $k \to \infty$. By (6.24), φ_k is a Cauchy sequence in $C_{L_\infty}^m(\mathbb{R}^n)$, and since this space is a Banach space, there is a limit $v \in C_{L_\infty}^m(\mathbb{R}^n)$. Both the convergence in H^s and the convergence in $C_{L_\infty}^m$ imply convergence in $\mathscr{S}'$, thus $u = v$ as elements of $\mathscr{S}'$, and thereby as locally integrable functions. This shows the injection (6.22), with (6.23). $\square$

The theorem will be illustrated by an application:

Theorem 6.12. *Let $u \in \mathscr{S}'(\mathbb{R}^n)$ with $\hat{u} \in L_{2,\mathrm{loc}}(\mathbb{R}^n)$. Then one has for $s \in \mathbb{R}$,*

$$\begin{aligned}
\Delta u \in H^s(\mathbb{R}^n) &\iff u \in H^{s+2}(\mathbb{R}^n)\,, \quad and \\
\Delta u \in C_{L_2}^\infty(\mathbb{R}^n) &\iff u \in C_{L_2}^\infty(\mathbb{R}^n)\,.
\end{aligned} \tag{6.25}$$

Here

$$\bigcap_{s \in \mathbb{R}} H^s(\mathbb{R}^n) = C^\infty_{L_2}(\mathbb{R}^n). \tag{6.26}$$

Proof. We start by showing the first line in (6.25). When $u \in H^{s+2}$, then $\Delta u \in H^s$, since $\langle \xi \rangle^s |\xi|^2 \le \langle \xi \rangle^{s+2}$. Conversely, when $\Delta u \in H^s$ and $\hat{u} \in L_{2,\mathrm{loc}}$, then

$$\langle \xi \rangle^s |\xi|^2 \hat{u}(\xi) \in L_2 \quad \text{and} \quad 1_{|\xi| \le 1} \hat{u} \in L_2,$$

which implies that

$$\langle \xi \rangle^{s+2} \hat{u}(\xi) \in L_2,$$

i.e., $u \in H^{s+2}$.

We now observe that

$$C^\infty_{L_2}(\mathbb{R}^n) \subset \bigcap_{s \in \mathbb{N}_0} H^s(\mathbb{R}^n) = \bigcap_{s \in \mathbb{R}} H^s(\mathbb{R}^n)$$

by definition, whereas

$$\bigcap_{s \in \mathbb{R}} H^s(\mathbb{R}^n) \subset C^\infty_{L_\infty}(\mathbb{R}^n) \cap C^\infty_{L_2}(\mathbb{R}^n) \subset C^\infty_{L_2}(\mathbb{R}^n)$$

follows by the Sobolev theorem. These inclusions imply (6.26), and then the validity of the first line in (6.25) for all $s \in \mathbb{R}$ implies the second line. $\quad\square$

Remark 6.13. Theorem 6.12 clearly shows that the Sobolev spaces are very well suited to describe the regularity of solutions of $-\Delta u = f$. The same *cannot* be said of the spaces of continuously differentiable functions, for here we have $u \in C^2(\mathbb{R}^n) \implies \Delta u \in C^0(\mathbb{R}^n)$ without the converse implication being true. An example in dimension $n = 3$ (found in N. M. Günther [G57] page 82 ff.) is the function

$$f(x) = \begin{cases} \dfrac{1}{\log |x|} \left(\dfrac{3x_1^2}{|x|^2} - 1 \right) \chi(x) & \text{for } x \ne 0, \\ 0 & \text{for } x = 0, \end{cases}$$

which is continuous with compact support, and is such that $u = \dfrac{1}{4\pi} \dfrac{1}{|x|} * f$ is in $C^1(\mathbb{R}^3) \setminus C^2(\mathbb{R}^3)$ and solves $-\Delta u = f$ in the distribution sense. (Here $u \in H^2_{\mathrm{loc}}(\mathbb{R}^n)$, cf. Theorem 6.29 later.)

There is another type of (Banach) spaces which is closer to the C^k- spaces than the Sobolev spaces and works well in the study of Δ, namely, the Hölder spaces $C^{k,\sigma}$ with $\sigma \in \,]0,1[$, where

$$C^{k,\sigma}(\Omega) = \left\{ u \in C^k(\Omega) \mid |D^\alpha u(x) - D^\alpha u(y)| \le C|x - y|^\sigma \text{ for } |\alpha| \le k \right\},$$

cf. also Exercise 4.18. Here one finds that $\Delta u \in C^{k,\sigma} \iff u \in C^{k+2,\sigma}$, at least locally. These spaces are useful also in studies of nonlinear problems (but are on the other hand not very easy to handle in connection with Fourier transformation). Elliptic differential equations in $C^{k,\sigma}$- spaces are treated for example in the books of R. Courant and D. Hilbert [CH62], D. Gilbarg and N. Trudinger [GT77]; the key word is "Schauder estimates".

The Sobolev theorem holds also for nice subsets of $\mathbb{R}^n$.

Corollary 6.14. *When* $\Omega = \mathbb{R}^n_+$, *or* Ω *is bounded, smooth and open, then one has for integer* m *and* $l \geq 0$, *with* $l > m + n/2$:

$$H^l(\Omega) \subset C^m_{L_\infty}(\overline{\Omega}), \ \text{ with } \ \sup\{\, |D^\alpha u(x)| \mid x \in \overline{\Omega}, |\alpha| \leq m \,\} \leq C_l \|u\|_l. \quad (6.27)$$

Proof. Here we use Theorem 4.12, which shows the existence of a continuous map $E : H^l(\Omega) \to H^l(\mathbb{R}^n)$ such that $u = (Eu)\big|_\Omega$. When $u \in H^l(\Omega)$, Eu is in $H^l(\mathbb{R}^n)$ and hence in $C^m_{L_\infty}(\mathbb{R}^n)$ by Theorem 6.11; then $u = (Eu)\big|_\Omega \in C^m_{L_\infty}(\overline{\Omega})$, and

$$\sup\big\{\, |D^\alpha u(x)| \mid x \in \overline{\Omega}, |\alpha| \leq m \,\big\} \leq \sup\big\{\, |D^\alpha Eu(x)| \mid x \in \mathbb{R}^n, |\alpha| \leq m \,\big\}$$
$$\leq C_l \|Eu\|_{l,\wedge} \leq C'_l \|u\|_{H^l(\Omega)}. \qquad \square$$

6.3 Dualities between Sobolev spaces. The Structure theorem

We shall now investigate the Sobolev spaces with negative exponents. The main point is that they will be viewed as *dual spaces* of the Sobolev spaces with positive exponent! For the $L_{2,s}$-spaces, this is very natural, and the corresponding interpretation is obtained for the H^s-spaces by application of F^{-1}. We here use the sesquilinear duality; i.e., the dual space is the space of continuous, *conjugate-linear* — also called *antilinear* or *semilinear* — functionals. (See also the discussion after Lemma 12.15. More precisely, we are working with the *antidual* space, but the prefix anti- is usually dropped.)

Theorem 6.15. *Let* $s \in \mathbb{R}$.

$1°$ $L_{2,-s}$ *can be identified with the dual space of* $L_{2,s}$ *by an isometric isomorphism, such that the function* $u \in L_{2,-s}$ *is identified with the functional* $\Lambda \in (L_{2,s})^*$ *precisely when*

$$\int u(\xi)\overline{\varphi}(\xi)\,d\xi = \Lambda(\varphi) \quad \text{for } \varphi \in \mathscr{S}. \qquad (6.28)$$

$2°$ $H^{-s}(\mathbb{R}^n)$ *can be identified with the dual space of* $H^s(\mathbb{R}^n)$, *by an isometric isomorphism, such that the distribution* $u \in H^{-s}(\mathbb{R}^n)$ *is identified with the functional* $\Lambda \in (H^s(\mathbb{R}^n))^*$ *precisely when*

$$\langle u, \overline{\varphi} \rangle = \Lambda(\varphi) \quad for \ \varphi \in \mathscr{S} \, . \tag{6.29}$$

Proof. 1°. When $u \in L_{2,-s}$, it defines a continuous antilinear functional Λ_u on $L_{2,s}$ by

$$\Lambda_u(v) = \int u(\xi)\overline{v}(\xi) \, d\xi \quad for \ v \in L_{2,s} \, ,$$

since

$$|\Lambda_u(v)| = \Big| \int \langle \xi \rangle^{-s} u(\xi) \langle \xi \rangle^s \overline{v}(\xi) \, d\xi \Big| \le \|u\|_{L_{2,-s}} \|v\|_{L_{2,s}}, \tag{6.30}$$

by the Cauchy-Schwarz inequality. Note here that

$$\|\Lambda_u\|_{L_{2,s}^*} = \sup_{v \in L_{2,s} \setminus \{0\}} \frac{|\Lambda_u(v)|}{\|v\|_{L_{2,s}}} = \sup_{\langle \xi \rangle^s v \in L_2 \setminus \{0\}} \frac{|\int \langle \xi \rangle^{-s} u(\xi) \langle \xi \rangle^s \overline{v}(\xi) \, d\xi|}{\|\langle \xi \rangle^s v\|_{L_2}}$$

$$= \|\langle \xi \rangle^{-s} u\|_{L_2} = \|u\|_{L_{2,-s}},$$

by the sharpness of the Cauchy-Schwarz inequality, so the mapping $u \mapsto \Lambda_u$ is an isometry, in particular injective. To see that it is an isometric isomorphism as stated in the theorem, we then just have to show its surjectiveness. So let Λ be given as a continuous functional on $L_{2,s}$; then we get by composition with the isometry $M_{\langle \xi \rangle^{-s}} : L_2 \to L_{2,s}$ a continuous functional

$$\Lambda' = \Lambda M_{\langle \xi \rangle^{-s}}$$

on L_2. Because of the identification of L_2 with its own dual space, there exists a function $f \in L_2$ such that $\Lambda'(v) = (f, v)$ for all $v \in L_2$. Then we have for $v \in L_{2,s}$,

$$\Lambda(v) = \Lambda(\langle \xi \rangle^{-s} \langle \xi \rangle^s v) = \Lambda'(\langle \xi \rangle^s v) = (f, \langle \xi \rangle^s v) = \int f(\xi) \langle \xi \rangle^s \overline{v}(\xi) d\xi \, ,$$

which shows that $\Lambda = \Lambda_u$ with $u = \langle \xi \rangle^s f \in L_{2,-s}$. Since $\mathscr{S}$ is dense in $L_{2,s}$, this identification of u with Λ is determined already by (6.28).

2°. The proof of this part now just consists of a "translation" of all the consideration under 1°, by use of F^{-1} and its isometry properties and homeomorphism properties. □

For the duality between H^{-s} and H^s we shall use the notation

$$\langle u, \overline{v} \rangle_{H^{-s} \, H^s}, \ \langle u, \overline{v} \rangle_{H^{-s},H^s} \text{ or just } \langle u, \overline{v} \rangle, \text{ for } u \in H^{-s}, \ v \in H^s, \tag{6.31}$$

since it coincides with the scalar product in $L_2(\mathbb{R}^n)$ and with the distribution duality, when these are defined. Note that we have shown (cf. (6.30)):

$$|\langle u, \overline{v} \rangle| \le \|u\|_{-s,\wedge} \|v\|_{s,\wedge} \text{ when } u \in H^{-s} \, , \ v \in H^s; \tag{6.32}$$

this is sometimes called the Schwartz inequality (with a "t") after Laurent Schwartz. Observe also (with the notation of 2°):

$$\|u\|_{-s,\wedge} = \|\Lambda_u\|_{(H^s)^*} = \sup\left\{\frac{|\Lambda_u(v)|}{\|v\|_{s,\wedge}} \mid v \in H^s \setminus \{0\}\right\} \qquad (6.33)$$

$$= \sup\left\{\frac{|\langle u, \overline{v}\rangle|}{\|v\|_{s,\wedge}} \mid v \in H^s \setminus \{0\}\right\} = \sup\left\{\frac{|\langle u, \varphi\rangle|}{\|\varphi\|_{s,\wedge}} \mid \varphi \in \mathscr{S} \setminus \{0\}\right\}.$$

Example 6.16. As an example of the importance of "negative Sobolev spaces", consider the variational construction from Theorem 12.18 and its corollary, applied to the situation where $H = L_2(\mathbb{R}^n)$, $V = H^1(\mathbb{R}^n)$ and $a(u, v) = \sum_{j=1}^n (\partial_j u, \partial_j v)_{L_2} = (u, v)_1 - (u, v)_0$. The embedding of H into V^* considered there corresponds exactly to the embedding of $L_2(\mathbb{R}^n)$ into $H^{-1}(\mathbb{R}^n)$! The operator $\widetilde{A}$ then goes from $H^1(\mathbb{R}^n)$ to $H^{-1}(\mathbb{R}^n)$ and restricts to A going from $D(A)$ to $L_2(\mathbb{R}^n)$. We know from the end of Chapter 4 that A acts like $-\Delta$ in the distribution sense, with domain $D(A) = H^1 \cap D(A_{\max})$ dense in H^1 (and clearly, $H^2 \subset D(A)$). Then $\widetilde{A}$, extending A to a mapping from H^1 to H^{-1}, likewise acts like $-\Delta$ in the distribution sense. Finally we have from Theorem 6.12 that $D(A) \subset H^2$, so in fact, $D(A) = H^2$. To sum up, we have inclusions

$$D(A) = H^2 \subset V = H^1 \subset H = L_2 \subset V^* = H^{-1},$$

for the variational realization of $-\Delta$ on the full space $\mathbb{R}^n$. This is the same operator as the one described in Corollary 6.5.

Having the full scale of Sobolev spaces available, we can apply differential operators (with smooth coefficients) without limitations:

Lemma 6.17. *Let $s \in \mathbb{R}$.*

1° For each $\alpha \in \mathbb{N}_0^n$, D^α is a continuous operator from $H^s(\mathbb{R}^n)$ into $H^{s-|\alpha|}(\mathbb{R}^n)$.

2° For each $f \in \mathscr{S}(\mathbb{R}^n)$, the multiplication by f is a continuous operator from $H^s(\mathbb{R}^n)$ into $H^s(\mathbb{R}^n)$.

Proof. 1°. That D^α maps $H^s(\mathbb{R}^n)$ continuously into $H^{s-|\alpha|}(\mathbb{R}^n)$ is seen from the fact that since $|\xi^\alpha| \le \langle\xi\rangle^{|\alpha|}$ (cf. (5.2)),

$$\|D^\alpha u\|_{s-|\alpha|,\wedge} = (2\pi)^{-n/2}\|\langle\xi\rangle^{s-|\alpha|}\xi^\alpha \hat{u}(\xi)\|_0$$

$$\le (2\pi)^{-n/2}\|\langle\xi\rangle^s \hat{u}(\xi)\|_0 = \|u\|_{s,\wedge} \text{ for } u \in H^s(\mathbb{R}^n).$$

2°. Let us first consider integer values of s. Let $s \in \mathbb{N}_0$, then it follows immediately from the Leibniz formula that one has for a suitable constant c'_s:

$$\|fu\|_s \le c'_s \sup\{|D^\alpha f(x)| \mid x \in \mathbb{R}^n, |\alpha| \le s\}\|u\|_s, \qquad (6.34)$$

which shows the continuity in this case. For $u \in H^{-s}(\mathbb{R}^n)$, we now use Theorem 6.15, (6.14) and (6.34):

$$
\begin{aligned}
|\langle fu, \varphi \rangle| = |\langle u, f\varphi \rangle| &\leq \|u\|_{-s,\wedge} \|f\varphi\|_{s,\wedge} \leq \|u\|_{-s,\wedge} C_s^{\frac{1}{2}} \|f\varphi\|_s \\
&\leq \|u\|_{-s,\wedge} C_s^{\frac{1}{2}} c_s' \sup \{\, |D^\alpha f(x)| \mid x \in \mathbb{R}^n \,,\, |\alpha| \leq s \,\} \|\varphi\|_s \\
&\leq \|u\|_{-s,\wedge} C_s^{\frac{1}{2}} c_s' \sup \{\, |D^\alpha f(x)| \mid x \in \mathbb{R}^n \,,\, |\alpha| \leq s \,\} \|\varphi\|_{s,\wedge} \\
&= C \|u\|_{-s,\wedge} \|\varphi\|_{s,\wedge},
\end{aligned}
$$

whereby $fu \in H^{-s}(\mathbb{R}^n)$ with

$$
\|fu\|_{-s,\wedge} \leq C \|u\|_{-s,\wedge}
$$

(cf. (6.33)). This shows the continuity in H^{-s} for s integer ≥ 0.

When s is noninteger, the proof is more technical. We can appeal to convolution (5.36) and use the Peetre inequality (6.12) in the following way: Let $u \in H^s$. Since $f \in \mathscr{S}$, there are inequalities

$$
|\hat{f}(\xi)| \leq C_N' \langle \xi \rangle^{-N}
$$

for all $N \in \mathbb{R}$. Then we get that

$$
\begin{aligned}
\|fu\|_{s,\wedge}^2 &= (2\pi)^{-n} \int_{\mathbb{R}^n} \langle \xi \rangle^{2s} |\widehat{fu}(\xi)|^2 \, d\xi \\
&\leq (2\pi)^{-3n} \int_{\mathbb{R}^n} \langle \xi \rangle^{2s} \left(\int_{\mathbb{R}^n} |\hat{f}(\xi - \eta) \hat{u}(\eta)| \, d\eta \right)^2 d\xi \\
&\leq (2\pi)^{-3n} (C_N')^2 \int_{\mathbb{R}^n} \left(\int_{\mathbb{R}^n} \langle \xi \rangle^s \langle \xi - \eta \rangle^{-N} |\hat{u}(\eta)| \, d\eta \right)^2 d\xi \\
&\leq (2\pi)^{-3n} (C_N')^2 c_s \int_{\mathbb{R}^n} \left(\int_{\mathbb{R}^n} \langle \xi - \eta \rangle^{|s|-N} \langle \eta \rangle^s |\hat{u}(\eta)| \, d\eta \right)^2 d\xi,
\end{aligned}
$$

where we choose N so large that $\langle \zeta \rangle^{|s|-N}$ is integrable, and apply the Cauchy-Schwarz inequality:

$$
\begin{aligned}
&\leq c' \int_{\mathbb{R}^n} \left(\int_{\mathbb{R}^n} \langle \xi - \eta \rangle^{|s|-N} \, d\eta \right) \left(\int_{\mathbb{R}^n} \langle \xi - \eta \rangle^{|s|-N} \langle \eta \rangle^{2s} |\hat{u}(\eta)|^2 \, d\eta \right) d\xi \\
&= c'' \int_{\mathbb{R}^n} \int_{\mathbb{R}^n} \langle \xi - \eta \rangle^{|s|-N} \langle \eta \rangle^{2s} |\hat{u}(\eta)|^2 \, d\eta \, d\xi \\
&= c'' \int_{\mathbb{R}^n} \int_{\mathbb{R}^n} \langle \zeta \rangle^{|s|-N} \langle \eta \rangle^{2s} |\hat{u}(\eta)|^2 \, d\eta \, d\zeta = c''' \|u\|_{s,\wedge}^2. \qquad \square
\end{aligned}
$$

It can sometimes be useful to observe that for m integer > 0, the proof shows that

$$
\|fu\|_m \leq \|f\|_{L_\infty} \|u\|_m + C \sup_{|\beta| \leq m-1} \|D^\beta f\|_{L_\infty} \|u\|_{m-1}. \tag{6.35}
$$

The spaces H^{-s}, $s > 0$, contain more proper distributions, the larger s is taken.

Example 6.18. The δ-distribution satisfies

$$\delta \in H^{-s}(\mathbb{R}^n) \iff s > n/2, \tag{6.36}$$

and its α-th derivative $D^\alpha \delta$ is in H^{-s} precisely when $s > |\alpha| + n/2$. This follows from the fact that $\mathscr{F}(D^\alpha \delta) = \xi^\alpha$ (cf. (5.39)) is in $L_{2,-s}$ if and only if $|\alpha| - s < -n/2$.

For more general distributions we have:

Theorem 6.19. *Let* $u \in \mathscr{E}'(\Omega)$, *identified with a subspace of* $\mathscr{E}'(\mathbb{R}^n)$ *by extension by* 0, *and let* N *be such that for some* C_N,

$$|\langle u, \varphi \rangle| \le C_N \sup \{|D^\alpha \varphi(x)| \mid x \in \mathbb{R}^n, |\alpha| \le N\}, \tag{6.37}$$

for all $\varphi \in C_0^\infty(\mathbb{R}^n)$; *so* u *is of order* N. *Then* $u \in H^{-s}(\mathbb{R}^n)$ *for* $s > N + n/2$.

Proof. We have by Theorem 3.12 and its proof that when $u \in \mathscr{E}'(\mathbb{R}^n)$, then u is of some finite order N, for which there exists a constant C_N such that (6.37) holds (regardless of the location of the support of φ), cf. (3.35). By (6.23) we now get that

$$|\langle u, \overline{\varphi} \rangle| \le C_s' \|\varphi\|_{s,\wedge} \text{ for } s > N + n/2, \text{ when } u \in C_0^\infty(\mathbb{R}^n), \tag{6.38}$$

whereby $u \in H^{-s}$ according to Theorem 6.15 (since $C_0^\infty(\mathbb{R}^n)$ is dense in H^s, cf. Lemma 6.10). $\qquad \square$

Note that both for $\mathscr{E}'$ and for the H^s spaces, the Fourier transformed space consists of (locally square integrable) *functions*. For $\mathscr{E}'$ this follows from Remark 5.19 or Theorem 6.19; for the H^s spaces it is seen from the definition. Then Theorem 6.12 can be applied directly to the elements of $\mathscr{E}'(\mathbb{R}^n)$, and more generally to the elements of $\bigcup_{t \in \mathbb{R}} H^t(\mathbb{R}^n)$.

We can now finally give an easy proof of the structure theorem that was announced in Chapter 3 (around formula (3.17)).

Theorem 6.20 (THE STRUCTURE THEOREM). *Let* Ω *be open* $\subset \mathbb{R}^n$ *and let* $u \in \mathscr{E}'(\Omega)$. *Let* V *be an open neighborhood of* supp u *with* $\overline{V}$ *compact* $\subset \Omega$, *and let* M *be an integer* $> (N + n)/2$, *where* N *is the order of* u *(as in Theorem 6.19). There exists a system of continuous functions* f_α *with support in* V *for* $|\alpha| \le 2M$ *such that*

$$u = \sum_{|\alpha| \le 2M} D^\alpha f_\alpha. \tag{6.39}$$

Moreover, there exists a continuous function g *on* $\mathbb{R}^n$ *such that* $u = (1 - \Delta)^M g$ *(and one can obtain that* $g \in H^{n/2 + 1 - \varepsilon}(\mathbb{R}^n)$ *for any* $\varepsilon > 0$).

Proof. We have according to Theorem 6.19 that $u \in H^{-s}$ for $s = N + n/2 + \varepsilon$ (for any $\varepsilon \in]0,1[$). Now $H^{-s} = \Xi^t H^{t-s}$ for all t. Taking $t = 2M > N + n$, we have that $t - s \geq N + n + 1 - N - n/2 - \varepsilon = n/2 + 1 - \varepsilon$, so that $H^{t-s} \subset C_{L_\infty}^0(\mathbb{R}^n)$, by the Sobolev theorem. Hence

$$H^{-s} = \Xi^{2M} H^{2M-s} = (1 - \Delta)^M H^{t-s} \subset (1 - \Delta)^M C_{L_\infty}^0(\mathbb{R}^n),$$

and then (by the bijectiveness of $I - \Delta = \Xi^2$) there exists a $g \in H^{t-s} \subset H^{n/2+1-\varepsilon} \subset C_{L_\infty}^0$ such that

$$u = (1 - \Delta)^M g = \sum_{|\alpha| \leq M} C_{M,\alpha} D^{2\alpha} g;$$

in the last step we used (5.2). Now let $\eta \in C_0^\infty(V)$ with $\eta = 1$ on a neighborhood of supp u. Then $u = \eta u$, so we have for any $\varphi \in C_0^\infty(\Omega)$:

$$\langle u, \varphi \rangle = \langle u, \eta \varphi \rangle = \langle \sum_{|\alpha| \leq M} C_{M,\alpha} D^{2\alpha} g, \eta \varphi \rangle = \sum_{|\alpha| \leq M} C_{M,\alpha} \langle g, (-D)^{2\alpha}(\eta \varphi) \rangle$$

$$= \sum_{|\alpha| \leq M} \sum_{\beta \leq 2\alpha} C_{M,\alpha} C_{2\alpha,\beta} \langle g, (-D)^{2\alpha-\beta} \eta \, (-D)^\beta \varphi \rangle$$

$$= \sum_{|\alpha| \leq M, \beta \leq 2\alpha} C_{M,\alpha} C_{2\alpha,\beta} \langle D^\beta [(-D)^{2\alpha-\beta} \eta \, g], \varphi \rangle,$$

by Leibniz' formula. This can be rearranged in the form $\langle \sum_{|\beta| \leq 2M} D^\beta f_\beta, \varphi \rangle$ with f_β continuous and supported in V since η and its derivatives are supported in V, and this shows (6.39). $\qquad\square$

As an immediate consequence we get the following result for arbitrary distributions:

Corollary 6.21. *Let Ω be open $\subset \mathbb{R}^n$, let $u \in \mathscr{D}'(\Omega)$ and let Ω' be an open subset of Ω with $\overline{\Omega'}$ compact $\subset \Omega$. Let $\zeta \in C_0^\infty(\Omega)$ with $\zeta = 1$ on Ω', and let N be the order of $\zeta u \in \mathscr{E}'(\Omega)$ (as in Theorem 6.19). When V is a neighborhood of supp ζ in Ω and M is an integer $> (N + n)/2$, then there exists a system of continuous functions with compact support in V such that $\zeta u = \sum_{|\alpha| \leq 2M} D^\alpha f_\alpha$; in particular,*

$$u = \sum_{|\alpha| \leq 2M} D^\alpha f_\alpha \quad on \ \Omega'. \tag{6.40}$$

Based on this corollary and a partition of unity as in Theorem 2.16 one can for any $u \in \mathscr{D}'(\Omega)$ construct a system $(g_\alpha)_{\alpha \in \mathbb{N}_0^n}$ of continuous functions g_α on Ω, which is *locally finite* (only finitely many functions are different from 0 on each compact subset of Ω), such that $u = \sum_{\alpha \in \mathbb{N}_0^n} D^\alpha g_\alpha$.

6.4 Regularity theory for elliptic differential equations

When $P(x, D)$ is an m-th order differential operator (6.5) with symbol (6.6), the part of order m is called the *principal part*:

$$P_m(x, D) = \sum_{|\alpha|=m} a_\alpha(x) D^\alpha , \tag{6.41}$$

and the associated symbol is called the *principal symbol*:

$$p_m(x, \xi) = \sum_{|\alpha|=m} a_\alpha(x) \xi^\alpha ; \tag{6.42}$$

the latter is also sometimes called the *characteristic polynomial*. The operator $P(x, D)$ is said to be *elliptic* on M ($M \subset \mathbb{R}^n$), when

$$p_m(x, \xi) \neq 0 \text{ for } \xi \in \mathbb{R}^n \setminus \{0\}, \text{ all } x \in M . \tag{6.43}$$

(This extends the definition given in (5.41) ff. for constant-coefficient operators.) We recall that the Laplace operator, whose symbol and principal symbol equal $-|\xi|^2$, is elliptic on $\mathbb{R}^n$.

The argumentation in Theorem 6.12 can easily be extended to general elliptic operators with constant coefficients a_α:

Theorem 6.22. 1° *Let $P(D) = \mathrm{Op}(p(\xi))$, where $p(\xi) \in \mathscr{O}_M$ and there exist $m \in \mathbb{R}$, $c > 0$ and $r \geq 0$ such that*

$$|p(\xi)| \geq c \langle \xi \rangle^m \text{ for } |\xi| \geq r. \tag{6.44}$$

For $s \in \mathbb{R}$ one then has: When $u \in \mathscr{S}'$ with $\hat{u} \in L_{2,\mathrm{loc}}$, then

$$P(D)u \in H^s(\mathbb{R}^n) \implies u \in H^{s+m}(\mathbb{R}^n). \tag{6.45}$$

2° *In particular, this holds when $P(D)$ is an elliptic differential operator of order $m \in \mathbb{N}$ with constant coefficients.*

Proof. 1°. That $P(D)u \in H^s(\mathbb{R}^n)$ means that $\langle \xi \rangle^s p(\xi) \hat{u}(\xi) \in L_2(\mathbb{R}^n)$. Therefore we have when $\hat{u}(\xi) \in L_{2,\mathrm{loc}}(\mathbb{R}^n)$, using (6.44):

$$1_{\{|\xi| \geq r\}} \langle \xi \rangle^{s+m} \hat{u}(\xi) \in L_2(\mathbb{R}^n), \quad 1_{\{|\xi| \leq r\}} \hat{u}(\xi) \in L_2(\mathbb{R}^n),$$

and hence that $\langle \xi \rangle^{s+m} \hat{u}(\xi) \in L_2(\mathbb{R}^n)$, i.e., $u \in H^{s+m}(\mathbb{R}^n)$.

2°. Now let $p(\xi)$ be the symbol of an elliptic differential operator of order $m \in \mathbb{N}$, i.e., $p(\xi)$ is a polynomial of degree m, where the principal part $p_m(\xi) \neq 0$ for all $\xi \neq 0$. Then $|p_m(\xi)|$ has a positive minimum on the unit sphere $\{ \xi \in \mathbb{R}^n \mid |\xi| = 1 \}$,

$$c_0 = \min\{ |p_m(\xi)| \mid |\xi| = 1 \} > 0,$$

and because of the homogeneity,

$$|p_m(\xi)| \geq c_0 |\xi|^m \text{ for all } \xi \in \mathbb{R}^n.$$

Since $p(\xi) - p_m(\xi)$ is of degree $\leq m - 1$,

$$\frac{|p(\xi) - p_m(\xi)|}{|\xi|^m} \to 0 \text{ for } |\xi| \to \infty.$$

Choose $r \geq 1$ so that this fraction is $\leq c_0/2$ for $|\xi| \geq r$. Since $\langle \xi \rangle^m \leq 2^{m/2}|\xi|^m$ for $|\xi| \geq 1$, we obtain that

$$|p(\xi)| \geq |p_m(\xi)| - |p(\xi) - p_m(\xi)| \geq \frac{c_0}{2}|\xi|^m \geq \frac{c_0}{2^{1+m/2}}\langle \xi \rangle^m, \text{ for } |\xi| \geq r.$$

This shows (6.44). $\qquad\qquad\qquad\qquad\qquad\qquad\qquad\qquad\qquad\qquad\qquad\square$

Corollary 6.23. *When $P(D)$ is an elliptic differential operator of order m with constant coefficients, one has for each $s \in \mathbb{R}$, when $u \in \mathscr{S}'$ with $\hat{u} \in L_{2,\mathrm{loc}}$:*

$$P(D)u \in H^s(\mathbb{R}^n) \iff u \in H^{s+m}(\mathbb{R}^n).$$

Proof. The implication $\Longleftarrow$ is an immediate consequence of Lemma 6.17, while $\Longrightarrow$ follows from Theorem 6.22. $\qquad\qquad\qquad\qquad\qquad\square$

We have furthermore for the minimal realization, in the case of constant coefficients:

Theorem 6.24. *Let $P(D)$ be elliptic of order m on $\mathbb{R}^n$, with constant coefficients. Let Ω be an open subset of $\mathbb{R}^n$. The minimal realization $P_{\min}$ of $P(D)$ in $L_2(\Omega)$ satisfies*

$$D(P_{\min}) = H_0^m(\Omega) . \qquad\qquad\qquad (6.46)$$

When $\Omega = \mathbb{R}^n$, $D(P_{\min}) = D(P_{\max}) = H^m(\mathbb{R}^n)$, with equivalent norms.

Proof. For $\Omega = \mathbb{R}^n$ we have already shown in Theorem 6.3 that $D(P_{\min}) = D(P_{\max})$, and the identification of this set with $H^m(\mathbb{R}^n)$ follows from Corollary 6.23. That the graph-norm and the H^m-norm are equivalent follows e.g. when we note that by the Parseval-Plancherel theorem,

$$\|u\|_0^2 + \|Pu\|_0^2 = (2\pi)^{-n}(\|\hat{u}\|_0^2 + \|\widehat{Pu}\|_0^2) = (2\pi)^{-n} \int_{\mathbb{R}^n} (1 + |p(\xi)|^2)|\hat{u}(\xi)|^2 \, d\xi \,,$$

and combine this with the estimates in Theorem 6.22, implying that there are positive constants c' and C' so that

$$c'\langle \xi \rangle^{2m} \leq 1 + |p(\xi)|^2 \leq C'\langle \xi \rangle^{2m} \,, \text{ for } \xi \in \mathbb{R}^n.$$

(One could also deduce the equivalence of norms from the easy fact that the graph-norm is dominated by the H^m-norm, and both norms define a

Hilbert space (since $P(D)_{\max}$ is closed). For then the identity mapping $\iota :$
$H^m(\mathbb{R}^n) \to D(P(D)_{\max})$ is both continuous and surjective, hence must be a
homeomorphism by the open mapping principle (Theorem B.15).)

For the assertion concerning the realization in $L_2(\Omega)$ we now observe that
the closures of $C_0^\infty(\Omega)$ in graph-norm and in H^m-norm must be identical;
this shows (6.46). □

For differential operators with variable coefficients it takes some further
efforts to show regularity of solutions of elliptic differential equations. We
shall here just give a relatively easy proof in the case where the principal
part has constant coefficients (a general result is shown in Corollary 7.20
later).

Here we need *locally defined* Sobolev spaces.

Definition 6.25. Let $s \in \mathbb{R}$, and let Ω be open $\subset \mathbb{R}^n$. The space $H^s_{\text{loc}}(\Omega)$ is
defined as the set of distributions $u \in \mathscr{D}'(\Omega)$ for which $\varphi u \in H^s(\mathbb{R}^n)$ for all
$\varphi \in C_0^\infty(\Omega)$ (where φu as usual is understood to be extended by zero outside
Ω).

Concerning multiplication by φ, see Lemma 6.17. The lemma implies that
in order to show that a distribution $u \in \mathscr{D}'(\Omega)$ belongs to $H^s_{\text{loc}}(\Omega)$, it suffices
to show e.g. that $\eta_l u \in H^s(\mathbb{R}^n)$ for each of the functions η_l introduced in
Corollary 2.14 (for a given $\varphi \in C_0^\infty(\Omega)$ one takes l so large that $\operatorname{supp}\varphi \subset K_l$;
then $\varphi u = \varphi \eta_l u$). It is also sufficient in order for $u \in \mathscr{D}'(\Omega)$ to lie in $H^s_{\text{loc}}(\Omega)$
that for any $x \in \Omega$ there exists a neighborhood ω and a nonnegative test
function $\psi \in C_0^\infty(\Omega)$ with $\psi = 1$ on ω such that $\psi u \in H^s(\mathbb{R}^n)$. To see
this, note that for each l, K_{l+1} can be covered by a finite system of such
neighborhoods $\omega_1, \ldots, \omega_N$, and

$$1 \leq \psi_1(x) + \cdots + \psi_N(x) \leq N \quad \text{for} \quad x \in K_{l+1} ,$$

so that

$$\eta_l u = \sum_{j=1}^{N} \frac{\eta_l}{\psi_1 + \cdots + \psi_N} \psi_j u \in H^s(\mathbb{R}^n) .$$

The space $H^s_{\text{loc}}(\Omega)$ is a Fréchet space with the topology defined by the
seminorms

$$p_l(u) = \|\eta_l u\|_{H^s(\mathbb{R}^n)} \quad \text{for} \quad l = 1, 2, \ldots . \tag{6.47}$$

Remark 6.26. For completeness we mention that $H^s_{\text{loc}}(\Omega)$ has the dual space
$H^{-s}_{\text{comp}}(\Omega)$ (which it is itself the dual space of), in a similar way as in Theorem
6.15 (and Exercises 2.4 and 2.8). Here

$$H^t_{\text{comp}}(\Omega) = \bigcup_{l=1}^{\infty} H^t_{K_l} , \tag{6.48}$$

where $H_{K_l}^t$ is the closed subspace of $H^t(\mathbb{R}^n)$ consisting of the elements with support in K_l; the space $H_{\mathrm{comp}}^t(\Omega)$ is provided with the inductive limit topology (Appendix B).

Using Lemma 6.17, we find:

Lemma 6.27. *Let $s \in \mathbb{R}$. When $f \in C^\infty(\Omega)$ and $\alpha \in \mathbb{N}_0^n$, then the operator $u \mapsto f D^\alpha u$ is a continuous mapping of $H_{\mathrm{loc}}^s(\Omega)$ into $H_{\mathrm{loc}}^{s-|\alpha|}(\Omega)$.*

Proof. When $u \in H_{\mathrm{loc}}^s(\Omega)$, one has for each $j = 1, \ldots, n$, each $\varphi \in C_0^\infty(\Omega)$, that

$$\varphi(D_j u) = D_j(\varphi u) - (D_j \varphi) u \in H^{s-1}(\mathbb{R}^n) \,,$$

since $\varphi u \in H^s(\mathbb{R}^n)$ implies $D_j(\varphi u) \in H^{s-1}(\mathbb{R}^n)$, and $D_j \varphi \in C_0^\infty(\Omega)$. Thus D_j maps the space $H_{\mathrm{loc}}^s(\Omega)$ into $H_{\mathrm{loc}}^{s-1}(\Omega)$, and it is found by iteration that D^α maps $H_{\mathrm{loc}}^s(\Omega)$ into $H_{\mathrm{loc}}^{s-|\alpha|}(\Omega)$. Since $f\varphi \in C_0^\infty(\Omega)$ when $\varphi \in C_0^\infty(\Omega)$, we see that $f D^\alpha u \in H_{\mathrm{loc}}^{s-|\alpha|}(\Omega)$. The continuity is verified in the usual way. $\quad\square$

Observe moreover the following obvious consequence of the Sobolev theorem:

Corollary 6.28. *For Ω open $\subset \mathbb{R}^n$ one has:*

$$\bigcap_{s \in \mathbb{R}} H_{\mathrm{loc}}^s(\Omega) = C^\infty(\Omega) \,. \tag{6.49}$$

Now we shall show the regularity theorem:

Theorem 6.29. *Let Ω be open $\subset \mathbb{R}^n$, and let $P = P(x, D)$ be an **elliptic** differential operator of order $m > 0$ on Ω, with constant coefficients in the principal part and C^∞-coefficients in the other terms. Then one has for any $s \in \mathbb{R}$, when $u \in \mathscr{D}'(\Omega)$:*

$$Pu \in H_{\mathrm{loc}}^s(\Omega) \iff u \in H_{\mathrm{loc}}^{s+m}(\Omega); \tag{6.50}$$

in particular,

$$Pu \in C^\infty(\Omega) \iff u \in C^\infty(\Omega) \,. \tag{6.51}$$

Proof. The implication $\impliedby$ in (6.51) is obvious, and it follows in (6.50) from Lemma 6.27. Now let us show $\implies$ in (6.50). It is given that P is of the form

$$P(x, D) = P_m(D) + Q(x, D) \,, \tag{6.52}$$

where $P_m(D) = \mathrm{Op}(p_m(\xi))$ is an elliptic m-th order differential operator with constant coefficients and Q is a differential operator of order $m-1$ with C^∞-coefficients.

Let u satisfy the left-hand side of (6.50), and let $x \in \Omega$. According to the descriptions of $H_{\mathrm{loc}}^t(\Omega)$ we just have to show that there is a neighborhood ω of x and a function $\psi \in C_0^\infty(\Omega)$ which is 1 on ω such that $\psi u \in H^{s+m}(\mathbb{R}^n)$.

We first choose $r > 0$ such that $\underline{B}(x,r) \subset \Omega$. Let $V_j = B(x,r/j)$ for $j = 1,2,\dots$. As in Corollary 2.14, we can for each j find a function $\psi_j \in C_0^\infty(V_j)$ with $\psi_j = 1$ on V_{j+1}. Then in particular, $\psi_j \psi_{j+1} = \psi_{j+1}$.

Since $\psi_1 u$ can be considered as a distribution on $\mathbb{R}^n$ with compact support, $\psi_1 u$ is of finite order, and there exists by Theorem 6.19 a number $M \in \mathbb{Z}$ such that $\psi_1 u \in H^{-M}(\mathbb{R}^n)$. We will show inductively that

$$\psi_{j+1}u \in H^{-M+j}(\mathbb{R}^n) \cup H^{s+m}(\mathbb{R}^n) \quad \text{for } j = 1,2,\dots . \tag{6.53}$$

When j gets so large that $-M + j \geq s + m$, then $\psi_{j+1}u \in H^{s+m}(\mathbb{R}^n)$, and the desired information has been obtained, with $\omega = V_{j+2}$ and $\psi = \psi_{j+1}$.

The induction step goes as follows: Let it be given that

$$\psi_j u \in H^{-M+j-1} \cup H^{s+m} , \text{ and } Pu \in H_{\text{loc}}^s(\Omega) . \tag{6.54}$$

Now we write

$$Pu = P_m(D)u + Q(x,D)u ,$$

and observe that in view of the Leibniz formula, we have for each l:

$$\psi_l Pu = P_m(D)\psi_l u + S_l(x,D)u , \tag{6.55}$$

where $S_l(x,D) = (\psi_l P_m - P_m \psi_l) + \psi_l Q$ is a differential operator *of order* $m-1$, which has *coefficients supported in* $\operatorname{supp}\psi_l \subset V_l$. We then get

$$P_m \psi_{j+1}u = \psi_{j+1}Pu - S_{j+1}u = \psi_{j+1}Pu - S_{j+1}\psi_j u , \tag{6.56}$$

since ψ_j is 1 on V_{j+1}, which contains the supports of ψ_{j+1} and the coefficients of S_{j+1}. According to the given information (6.54) and Lemma 6.27,

$$S_{j+1}\psi_j u \in H^{-M+j-1-m+1} \cup H^{s+m-m+1} = H^{-M+j-m} \cup H^{s+1} ,$$

and $\psi_{j+1}Pu \in H^s$, so that, all taken together,

$$P_m \psi_{j+1}u \in H^{-M+j-m} \cup H^s . \tag{6.57}$$

Now we can apply Corollary 6.23 to P_m, which allows us to conclude that

$$\psi_{j+1}u \in H^{-M+j} \cup H^{s+m} .$$

This shows that (6.54) implies (6.53), and the induction works as claimed.

The last implication in (6.51) now follows from Corollary 6.28.   □

An argumentation as in the above proof is often called a "bootstrap argument", which relates the method to one of the adventures of Münchhausen, where he (on horseback) was stuck in a swamp and dragged himself and the horse up step by step by pulling at his bootstraps.

We get in particular from the case $s = 0$:

Corollary 6.30. *When P is an elliptic differential operator on Ω of order m, with constant coefficients in the principal symbol, then*

$$D(P_{\max}) \subset H^m_{\text{loc}}(\Omega) . \tag{6.58}$$

The corollary implies that the realizations T and T_1 of $-\Delta$ introduced in Theorems 4.27 and 4.28 have domains contained in $H^2_{\text{loc}}(\Omega)$; the so-called "interior regularity". There remains the question of "regularity up to the boundary", which can be shown for nice domains by a larger effort.

The theorem and its corollary can also be shown for elliptic operators with all coefficients variable. Classical proofs in positive integer-order Sobolev spaces use approximation of u by difference quotients (and allow some relaxation of the smoothness assumptions on the coefficients, depending on how high a regularity one wants to show). There is also an elegant modern proof that involves construction of an approximate inverse operator (called a parametrix) by the help of pseudodifferential operator theory. This is taken up in Chapter 7, see Corollary 7.20.

One finds in general that $D(P_{\max})$ is *not* contained in $H^m(\Omega)$ when $\Omega \neq \mathbb{R}^n$ (unless the dimension n is equal to 1); see Exercises 4.5 and 6.2 for examples.

Besides the general *regularity* question for solutions of elliptic differential equations treated above, the question of *existence* of solutions can be conveniently discussed in the framework of Sobolev spaces and Fourier integrals. There is a fairly elementary introduction to partial differential equations building on distribution theory in F. Treves [T75]. The books of L. Hörmander [H83], [H85] (vol. I–IV) can be recommended for those who want a much deeper knowledge of the modern theory of linear differential operators. Let us also mention the books of J.-L. Lions and E. Magenes [LM68] on elliptic and parabolic boundary value problems, the book of D. Gilbarg and N. Trudinger [GT77] on linear and nonlinear elliptic problems in general spaces, and the book of L. C. Evans [E98] on PDE in general; the latter starts from scratch and uses only distribution theory in disguise (speaking instead of weak solvability), and has a large section on nonlinear questions.

Remark 6.31. The theory of elliptic problems has further developments in several directions. Let us point to the following two:

$1°$ *The Schrödinger operator.* Hereby is usually meant a realization of the differential operator $P_V = -\Delta + V$ on $\mathbb{R}^n$, where V is a multiplication operator (by a function $V(x)$ called the potential function). As we have seen (for $V = 0$), $P_0|_{C_0^\infty(\mathbb{R}^n)}$ is essentially selfadjoint in $L_2(\mathbb{R}^n)$ (Corollary 6.5). It is important to define classes of potentials V for which P_V with domain $C_0^\infty(\mathbb{R}^n)$ is essentially selfadjoint too, and to describe numerical ranges, spectra and other properties of these operators. The operators enter in quantum mechanics and in particular in *scattering theory*, where one investigates the connection between $\exp(itP_0)$ and $\exp(itP_V)$ (defined by functional analysis).

2° *Boundary value problems* in dimension $n \geq 2$. One here considers the Laplace operator and other elliptic operators on smooth open subsets Ω of $\mathbb{R}^n$. The statements in Chapter 4 give a beginning of this theory.

One can show that the boundary mapping (also called a trace operator)

$$\gamma_j : u \mapsto \left(\frac{\partial}{\partial n}\right)^j u\big|_{\partial\Omega} ,$$

defined on $C^m(\overline{\Omega})$, can be extended to a continuous map from the Sobolev space $H^m(\Omega)$ to the space $H^{m-j-\frac{1}{2}}(\partial\Omega)$ when $m > j$; here $H^s(\partial\Omega)$ is defined as in Section 6.2 when $\partial\Omega = \mathbb{R}^{n-1}$, and is more generally defined by the help of local coordinates. Theorems 4.17 and 4.25 have in the case $n \geq 2$ the generalization that $H_0^m(\Omega)$ consists of those H^m-functions u for which $\gamma_j u = 0$ for $j = 0, 1, \ldots, m - 1$. Equations for these boundary values can be given a sense when $u \in H^m(\Omega)$. As indicated briefly at the end of Section 4.4 for second-order operators, one can develop a theory of selfadjoint or variational realizations of elliptic operators on Ω determined by boundary conditions. More on this in Chapter 9 for a constant-coefficient case, and in Chapters 7 and 11 for variable-coefficient cases.

For a second-order elliptic operator A we have from Corollary 6.30 that the domains of its realizations are contained in $H_{\text{loc}}^2(\Omega)$. Under special hypotheses concerning the boundary condition and the smoothness of Ω, one can show with a greater effort that the domains are in fact contained in $H^2(\Omega)$; this belongs to the regularity theory for boundary value problems. A particular case is treated in Chapter 9; a technique for general cases is developed in Chapters 10 and 11.

Having such realizations available, one can furthermore discuss evolution equations with a time parameter:

$$\partial_t u(x,t) + Au(x,t) = f(x,t) \text{ for } t > 0,$$
$$u(x,0) = g(x)$$

(with boundary conditions); here the semiboundedness properties of variational operators allow a construction of solutions by use of the semigroup theory established in functional analysis (more about this e.g. in books of K. Yoshida [Y68] and A. Friedman [F69]). Semigroups are in the present book taken up in Chapter 14.

Exercises for Chapter 6

6.1. Show that when $u \in \mathscr{S}'$ with $\hat{u} \in L_{2,\text{loc}}(\mathbb{R}^n)$, then

$$u \in H^s(\mathbb{R}^n) \iff \Delta^2 u \in H^{s-4}(\mathbb{R}^n).$$

(Δ^2 is called the biharmonic operator.)

6.2. For $\varphi(x') \in \mathscr{S}(\mathbb{R}^{n-1})$, we can define the function (with notation as in (A.1) and (A.2))

$$u_\varphi(x', x_n) = \mathscr{F}^{-1}_{\xi' \to x'}(\hat\varphi(\xi')e^{-\langle\xi'\rangle x_n}) \text{ for } x = (x', x_n) \in \mathbb{R}^n_+,$$

by use of Fourier transformation in the x'-variable only.

(a) Show that $(\langle\xi'\rangle^2 - \partial^2_{x_n})(\hat\varphi(\xi')e^{-\langle\xi'\rangle x_n}) = 0$, and hence that $(I-\Delta)u_\varphi = 0$ on $\mathbb{R}^n_+$.

(b) Show that if a sequence φ_k in $\mathscr{S}(\mathbb{R}^{n-1})$ converges in $L_2(\mathbb{R}^{n-1})$ to a function ψ, then u_{φ_k} is a Cauchy sequence in $L_2(\mathbb{R}^n_+)$. (Calculate the norm of $u_{\varphi_k} - u_{\varphi_l}$ by use of the Parseval-Plancherel theorem in the x'-variable.)

(c) Denoting the limit of u_{φ_k} in $L_2(\mathbb{R}^n_+)$ by v, show that v is in the maximal domain for $I - \Delta$ (and for Δ) on $\Omega = \mathbb{R}^n_+$.

(*Comment.* One can show that ψ is the boundary value of v in a general sense, consistent with that of Theorem 4.24. Then if $v \in H^2(\mathbb{R}^n_+)$, ψ must be in $H^1(\mathbb{R}^{n-1})$, cf. Exercise 4.22. So if ψ is taken $\notin H^1(\mathbb{R}^{n-1})$, then $v \notin H^2(\mathbb{R}^n_+)$, and we have an example of a function in the maximal domain which is not in $H^2(\mathbb{R}^n_+)$. The tools for a complete clarification of these phenomena are given in Chapter 9).

6.3. (a) Show that when $u \in \mathscr{E}'(\mathbb{R}^n)$ (or $\bigcup_t H^t(\mathbb{R}^n)$) and $\psi \in C_0^\infty(\mathbb{R}^n)$, then $u * \psi \in C^\infty(\mathbb{R}^n)$. (One can use (5.36).)

(b) Show that when $u \in \mathscr{D}'(\mathbb{R}^n)$ and $\psi \in C_0^\infty(\mathbb{R}^n)$, then $u * \psi \in C^\infty(\mathbb{R}^n)$. (One can write

$$u = \eta_1 u + \sum_{j=1}^\infty (\eta_{j+1} - \eta_j)u,$$

where η_j is as in Corollary 2.14; the sum is locally finite, i.e., finite on compact subsets. Then $u * \psi = \eta_1 u * \psi + \sum(\eta_{j+1} - \eta_j)u * \psi$ is likewise locally finite.)

6.4. Show that the heat equation for $x \in \mathbb{R}^n$,

$$\frac{\partial u(x,t)}{\partial t} - \Delta_x u(x,t) = 0, \qquad t > 0,$$

$$u(x,0) = \varphi(x),$$

for each $\varphi \in \mathscr{S}(\mathbb{R}^n)$ has a solution of the form

$$u(x,t) = c_1 t^{-n/2} \int_{\mathbb{R}^n} \exp(-c_2|x-y|^2/t)\varphi(y)\,dy;$$

determine the constants c_1 and c_2.

6.5. (a) Show that $\delta * f = f$ for $f \in \mathscr{S}(\mathbb{R}^n)$.

(b) Show that the function $H(s)H(t)$ on $\mathbb{R}^2$ (with points (s,t)) satisfies

$$\frac{\partial^2}{\partial s \partial t}H(s)H(t) = \delta \quad \text{in} \quad \mathscr{S}'(\mathbb{R}^2) . \tag{6.59}$$

(c) Show that the function $U(x,y) = H(x+y)H(x-y)$ on $\mathbb{R}^2$ (with points (x,y)) is a solution of the differential equation

$$\frac{\partial^2 U}{\partial x^2} - \frac{\partial^2 U}{\partial y^2} = 2\delta \quad \text{in} \quad \mathscr{S}'(\mathbb{R}^2) . \tag{6.60}$$

(A coordinate change $s = x + y$, $t = x - y$, may be useful.)

(d) Show that when $f(x,y) \in \mathscr{S}(\mathbb{R}^2)$, then $u = \frac{1}{2}U * f$ is a C^∞-solution of

$$\frac{\partial^2 u}{\partial x^2} - \frac{\partial^2 u}{\partial y^2} = f \quad \text{on} \quad \mathbb{R}^2 . \tag{6.61}$$

6.6.   Let $P(D)$ be a differential operator with constant coefficients. A distribution $E \in \mathscr{D}'(\mathbb{R}^n)$ is called a fundamental solution (or an elementary solution) of $P(D)$ if E satisfies

$$P(D)E = \delta .$$

(a) Show that when E is a fundamental solution of $P(D)$ in $\mathscr{S}'$, then

$$P(D)(E * f) = f \quad \text{for} \ \ f \in \mathscr{S}$$

(cf. Exercise 6.5(a)), i.e., the equation $P(D)u = f$ has the solution $u = E * f$ for $f \in \mathscr{S}$.

(b) Find fundamental solutions of $-\Delta$ and of $-\Delta + 1$ in $\mathscr{S}'(\mathbb{R}^3)$ (cf. Section 5.4 and Exercise 5.4).

(c) Show that on $\mathbb{R}^2$, $\frac{1}{2}H(x+y)H(x-y)$ is a fundamental solution of $P(D) = \frac{\partial^2}{\partial x^2} - \frac{\partial^2}{\partial y^2}$. (Cf. Exercise 6.5.)

(*Comment.* Point (c) illustrates the fact that fundamental solutions exist for much more general operators than those whose symbol is invertible, or is so outside a bounded set (e.g., elliptic operators). In fact, Ehrenpreis and Malgrange showed in the 1950s that any nontrivial constant-coefficient partial differential operator has a fundamental solution, see proofs in [R74, Sect. 8.1], or [H83, Sect. 7.3]. The latter book gives many important examples.)

Miscellaneous exercises (exam problems)

The following problems have been used for examinations at Copenhagen University in courses in "Modern Analysis" since the 1980s, drawing on material from Chapters 1–6, 12 and the appendices.

6.7. (Concerning the definition of fundamental solution, see Exercise 6.6.)
(a) Show that when f and g are locally integrable functions on $\mathbb{R}$ with supports satisfying

$$\operatorname{supp} f \subset [a, \infty[\,, \quad \operatorname{supp} g \subset [b, \infty[\,,$$

where a and $b \in \mathbb{R}$, then $f * g$ is a locally integrable function on $\mathbb{R}$, with $\operatorname{supp}(f * g) \subset [a + b, \infty[$. (Write the convolution integral.)
(b) Let λ_1 and $\lambda_2 \in \mathbb{C}$. Find

$$E(x) = (H(x)e^{\lambda_1 x}) * (H(x)e^{\lambda_2 x}),$$

where $H(x)$ is the Heaviside function ($H(x) = 1$ for $x > 0$, $H(x) = 0$ for $x \le 0$).
(c) Let $P(t)$ be a second-order polynomial with the factorization $P(t) = (t - \lambda_1)(t - \lambda_2)$. Show that

$$[(\delta' - \lambda_1 \delta) * (\delta' - \lambda_2 \delta)] * E(x) = \delta,$$

and thereby that E is a fundamental solution of the operator

$$P\left(\frac{d}{dx}\right) = \frac{d^2}{dx^2} - (\lambda_1 + \lambda_2)\frac{d}{dx} + \lambda_1 \lambda_2.$$

(d) Find out whether there exists a fundamental solution of $P\left(\dfrac{d}{dx}\right)$ with support in $]-\infty, 0]$.
(e) Find the solution of the problem

$$(*) \qquad \begin{cases} (P(\frac{d}{dx})u)(x) = f(x) & \text{for } x > 0, \\ u(0) = 0, \\ u'(0) = 0, \end{cases}$$

where f is a given continuous function on $[0, \infty[$.

6.8. Let t denote the vector space of real sequences $\underline{a} = (a_k)_{k \in \mathbb{N}}$. For each $N \in \mathbb{Z}$ one defines $\ell_{2,N}$ as the subspace of t consisting of sequences $\underline{a}$ for which

$$\|a\|_N = \left(\sum_{k \in \mathbb{N}} k^{2N} |a_k|^2\right)^{\frac{1}{2}} < \infty. \tag{6.62}$$

Let $\mathfrak{s}$ denote the set $\bigcap_{N \geq 0} \ell_{2,N}$. Moreover, write

$$\langle \underline{a}, \underline{b} \rangle = \sum_{k \in \mathbb{N}} a_k b_k , \qquad (6.63)$$

when this series is convergent.

(a) Let $\ell_{2,N}$ be provided with the topology determined by the norm $\| \cdot \|_N$, and investigate which of the following properties hold for the topological vector space $\ell_{2,N}$: locally convex, locally bounded, complete, Banach space, Fréchet space.

(b) Let $\mathfrak{s}$ be provided with the topology determined by the sequence of norms $\| \cdot \|_N$, $N = 0, 1, 2, \ldots$, and investigate which of the following properties hold for the topological vector space $\mathfrak{s}$: locally convex, locally bounded, complete, Banach space, Fréchet space.

(c) Let N be an integer ≥ 0. Show that $(\ell_{2,N})^*$ can be identified with the space $\ell_{2,-N}$ in such a way that when $\Lambda \in (\ell_{2,N})^*$ is identified with the sequence $\underline{a} = (a_k)_{k \in \mathbb{N}}$, then

$$\Lambda(\underline{b}) = \langle \underline{a}, \underline{b} \rangle \qquad (6.64)$$

for all $\underline{b} = (b_k)_{k \in \mathbb{N}}$ in $\ell_{2,N}$.

(d) Show that the dual space $\mathfrak{s}^*$ of $\mathfrak{s}$ can be identified with the space $\bigcup_{N \geq 0} \ell_{2,-N}$.

(e) Show that the operator T from $\mathfrak{t}$ into $\mathfrak{t}$ defined by

$$T[(a_k)_{k \in \mathbb{N}}] = \left(\frac{1}{k} a_k + k^3 a_{k+1} \right)_{k \in \mathbb{N}} ,$$

defines a continuous operator from $\mathfrak{s}$ into $\mathfrak{s}$.

6.9.   Let u denote the distribution on $\mathbb{R}$:

$$u = \delta_0 - \delta_1.$$

(a) Show that there exists a continuous function f on $\mathbb{R}$, for which

$$u = f'',$$

and indicate such one.

(b) Show that there exists a triple of continuous functions g_0, g_1 and g_2 on $\mathbb{R}$ with compact support such that

$$u = g_0 + g_1' + g_2'',$$

and find such a triple.

6.10. Let $a(x)$ be a real C^∞-function on $\mathbb{R}$, satisfying

$$c_1 \geq a(x) \geq c_2, \qquad |a'(x)| \leq c_3,$$

for all $x \in \mathbb{R}$, with positive constants c_1, c_2 and c_3. Let S_0 be the operator $-\frac{d}{dx}a\frac{d}{dx} : u \mapsto -(au')'$ with domain $D(S_0) = C_0^\infty(\mathbb{R})$.

(a) Show that S_0 is a symmetric operator in $L_2(\mathbb{R})$ with lower bound 0.

(b) Show that the Friedrichs extension S of S_0 is the operator $-\frac{d}{dx}a\frac{d}{dx}$ with domain $D(S) = H^2(\mathbb{R})$.

(c) Let furthermore $b(x)$ be a real C^∞-function, with $|b(x)| \leq a(x)$ for all x and $b'(x)$ bounded. Let $s_1(u, v)$ be the sesquilinear form

$$s_1(u, v) = \int_\mathbb{R} (a(x) + ib(x))u'(x)\overline{v'(x)}\, dx,$$

defined on $H^1(\mathbb{R}) \subset L_2(\mathbb{R})$. Show that $s_1(u, v)$ satisfies the conditions for application of the Lax–Milgram theorem (with $H = L_2(\mathbb{R})$ and $V = H^1(\mathbb{R})$), and determine the associated operator S_1. Show that its numerical range satisfies

$$\nu(S_1) \subset \{\, z \in \mathbb{C} \mid |\operatorname{Im} z| \leq \operatorname{Re} z \,\}.$$

6.11. Let a and b be real numbers, and let $u(x, y)$ be a function in $L_2(\mathbb{R}^2)$ satisfying the differential equation

$$(a\frac{\partial}{\partial x} + b\frac{\partial^2}{\partial x^2})u + \frac{\partial^2}{\partial y^2}u = f, \qquad (6.65)$$

where f is a function in $L_2(\mathbb{R}^2)$.

(a) Show that if $b > 0$, then $u \in H^2(\mathbb{R}^2)$.

(b) Show that if $b = 0$ and $a \neq 0$, then $u \in H^1(\mathbb{R}^2)$.

From here on, consider (6.65) for u and f in $L_{2,\mathrm{loc}}(\mathbb{R}^2)$.

(c) Let $a = 0$ and $b = -1$. Show that the function $u(x, y) = H(x - y)$ is a solution of (6.65) with $f = 0$, for which $\partial_x u$ and $\partial_y u$ do not belong to $L_{2,\mathrm{loc}}(\mathbb{R}^2)$, and hence $u \notin H^1_{\mathrm{loc}}(\mathbb{R}^2)$. ($H$ denotes the Heaviside function.)

6.12. Let λ denote the topology on $C_0^\infty(\mathbb{R}^n)$ defined by the seminorms $\varphi \mapsto \sup_{x \in \mathbb{R}^n, |\alpha| \leq m} |\partial^\alpha \varphi(x)|$, $\varphi \in C_0^\infty(\mathbb{R}^n)$, with $m = 0, 1, 2, \ldots$.

(a) Show that for $\beta \in \mathbb{N}_0^n$, ∂^β is a continuous mapping of $(C_0^\infty(\mathbb{R}^n), \lambda)$ into $(C_0^\infty(\mathbb{R}^n), \lambda)$.

Let $\mathscr{D}_\lambda'(\mathbb{R}^n)$ denote the dual space of $(C_0^\infty(\mathbb{R}^n), \lambda)$.

(b) Show that $\mathscr{D}_\lambda'(\mathbb{R}^n) \subset \mathscr{D}'(\mathbb{R}^n)$.

(c) Show that for any function in $L_1(\mathbb{R}^n)$, the corresponding distribution belongs to $\mathscr{D}_\lambda'(\mathbb{R}^n)$.

(d) Show that any distribution with compact support on $\mathbb{R}^n$ belongs to $\mathscr{D}'_\lambda(\mathbb{R}^n)$.

(e) Show that every distribution in $\mathscr{D}'_\lambda(\mathbb{R}^n)$ is temperate and even belongs to one of the Sobolev spaces $H^t(\mathbb{R}^n)$, $t \in \mathbb{R}$.

(f) Show that the distribution given by the function 1 is temperate, but does not belong to any of the Sobolev spaces $H^t(\mathbb{R}^n)$, $t \in \mathbb{R}$.

6.13. 1. Let there be given two Hilbert spaces V and H and a continuous linear map J of V into H. Let V_0 be a dense subspace of V. Assume that $J(V)$ is dense in H, and that for any u in V_0 there exists a constant $c_u \in [0, \infty[$ such that $|(u,v)_V| \leq c_u\|Jv\|_H$ for all v in V_0.

(a) Show that J and J^* both are injective.

(b) Show that $J^{*-1}J^{-1}$ is a selfadjoint operator on H.

(c) Show that for u in V_0, $y \mapsto (J^{-1}y, u)_V$ is a continuous linear functional on $J(V)$.

(d) Show that $J(V_0)$ is contained in the domain of $J^{*-1}J^{-1}$.

2. Consider the special case where $H = L_2(\mathbb{R}^2)$, $V = L_{2,1}(\mathbb{R}^2)$, $V_0 = L_{2,2}(\mathbb{R}^2)$ and J is the identity map of $L_{2,1}(\mathbb{R}^2)$ into $L_2(\mathbb{R}^2)$. (Recall Definition 6.6.) Show that $J^*(L_2(\mathbb{R}^2)) = L_{2,2}(\mathbb{R}^2)$, and find $J^{*-1}J^{-1}$.

3. Consider finally the case where $H = L_2(\mathbb{R}^2)$, $V = H^1(\mathbb{R}^2)$, $V_0 = H^2(\mathbb{R}^2)$ and J is the identity map of $H^1(\mathbb{R}^2)$ into $L_2(\mathbb{R}^2)$.
Find $J^{*-1}J^{-1}$.

6.14. Let b denote the function $b(x) = (2\pi)^{-\frac{1}{2}}e^{-\frac{x^2}{2}}$, $x \in \mathbb{R}$.

Let A denote the differential operator $A = \frac{1}{i}\frac{d}{dx} + b$ on the interval $I =]-a, a[\subset \mathbb{R}, a \in]0, \infty]$.

(a) Find the domain of the maximal realization A_{max}.

(b) Find the domain of the minimal realization A_{min}.

(c) Show that A has a selfadjoint realization.

(d) Show that if $\lambda \in \mathbb{R}$, $f \in D(A_{max})$ and $A_{max}f = \lambda f$, then $\bar{f}f$ is a constant function.

(e) Assume that $a = \infty$, i.e., $I = \mathbb{R}$. Show that A_{max} has no eigenvalues.

6.15. Let $n \in \mathbb{N}$ and an open nonempty subset Ω of $\mathbb{R}^n$ be given.
Let $\mathscr{D}'_F(\Omega)$ denote the set of distributions of finite order on Ω.

(a) Show that for β in $\mathbb{N}_0^n$ and Λ in $\mathscr{D}'_F(\Omega)$, $\partial^\beta \Lambda$ is in $\mathscr{D}'_F(\Omega)$.

(b) Show that for f in $C^\infty(\Omega)$ and Λ in $\mathscr{D}'_F(\Omega)$, $f\Lambda$ is in $\mathscr{D}'_F(\Omega)$.

(c) Show that any temperate distribution on $\mathbb{R}^n$ is of finite order.

(d) Give an example of a distribution in $\mathscr{D}'_F(\mathbb{R})$, which is not temperate.

(e) Show that when φ belongs to $C_0^\infty(\mathbb{R}^n)$ and Λ is a distribution on $\mathbb{R}^n$, then $\varphi * \Lambda$ is a distribution of order 0 on $\mathbb{R}^n$.

6.16.   Let b denote the function $b(x) = e^{-\frac{x^2}{2}}$, $x \in [0, 2]$.

· Let A_0 denote the operator in $H = L_2([0, 2])$ with domain

$$D(A_0) = \{f \in C^2([0, 2]) \mid f(0) - 2f'(0) - f'(2) = 0, \, f(2) + e^2 f'(0) + 5f'(2) = 0\}$$

and action $A_0 f = -bf'' + xbf' + f$ for f in $D(A_0)$.

Let V denote the subspace of $\mathbb{C}^4$ spanned by the vectors

$$\begin{pmatrix} \frac{2}{-e^2} \\ 1 \\ 0 \end{pmatrix} \quad \text{and} \quad \begin{pmatrix} 1 \\ -5 \\ 0 \\ 1 \end{pmatrix}.$$

(a) Show that

$$D(A_0) = \left\{ f \in C^2([0, 2]) \;\middle|\; \begin{pmatrix} f(0) \\ f(2) \\ f'(0) \\ f'(2) \end{pmatrix} \in V \right\}.$$

(b) Show that A_0 can be extended to a selfadjoint operator A in H.

6.17.   (a) Show that the equations

$$\mathrm{Pf}\left(\frac{1}{|x|}\right)(\varphi) = \lim_{\varepsilon \to 0+} \left[\int_{-\infty}^{-\varepsilon} \frac{\varphi(x)}{|x|} \, dx + \int_{\varepsilon}^{\infty} \frac{\varphi(x)}{|x|} \, dx + 2\varphi(0) \log \varepsilon \right],$$

for $\varphi \in C_0^\infty(\mathbb{R})$, define a distribution $\mathrm{Pf}(\frac{1}{|x|})$ on $\mathbb{R}$.

(b) Show that $\mathrm{Pf}(\frac{1}{|x|})$ is a temperate distribution of order ≤ 1.

(c) Show that the restriction of $\mathrm{Pf}(\frac{1}{|x|})$ to $\mathbb{R} \setminus \{0\}$ is a distribution given by a locally integrable function on $\mathbb{R} \setminus \{0\}$.

(d) Find the distribution $x \, \mathrm{Pf}(\frac{1}{|x|})$ on $\mathbb{R}$. (Pf is short for pseudo-function.)

(e) Show that there is a constant C such that the Fourier transform of $\mathrm{Pf}(\frac{1}{|x|})$ is the distribution given by the locally integrable function $C - 2\log|\xi|$ on $\mathbb{R}$. (*Comment.* Do not try to guess C, it is not easy. There is more information in Exercise 5.12.)

6.18.   In this exercise we consider the Laplace operator $\Delta = \partial_1^2 + \partial_2^2$ on $\mathbb{R}^2$.

(a) Show that $H_{\mathrm{loc}}^2(\mathbb{R}^2)$ is contained in $C^0(\mathbb{R}^2)$.

(b) Let u be a distribution on $\mathbb{R}^2$. Assume that there exists a continuous function h on $\mathbb{R}^2$ such that $\langle u, \Delta \varphi \rangle = \int_{\mathbb{R}^2} h\varphi \, dx$ for all $\varphi \in C_0^\infty(\mathbb{R}^2)$. Show that there exists a continuous function k on $\mathbb{R}^2$ such that $\langle u, \varphi \rangle = \int_{\mathbb{R}^2} k\varphi \, dx$ for all $\varphi \in C_0^\infty(\mathbb{R}^2)$.

6.19.   Let I denote the interval $]-\pi, \pi[$, and — as usual — let $\mathscr{D}'(I)$ be the space of distributions on I with the weak* topology.

(a) Show that for any given $r \in \,]0, 1]$, the sequence

$$\Big(\tfrac{1}{2\pi} \sum_{n=-N}^{N} r^{|n|} e^{-int}\Big)_{N \in \mathbb{N}}$$

converges to a limit P_r in $\mathscr{D}'(I)$, and that $P_1(\varphi) = \langle P_1, \varphi \rangle = \varphi(0)$, $\varphi \in C_0^\infty(I)$.

(*Hint* for (a) and (b): Put $c_n(\varphi) = \tfrac{1}{2\pi} \int_{-\pi}^{\pi} e^{-in\theta} \varphi(\theta) d\theta$, $\varphi \in C_0^\infty(I)$; you can utilize that $\sum_{n=-\infty}^{\infty} |c_n(\varphi)| < \infty$ when $\varphi \in C_0^\infty(I)$.)

(b) Show that $r \mapsto P_r$ is a continuous map of $]0, 1]$ into $\mathscr{D}'(I)$.

(c) Show that when r converges to 1 from the left, then

$$\frac{1}{2\pi} \int_{-\pi}^{\pi} \frac{1 - r^2}{1 - 2r \cos \theta + r^2} \varphi(\theta) d\theta$$

converges to $\varphi(0)$ for each φ in $C_0^\infty(I)$.

6.20.   Let Λ denote a distribution on $\mathbb{R}$. For any given f in $C(\mathbb{R})$ and x in $\mathbb{R}$ we define $\tau(x)f$ in $C(\mathbb{R})$ by

$$(\tau(x)f)(y) = f(y + x), \quad y \in \mathbb{R}.$$

Define

$$(T\varphi)(x) = \Lambda(\tau(x)\varphi) = \langle \Lambda, \tau(x)\varphi \rangle, \quad x \in \mathbb{R}, \varphi \in C_0^\infty(\mathbb{R}).$$

(a) Show that $T\varphi$ is a continuous function on $\mathbb{R}$ for each φ in $C_0^\infty(\mathbb{R})$.

(b) The space $C(\mathbb{R})$ of continuous functions on $\mathbb{R}$ is topologized by the increasing sequence $(p_n)_{n \in \mathbb{N}}$ of seminorms defined by

$$p_n(f) = \sup_{|x| \le n} |f(x)|, \ n \in \mathbb{N}, \quad f \in C(\mathbb{R}).$$

Show that T is a continuous linear map of $C_0^\infty(\mathbb{R})$ into $C(\mathbb{R})$.

(c) Show that $T(\tau(y)\varphi) = \tau(y)(T\varphi)$ for y in $\mathbb{R}$ and φ in $C_0^\infty(\mathbb{R})$.

(d) Show that every continuous linear map S of $C_0^\infty(\mathbb{R})$ into $C(\mathbb{R})$ with the property that $S(\tau(y)\varphi) = \tau(y)(S\varphi)$ for all y in $\mathbb{R}$ and φ in $C_0^\infty(\mathbb{R})$, is given by $(S\varphi)(x) = \langle M, \tau(x)\varphi \rangle$, $\varphi \in C_0^\infty(\mathbb{R})$, $x \in \mathbb{R}$, for some distribution M on $\mathbb{R}$.

6.21. Let n be a natural number.

The space $C_{L_2}^\infty(\mathbb{R}^n)$ of functions f in $C^\infty(\mathbb{R}^n)$ with $\partial^\alpha f$ in $L_2(\mathbb{R}^n)$ for each multiindex $\alpha \in \mathbb{N}_0^n$ is topologized by the increasing sequence $(\|\cdot\|_k)_{k\in\mathbb{N}_0}$ of norms defined by

$$\|f\|_0^2 = \int_{\mathbb{R}^n} |f(x)|^2\, dx \text{ and } \|f\|_k^2 = \sum_{|\alpha|\le k} \|\partial^\alpha f\|_0^2\,, \; k \in N\,, \; f \in C_{L_2}^\infty(\mathbb{R}^n)\,.$$

(a) Show that any distribution Λ in one of the Sobolev spaces $H^t(\mathbb{R}^n)$, $t \in \mathbb{R}$, by restriction defines a continuous linear functional on $C_{L_2}^\infty(\mathbb{R}^n)$.

(b) Let M be a continuous linear functional on $C_{L_2}^\infty(\mathbb{R}^n)$. Show that there exists one and only one distribution Λ in $\bigcup_{t\in\mathbb{R}} H^t(\mathbb{R}^n)$ such that $\Lambda(\varphi) = M(\varphi)$ when $\varphi \in C_{L_2}^\infty(\mathbb{R}^n)$.

(c) Let M be a linear functional on $C_{L_2}^\infty(\mathbb{R}^n)$. Show that M is continuous if and only if there exists a finite family $(f_i, \alpha_i)_{i\in I}$ of pairs, with $f_i \in L_2(\mathbb{R}^n)$ and $\alpha_i \in \mathbb{N}_0^n$, $i \in I$, such that $M(\varphi) = \sum_{i\in I} \int_{\mathbb{R}^n} f_i \partial^{\alpha_i} \varphi\, dx$ for φ in $C_{L_2}^\infty(\mathbb{R}^n)$.

6.22. Let a be a real number. Let A denote the differential operator on $\mathbb{R}^2$ given by
$$A = 2D_1^4 + a(D_1^3 D_2 + D_1 D_2^3) + 2D_2^4.$$

(a) Show that for an appropriate choice of a, the operator is not elliptic.

In the rest of the problem $a = 1$.

(b) Show that A is elliptic.

(c) Show that the equation $u + Au = f$ has a unique solution in $S'(\mathbb{R}^2)$ for each f in $L_2(\mathbb{R}^2)$, and that the solution is a function in $C^2(\mathbb{R}^2)$.

(d) What is the domain of definition of the maximal realization $A_{\max}$ of A in $L_2(\mathbb{R}^2)$?

6.23. Let — as usual — χ denote a function in $C_0^\infty(\mathbb{R})$ taking values in $[0,1]$ and satisfying

$$\chi(x) = 0 \text{ for } x \notin\,]-2,2[,\quad \chi(x) = 1 \text{ for } x \in [-1,1]\,.$$

Define
$$\kappa_n(x) = \begin{cases} \chi(nx-3), & x < \frac{4}{n}, \\ 1, & \frac{3}{n} < x < 6, \\ \chi(x-6), & 5 < x, \end{cases}$$

for $n = 1,2,3,\dots$.

(a) Explain why κ_n is a well-defined function in $C_0^\infty(\mathbb{R})$ for each n in $\mathbb{N}$. Show that the sequence of functions $(e^{-n}\kappa_n)_{n\in\mathbb{N}}$ converges to 0 in $C_0^\infty(\mathbb{R})$.

(b) Show that there exists no distribution u on $\mathbb{R}$ with the property that the restriction of u to $]0,\infty[$ equals the distribution given by the locally integrable function $x \mapsto e^{\frac{6}{x}}$ on $]0,\infty[$.

6.24. Let $n \in \mathbb{N}$ be given. For an arbitrary function φ on $\mathbb{R}^n$, define $\check{\varphi} = S\varphi$ by $\check{\varphi}(x) = \varphi(-x)$, $x \in \mathbb{R}^n$. For u in $\mathscr{D}'(\mathbb{R}^n)$, define $\check{u}$ by $\langle \check{u}, \varphi \rangle = \langle u, \check{\varphi} \rangle$, $\varphi \in C_0^\infty(\mathbb{R}^n)$. For $\varphi \in C_0^\infty(\mathbb{R}^n)$ and $u \in \mathscr{D}'(\mathbb{R}^n)$ set $u * \varphi = \varphi * u$. Similarly, set $u * \varphi = \varphi * u$ for $\varphi \in \mathscr{S}(\mathbb{R}^n)$ and u in $\mathscr{S}'(\mathbb{R}^n)$. For f in $L_{1,\mathrm{loc}}(\mathbb{R}^n)$, denote the corresponding distribution on $\mathbb{R}^n$ by Λ_f or f.

(a) Show that $(\partial^\alpha \Lambda_f)^\vee = (-\partial)^\alpha \Lambda_{\check{f}}$ for f continuous on $\mathbb{R}^n$ and $\alpha \in \mathbb{N}_0^n$. Show that for u in $\mathscr{E}'(\mathbb{R}^n)$, $\check{u}$ is in $\mathscr{E}'(\mathbb{R}^n)$ with support supp $(\check{u}) = -\mathrm{supp}\,(u)$.

(b) Explain the fact that when f is continuous with compact support in $\mathbb{R}^n$ and α in $\mathbb{N}_0^n$, and φ in $C_0^\infty(\mathbb{R}^n)$, then the distribution $\varphi * \partial^\alpha \Lambda_f$ is given by the function $(\partial^\alpha \varphi) * f$ in $C_0^\infty(\mathbb{R}^n)$. Show that for u in $\mathscr{E}'(\mathbb{R}^n)$ and φ in $C_0^\infty(\mathbb{R}^n)$, $\varphi * u$ is given by a function (that we shall also denote $\varphi * u = u * \varphi$); show that $\varphi \mapsto \varphi * u$ defines a continuous mapping of $C_0^\infty(\mathbb{R}^n)$ into $C_0^\infty(\mathbb{R}^n)$.

(c) Show that for $\varphi \in C_0^\infty(\mathbb{R}^n)$ and $v \in \mathscr{D}'(\mathbb{R}^n)$,

$$\langle (\varphi * v), \psi \rangle = \langle v, \psi * (\Lambda_\varphi)^\vee \rangle$$

when $\psi \in C_0^\infty(\mathbb{R}^n)$. Show that for $u \in \mathscr{E}'(\mathbb{R}^n)$ and $\varphi \in \mathscr{S}(\mathbb{R}^n)$,

$$\langle \varphi * u, \psi \rangle = \langle \Lambda_\varphi, \psi * \check{u} \rangle$$

for $\psi \in C_0^\infty(\mathbb{R}^n)$.

(d) Show that for $u \in \mathscr{E}'(\mathbb{R}^n)$ and $v \in \mathscr{D}'(\mathbb{R}^n)$, the expression $\langle u * v, \psi \rangle = \langle v, \psi * \check{u} \rangle$, $\psi \in C_0^\infty(\mathbb{R}^n)$, defines a distribution $u * v$ in $\mathscr{D}'(\mathbb{R}^n)$, the convolution of u and v; moreover, $v \mapsto u * v$ defines a continuous linear map of $\mathscr{D}'(\mathbb{R}^n)$ into $\mathscr{D}'(\mathbb{R}^n)$.

(e) Show that for $u \in \mathscr{E}'(\mathbb{R}^n)$, $v \in \mathscr{D}'(\mathbb{R}^n)$ and $\alpha \in \mathbb{N}_0^n$,

$$\partial^\alpha (u * v) = (\partial^\alpha u) * v = u * (\partial^\alpha v).$$

(f) Assume in this question that $n = 1$. Find, for $j \in \mathbb{N}_0$, the convolution of the j-th derivative of the distribution $\delta : \varphi \mapsto \varphi(0)$, $\varphi \in C_0^\infty(\mathbb{R})$, and the distribution corresponding to the Heaviside function $H = 1_{]0,\infty[}$.

(g) Show that for u and v in $\mathscr{E}'(\mathbb{R}^n)$, $u * v$ is in $\mathscr{E}'(\mathbb{R}^n)$ with supp $(u * v) \subset$ supp $(u) +$ supp (v). Moreover, the Fourier transformation carries convolution into a product:

$$\mathscr{F}(u * v) = \mathscr{F}(u)\mathscr{F}(v).$$

(One can use here that for u in $\mathscr{E}'(\mathbb{R}^n)$, $\mathscr{F}u$ is given by a function (also denoted $\mathscr{F}u$) in $C^\infty(\mathbb{R}^n)$.)

(h) Show that for u and v in $\mathscr{E}'(\mathbb{R}^n)$ and w in $\mathscr{D}'(\mathbb{R}^n)$,

$$(u * v) * w = u * (v * w),$$

and $\delta * w = w$.

(i) Let $P(D)$ denote a partial differential operator with constant coefficients on $\mathbb{R}^n$. Assume that the distribution v on $\mathbb{R}^n$ is a fundamental solution, i.e., $P(D)v = \delta$. Show that if f is a distribution on $\mathbb{R}^n$, and f — or v — has compact support, then the distribution $f * v$ — or $v * f$ — is a solution u of the equation $P(D)u = f$.

6.25. Let Ω denote $\{(x, y) \in \mathbb{R}^2 \mid x^2 + y^2 < 1\}$. Consider the differential operator A on Ω given by

$$A\varphi = -(1 + \cos^2 x)\varphi''_{x,x} - (1 + \sin^2 x + i\cos^2 y)\varphi''_{y,y}$$
$$+ (2\cos x \sin x)\varphi'_x + (i2\cos y \sin y)\varphi'_y + \varphi,$$

when $\varphi \in C_0^\infty(\Omega)$.

(a) Show that $H^2(\Omega)$ is contained in the domain of the maximal realization $A_{\max}$ of A on $L_2(\Omega)$, and that $H_0^2(\Omega)$ is contained in the domain of the minimal realization $A_{\min}$ of A on $L_2(\Omega)$.

(b) Show that the sesquilinear form

$$\{\varphi, \psi\} \mapsto (A\varphi, \psi), \quad \varphi, \psi \in C_0^\infty(\Omega),$$

has one and only one extension to a bounded sesquilinear form on $H_0^1(\Omega)$, and that this form is $H_0^1(\Omega)$-coercive.

(c) Show that $H^2(\Omega) \cap H_0^1(\Omega)$ is contained in the domain of the corresponding variational operator $\widetilde{A}$.

(d) Show that for (a, b) in $\mathbb{R}^2$ satisfying $|b| > 3a$, $\widetilde{A} - a - ib$ is a bijective map of the domain of $\widetilde{A}$ onto $L_2(\Omega)$, with a bounded inverse.

6.26. Let $Q =]0, 1[\times]0, 1[\subset \mathbb{R}^2$, and consider the sesquilinear form

$$a(u, v) = \int_Q (\partial_1 u \partial_1 \bar{v} + \partial_2 u \partial_2 \bar{v} + u \partial_1 \bar{v}) \, dx_1 dx_2.$$

Let $H = L_2(Q)$, $V_0 = H_0^1(Q)$ and $V_1 = H^1(Q)$, and let, respectively, a_0 and a_1 denote a defined on V_0 resp. V_1. One considers the triples (H, V_0, a_0) and (H, V_1, a_1).

(a) Show that a_0 is V_0-elliptic and that a_1 is V_1-coercive, and explain why the Lax-Milgram construction can be applied to the triples (H, V_0, a_0) and (H, V_1, a_1). The hereby defined operators will be denoted A_0 and A_1.
(*Hint.* One can show that $(u, \partial_1 u)_H$ is purely imaginary, when $u \in C_0^\infty(Q)$.)

(b) Show that A_0 acts like $-\Delta - \partial_1$ in the distribution sense, and that functions $u \in D(A_0) \cap C(\overline{Q})$ satisfy $u|_{\partial Q} = 0$.
(*Hint.* Let $u_k \to u$ in $H_0^1(Q)$, $u_k \in C_0^\infty(Q)$. For a boundary point x which is not a corner, one can by a suitable choice of truncation function reduce the u_k's and u to have support in a small neighborhood $B(x, \delta) \cap \overline{Q}$ and show

that u is 0 as an L_2-function on the interval $B(x, \delta/2) \cap \partial Q$, by inequalities as in Theorem 4.24.)

(c) Show that A_1 acts like $-\Delta - \partial_1$ in the distribution sense, and that functions $u \in D(A_1) \cap C^2(\overline{Q})$ satisfy certain first-order boundary conditions on the edges of Q; find them. (Note that the Gauss and Green's formulas hold on Q, with a piecewise continuous definition of the normal vector at the boundary.)

(d) Show that A_0 has its numerical range (and hence its spectrum) contained in the set

$$\{\lambda \in \mathbb{C} \mid \operatorname{Re}\lambda \geq 2, \quad |\operatorname{Im}\lambda| \leq \sqrt{\operatorname{Re}\lambda}\}.$$

(*Hint.* Note that for $u \in V_0$ with $\|u\|_H = 1$, $|\operatorname{Im} a_0(u, u)| \leq \|D_1 u\|_H$, while $\operatorname{Re} a_0(u, u) \geq \|D_1 u\|_H^2$.)

(e) Investigate the numerical range (and spectrum) of A_1. (One should at least find a convex set as in Corollary 12.21. One may possibly improve this to the set

$$\{\lambda \in \mathbb{C} \mid \operatorname{Re}\lambda \geq -\tfrac{1}{2}, \quad |\operatorname{Im}\lambda| \leq \sqrt{2\operatorname{Re}\lambda + 1}\}.)$$

6.27. Let $\mathcal{L}$ denote the differential operator defined by

$$\mathcal{L}u = -\partial_x(x\partial_x u) + (x+1)u = (1+\partial_x)[x(1-\partial_x)u],$$

for $u \in \mathscr{S}'(\mathbb{R})$.

(a) Find the operator $\widehat{\mathcal{L}}$ that $\mathcal{L}$ carries over to by Fourier transformation, in other words, $\widehat{\mathcal{L}} = \mathscr{F}\mathcal{L}\mathscr{F}^{-1}$.

(b) Show that the functions

$$g_k(\xi) = \frac{(1 - i\xi)^k}{(1 + i\xi)^{k+1}}, \quad k \in \mathbb{Z},$$

satisfy

$$\widehat{\mathcal{L}} g_k = 2(k+1)g_k$$

(hence are eigenfunctions for $\widehat{\mathcal{L}}$ with eigenvalues $2(k+1)$), and that the system $\{\frac{1}{\sqrt{\pi}}g_k\}_{k \in \mathbb{Z}}$ is orthonormal in $L_2(\mathbb{R})$.

(c) Show that $H(x)e^{-x}$ by convolution with itself m times gives

$$H(x)e^{-x} * \cdots * H(x)e^{-x} = \frac{x^m}{m!}H(x)e^{-x} \quad (m+1 \text{ factors}).$$

(d) Show that $\mathcal{L}$ has a system of eigenfunctions

$$f_k(x) = \mathscr{F}^{-1}g_k = p_k(x)H(x)e^{-x}, \quad k \in \mathbb{N}_0,$$

belonging to eigenvalues $2(k+1)$, where each p_k is a polynomial of degree k. Calculate p_k for $k = 0, 1, 2$.

(*Hint.* One can for example calculate $\mathscr{F}^{-1}\frac{1}{1+i\xi}$ and use point (c).)

(e) Show that further eigenfunctions for $\mathcal{L}$ (with eigenalues $2(-m+1)$) are obtained by taking

$$f_{-m}(x) = f_{m-1}(-x), \quad m \in \mathbb{N},$$

and show that the whole system $\{\sqrt{2}\,f_k\}_{k\in\mathbb{Z}}$ is an orthonormal system in $L_2(\mathbb{R})$.

(f) Show that when $\varphi \in \mathscr{S}(\mathbb{R})$, then

$$\sum_{k\in\mathbb{N}_0} |(f_k, \varphi)_{L_2}| \le C\|\mathcal{L}\varphi\|_{L_2},$$

for a suitable constant C.

(*Hint.* A useful inequality can be obtained by applying the Bessel inequality to $\mathcal{L}\varphi$ and observe that $(f_k, \mathcal{L}\varphi) = (\mathcal{L}f_k, \varphi)$.)

(g) Show that $\Lambda = \sum_{k\in\mathbb{N}_0} f_k$ defines a distribution in $\mathscr{S}'(\mathbb{R})$ by the formula

$$\langle \Lambda, \varphi \rangle = \lim_{N\to\infty} \sum_{k=0}^{N} \langle f_k, \varphi \rangle,$$

and that this distribution is supported in $[0, \infty[$.

(*Comment:* The system $\{\sqrt{2}f_k \mid k \in \mathbb{N}_0\}$ on $\mathbb{R}_+$ is a variant of what is usually called the Laguerre orthonormal system; it is complete in $L_2(\mathbb{R}_+)$. It is used for example in the calculus of pseudodifferential boundary operators, see Section 10.2.)

6.28. For $a \in \mathbb{R}_+$, let

$$f_a(x) = \frac{a}{\pi}\frac{1}{x^2 + a^2} \quad \text{for } x \in \mathbb{R}.$$

Show that $f_a \to \delta$ in $H^{-1}(\mathbb{R})$ for $a \to 0+$.

6.29. Consider the partial differential operator in two variables

$$A = D_1^4 + D_2^4 + bD_1^2D_2^2,$$

where b ia a complex constant.

(a) Show that A is elliptic if and only if $b \in \mathbb{C}\backslash\,]-\infty, -2]$. (One can investigate the cases $b \in \mathbb{R}$ and $b \in \mathbb{C}\setminus\mathbb{R}$ separately.)

(b) Show (by reference to the relevant theorems) that the maximal realization and the minimal realization of A on $\mathbb{R}^2$ coincide, and that they in the elliptic cases have domain $H^4(\mathbb{R}^2)$.

(c) Show that $A_{\max}$ can be defined by the Lax-Milgram construction from a sesquilinear form $a(u, v)$ on a suitable subspace V of $H = L_2(\mathbb{R}^2)$ (indicate a and V).

Describe polygonal sets containing the numerical range and the spectrum of $A_{\max}$, on one hand when $b \in\,]-2, \infty[$, on the other hand when $b = \alpha + i\beta$ with $\alpha, \beta \in \mathbb{R}$, $\beta \neq 0$.

6.30.   In the following, $\varphi(x)$ denotes a given function in $L_2(\mathbb{R})$ with compact support.

(a) Show that $\hat{\varphi} \in H^m(\mathbb{R})$ for all $m \in \mathbb{N}$, and that $\hat{\varphi} \in C^\infty(\mathbb{R})$.

(b) Show that when $\psi \in L_1(\mathbb{R})$, then $\sum_{l \in \mathbb{Z}} \psi(\xi + 2\pi l)$ defines a function in $L_1(\mathbb{T})$, and

$$\int_{\mathbb{R}} \psi(\xi)\, d\xi = \int_0^{2\pi} \sum_{l \in \mathbb{Z}} \psi(\xi + 2\pi l)\, d\xi.$$

(We recall that $L_p(\mathbb{T})$ ($1 \leq p < \infty$) denotes the space of (equivalence classes of) functions in $L_{p,\mathrm{loc}}(\mathbb{R})$ with period 2π; it is a Banach space when provided with the norm $\left(\frac{1}{2\pi} \int_0^{2\pi} |\psi(\xi)|^p\, d\xi\right)^{1/p}$.)

Show that the series $\sum_{l \in \mathbb{Z}} |\hat{\varphi}(\xi + 2\pi l)|^2$ converges uniformly toward a continuous function $g(\xi)$ with period 2π.

(*Hint.* Using an inequality from Chapter 4 one can show that

$$\sup_{\xi \in [2\pi l, 2\pi(l+1)]} |\hat{\varphi}(\xi)|^2 \leq c \|1_{[2\pi l, 2\pi(l+1)]}\hat{\varphi}\|^2_{H^1([2\pi l, 2\pi(l+1)])}.)$$

(c) Show the identities, for $n \in \mathbb{Z}$,

$$\int_{\mathbb{R}} \varphi(x - n)\overline{\varphi(x)}\, dx = \frac{1}{2\pi} \int_{\mathbb{R}} |\hat{\varphi}(\xi)|^2 e^{-in\xi}\, d\xi$$

$$= \frac{1}{2\pi} \int_0^{2\pi} \sum_{l \in \mathbb{R}} |\hat{\varphi}(\xi + 2\pi l)|^2 e^{-in\xi}\, d\xi.$$

(d) Show that the following statements (i) and (ii) are equivalent:

 (i) The system of functions $\{\, \varphi(x - n) \mid n \in \mathbb{Z}\,\}$ is an orthonormal system in $L_2(\mathbb{R})$.
 (ii) The function $g(\xi)$ defined in (b) is constant $= 1$.

(*Hint.* Consider the Fourier series of g.)

(*Comment.* The results of this exercise are used in the theory of wavelets.)

6.31.   For each $j \in \mathbb{N}$, define the distribution u_j by

$$u_j = \sum_{k=1}^{2^j - 1} \frac{1}{2^j} \delta_{\frac{k}{2^j}}.$$

Show that $u_j \in \mathscr{D}'(\,]0,1[\,)$, and that $u_j \to 1$ in $\mathscr{D}'(\,]0,1[\,)$ for $j \to \infty$.

6.32. Considering $\mathbb{R}^2$ with coordinates (x,y), denote $(x^2+y^2)^{1/2} = r$. Let $v_1(x,y) = \log r$ and $v_2(x,y) = \partial_x \log r$ for $(x,y) \neq (0,0)$, setting them equal to 0 for $(x,y) = (0,0)$; show that both functions belong to $L_{1,\mathrm{loc}}(\mathbb{R}^2)$. Identifying $\log r$ with v_1 as an element of $\mathscr{D}'(\mathbb{R}^2)$, show that the derivative in the distribution sense $\partial_x \log r$ can be identified with the function v_2, and that both distributions have order 0.

6.33. Let $f \in L_1(\mathbb{R}^n)$ with $\int_{\mathbb{R}^n} f(x)\,dx = 1$. For each $j \in \mathbb{N}$, define f_j by

$$f_j(x) = j^n f(jx).$$

Show that $f_j \to \delta$ in $\mathscr{S}'(\mathbb{R}^n)$.

6.34. Consider the differential operator A in $H = L_2(\mathbb{R}^3_+)$ defined by

$$A = -\partial_1^2 - \partial_2^2 - \partial_3^2 + \partial_2\partial_3 + 1,$$

and the sesquilinear form $a(u,v)$ on $V = H^1(\mathbb{R}^3_+)$ defined by

$$a(u,v) = (\partial_1 u, \partial_1 v) + (\partial_2 u, \partial_2 v) + (\partial_3 u, \partial_3 v) - (\partial_2 u, \partial_3 v) + (u,v),$$

where $\mathbb{R}^3_+ = \{(x_1,x_2,x_3) \mid x_3 > 0\}$.
(a) Show that A is elliptic of order 2, and that a is V-elliptic.
(b) Let A_1 be the variational operator defined from the triple (H,V,a). Show that A_1 is a realization of A.
(c) What is the boundary condition satisfied by the functions $u \in D(A_1) \cap C^\infty_{(0)}(\overline{\mathbb{R}^3_+})$?

6.35. With B denoting the unit ball in $\mathbb{R}^n$, consider the two functions

$$u = 1_B, \quad v = 1_{\mathbb{R}^n \setminus B}.$$

For each of the distributions u, $\partial_1 u$, v and $\partial_1 v$, find out whether it belongs to a Sobolev space $H^s(\mathbb{R}^n)$, and indicate such a space in the affirmative cases.

6.36. Let f be the function on $\mathbb{R}$ defined by

$$f(x) = \begin{cases} 1 & \text{for } |x| > \pi/2, \\ 1 + \cos x & \text{for } |x| \le \pi/2. \end{cases}$$

(a) Find f', f'', f''' and $\hat{f}$. (Recall that $\cos x = \frac{1}{2}(e^{ix} + e^{-ix})$.)
(b) For each of these distributions, determine whether it is an $L_{1,\mathrm{loc}}(\mathbb{R})$-function, and in case not, find what the order of the distribution is.

6.37. Let Ω be the unit disk $B(0,1)$ in $\mathbb{R}^2$ with points denoted (x,y), let $H = L_2(\Omega)$ and let $V = H_0^1(\Omega)$. Consider the sesquilinear form $a(u,v)$ with domain V, defined by

$$a(u,v) = \int_\Omega ((2+x)\partial_x u\, \partial_x \bar{v} + (2+y)\partial_y u\, \partial_y \bar{v})\, dx dy.$$

(a) Show that a is bounded on V and V-coercive. Is it V-elliptic?

(b) Show that the variational operator A_0 defined from the triple (H,V,a) is selfadjoint in H.

(c) Show that A_0 is a realization of a partial differential operator; which one is it?

(d) The functions $u \in D(A_0)$ have boundary value zero; indicate why.

6.38. For each nonnegative integer m, define the space $K^m(\mathbb{R})$ of distributions on $\mathbb{R}$ by

$$K^m(\mathbb{R}) = \{u \in L_2(\mathbb{R}) \mid x^j D^k u(x) \in L_2(\mathbb{R}) \text{ for } j + k \leq m\};$$

here j and k denote nonnegative integers.

(a) Provided with the norm

$$\|u\|_{K^m} = \Big(\sum_{j+k \leq m} \|x^j D^k u(x)\|_{L_2(\mathbb{R})}^2 \Big)^{\frac{1}{2}},$$

$K^m(\mathbb{R})$ is a Hilbert space; indicate why.

(b) Show that $\mathscr{F}(K^m(\mathbb{R})) = K^m(\mathbb{R})$.

(c) Show that $\bigcap_{m \geq 0} K^m(\mathbb{R}) = \mathscr{S}(\mathbb{R})$.

(*Hint.* One can for example make use of Theorem 4.18.)

Part III
Pseudodifferential operators

Chapter 7
Pseudodifferential operators on open sets

7.1 Symbols and operators, mapping properties

The Fourier transform is an important tool in the theory of PDE because of its very convenient property of *replacing differentiation by multiplication* by a polynomial:

$$\mathscr{F}(D^\alpha u) = \xi^\alpha \hat{u},$$

and the fact that $(2\pi)^{-n/2}\mathscr{F}$ defines a unitary operator in $L_2(\mathbb{R}^n)$ with a similar inverse $(2\pi)^{-n/2}\overline{\mathscr{F}}$. We have exploited this for example in the definition of Sobolev spaces of all orders

$$H^s(\mathbb{R}^n) = \{\, u \in \mathscr{S}'(\mathbb{R}^n) \mid \langle\xi\rangle^s \hat{u} \in L_2(\mathbb{R}^n) \,\},$$

used in Chapter 6 to discuss the regularity of the distribution solutions of elliptic equations. For constant-coefficient elliptic operators the Fourier transform is easy to use, for example in the simple case of the operator $I - \Delta$ that has the solution operator

$$\mathrm{Op}\Big(\frac{1}{1+|\xi|^2}\Big) = \mathscr{F}^{-1}\frac{1}{1+|\xi|^2}\mathscr{F}.$$

When we more generally define the operator $\mathrm{Op}(p(\xi))$ with symbol $p(\xi)$ by the formula

$$\mathrm{Op}(p(\xi))u = \mathscr{F}^{-1}(p(\xi)\mathscr{F}u),$$

we have a straightforward composition rule

$$\mathrm{Op}(p(\xi))\,\mathrm{Op}(q(\xi)) = \mathrm{Op}(p(\xi)q(\xi)), \tag{7.1}$$

where composition of operators is turned into multiplication of symbols.

However, these simple mechanisms hold only for x-independent ("constant-coefficient") operators. As soon as one has to deal with differential operators with variable coefficients, the situation becomes much more complicated.

Pseudodifferential operators (ψdo's) were introduced as a tool to handle this, and to give a common framework for partial differential operators and their solution integral operators. A symbol p is now taken to depend (smoothly) on x also, and we define $P \equiv \mathrm{Op}(p(x, \xi)) \equiv p(x, D)$ by

$$
\begin{aligned}
\mathrm{Op}(p(x, \xi))u &= \int_{\mathbb{R}^n} e^{ix \cdot \xi} p(x, \xi) \hat{u} \, d\xi \\
&= \int_{\mathbb{R}^n} \int_{\mathbb{R}^n} e^{i(x-y) \cdot \xi} p(x, \xi) u(y) \, dy d\xi.
\end{aligned}
\tag{7.2}
$$

We here (and in the following) use the notation

$$
d\xi = (2\pi)^{-n} d\xi,
\tag{7.3}
$$

which was first introduced in the Russian literature on the subject. In the second line, the expression for the Fourier transform of u has been inserted. This formula will later be generalized, allowing p to depend on y also, see (7.16). Note that $(Pu)(x) = \{\mathscr{F}_{\xi \to x}^{-1}[p(z, \xi)(\mathscr{F}u)(\xi)]\}_{z=x}$.

With this notation, a differential operator $A = \sum_{|\alpha| \leq d} a_\alpha(x) D^\alpha$ with C^∞ coefficients $a_\alpha(x)$ on $\mathbb{R}^n$ can be written as

$$
A = \sum_{|\alpha| \leq d} a_\alpha(x) D^\alpha = \sum_{|\alpha| \leq d} a_\alpha(x) \mathscr{F}^{-1} \xi^\alpha \mathscr{F} = \mathrm{Op}(a(x, \xi)),
$$

$$
\text{where } a(x, \xi) = \sum_{|\alpha| \leq d} a_\alpha(x) \xi^\alpha, \text{ the symbol of } A.
$$

For operators as in (7.2) we do not have a simple product rule like (7.1). But it is important here that for "reasonable choices" of symbols, one can show something that is approximately as good:

$$
\mathrm{Op}(p(x, \xi)) \, \mathrm{Op}(q(x, \xi)) = \mathrm{Op}(p(x, \xi)q(x, \xi)) + \mathcal{R},
\tag{7.4}
$$

where $\mathcal{R}$ is an operator that is "of lower order" than $\mathrm{Op}(pq)$.

We shall now describe a couple of the reasonable choices, namely, the space S^d of so-called *classical* (or polyhomogeneous) symbols of order d (as systematically presented by Kohn and Nirenberg in [KN65]), and along with it the space $S_{1,0}^d$ (of Hörmander [H67]). We shall go rapidly through the main points in the classical theory without explaining everything in depth; detailed introductions are found e.g. in Seeley [S69], Hörmander [H71], [H85], Taylor [T81], Treves [T80].

In the next two definitions, n and n' are positive integers, Σ is an open subset of $\mathbb{R}^{n'}$ whose points are denoted X, and $d \in \mathbb{R}$. As usual, $\langle \xi \rangle = (1 + |\xi|^2)^{\frac{1}{2}}$.

Definition 7.1. The space $S_{1,0}^d(\Sigma, \mathbb{R}^n)$ of symbols of degree d and type $1, 0$ is defined as the set of functions $p(X, \xi) \in C^\infty(\Sigma \times \mathbb{R}^n)$ such that for all indices $\alpha \in \mathbb{N}_0^n$ and $\beta \in \mathbb{N}_0^{n'}$ and any compact set $K \subset \Sigma$, there is a constant $c_{\alpha,\beta,K}$ such that

$$|D_X^\beta D_\xi^\alpha p(X,\xi)| \le c_{\alpha,\beta,K}\langle\xi\rangle^{d-|\alpha|}. \tag{7.5}$$

When $p \in S_{1,0}^{m_0}(\Sigma,\mathbb{R}^n)$ and there exists a sequence of symbols p_{m_j}, $j \in \mathbb{N}_0$, with $p_{m_j} \in S_{1,0}^{m_j}(\Sigma,\mathbb{R}^n)$, $m_j \searrow -\infty$, such that $p - \sum_{j<M} p_{m_j} \in S_{1,0}^{m_M}(\Sigma,\mathbb{R}^n)$ for all M, we say that p has the asymptotic expansion $\sum_{j\in\mathbb{N}_0} p_{m_j}$, in short, $p \sim \sum_j p_{m_j}$ in $S_{1,0}^{m_0}(\Sigma,\mathbb{R}^n)$.

Actually, [H67] also introduces some more general symbol spaces $S_{\varrho,\delta}^d$, where $0 \le \delta \le 1$, $0 \le \varrho \le 1$, and the estimates (7.5) are replaced by estimates

$$|D_X^\beta D_\xi^\alpha p(X,\xi)| \le c_{\alpha,\beta,K}\langle\xi\rangle^{d-\varrho|\alpha|+\delta|\beta|}.$$

Many of the results we discuss in the following are also valid for these spaces, when $0 \le 1-\varrho \le \delta < \varrho \le 1$. We shall not pursue the study of symbol spaces of type ϱ,δ here.

A prominent example of a function in $S_{1,0}^d(\Sigma,\mathbb{R}^n)$ is a function $p_d(X,\xi) \in C^\infty(\Sigma \times \mathbb{R}^n)$, which is *positively homogeneous of degree d in ξ for $|\xi| \ge 1$*, i.e., satisfies

$$p_d(X,t\xi) = t^d p_d(X,\xi) \text{ for } |\xi| \ge 1,\ t \ge 1. \tag{7.6}$$

For such a function we have:

$$|p_d(X,\xi)| = |\xi|^d |p_d(X,\xi/|\xi|)| \le c(X)\langle\xi\rangle^d \text{ for } |\xi| \ge 1,$$

and its α-th derivative in ξ is homogeneous of degree $d-|\alpha|$, hence bounded by $c(X)\langle\xi\rangle^{d-|\alpha|}$, for $|\xi| \ge 1$. (For the latter homogeneity, note that $\partial_{\xi_j} p_d(X,\xi)$ $= \partial_{\xi_j}(t^{-d} p_d(X,t\xi)) = t^{-d+1}(\partial_{\xi_j} p_d)(X,t\xi)$.)

Definition 7.2. The space $S^d(\Sigma,\mathbb{R}^n)$ of **polyhomogeneous symbols of degree d** is defined as the set of symbols $p(X,\xi) \in S_{1,0}^d(\Sigma,\mathbb{R}^n)$ for which there exists a sequence of functions $p_{d-l}(X,\xi) \in C^\infty(\Sigma \times \mathbb{R}^n)$ for $l \in \mathbb{N}_0$, satisfying (i) and (ii):

(i) Each p_{d-l} is positively homogeneous of degree $d-l$ in ξ for $|\xi| \ge 1$,

(ii) p has the asymptotic expansion

$$p(X,\xi) \sim \sum_{l\in\mathbb{N}_0} p_{d-l}(X,\xi) \text{ in } S_{1,0}^d(\Sigma,\mathbb{R}^n); \tag{7.7}$$

in other words, for any compact set $K \subset \Sigma$, any multiindices $\alpha \in \mathbb{N}_0^n$, $\beta \in \mathbb{N}_0^{n'}$ and any $M \in \mathbb{N}_0$, there is a constant $c_{\alpha,\beta,M,K}$ such that for all $(X,\xi) \in K \times \mathbb{R}^n$,

$$|D_X^\beta D_\xi^\alpha [p(X,\xi) - \sum_{0\le l<M} p_{d-l}(X,\xi)]| \le c_{\alpha,\beta,M,K}\langle\xi\rangle^{d-|\alpha|-M}. \tag{7.8}$$

These symbols are also often called *classical* in the literature. Some authors call them *one-step polyhomogeneous* to underline that the degree of homogeneity fall in steps of length 1 (noninteger or varying steps could be needed in other contexts).

The leading term $p_d(X,\xi)$ is called *the principal symbol*, also denoted $p^0(X,\xi)$; the term $p_{d-l}(X,\xi)$ is called the symbol (or term) of degree $d-l$; and the series $\sum_{l=M}^{\infty} p_{d-l}(X,\xi)$ is called the symbol of degree $\leq d - M$ (of p). (For ψdo's we use the word *degree* interchangeably with *order*, where the latter reflects their continuity properties, see Theorem 7.5 below.) From a given symbol $p(X,\xi) \in S^d(\Sigma,\mathbb{R}^n)$ one can determine the terms of degree $d-l$ successively by the formulas

$$p_{d-l}(X,\xi) = \lim_{t\to\infty} (t^{-d+l}[p(X,t\xi) - \sum_{j<l} p_{d-j}(X,t\xi)]), \text{ for } |\xi| \geq 1. \quad (7.9)$$

In view of the estimates (7.8), this convergence is uniform, locally in X and in ξ.

Observe that the series $\sum_{l\in\mathbb{N}_0} p_{d-l}$ is by no means assumed to be convergent; it is an *asymptotic series*, and its connection with p is described in a precise way in (7.8). It is important to know that there holds the following "reconstruction lemma" for general $S^d_{1,0}(\Sigma,\mathbb{R}^n)$ symbols:

Lemma 7.3. *For any sequence of symbols $p_{m_j}(X,\xi)$ in $S^{m_j}_{1,0}(\Sigma,\mathbb{R}^n)$, $m_j \searrow -\infty$, there exists a function $p(X,\xi)$ such that $p \sim \sum_j p_{m_j}$ in $S^{m_0}_{1,0}(\Sigma,\mathbb{R}^n)$.*

For the proof, one takes

$$p(X,\xi) = \sum_{j\in\mathbb{N}_0} p_{m_j}(X,\xi)(1 - \chi(\varepsilon_j\xi)), \quad (7.10)$$

where χ is our usual cut-off function, and ε_j goes to zero sufficiently rapidly for $j \to \infty$. Details are given in [S69] and [H71], [H85], see e.g. [H85, Prop. 18.1.3]. There is also a proof in [S91, Lemma 2.2]. The construction is a generalization of an old construction by Borel of a C^∞-function with arbitrarily given Taylor coefficients at a point.

A simple but important example of a symbol in $S^d(\mathbb{R}^n,\mathbb{R}^n)$ with $d=1$ is the function $\langle\xi\rangle = (1+|\xi|^2)^{\frac{1}{2}}$, which for $|\xi| > 1$ is the sum of a convergent series

$$\langle\xi\rangle = |\xi|(1 + \tfrac{1}{2}|\xi|^{-2} - \tfrac{1}{8}|\xi|^{-4} + \cdots + \binom{\frac{1}{2}}{j}|\xi|^{-2j} + \cdots), \quad (7.11)$$

where $\binom{s}{j} = s(s-1)\cdots(s-j+1)/j!$. Then $\langle\xi\rangle$ has the asymptotic expansion

$$\langle\xi\rangle \sim \eta(\xi)|\xi| + \tfrac{1}{2}\eta(\xi)|\xi|^{-1} - \tfrac{1}{8}\eta(\xi)|\xi|^{-3} + \cdots + \binom{\frac{1}{2}}{j}\eta(\xi)|\xi|^{1-2j} + \cdots, \quad (7.12)$$

where $\eta(\xi) = 1 - \chi(2\xi)$ was inserted to make the terms smooth near 0.

The space $S_{1,0}^d(\Sigma, \mathbb{R}^n)$ is a Fréchet space with the seminorms defined as the least constants entering in (7.5) for each choice of α, β and K (the K can be replaced by an exhausting sequence $K_j \to \Sigma$ defined as in Lemma 2.2). Similarly, the space $S^d(\Sigma, \mathbb{R}^n)$ is a Fréchet space with the seminorms defined as the least constants entering in (7.8) for each choice of α, β, M and K. Clearly,

$$S_{1,0}^d(\Sigma, \mathbb{R}^n) \subset S_{1,0}^{d'}(\Sigma, \mathbb{R}^n) \text{ when } d' > d,$$

$$S^d(\Sigma, \mathbb{R}^n) \subset S^{d'}(\Sigma, \mathbb{R}^n) \text{ when } d' - d \in \mathbb{N}_0.$$

We can define

$$S_{1,0}^\infty(\Sigma, \mathbb{R}^n) = \bigcup_{d \in \mathbb{R}} S_{1,0}^d(\Sigma, \mathbb{R}^n), \quad S^\infty(\Sigma, \mathbb{R}^n) = \bigcup_{d \in \mathbb{R}} S^d(\Sigma, \mathbb{R}^n),$$

$$S_{1,0}^{-\infty}(\Sigma, \mathbb{R}^n) = \bigcap_{d \in \mathbb{R}} S_{1,0}^d(\Sigma, \mathbb{R}^n) = \bigcap_{d \in \mathbb{R}} S^d(\Sigma, \mathbb{R}^n) = S^{-\infty}(\Sigma, \mathbb{R}^n). \tag{7.13}$$

The symbols can be $(N' \times N)$-matrix formed; then the symbol space will be indicated by $S^d(\Sigma, \mathbb{R}^n) \otimes \mathcal{L}(\mathbb{C}^N, \mathbb{C}^{N'})$ or just $S^d(\Sigma, \mathbb{R}^n)$. The norm in (7.5) or (7.8) then stands for a matrix norm (some convenient norm on $\mathcal{L}(\mathbb{C}^N, \mathbb{C}^{N'})$, chosen once and for all).

In Definition 7.2, condition (ii) can be replaced by the following equivalent condition (ii'):

(ii') For any indices $\alpha \in \mathbb{N}_0^n, \beta \in \mathbb{N}_0^{n'}$ and $M \in \mathbb{N}_0$, there is a continuous function $c(X)$ on Σ (depending on α, β and M but not on ξ) so that

$$|D_X^\beta D_\xi^\alpha [p(X, \xi) - \sum_{l < M} p_{d-l}(X, \xi)]| \leq c(X)\langle\xi\rangle^{d-|\alpha|-M}; \tag{7.14}$$

a formulation we shall often use. Similarly, we can reformulate the estimates (7.5) in the form

$$|D_X^\beta D_\xi^\alpha p(X, \xi)| \leq c(X)\langle\xi\rangle^{d-|\alpha|}. \tag{7.15}$$

In the case where $\Sigma = \mathbb{R}^{n'}$, one can instead work with more restrictive symbol classes where the estimates in (7.5), (7.8) or (7.14) (local in X) are replaced by global estimates on $\mathbb{R}^{n'}$ (with constants independent of K or X); this is done in [H85, Sect. 18.1], in [S91] and in [G96]. The basic calculations such as proofs of composition rules are somewhat harder in that case than in the case we consider here, but the global calculus has the advantage that the rules can be made exact, without remainder terms. One can get the local calculus from the global calculus by use of cut-off functions.

Besides the need to construct a symbol with a given asymptotic series, we shall also sometimes need to *rearrange* an asymptotic series. For example, let $p \sim \sum_{l \in \mathbb{N}_0} p_{d-l}$ in $S_{1,0}^d$, where $p_{d-l} \in S_{1,0}^{d-l}$, and assume that each p_{d-l} has

an asymptotic expansion $p_{d-l} \sim \sum_{k \in \mathbb{N}_0} p_{d-l,k}$ in $S_{1,0}^{d-l}$ with $p_{d-l,k} \in S_{1,0}^{d-l-k}$.
Then we also have that

$$p \sim \sum_{l \in \mathbb{N}_0} q_{d-l}, \text{ where } q_{d-l} = \sum_{j+k=l} p_{d-j,k}.$$

In fact, $q_{d-l} = \sum_{j+k=l} p_{d-j,k}$ is a finite sum of terms in $S_{1,0}^{d-l}$ for each l, and
$p - \sum_{l<M} q_{d-l}$ is the sum of $p - \sum_{l<M} p_{d-l} \in S_{1,0}^{d-M}$ and finitely many "tails"
$p_{d-j} - \sum_{k<M-j} p_{d-j,k} \in S_{1,0}^{d-M}$. This is useful e.g. if p is given as a series
of polyhomogeneous symbols of decreasing orders, and we want to rearrange
the terms, collecting those that have the same degree of homogeneity.

We now specialize Σ somewhat. When $n = n'$, i.e., Σ is an open sub-
set of $\mathbb{R}^n$ (with points x) and $p(x, \xi) \in S_{1,0}^d(\Sigma, \mathbb{R}^n)$, then $p(x, \xi)$ defines a
pseudodifferential operator $P \equiv \mathrm{Op}(p(x, \xi)) \equiv p(x, D)$ by the formula (7.2),
considered e.g. for $u \in C_0^\infty(\Sigma)$ or $u \in \mathscr{S}(\mathbb{R}^n)$.

Another interesting case is when $\Sigma = \Omega_1 \times \Omega_2$, Ω_1 and Ω_2 open $\subset \mathbb{R}^n$ (so
$n' = 2n$), the points in Σ denoted (x, y). Here a symbol $p(x, y, \xi)$ defines an
operator P, also denoted $\mathrm{Op}(p(x, y, \xi))$, by the formula

$$(Pu)(x) = \int e^{i(x-y) \cdot \xi} p(x, y, \xi) u(y) \, dy d\xi, \tag{7.16}$$

for $u \in C_0^\infty(\Omega_2)$. This generalizes the second line in (7.2). (The functions
$p(x, y, \xi)$ are in some texts called amplitude functions, see Remark 7.4 below.)

The integration over ξ is defined in the sense of oscillatory integrals (cf.
[H71], [H85], [S91]). A brief explanation goes as follows: When $d < -n$, the
integrand in (7.16) is in L_1 since it is $O(\langle\xi\rangle^d)$, so the integral has the usual
meaning. Otherwise, insert a convergence factor $\chi(\varepsilon\xi)$, and let $\varepsilon \to 0$ (note
that then $\chi(\varepsilon x) \to 1$ pointwise). The limit exists and can be found as follows:
Inserting

$$e^{-iy \cdot \xi} = (1 + |\xi|^2)^{-N} (1 - \Delta_y)^N e^{-iy \cdot \xi} \tag{7.17}$$

(with N so large that $d - 2N < -n$) and integrating by parts with respect
to y, we see that when $u \in C_0^\infty(\Omega_2)$,

$$
\begin{aligned}
(Pu)(x) &= \lim_{\varepsilon \to 0} \int \chi(\varepsilon\xi) e^{i(x-y) \cdot \xi} p(x, y, \xi) u(y) \, dy d\xi \\
&= \lim_{\varepsilon \to 0} \int \chi(\varepsilon\xi) \langle\xi\rangle^{-2N} [(1 - \Delta_y)^N e^{i(x-y) \cdot \xi}] p(x, y, \xi) u(y) \, dy d\xi \\
&= \lim_{\varepsilon \to 0} \int \chi(\varepsilon\xi) e^{i(x-y) \cdot \xi} \langle\xi\rangle^{-2N} (1 - \Delta_y)^N [p(x, y, \xi) u(y)] \, dy d\xi \\
&= \int e^{i(x-y) \cdot \xi} \langle\xi\rangle^{-2N} (1 - \Delta_y)^N [p(x, y, \xi) u(y)] \, dy d\xi,
\end{aligned}
\tag{7.18}
$$

where the limit exists since the integrand in the third line equals $\chi(\varepsilon\xi)$ times an L_1-function, and $\chi(\varepsilon\xi) \to 1$ boundedly (equals 1 for $|\xi| \leq 1/\varepsilon$). The last expression defines a continuous function of x; it is independent of N since N was arbitrarily chosen ($> \frac{1}{2}(d+n)$). Thus we can use the last expression as the definition of Pu (for any large N). For $2N > n+d+m$, it allows differentiation with respect to x of order up to m, carried through the integral sign; since m can be taken arbitrarily large, we can conclude that $Pu \in C^\infty(\Omega_1)$ (when $u \in C_0^\infty(\Omega_2)$). One can also verify that P is continuous from $C_0^\infty(\Omega_2)$ to $C^\infty(\Omega_1)$ on the basis of these formulas.

Remark 7.4. Symbols of the form $p(x,y,\xi)$ are sometimes called amplitude functions, to distinguish them from the sharper notion of symbols $p(x,\xi)$, since they are far from uniquely determined by the operator (as shown below around (7.28)). We shall stay with the more vague terminology where everything is called a symbol, but distinguish by speaking of symbols "in x-form" (symbols $p(x,\xi)$), "in y-form" (symbols $p(y,\xi)$), "in (x,y)-form" (symbols $p(x,y,\xi)$). Moreover, we can speak of for example symbols "in (x',y_n)-form" (symbols $p(x',y_n,\xi)$), etc.

In the following we shall use (without further argumentation) that the occurring integrals all have a sense as oscillatory integrals. Oscillatory integrals have many of the properties of usual integrals, allowing change of variables, change of order of integration (Fubini theorems), etc.

When Ω_1 and $\Omega_2 \subset \mathbb{R}^n$, the notation $S^d(\Omega_1, \mathbb{R}^n)$ and $S^d(\Omega_1 \times \Omega_2, \mathbb{R}^n)$ is often abbreviated to $S^d(\Omega_1)$ resp. $S^d(\Omega_1 \times \Omega_2)$, and the space of operators defined from these symbols is denoted $\operatorname{Op} S^d(\Omega_1)$ resp. $\operatorname{Op} S^d(\Omega_1 \times \Omega_2)$ (with similar notation for $S_{1,0}^d$).

The pseudodifferential operators have the continuity property with respect to Sobolev spaces:

$$
\begin{aligned}
&P : H_{\text{comp}}^s(\Omega_2) \to H_{\text{loc}}^{s-d}(\Omega_1) \quad \text{continuously, when} \\
&P = \operatorname{Op}\big(p(x,y,\xi)\big)\,,\ p \in S_{1,0}^d(\Omega_1 \times \Omega_2, \mathbb{R}^n).
\end{aligned}
\tag{7.19}
$$

This follows from the next theorem, when we use that for φ and ψ in $C_0^\infty(\Omega_1)$ resp. $C_0^\infty(\Omega_2)$,

$$
\varphi P(\psi u) = \operatorname{Op}\big(\varphi(x)p(x,y,\xi)\psi(y)\big)u.
$$

Theorem 7.5. *Let $p(x,y,\xi) \in S_{1,0}^d(\mathbb{R}^n \times \mathbb{R}^n, \mathbb{R}^n)$, vanishing for $|x| > a$ and for $|y| \geq a$, for some $a > 0$. For each $s \in \mathbb{R}$ there is a constant C, depending only on n, d, a and s; such that the norm of $P = \operatorname{Op}\big(p(x,y,\xi)\big)$ as an operator from $H^s(\mathbb{R}^n)$ to $H^{s-d}(\mathbb{R}^n)$ satisfies*

$$
\|P\|_{s,s-d} \equiv \sup\{\,\|Pu\|_{s-d} \mid u \in \mathscr{S}(\mathbb{R}^n), \|u\|_s = 1\,\}
\tag{7.20}
$$

$$
\leq C \sup\{\,|\langle\xi\rangle^{-d} D_{x,y}^\beta p| \mid x, y, \xi \in \mathbb{R}^n, |\beta| \leq 2(\max\{|d-s|, |s|\} + n + 2)\,\}.
$$

Proof. Let $u, v \in \mathscr{S}(\mathbb{R}^n)$. By Fourier transformation, we find

$$|(Pu, v)| = |\int_{\mathbb{R}^{3n}} e^{i(x-y)\cdot\xi} p(x, y, \xi) u(y) \overline{v}(x) \, dy d\xi dx|$$

$$= |\int_{\mathbb{R}^{5n}} e^{ix\cdot(\xi-\theta)-iy\cdot(\xi-\eta)} p(x, y, \xi) \hat{u}(\eta) \overline{\hat{v}}(\theta) \, d\xi d\eta d\theta dx dy| \qquad (7.21)$$

$$\leq \int_{\mathbb{R}^{3n}} |p(\widehat{\theta - \xi}, \widehat{\xi - \eta}, \xi) \hat{u}(\eta) \overline{\hat{v}}(\theta)| \, d\xi d\eta d\theta,$$

where $p(\hat{\theta}, \hat{\eta}, \xi) = \mathscr{F}_{x\to\theta} \mathscr{F}_{y\to\eta} p(x, y, \xi)$. Now for any $N \in \mathbb{N}_0$,

$$\langle\xi\rangle^{-d} |\langle\theta\rangle^{2N} \langle\eta\rangle^{2N} p(\hat{\theta}, \hat{\eta}, \xi)|$$

$$\leq \langle\xi\rangle^{-d} \|(1 - \Delta_x)^N (1 - \Delta_y)^N p(x, y, \xi)\|_{L_{1,x,y}(B(0,a)\times B(0,a))}$$

$$\leq C_{a,N} \sup\{ |\langle\xi\rangle^{-d} D_{x,y}^\beta p| \mid x, y, \xi \in \mathbb{R}^n, |\beta| \leq 4N \}$$

$$\equiv M,$$

so the symbol p satisfies

$$|p(\hat{\theta}, \hat{\eta}, \xi)| \leq M \langle\xi\rangle^d \langle\theta\rangle^{-2N} \langle\eta\rangle^{-2N}. \qquad (7.22)$$

By the Peetre inequality (6.12), we have that

$$\langle\xi\rangle^d = \langle\xi\rangle^s \langle\xi\rangle^{d-s} \leq C'_{s,d} \langle\eta\rangle^s \langle\xi - \eta\rangle^{|s|} \langle\theta\rangle^{d-s} \langle\xi - \theta\rangle^{|d-s|}.$$

Then we find from (7.21), by applying the Schwarz inequality (and absorbing universal constants in c and c'):

$$|(Pu, v)| \leq M \int \langle\xi\rangle^d \langle\xi - \eta\rangle^{-2N} \langle\xi - \theta\rangle^{-2N} |\hat{u}(\eta) \hat{v}(\theta)| \, d\xi d\eta d\theta$$

$$\leq cM \int \langle\xi - \eta\rangle^{|s|-2N} \langle\xi - \theta\rangle^{|d-s|-2N} \langle\eta\rangle^s \langle\theta\rangle^{d-s} |\hat{u}(\eta) \hat{v}(\theta)| \, d\xi d\eta d\theta \qquad (7.23)$$

$$\leq c'M \| \langle\xi - \eta\rangle^{\frac{1}{2}|s|-N} \langle\xi - \theta\rangle^{\frac{1}{2}|d-s|-N} \langle\eta\rangle^s \hat{u}(\eta) \|_{L^2_{\xi,\eta,\theta}} \cdot$$

$$\cdot \| \langle\xi - \eta\rangle^{\frac{1}{2}|s|-N} \langle\xi - \theta\rangle^{\frac{1}{2}|d-s|-N} \langle\theta\rangle^{d-s} \hat{v}(\theta) \|_{L^2_{\xi,\eta,\theta}}.$$

Using a change of variables, we calculate e.g.

$$\| \langle\xi - \eta\rangle^{\frac{1}{2}|s|-N} \langle\xi - \theta\rangle^{\frac{1}{2}|d-s|-N} \langle\eta\rangle^s \hat{u}(\eta) \|^2_{L^2_{\xi,\eta,\theta}}$$

$$= \int \langle\varrho\rangle^{|s|-2N} \langle\sigma\rangle^{|d-s|-2N} \langle\eta\rangle^{2s} |\hat{u}(\eta)|^2 \, d\varrho d\sigma d\eta$$

$$= c'' \int \langle\eta\rangle^{2s} |\hat{u}(\eta)|^2 \, d\eta = c'' \|u\|^2_s,$$

when $2N > \max\{|s|, |d - s|\} + n + 1$. It follows that

$$|(Pu, v)| \leq CM \|u\|_s \|v\|_{d-s},$$

with C depending only on a, N, s, d, n. Then

$$\|P\|_{s,s-d} = \sup\{\,|(Pu, v)|\,\mid\,\|u\|_s = 1, \|v\|_{d-s} = 1\,\}$$

satisfies (7.20).   □

Approximating the elements in $H^s_{\mathrm{comp}}(\Omega_2)$ by $C^\infty_0(\Omega_2)$-functions, we extend P by continuity to a mapping from $H^s_{\mathrm{comp}}(\Omega_2)$ to $H^{s-d}_{\mathrm{loc}}(\Omega_1)$ (it is continuous, since it is clearly continuous from $H^s_{K_j}$ to $H^{s-d}_{\mathrm{loc}}(\Omega_1)$ for each K_j compact $\subset \Omega_2$). Since s can be arbitrary in $\mathbb{R}$, an application of the Sobolev theorem confirms that P maps $C^\infty_0(\Omega_2)$ continuously into $C^\infty(\Omega_1)$. Note also that since each element of $\mathscr{E}'(\Omega_2)$ lies in $H^t_{K_j}$ for some t, some K_j compact $\subset \Omega_2$ (cf. Theorem 6.19), P maps $\mathscr{E}'(\Omega_2)$ into $\mathscr{D}'(\Omega_1)$ (continuously).

One advantage of including symbols in (x, y)-form (by (7.16)) is that the formal adjoint $P^\times$ of P is easy to describe; where $P^\times$ stands for the operator from $C^\infty_0(\Omega_1)$ to $\mathscr{D}'(\Omega_2)$ for which

$$\langle P^\times u, \overline{v}\rangle_{\Omega_2} = \langle u, \overline{Pv}\rangle_{\Omega_1} \text{ for } u \in C^\infty_0(\Omega_1), v \in C^\infty_0(\Omega_2). \tag{7.24}$$

We here find that when $P = \mathrm{Op}\big(p(x, y, \xi)\big)$, possibly matrix-formed, then $P^\times$ is simply $\mathrm{Op}\big(p_1(x, y, \xi)\big)$, where $p_1(x, y, \xi)$ is defined as the conjugate transpose of $p(y, x, \xi)$:

$$p_1(x, y, \xi) = {}^t\overline{p}(y, x, \xi) \equiv p(y, x, \xi)^*; \tag{7.25}$$

this is seen by writing out the integrals (7.24) and interchanging integrations (all is justified in the sense of oscillatory integrals). In particular, if $P = \mathrm{Op}\big(p(x, \xi)\big)$, then $P^\times = \mathrm{Op}\big(p(y, \xi)^*\big)$. (The transposition is only relevant when p is matrix-formed.) Note that $P^\times$ in fact maps $C^\infty_0(\Omega_1)$ into $C^\infty(\Omega_2)$, since it is a ψdo.

We can relate this to the extension of P to distributions: The operator

$$P^\times : C^\infty_0(\Omega_1) \to C^\infty(\Omega_2)$$

has an adjoint (not just a formal adjoint)

$$(P^\times)^* : \mathscr{E}'(\Omega_2) \to \mathscr{D}'(\Omega_1),$$

that coincides with P on the set $C^\infty_0(\Omega_2)$. Since $C^\infty_0(\Omega_2)$ is dense in $\mathscr{E}'(\Omega_2)$ (a distribution $u \in \mathscr{E}'(\Omega_2)$ is the limit of $h_j * u$ lying in $C^\infty_0(\Omega_2)$ for j sufficiently large, cf. Lemma 3.17), there can be at most one extension of P to a continuous operator from $\mathscr{E}'(\Omega_2)$ to $\mathscr{D}'(\Omega_1)$. Thus $(P^\times)^*$ acts in the same way as $P : \mathscr{E}'(\Omega_2) \to \mathscr{D}'(\Omega_1)$. It is also consistent with the extensions of P to maps between Sobolev spaces. We use the notation P for all the extensions, since they are consistent with each other.

It is customary in the pseudodifferential theory to use the notation P^* for any true or formal adjoint of any of the versions of P. We shall follow this custom, as long as it does not conflict with the strict rules for adjoints of Hilbert space operators recalled in Chapter 12. (The notation P' can then be used more liberally for other purposes.)

7.2 Negligible operators

Another advantage of including symbols in (x, y)-form, working with the formulation (7.16) and not just (7.2), is that in this way, the so-called *negligible pseudodifferential operators* are included in the theory in a natural way:
When $p(x, y, \xi) \in S_{1,0}^{-\infty}(\Omega_1 \times \Omega_2, \mathbb{R}^n)$, then

$$\mathrm{Op}(p)u(x) = \int_{\Omega_2} K_p(x, y)u(y)\, dy, \text{ with kernel}$$

$$K_p(x, y) = \int_{\mathbb{R}^n} e^{i(x-y)\cdot\xi} p(x, y, \xi)\, d\xi = \mathscr{F}_{\xi\to z}^{-1} p(x, y, \xi)|_{z=x-y},$$

(7.26)

via an interpretation in terms of oscillatory integrals. Since this p is in $\mathscr{S}(\mathbb{R}^n)$ as a function of ξ, $\mathscr{F}_{\xi\to z}^{-1}p$ is in $\mathscr{S}(\mathbb{R}^n)$ as a function of z. Taking the smooth dependence of p on x and y into account, one finds that $K_p(x, y) \in C^\infty(\Omega_1 \times \Omega_2)$. So in fact $\mathrm{Op}(p)$ is an integral operator with C^∞ kernel; we call such operators *negligible*. Conversely, if $\mathcal{R}$ is an integral operator from Ω_2 to Ω_1 with kernel $K(x, y) \in C^\infty(\Omega_1 \times \Omega_2)$, then there is a symbol $r(x, y, \xi) \in S_{1,0}^{-\infty}(\Omega_1 \times \Omega_2, \mathbb{R}^n)$ such that $\mathcal{R} = \mathrm{Op}\big(r(x, y, \xi)\big)$, namely,

$$r(x, y, \xi) = ce^{i(y-x)\cdot\xi} K(x, y)\chi(\xi),$$

(7.27)

where the constant c equals $\left(\int \chi(\xi)d\xi\right)^{-1}$. (Integral operators with C^∞ kernels are in some other texts called smoothing operators, or regularizing operators.)

The reason that we need to include the negligible operators in the calculus is that there is a certain vagueness in the definition. For example, for the polyhomogeneous symbols there is primarily a freedom of choice in how each term $p_{d-l}(x, \xi)$ (or $p_{d-l}(x, y, \xi)$) is extended as a C^∞-function into the set $|\xi| \leq 1$ where it is not assumed to be homogeneous; and second, p is only associated with the series $\sum_{l\in\mathbb{N}_0} p_{d-l}$ in an asymptotic sense (cf. Definition 7.2), which also leaves a free choice of the value of p, as long as the estimates are respected. These choices are free precisely modulo symbols of order $-\infty$. *Moreover*, we shall find that when the *composition* of two operators $P' = \mathrm{Op}\big(p'(x, \xi)\big)$ and $P'' = \mathrm{Op}\big(p''(x, \xi)\big)$ with symbols in $S_{1,0}^\infty(\Omega, \mathbb{R}^n)$ is defined, the resulting operator $P = P'P''$ need not be of the exact form $\mathrm{Op}\big(p(x, \xi)\big)$, but does have the form

$$P = \mathrm{Op}\big(p(x, \xi)\big) + \mathcal{R},$$

for some negligible operator $\mathcal{R}$ (in $\operatorname{Op} S^{-\infty}(\Omega \times \Omega)$), see below. (As mentioned earlier, one can get a more exact calculus on $\mathbb{R}^n$ by working with the more restrictive class of globally estimated symbols introduced in [H85, 18.1], see also e.g. Saint Raymond [S91].)

When p and p' are symbols in $S_{1,0}^\infty(\Sigma, \mathbb{R}^n)$ with $p - p' \in S^{-\infty}(\Sigma, \mathbb{R}^n)$, we say that $p \sim p'$. When P and P' are linear operators from $C_0^\infty(\Omega_2)$ to $C^\infty(\Omega_1)$ with $P - P'$ a negligible ψdo, we say that $P \sim P'$.

From now on, we restrict the attention to cases where $\Omega_1 = \Omega_2 = \Omega$.

We shall now discuss the question of whether a symbol is determined from a given ψdo. There are the following facts: On one hand, the (x, y)-dependent symbols $p(x, y, \xi)$ are very far from uniquely determined from $\operatorname{Op}(p(x, y, \xi))$; for example, the symbol (where $a(x) \in C_0^\infty(\Omega) \setminus \{0\}$)

$$p(x, y, \xi) = \xi_j a(y) - a(x)\xi_j - D_{x_j} a(x) \tag{7.28}$$

is an element of $S^1(\Omega \times \Omega) \setminus S^0(\Omega \times \Omega)$, for which $\operatorname{Op}(p)$ is *the zero operator*. On the other hand, it can be shown that an x-dependent symbol $p(x, \xi)$ *is uniquely determined from* $P = \operatorname{Op}(p)$ modulo $S^{-\infty}(\Omega)$, in a sense that we shall explain below. First we need to introduce a restricted class of operators:

Definition 7.6. A ψdo P will be said to be *properly supported* in Ω when both P and P^* have the property: For each compact $K \subset \Omega$ there is a compact $K' \subset \Omega$ such that distributions supported in K are mapped into distributions supported in K'.

When P is properly supported, P and P^* map $C_0^\infty(\Omega)$ into itself, and hence P extends to a mapping from $\mathscr{D}'(\Omega)$ to itself, as the adjoint of P^* on $C_0^\infty(\Omega)$. Moreover, P maps $C^\infty(\Omega)$ to itself (since P^* maps $\mathscr{E}'(\Omega)$ to itself), and it maps $H_{comp}^s(\Omega)$ to $H_{comp}^{s-d}(\Omega)$ and $H_{loc}^s(\Omega)$ to $H_{loc}^{s-d}(\Omega)$ for all s when of order d. — Note that *differential* operators are always properly supported.

Consider a properly supported ψdo P in Ω. If $u \in C^\infty(\Omega)$ and we want to evaluate Pu at $x \in \Omega$, we can replace Pu by ϱPu, where $\varrho = 1$ at x and is C^∞, supported in a compact set $K \subset \Omega$. Then if K' is chosen for P^* according to Definition 7.6, and $\psi = 1$ on K', $\psi \in C_0^\infty(\Omega)$, we have for any $\varphi \in C_0^\infty(\Omega)$ with $\operatorname{supp} \varphi \subset K$:

$$\langle Pu, \bar{\varphi} \rangle = \langle u, \overline{P^*\varphi} \rangle = \langle u, \psi \overline{P^*\varphi} \rangle = \langle P(\psi u), \bar{\varphi} \rangle.$$

So $Pu = P\psi u$ on K°, and $(\varrho Pu)(x) = (\varrho P\psi u)(x)$. This allows us to give a meaning to $e^{-ix\cdot\xi} P(e^{i(\cdot)\cdot\xi})$, namely, as

$$e^{-ix\cdot\xi} P(e^{i(\cdot)\cdot\xi}) = e^{-ix\cdot\xi} \varrho(x) P(\psi(y)e^{iy\cdot\xi}),$$

for any pair of ϱ and ψ chosen as just described; it is independent of the choice of ϱ and ψ. With a certain abuse of notation, this function of x and ξ is often denoted $e^{-ix\cdot\xi} P(e^{ix\cdot\xi})$.

Now we claim that if $P = \mathrm{Op}(p(x,\xi))$ and is properly supported, then p is determined from P by

$$p(x,\xi) = e^{-ix\cdot\xi}P(e^{ix\cdot\xi}), \tag{7.29}$$

for $x \in \Omega$. For then, by reading the integrals as forwards or backwards Fourier transforms,

$$\begin{aligned}
\mathrm{Op}(p)(e^{i(\cdot)\cdot\xi}) &= \int_{\mathbb{R}^{2n}} e^{i(x-y)\cdot\eta}p(x,\eta)e^{iy\cdot\xi}\,dy d\eta \\
&= \int_{\mathbb{R}^n} p(x,\widetilde{x-y})e^{iy\cdot\xi}\,dy \\
&= e^{ix\cdot\xi}\int_{\mathbb{R}^n} p(x,\widetilde{x-y})e^{-i(x-y)\cdot\xi}\,dy = e^{ix\cdot\xi}p(x,\xi),
\end{aligned}$$

for all $x \in \Omega$, $\xi \in \mathbb{R}^n$. We have here used the notation $p(x,\tilde{z})$ for the inverse Fourier transform with respect to the last variable, $\mathscr{F}^{-1}_{\eta\to z}p(x,\eta)$. This shows that $p(x,\xi)$ is uniquely determined from $\mathrm{Op}(p)$.

On the other hand, if p is defined from P by (7.29), then when $u \in C^\infty_0(\Omega)$, one can justify the calculation

$$Pu = P\Big(\int_{\mathbb{R}^n} e^{i(\cdot)\cdot\xi}\hat{u}(\xi)\,d\xi\Big) = \int_{\mathbb{R}^n} P(e^{i(\cdot)\cdot\xi})\hat{u}(\xi)\,d\xi = \int_{\mathbb{R}^n} e^{ix\cdot\xi}p(x,\xi)\hat{u}(\xi)\,d\xi,$$

by inserting the Fourier transform of u and writing the integral as a limit of Riemann sums, such that the linearity and continuity of P allows us to pull it through the integration; this shows that $P = \mathrm{Op}(p(x,\xi))$.

All this implies:

Lemma 7.7. *When P is properly supported in Ω, there is a unique symbol $p(x,\xi) \in S^\infty(\Omega)$ such that $P = \mathrm{Op}(p(x,\xi))$, namely, the one determined by (7.29).*

As we shall see below in Theorem 7.10, an operator $P = \mathrm{Op}(p(x,\xi))$ can always be written as a sum $P = P' + \mathcal{R}$, where $P' = \mathrm{Op}(p'(x,y,\xi))$ is properly supported and $\mathcal{R}$ is negligible. By the preceding remarks there is then also a symbol $p''(x,\xi)$ (by (7.29)) so that $P = \mathrm{Op}(p''(x,\xi)) + \mathcal{R}$, and then $\mathcal{R} = \mathrm{Op}(r(x,\xi))$ with $r(x,\xi) = p(x,\xi) - p''(x,\xi)$. Moreover, one can show that when $r(x,\xi)$ defines a negligible operator, then necessarily $r(x,\xi) \in S^{-\infty}(\Omega)$ (for example by use of Remark 7.16 below). We conclude:

Proposition 7.8. *The symbol $p(x,\xi)$ in a representation*

$$P = \mathrm{Op}(p(x,\xi)) + \mathcal{R}, \tag{7.30}$$

with $\mathrm{Op}(p(x,\xi))$ properly supported and $\mathcal{R}$ negligible, is determined from P uniquely modulo $S^{-\infty}(\Omega)$.

It remains to establish Theorem 7.10. For an open set $\Omega \subset \mathbb{R}^n$, denote $\operatorname{diag}(\Omega \times \Omega) = \{ (x,y) \in \Omega \times \Omega \mid x = y \}$.

Lemma 7.9. *Let* $p(x,y,\xi) \in S_{1,0}^d(\Omega \times \Omega, \mathbb{R}^n)$. *When* $\varphi(x,y) \in C^\infty(\Omega \times \Omega)$ *with* $\operatorname{supp}\varphi \subset (\Omega \times \Omega) \setminus \operatorname{diag}(\Omega \times \Omega)$, *then* $\operatorname{Op}(\varphi(x,y)p(x,y,\xi))$ *is negligible.*

Proof. Since $\varphi(x,y)$ vanishes on a neighborhood of the diagonal $\operatorname{diag}(\Omega \times \Omega)$, $\varphi(x,y)/|y-x|^{2N}$ is C^∞ for any $N \in \mathbb{N}_0$, so we may write $\varphi(x,y)$ as

$$\varphi(x,y) = |y-x|^{2N}\varphi_N(x,y), \tag{7.31}$$

where also the $\varphi_N(x,y)$ are in $C^\infty(\Omega \times \Omega)$ with support in $(\Omega \times \Omega) \setminus \operatorname{diag}(\Omega \times \Omega)$. Then an integration by parts (in the oscillatory integrals) gives

$$
\begin{aligned}
\operatorname{Op}&(\varphi(x,y)p(x,y,\xi))u \\
&= \int e^{i(x-y)\cdot\xi}|y-x|^{2N}\varphi_N(x,y)p(x,y,\xi)u(y)\,dyd\xi \\
&= \int [(-\Delta_\xi)^N e^{i(x-y)\cdot\xi}]\varphi_N(x,y)p(x,y,\xi)u(y)\,dyd\xi \qquad (7.32)\\
&= \int e^{i(x-y)\cdot\xi}\varphi_N(x,y)(-\Delta_\xi)^N p(x,y,\xi)u(y)\,dyd\xi \\
&= \operatorname{Op}(\varphi_N(x,y)(-\Delta_\xi)^N p(x,y,\xi))u,
\end{aligned}
$$

where the symbol is in $S_{1,0}^{d-2N}(\Omega \times \Omega, \mathbb{R}^n)$. Calculating the kernel of this operator as in (7.26), we get a function of (x,y) with more continuous derivatives the larger N is taken. Since the original expression is independent of N, we conclude that $\operatorname{Op}(\varphi p)$ is an integral operator with kernel in $C^\infty(\Omega \times \Omega)$, i.e., is a negligible ψdo. $\qquad\square$

Theorem 7.10. *Any* $P = \operatorname{Op}(p(x,y,\xi))$ *with* $p \in S_{1,0}^d(\Omega \times \Omega)$ *can be written as the sum of a properly supported operator* P' *and a negligible operator* $\mathcal{R}$.

Proof. The basic idea is to obtain the situation of Lemma 7.9 with $\varphi(x,y) = 1 - \varrho(x,y)$, where ϱ has the following property: Whenever M_1 and M_2 are compact $\subset \Omega$, then the sets

$$
\begin{aligned}
M_{12} &= \{ y \in \Omega \mid \exists x \in M_1 \text{ with } (x,y) \in \operatorname{supp}\varrho \} \\
M_{21} &= \{ x \in \Omega \mid \exists y \in M_2 \text{ with } (x,y) \in \operatorname{supp}\varrho \}
\end{aligned}
$$

are compact. We then say that $\varrho(x,y)$ is *properly supported.*

Once we have such a function, we can take

$$p(x,y,\xi) = \varrho(x,y)p(x,y,\xi) + (1 - \varrho(x,y))p(x,y,\xi); \tag{7.33}$$

here the first term defines a properly supported operator $P = \operatorname{Op}(\varrho p)$ and the second term defines, by Lemma 7.9, a negligible operator $\mathcal{R} = \operatorname{Op}((1-\varrho)p) = \operatorname{Op}(\varphi p)$. Then the statement in the theorem is obtained.

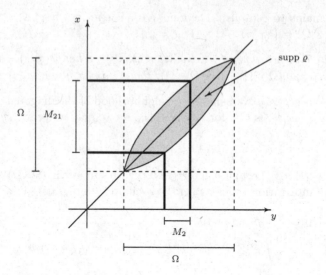

To construct the function ϱ, we can use a partition of unity $1 = \sum_{j\in\mathbb{N}_0}\psi_j$ for Ω as in Theorem 2.16. Take

$$J = \{(j,k)\in\mathbb{N}_0^2 \mid \operatorname{supp}\psi_j\cap\operatorname{supp}\psi_k = \emptyset\}, \quad J' = \mathbb{N}_0^2\setminus J,$$

$$\varphi(x,y) = \sum_{(j,k)\in J}\psi_j(x)\psi_k(y), \quad \varrho(x,y) = \sum_{(j,k)\in J'}\psi_j(x)\psi_k(y).$$

In the proof that φ and ϱ are as asserted it is used again and again that any compact subset of Ω meets only finitely many of the supports of the ψ_j:

To see that ϱ is properly supported, let M_2 be a compact subset of Ω. Then there is a finite set $I_2\subset\mathbb{N}_0$ such that $\operatorname{supp}\psi_k\cap M_2 = \emptyset$ for $k\notin I_2$, and hence

$$\varrho(x,y) = \sum_{(j,k)\in J',k\in I_2}\psi_j(x)\psi_k(y) \text{ for } y\in M_2.$$

By definition of J', the indices j that enter here are at most those for which $\operatorname{supp}\psi_j\cap M_2' \neq \emptyset$, where M_2' is the compact set $M_2' = \bigcup_{k\in I_2}\operatorname{supp}\psi_k$ in Ω. There are only finitely many such j; let I_1 denote the set of these j. Then $\varrho(x,y)$ vanishes for $x\notin M_{21} = \bigcup_{j\in I_1}\operatorname{supp}\psi_j$, when $y\in M_2$. — There is a similar proof with the roles of x and y exchanged.

To see that φ vanishes on a neighborhood of the diagonal, let $x_0\in\Omega$, and let $B\subset\Omega$ be a closed ball around x_0. There is a finite set $I_0\subset\mathbb{N}_0$ such that $\operatorname{supp}\psi_k\cap B = \emptyset$ for $k\notin I_0$, so

$$\varphi(x,y) = \sum_{(j,k)\in J, j\in I_0, k\in I_0}\psi_j(x)\psi_k(y) \text{ for } (x,y)\in B\times B.$$

This is a finite sum, and we examine each term. Consider $\varphi_j(x)\varphi_k(y)$. The supports of φ_j and φ_k have a positive distance r_{jk}, by definition of J, so $\psi_j(x)\varphi_k(y) = 0$ for $|x - y| < r_{jk}$. Take as r the smallest occurring r_{jk}, then φ vanishes on $\{(x,y) \in B \times B \mid |x - y| < r\}$, a neighborhood of (x_0, x_0). $\square$

A further consequence of Lemma 7.9 is the "pseudolocal" property of pseudodifferential operators. For a $u \in \mathscr{D}'(\Omega)$, define

$$\Omega_\infty(u) = \bigcup \{\, \omega \text{ open } \subset \Omega \mid u|_\omega \in C^\infty(\omega) \,\}; \qquad (7.34)$$

it is the largest open subset of Ω where u coincides with a C^∞-function. Define the singular support of u as the complement

$$\operatorname{sing\,supp} u = \Omega \setminus \Omega_\infty(u), \qquad (7.35)$$

it is clearly a closed subset of $\operatorname{supp} u$. Differential operators preserve supports

$$\operatorname{supp} Pu \subset \operatorname{supp} u, \quad \text{when } P \text{ is a differential operator;} \qquad (7.36)$$

in short: They are *local*. Pseudodifferential operators do not in general have the property in (7.36), but Lemma 7.9 implies that they are *pseudolocal*:

Proposition 7.11. *A ψdo P preserves singular supports:*

$$\operatorname{sing\,supp} Pu \subset \operatorname{sing\,supp} u. \qquad (7.37)$$

Proof. Let $u \in \mathscr{E}'(\Omega)$ and write $u = u_\varepsilon + v_\varepsilon$ where $\operatorname{supp} u_\varepsilon \subset \operatorname{sing\,supp} u + B(0,\varepsilon)$, and $v_\varepsilon \in C_0^\infty(\Omega)$. Using the decomposition (7.33) with ϱ supported in $\operatorname{diag}(\Omega \times \Omega) + B(0,\varepsilon)$, $\varrho = 1$ on a neighborhood of $\operatorname{diag}(\Omega \times \Omega)$, we find

$$Pu = \operatorname{Op}(\varrho p)(u_\varepsilon + v_\varepsilon) + \operatorname{Op}\big((1 - \varrho)p\big)u$$
$$= \operatorname{Op}(\varrho p)u_\varepsilon + f_\varepsilon,$$

where $f_\varepsilon \in C^\infty(\Omega)$ and $\operatorname{supp}\operatorname{Op}(\varrho p)u_\varepsilon \subset \operatorname{sing\,supp} u + B(0, 2\varepsilon)$. Since ε can be taken arbitrarily small, this implies (7.37). $\square$

In preparation for general composition rules, we observe:

Lemma 7.12. *When $P = \operatorname{Op}(p(x,\xi))$ is properly supported and $\mathcal{R}$ is negligible, then $P\mathcal{R}$ and $\mathcal{R}P$ are negligible.*

Proof (sketch). Let $K(x,y)$ be the kernel of $\mathcal{R}$. Then for $u \in C_0^\infty(\Omega)$,

$$P\mathcal{R}u = \int_{\Omega \times \Omega \times \mathbb{R}^n} e^{i(x-z)\cdot\xi} p(x,\xi) K(z,y) u(y)\, dy\,dz\,d\xi$$

$$= \int_\Omega K'(x,y) u(y)\, dy, \quad \text{with}$$

$$K'(x,y) = \int_{\Omega \times \mathbb{R}^n} e^{i(x-z)\cdot\xi} p(x,\xi) K(z,y)\, dz\,d\xi$$

(oscillatory integrals). For each y, $K'(x,y)$ is $\mathrm{Op}(p)$ applied to the C^∞-function $K(\cdot,y)$, hence is C^∞ in x. Moreover, since $K(\cdot,y)$ depends smoothly on y, so does $K'(\cdot,y)$. So $K'(x,y)$ is C^∞ in (x,y); this shows the statement for $P\mathcal{R}$. For $\mathcal{R}P$ one can use that $(\mathcal{R}P)^* = P^*\mathcal{R}^*$ is of the type already treated.   □

7.3 Composition rules

Two pseudodifferential operators can be composed for instance when one of them is properly supported, or when the ranges and domains fit together in some other way. The "rules of calculus" are summarized in the following theorem (where $\overline{D}$ stands for $+i\partial$).

Theorem 7.13. *In the following, Ω is an open subset of $\mathbb{R}^n$, and d and $d' \in \mathbb{R}$.*

1° *Let $p(x,y,\xi) \in S_{1,0}^d(\Omega \times \Omega)$. Then*

$$\mathrm{Op}\big(p(x,y,\xi)\big) \sim \mathrm{Op}\big(p_1(x,\xi)\big) \sim \mathrm{Op}\big(p_2(y,\xi)\big) \, , \text{ where}$$

$$p_1(x,\xi) \sim \sum_{\alpha \in \mathbb{N}_0^n} \tfrac{1}{\alpha!} \partial_y^\alpha D_\xi^\alpha p(x,y,\xi) \, |_{y=x} \, , \text{ and} \tag{7.38}$$

$$p_2(y,\xi) \sim \sum_{\alpha \in \mathbb{N}_0^n} \tfrac{1}{\alpha!} \partial_x^\alpha \overline{D}_\xi^\alpha p(x,y,\xi) \, |_{x=y}, \tag{7.39}$$

as symbols in $S_{1,0}^d(\Omega)$.

2° *When $p(x,\xi) \in S_{1,0}^d(\Omega)$, then*

$$\mathrm{Op}\big(p(x,\xi)\big)^* \sim \mathrm{Op}\big(p_3(x,\xi)\big) \, , \text{ where}$$

$$p_3(x,\xi) \sim \sum_{\alpha \in \mathbb{N}_0^n} \tfrac{1}{\alpha!} \partial_x^\alpha D_\xi^\alpha p(x,\xi)^* \quad \text{in } S_{1,0}^d(\Omega). \tag{7.40}$$

3° *When $p(x,\xi) \in S_{1,0}^d(\Omega)$ and $p'(x,\xi) \in S_{1,0}^{d'}(\Omega)$, with $\mathrm{Op}(p)$ or $\mathrm{Op}(p')$ properly supported, then*

$$\mathrm{Op}\big(p(x,\xi)\big) \, \mathrm{Op}\big(p'(x,\xi)\big) \sim \mathrm{Op}\big(p''(x,\xi)\big) \, , \text{ where}$$

$$p''(x,\xi) \sim \sum_{\alpha \in \mathbb{N}_0^n} \tfrac{1}{\alpha!} D_\xi^\alpha p(x,\xi) \partial_x^\alpha p'(x,\xi) \quad \text{in } S_{1,0}^{d+d'}(\Omega). \tag{7.41}$$

In each of the rules, polyhomogeneous symbols give rise to polyhomogeneous symbols (when rearranged in series of terms according to degree of homogeneity).

Proof. The principal step is the proof of $1°$. Inserting the Taylor expansion of order N in y of $p(x, y, \xi)$ at $y = x$, we find

$$\text{Op}(p)u = \int e^{i(x-y)\cdot\xi} p(x, y, \xi) u(y) \, dy d\xi$$

$$= \int e^{i(x-y)\cdot\xi} \sum_{|\alpha|<N} \tfrac{1}{\alpha!}(y-x)^\alpha \partial_y^\alpha p(x, x, \xi) u(y) \, dy d\xi$$

$$+ \int e^{i(x-y)\cdot\xi} \sum_{|\alpha|=N} \tfrac{N(y-x)^\alpha}{\alpha!} \int_0^1 (1-h)^{N-1} \partial_y^\alpha p(x, x+(y-x)h, \xi) u(y) \, dh dy d\xi.$$

The first integral gives the terms for $|\alpha| < N$ in the series (7.38), by an integration by parts like in (7.32). Also in the second integral, an integration by parts transforms the factor $(y-x)^\alpha$ to a derivation D_ξ^α on p; then for sufficiently large N the integral equals

$$\int K_N(x, y) u(y) dy,$$

with continuous kernel

$$K_N(x, y) = \sum_{|\alpha|=N} c_\alpha \int_{\mathbb{R}^n} e^{i(x-y)\xi} \int_0^1 (1-h)^{N-1} \partial_y^\alpha D_\xi^\alpha p(x, x+(y-x)h, \xi) \, dh d\xi.$$

More precisely, this integral defines a continuous function when $N > d+n$ because $\partial_y^\alpha D_\xi^\alpha p$ is integrable in ξ then; and K_N has continuous derivatives in (x, y) up to order k, when $N > d+n+k$. Let $p_1(x, \xi)$ be a symbol satisfying (7.38), then $\text{Op}(p_1 - \sum_{|\alpha|<N} \tfrac{1}{\alpha!}\partial_y^\alpha D_\xi^\alpha p(x, y, \xi) |_{y=x})$ has a continuous kernel $K_{1,N}(x, y)$ with similar properties for large N. Altogether, $\text{Op}(p)$ differs from $\text{Op}(p_1)$ by an operator with a kernel $K_N - K_{1,N}$, which is C^k when $N > d+n+k$. Since $\text{Op}(p(x, y, \xi)) - \text{Op}(p_1(x, \xi))$ is independent of N, its kernel is independent of N; then since we can take N arbitrarily large, it must be C^∞. Hence $\text{Op}(p(x, y, \xi)) - \text{Op}(p_1(x, \xi))$ is negligible. This shows (7.38).

(7.39) is obtained by considering the adjoints (cf. (7.25)):

$$\text{Op}(p(x, y, \xi))^* = \text{Op}(^t\overline{p}(y, x, \xi)) \sim \text{Op}(p_2'(x, \xi)),$$

where

$$p_2'(x, \xi) \sim \sum_{\alpha \in \mathbb{N}_0^n} \tfrac{1}{\alpha!}\partial_w^\alpha D_\xi^\alpha {}^t\overline{p}(w, z, \xi) |_{z=w=x},$$

by (7.38). Here $\text{Op}(p_2'(x, \xi))^* = \text{Op}(p_2(y, \xi))$, where

$$p_2(y, \xi) = {}^t\overline{p_2'}(y, \xi) \sim \sum_{\alpha \in \mathbb{N}_0^n} \tfrac{1}{\alpha!}\partial_x^\alpha \overline{D}_\xi^\alpha p(x, y, \xi) |_{x=y}.$$

This completes the proof of 1°; and 2° is obtained as a simple corollary, where we apply 1° to the symbol ${}^t\overline{p}(y,\xi)$.

Point 3° can likewise be obtained as a corollary, when we use 1° twice. Assume first that both $\mathrm{Op}(p)$ and $\mathrm{Op}(p')$ are properly supported. By 1°, we can replace $\mathrm{Op}(p'(x,\xi))$ by an operator in y-form, cf. (7.39):

$$\mathrm{Op}\big(p'(x,\xi)\big) \sim \mathrm{Op}\big(p_2(y,\xi)\big), \text{ where}$$
$$p_2(y,\xi) \sim \sum_{\alpha \in \mathbb{N}_0^n} \tfrac{1}{\alpha!}\partial_y^\alpha \overline{D}_\xi^\alpha p'(y,\xi).$$

So $P' = \mathrm{Op}(p_2(y,\xi)) + \mathcal{R}$ with $\mathcal{R}$ negligible, and then

$$PP' \sim \mathrm{Op}(p_1(x,\xi))\,\mathrm{Op}(p_2(y,\xi)),$$

in view of Lemma 7.12. Now we find

$$\mathrm{Op}\big(p(x,\xi)\big)\,\mathrm{Op}\big(p_2(y,\xi)\big)u(x)$$
$$= \int e^{i(x-y)\cdot\xi}p(x,\xi)e^{i(y-z)\cdot\theta}p_2(z,\theta)u(z)\,dz\,d\theta\,dy\,d\xi$$
$$= \int e^{i(x-z)\cdot\xi}p(x,\xi)p_2(z,\xi)u(z)\,dz\,d\xi$$
$$= \mathrm{Op}\big(p(x,\xi)p_2(y,\xi)\big)u(x),$$

since the integrations in θ and y represent a backwards and a forwards Fourier transform:

$$\int e^{i(y-z)\cdot\theta}p_2(z,\theta)\,d\theta = \mathscr{F}_{\theta\to y-z}^{-1}p_2(z,\theta) = p_2(z,\widetilde{y-z}),$$

and then (with $e^{i(x-y)\cdot\xi} = e^{i(x-z)\cdot\xi}e^{-i(y-z)\cdot\xi}$)

$$\int e^{-i(y-z)\cdot\xi}p_2(z,\widetilde{y-z})\,dy = p_2(z,\xi).$$

The resulting symbol of the composed operator is a simple product $p(x,\xi)p_2(y,\xi)$! One checks by use of the Leibniz formula, that it is a symbol of order $d + d'$.

Next, we use 1° to reduce $\mathrm{Op}\big(p(x,\xi)p_2(y,\xi)\big)$ to x-form. This gives $PP' \sim \mathrm{Op}\big(p''(x,\xi)\big)$, where

$$p''(x,\xi) \sim \sum_{\beta \in \mathbb{N}_0^n} \tfrac{1}{\beta!}D_\xi^\beta\Big[p(x,\xi)\partial_x^\beta \sum_{\alpha \in \mathbb{N}_0^n} \tfrac{1}{\alpha!}\partial_x^\alpha \overline{D}_\xi^\alpha p'(x,\xi)\Big].$$

This can be further reduced by use of the "backwards Leibniz formula" shown in Lemma 7.14 below. Doing a rearrangement as mentioned after (7.15), and applying (7.42), we find

$$p''(x,\xi) \sim \sum_{\beta \in \mathbb{N}_0^n} \sum_{\alpha \in \mathbb{N}_0^n} \frac{1}{\alpha! \beta!} D_\xi^\beta [p \overline{D}_\xi^\alpha \partial_x^{\alpha+\beta} p']$$

$$\sim \sum_{\theta \in \mathbb{N}_0^n} \frac{1}{\theta!} \sum_{\alpha+\beta=\theta} \frac{\theta!}{\alpha! \beta!} D_\xi^\beta [p \overline{D}_\xi^\alpha \partial_x^\theta p'] = \sum_{\theta \in \mathbb{N}_0^n} \frac{1}{\theta!} (D_\xi^\theta p) \partial_x^\theta p',$$

which gives the formula (7.41).

In all the formulas one makes rearrangements, when one starts with a polyhomogeneous symbol and wants to show that the resulting symbol is likewise polyhomogeneous.

Point 3° could alternatively have been shown directly by inserting a Taylor expansion for p in the ξ-variable, integrating by parts and using inequalities like in the proof of Theorem 7.5.

Finally, if only one of the operators $\operatorname{Op}(p)$ and $\operatorname{Op}(p')$, say $\operatorname{Op}(p')$, is properly supported, we reduce to the first situation by replacing the other operator $\operatorname{Op}(p(x,\xi))$ by $\operatorname{Op}(p_1(x,\xi)) + \mathcal{R}'$, where $\operatorname{Op}(p_1)$ is properly supported and $\mathcal{R}'$ is negligible; then $\operatorname{Op}(p)\operatorname{Op}(p') \sim \operatorname{Op}(p_1)\operatorname{Op}(p')$ in view of Lemma 7.12. □

Lemma 7.14 (BACKWARDS LEIBNIZ FORMULA). *For $u, v \in C^\infty(\Omega)$,*

$$(D^\theta u)v = \sum_{\alpha,\beta \in \mathbb{N}_0, \alpha+\beta=\theta} \frac{\theta!}{\alpha! \beta!} D^\beta(u\overline{D}^\alpha v). \tag{7.42}$$

Proof. This is deduced from the usual Leibniz formula by noting that

$$\langle (D^\theta u)v, \varphi \rangle = (-1)^{|\theta|} \langle u, D^\theta(v\varphi) \rangle = (-1)^{|\theta|} \langle u, \sum_{\alpha+\beta=\theta} \frac{\theta!}{\alpha! \beta!} D^\alpha v D^\beta \varphi \rangle$$

$$= \sum_{\alpha+\beta=\theta} (-1)^{|\theta|-|\beta|} \frac{\theta!}{\alpha! \beta!} \langle D^\beta(u D^\alpha v), \varphi \rangle,$$

where $|\theta| - |\beta| = |\alpha|$. □

Formula (7.42) extends of course to cases where u or v is in $\mathscr{D}'$.

Definition 7.15. The symbol (defined modulo $S^{-\infty}(\Omega)$) in the right-hand side of (7.41) is denoted $p(x,\xi) \circ p'(x,\xi)$ and is called the Leibniz product of p and p'.

The symbol in the right-hand side of (7.40) is denoted $p^{\circ*}(x,\xi)$.

The rule for $p \circ p'$ is a generalization of the usual (Leibniz) rule for composition of differential operators with variable coefficients. The notation $p \# p'$ is also often used.

Note that (7.41) shows

$$p(x,\xi) \circ p'(x,\xi) \sim p(x,\xi)p'(x,\xi) + r(x,\xi) \text{ with}$$

$$r(x,\xi) \sim \sum_{|\alpha| \geq 1} \tfrac{1}{\alpha!} D_\xi^\alpha p(x,\xi) \partial_x^\alpha p'(x,\xi), \qquad (7.43)$$

where r is of order $d + d' - 1$. Thus (7.4) has been obtained for these symbol classes with $\mathrm{Op}(pq)$ of order $d + d'$ and $\mathcal{R}$ of order $d + d' - 1$.

In calculations concerned with elliptic problems, this information is sometimes sufficient; one does not need the detailed information on the structure of $r(x,\xi)$. But there are also applications where the terms in r are important, first of all the term of order $d + d' - 1$, $\sum_{j=1}^n D_{\xi_j} p \, \partial_{x_j} p'$. For example, the commutator of $\mathrm{Op}(p)$ and $\mathrm{Op}(p')$ (for scalar operators) has the symbol

$$p \circ p' - p' \circ p \sim -i \sum_{j=1}^n \left(\partial_{\xi_j} p \, \partial_{x_j} p' - \partial_{x_j} p \, \partial_{\xi_j} p' \right) + r',$$

with r' of order $d + d' - 2$. The sum over j is called the Poisson bracket of p and p'; it plays a role in many considerations.

Remark 7.16. We here *sketch* a certain spectral property. Let $p(x,\xi) \in S^0(\Omega, \mathbb{R}^n)$ and let (x_0, ξ_0) be a point with $x_0 \in \Omega$ and $|\xi_0| = 1$; by translation and dilation we can obtain that $x_0 = 0$ and $B(0,2) \subset \Omega$. The sequence

$$u_k(x) = k^{n/2} \chi(kx) e^{ik^2 x \cdot \xi_0}, \quad k \in \mathbb{N}_0, \qquad (7.44)$$

has the following properties:

(i) $\|u_k\|_0 = \|\chi\|_0 \ (\neq 0)$ for all k,

(ii) $(u_k, v) \to 0$ for $k \to \infty$, all $v \in L_2(\mathbb{R}^n)$, $\qquad$ (7.45)

(iii) $\|\chi \, \mathrm{Op}(p) u_k - p^0(0, \xi_0) \cdot u_k\|_0 \to 0$ for $k \to \infty$.

Here (i) and (ii) imply that u_k has no convergent subsequence in $L_2(\mathbb{R}^n)$. It is used to show that if $\mathrm{Op}(p)$ is continuous from $H^0_{\mathrm{comp}}(\Omega)$ to $H^1_{\mathrm{loc}}(\Omega)$, then $p^0(0, \xi_0)$ must equal zero, for the compactness of the injection $H^1(B(0,2)) \hookrightarrow H^0(B(0,2))$ (cf. Section 8.2) then shows that $\chi \, \mathrm{Op}(p) u_k$ has a convergent subsequence in $H^0(B(0,2))$, and then, unless $p(0, \xi_0) = 0$, (iii) will imply that u_k has a convergent subsequence in L_2 in contradiction to (i) and (ii). Applying this argument to every point (x_0, ξ_0) with $|\xi_0| = 1$, we conclude that *if* $\mathrm{Op}(p)$ *is of order* 0 *and maps* $H^0_{\mathrm{comp}}(\Omega)$ *continuously into* $H^1_{\mathrm{loc}}(\Omega)$, *its principal symbol must equal* 0. (The proof is found in [H67] and is sometimes called Hörmander's variant of Gohberg's lemma, referring to a version given by Gohberg in [G60].)

The properties (i)–(iii) mean that u_k is a *singular sequence* for the operator $P_1 - a$ with $P_1 = \chi \, \mathrm{Op}(p)$ and $a = p^0(0, \xi_0)$; this implies that a belongs to the *essential spectrum* of P_1, namely, the set

$$\text{ess spec}(P_1) = \bigcap \{ \text{spec}(P_1 + K) \mid K \text{ compact in } L_2(\Omega) \}.$$

Since the operator norm is $\geq$ the spectral radius, the operator norm of P_1 (and of any $P_1 + K$) must be $\geq |a|$. It follows that the operator norm in $L_2(\Omega)$ of ψP for any $\psi \in C_0^\infty$ is $\geq \sup_{x \in \Omega, |\xi|=1} |\psi(x) p^0(x, \xi)|$. (So if we know that the norm of ψP is $\leq C$ for all $|\psi| \leq 1$, then also $\sup |p^0| \leq C$; a remark that can be very useful.)

By compositions with properly supported versions of $\text{Op}(\langle \xi \rangle^t)$ (for suitable t), it is seen more generally that *if $P = \text{Op}(p(x, \xi))$ is of order d and maps $H_{\text{comp}}^s(\Omega)$ into $H_{\text{loc}}^{s-d+1}(\Omega)$, then its principal symbol equals zero.*

In particular, if $P = \text{Op}(r(x, \xi))$ where $r \in S^{+\infty}(\Omega, \mathbb{R}^n)$, and maps $\mathscr{E}'(\Omega)$ into $C^\infty(\Omega)$, then all the homogeneous terms in each asymptotic series for $r(x, \xi)$ are zero, i.e., $r(x, \xi) \sim 0$ (hence is in $S^{-\infty}(\Omega, \mathbb{R}^n)$). This gives a proof that a symbol in x-form is determined from the operator it defines, uniquely modulo $S^{-\infty}(\Omega, \mathbb{R}^n)$.

7.4 Elliptic pseudodifferential operators

One of the most important applications of Theorem 7.13 is the construction of a *parametrix* (an almost-inverse) to an *elliptic* operator. We shall now define ellipticity, and here we include matrix-formed symbols and operators.

The space of $(N' \times N)$-matrices of symbols in $S^d(\Sigma, \mathbb{R}^n)$ (resp. $S_{1,0}^d(\Sigma, \mathbb{R}^n)$) is denoted $S^d(\Sigma, \mathbb{R}^n) \otimes \mathcal{L}(\mathbb{C}^N, \mathbb{C}^{N'})$ (resp. $S_{1,0}^d(\Sigma, \mathbb{R}^n) \otimes \mathcal{L}(\mathbb{C}^N, \mathbb{C}^{N'})$) since complex $(N' \times N)$-matrices can be identified with linear maps from $\mathbb{C}^N$ to $\mathbb{C}^{N'}$ (i.e., elements of $\mathcal{L}(\mathbb{C}^N, \mathbb{C}^{N'})$). The symbols in these classes of course define $(N' \times N)$-matrices of operators (notation: when $p \in S_{1,0}^d(\Omega \times \Omega, \mathbb{R}^n) \otimes \mathcal{L}(\mathbb{C}^N, \mathbb{C}^{N'})$, then $P = \text{Op}(p)$ sends $C_0^\infty(\Omega)^N$ into $C^\infty(\Omega)^{N'}$). Ellipticity is primarily defined for square matrices (the case $N = N'$), but has a natural extension to general matrices.

Definition 7.17. 1° Let $p \in S^d(\Omega, \mathbb{R}^n) \otimes \mathcal{L}(\mathbb{C}^N, \mathbb{C}^N)$. Then p, and $P = \text{Op}(p)$, and any ψdo P' with $P' \sim P$, is said to be **elliptic** of order d, when the principal symbol $p_d(x, \xi)$ is invertible for all $x \in \Omega$ and all $|\xi| \geq 1$.

2° Let $p \in S^d(\Omega, \mathbb{R}^n) \otimes \mathcal{L}(\mathbb{C}^N, \mathbb{C}^{N'})$. Then p (and $\text{Op}(p)$ and any $P' \sim \text{Op}(p)$) is said to be **injectively elliptic** of order d, resp. **surjectively elliptic** of order d, when $p_d(x, \xi)$ is injective, resp. surjective, from $\mathbb{C}^N$ to $\mathbb{C}^{N'}$, for all $x \in \Omega$ and $|\xi| \geq 1$. (In particular, $N' \geq N$ resp. $N' \leq N$.)

Note that since p_d is homogeneous of degree d for $|\xi| \geq 1$, it is only necessary to check the invertibility for $|\xi| = 1$. The definition (and its usefulness) extends to the classes $S_{\varrho, \delta}^d$, for the symbols that have a principal part in a suitable sense, cf. e.g. [H67], [T80] and [T81].

184 Pseudodifferential operators on open sets

Surjectively elliptic systems are sometimes called underdetermined elliptic or right elliptic systems, and injectively elliptic systems are sometimes called overdetermined elliptic or left elliptic systems. Elliptic systems may for precision be called two-sided elliptic.

When $P = \mathrm{Op}(p)$ and $Q = \mathrm{Op}(q)$ are pseudodifferential operators on Ω, we say that Q is a *right parametrix* for P if PQ can be defined and

$$PQ \sim I, \tag{7.46}$$

and q is a *right parametrix symbol* for p if

$$p \circ q \sim 1 \text{ (read as the identity, for matrix-formed operators)},$$

in the sense of the equivalences and the composition rules introduced above. Similarly, Q is a *left parametrix*, resp. q is a *left parametrix symbol* for p, when

$$QP \sim I \text{ resp. } q \circ p \sim 1.$$

When Q is both a right and a left parametrix, it is called a two-sided parametrix or simply a parametrix. (When P is of order d and Q is a one-sided parametrix of order $-d$, then it is two-sided if $N' = N$, as we shall see below.)

Theorem 7.18. 1° *Let* $p(x,\xi) \in S^d(\Omega, \mathbb{R}^n) \otimes \mathcal{L}(\mathbb{C}^N, \mathbb{C}^N)$. *Then* p *has a parametrix symbol* $q(x,\xi)$ *belonging to* $S^{-d}(\Omega, \mathbb{R}^n) \otimes \mathcal{L}(\mathbb{C}^N, \mathbb{C}^N)$ *if and only if* p *is elliptic.*

2° *Let* $p(x,\xi) \in S^d(\Omega, \mathbb{R}^n) \otimes \mathcal{L}(\mathbb{C}^N, \mathbb{C}^{N'})$. *Then* p *has a right (left) parametrix symbol* $q(x,\xi)$ *belonging to* $S^{-d}(\Omega, \mathbb{R}^n) \otimes \mathcal{L}(\mathbb{C}^{N'}, \mathbb{C}^N)$ *if and only if* p *is surjectively (resp. injectively) elliptic.*

Proof. 1° (The case of square matrices.) Assume that p is elliptic. Let $q_{-d}(x,\xi)$ be a C^∞-function on $\Omega \times \mathbb{R}^n$ (($N \times N$)-matrix-formed), that coincides with $p_d(x,\xi)^{-1}$ for $|\xi| \geq 1$ (one can extend p_d to be homogeneous for all $\xi \neq 0$ and take $q_{-d}(x,\xi) = [1 - \chi(2\xi)]p_d(x,\xi)^{-1}$). By Theorem 7.13 3° (cf. (7.43)),

$$p(x,\xi) \circ q_{-d}(x,\xi) \sim 1 - r(x,\xi), \tag{7.47}$$

for some $r(x,\xi) \in S^{-1}(\Omega)$. For each M, let

$$r^{\circ M}(x,\xi) \sim r(x,\xi) \circ r(x,\xi) \circ \cdots \circ r(x,\xi) \ (M \text{ factors}),$$

it lies in $S^{-M}(\Omega)$. Now

$$
\begin{aligned}
p(x,\xi) &\circ q_{-d}(x,\xi) \circ \big(1 + r(x,\xi) + r^{\circ 2}(x,\xi) + \cdots + r^{\circ M}(x,\xi)\big) \\
&\sim \big(1 - r(x,\xi)\big) \circ \big(1 + r(x,\xi) + r^{\circ 2}(x,\xi) + \cdots + r^{\circ M}(x,\xi)\big) \\
&\sim 1 - r^{\circ(M+1)}(x,\xi).
\end{aligned}
\tag{7.48}
$$

Here each term $r^{\circ M}(x,\xi)$ has an asymptotic development in homogeneous terms of degree $-M-j$, $j=0,1,2,\dots$,

$$r^{\circ M}(x,\xi) \sim r^{\circ M}_{-M}(x,\xi) + r^{\circ M}_{-M-1}(x,\xi) + \cdots,$$

and there exists a symbol $r'(x,\xi) \in S^{-1}(\Omega)$ with

$$r'(x,\xi) \sim \sum_{M \geq 1} r^{\circ M}(x,\xi), \quad \text{defined as} \quad \sum_{M \geq 1} \Big(\sum_{1 \leq j \leq M} r^{\circ j}_{-M} \Big)$$

(here we use rearrangement; the point is that there are only M terms of each degree $-M$). Finally,

$$p \circ q_{-d} \circ (1 + r') \sim 1, \tag{7.49}$$

which is seen as follows: For each M, there is a symbol $r'_{(M+1)}$ such that

$$r'_{(M+1)}(x,\xi) \sim \sum_{k \geq M+1} r^{\circ k}(x,\xi);$$

it is in $S^{-M-1}(\Omega)$. Then (cf. also (7.48))

$$p \circ q_{-d} \circ (1+r') \sim p \circ q_{-d} \circ (1 + \cdots + r^{\circ M}) + p \circ q_{-d} \circ r'_{(M+1)} \sim 1 + r''_{(M+1)},$$

where $r''_{(M+1)} = -r^{\circ(M+1)} + p \circ q_{-d} \circ r'_{(M+1)}$ is in $S^{-M-1}(\Omega)$. Since this holds for any M, $p \circ q_{-d} \circ (1 + r') - 1$ is in $S^{-\infty}(\Omega)$. In other words, (7.49) holds. This gives a right parametrix of p, namely,

$$q \sim q_{-d} \circ (1 + r') \in S^{-d}(\Omega).$$

Similarly, there exists a left parametrix q' for p. Finally, $q' \sim q$, since

$$q' - q \sim q' \circ (p \circ q) - (q' \circ p) \circ q \sim 0,$$

so q itself is also a left parametrix. We have then shown that when p is elliptic of order d, it has a parametrix symbol q of order $-d$, and any left/right parametrix symbol is also a right/left parametrix symbol.

Conversely, if $q \in S^{-d}(\Omega)$ is such that $p \circ q \sim 1$, then the principal symbols satisfy

$$p_d(x,\xi) q_{-d}(x,\xi) = 1 \quad \text{for } |\xi| \geq 1,$$

in view of (7.9) (since $p \circ q \sim p_d q_{-d} +$ terms of degree ≤ -1), so p_d is elliptic.

$2°$ We now turn to the case where p is not necessarily a square matrix; assume for instance that $N \geq N'$. Here $p_d(x,\xi)$ is, when p is surjectively elliptic of order d, a matrix defining a surjective operator from $\mathbb{C}^N$ to $\mathbb{C}^{N'}$ for each (x,ξ) with $|\xi| \geq 1$; and hence

$$\tilde{p}(x,\xi) = p_d(x,\xi) p_d(x,\xi)^* : \mathbb{C}^{N'} \to \mathbb{C}^{N'}$$

is bijective. Let $\tilde{q}(x,\xi) = \tilde{p}(x,\xi)^{-1}$ for $|\xi| \geq 1$, extended to a C^∞-function for $|\xi| \leq 1$; and note that $\tilde{q} \in S^{-2d}(\Omega, \mathbb{R}^n) \otimes \mathcal{L}(\mathbb{C}^{N'}, \mathbb{C}^{N'})$. Now, by Theorem 7.13 3°,

$$p(x,\xi) \circ p_d(x,\xi)^* \circ \tilde{q}(x,\xi) \sim 1 - r(x,\xi),$$

where $r(x,\xi) \in S^{-1}(\Omega, \mathbb{R}^n) \otimes \mathcal{L}(\mathbb{C}^{N'}, \mathbb{C}^{N'})$. We can then proceed exactly as under 1°, and construct the complete right parametrix symbol $q(x,\xi)$ as

$$q(x,\xi) \sim p_d(x,\xi)^* \circ \tilde{q}(x,\xi) \circ \big(1 + \sum_{M \geq 1} r^{\circ M}(x,\xi)\big). \tag{7.50}$$

(One could instead have taken $\tilde{p}' \sim p \circ p_d{}^*$, observed that it has principal symbol $p_d p_d{}^*$ (in S^{2d}) in view of the composition formula (7.41), thus is elliptic, and applied 1° to this symbol. This gives a parametrix symbol $\tilde{q}'$ such that $p \circ p_d{}^* \circ \tilde{q}' \sim 1$, and then $p_d{}^* \circ \tilde{q}'$ is a right parametrix symbol for p.)

The construction of a left parametrix symbol in case $N \leq N'$ is analogous. The necessity of ellipticity is seen as under 1°.   □

The above proof is *constructive*; it shows that p has the parametrix symbol

$$q(x,\xi) \sim q_{-d}(x,\xi) \circ (1 + \sum_{M \geq 1} (1 - p(x,\xi) \circ q_{-d}(x,\xi))^{\circ M}),$$

$$\text{with } q_{-d}(x,\xi) = p_d(x,\xi)^{-1} \text{ for } |\xi| \geq 1, \tag{7.51}$$

in the square matrix case, and (7.50) in case $N \geq N'$ (with a related left parametrix in case $N \leq N'$). Sometimes one is interested in the precise structure of the lower order terms, and they can be calculated from the above formulas.

Corollary 7.19. 1° *When P is a square matrix-formed ψdo on Ω that is elliptic of order d, then it has a properly supported parametrix Q that is an elliptic ψdo of order $-d$. The parametrix Q is unique up to a negligible term.*

2° When P is a surjectively elliptic ψdo of order d on Ω, then it has a properly supported right parametrix Q, which is an injectively elliptic ψdo of order $-d$. When P is an injectively elliptic ψdo of order d on Ω, then it has a properly supported left parametrix Q, which is a surjectively elliptic ψdo of order $-d$.

Proof. That P is surjectively/injectively elliptic of order d means that $P = \text{Op}\big(p(x,\xi)\big) + \mathcal{R}$, where $\mathcal{R}$ is negligible and $p \in S^d(\Omega) \otimes \mathcal{L}(\mathbb{C}^N, \mathbb{C}^{N'})$ with p_d surjective/injective for $|\xi| \geq 1$. Let $q(x,\xi)$ be the parametrix symbol constructed according to Theorem 7.18. For Q we can then take any properly supported operator $Q = \text{Op}\big(q(x,\xi)\big) + \mathcal{R}'$ with $\mathcal{R}'$ negligible.

In the square matrix-formed case, we have that when Q is a properly supported right parametrix, and Q' is a properly supported left parametrix, then since $PQ = I - \mathcal{R}_1$ and $Q'P = I - \mathcal{R}_2$,

$$Q' - Q = Q'(PQ + \mathcal{R}_1) - (Q'P + \mathcal{R}_2)Q = Q'\mathcal{R}_1 - \mathcal{R}_2 Q$$

is negligible; and Q and Q' are, both of them, two-sided parametrices.   □

There are some immediate consequences for the solvability of the equation $Pu = f$, when P is elliptic.

Corollary 7.20. *When P is injectively elliptic of order d on Ω, and properly supported, then any solution $u \in \mathscr{D}'(\Omega)$ of the equation*

$$Pu = f \quad in\ \Omega, \quad with\ f \in H^s_{\mathrm{loc}}(\Omega), \tag{7.52}$$

satisfies $u \in H^{s+d}_{\mathrm{loc}}(\Omega)$.

Proof. Let Q be a properly supported left parametrix of P, it is of order $-d$. Then $QP = I - \mathcal{R}$, with $\mathcal{R}$ negligible, so u satisfies

$$u = QPu + \mathcal{R}u = Qf + \mathcal{R}u.$$

Here Q maps $H^s_{\mathrm{loc}}(\Omega)$ into $H^{s+d}_{\mathrm{loc}}(\Omega)$ and $\mathcal{R}$ maps $\mathscr{D}'(\Omega)$ into $C^\infty(\Omega)$, so $u \in H^{s+d}_{\mathrm{loc}}(\Omega)$.   □

This is a far-reaching generalization of the regularity result in Theorem 6.29.

Without assuming that P is properly supported, we get the same conclusion for $u \in \mathscr{E}'(\Omega)$.

Corollary 7.20 is a *regularity* result, that is interesting also for uniqueness questions: If one knows that there is uniqueness of "smooth" solutions, then injective ellipticity gives uniqueness of any kind of solution.

Another consequence of injective ellipticity is that there is always *uniqueness modulo C^∞ solutions*: If u and u' satisfy $Pu = Pu' = f$, then (with notations as above)

$$u - u' = (QP + \mathcal{R})(u - u') = \mathcal{R}(u - u') \in C^\infty(\Omega). \tag{7.53}$$

Note also that there is an *estimate* for $u \in H^s_{\mathrm{comp}}(\Omega)$ with support in $K \subset \Omega$:

$$\|u\|_s = \|QPu + \mathcal{R}u\|_s \leq C(\|Pu\|_{s-d} + \|u\|_{s-1}), \tag{7.54}$$

since $\mathcal{R}$ is of order ≤ -1.

For surjectively elliptic differential operators we can show an *existence* result for data with small support.

Corollary 7.21. *Let P be a surjectively elliptic differential operator of order d in Ω, let $x_0 \in \Omega$ and let $r > 0$ be such that $\overline{B}(x_0, r) \subset \Omega$. Then there exists $r_0 \leq r$ so that for any $f \in L_2(B(x_0, r_0))$, there is a $u \in H^d(B(x_0, r_0))$ satisfying $Pu = f$ on $B(x_0, r_0)$.*

Proof. Let Q be a properly supported right parametrix of P. Then $PQ = I - \mathcal{R}$ where $\mathcal{R}$ is negligible, and

$$PQf = f - \mathcal{R}f \quad \text{for } f \in L_{2,\text{comp}}(\Omega).$$

Let $r_0 \leq r$ (to be fixed later), and denote $B(x_0, r_0) = B_0$. Similarly to earlier conventions, a function in $L_2(B_0)$ will be identified with its extension by 0 on $\Omega \setminus B_0$, which belongs to $L_2(\Omega)$ and has support in $\overline{B}_0$. We denote by r_{B_0} the operator that restricts to B_0.

When f is supported in $\overline{B}_0$,

$$r_{B_0} PQf = f - 1_{B_0} \mathcal{R}f \text{ on } B_0.$$

We know that $\mathcal{R}$ has a C^∞ kernel $K(x, y)$, so that

$$(\mathcal{R}f)(x) = \int_{B_0} K(x, y) f(y) \, dy, \text{ when } f \in L_2(B_0);$$

moreover,

$$\begin{aligned}
\|r_{B_0} \mathcal{R}f\|_{L_2(B_0)}^2 &= \int_{B_0} \left| \int_{B_0} K(x, y) f(y) \, dy \right|^2 dx \\
&\leq \int_{B_0} \left(\int_{B_0} |K(x, y)|^2 \, dy \right) \left(\int_{B_0} |f(y)|^2 \, dy \right) dx \\
&= \int_{B_0 \times B_0} |K(x, y)|^2 \, dx dy \, \|f\|_{L_2(B_0)}^2.
\end{aligned}$$

When $r_0 \to 0$, $\int_{B_0 \times B_0} |K(x, y)|^2 \, dx dy \to 0$, so there is an $r_0 > 0$ such that the integral is $\leq \frac{1}{4}$; then

$$\|r_{B_0} \mathcal{R}f\|_{L_2(B_0)} \leq \tfrac{1}{2} \|f\|_{L_2(B_0)} \text{ for } f \in L_2(B_0).$$

So the norm of the operator S in $L_2(B_0)$ defined from $r_{B_0} \mathcal{R}$,

$$S : f \mapsto r_{B_0} \mathcal{R}f,$$

is $\leq \frac{1}{2}$. Then $I - S$ can be inverted by a Neumann series

$$(I - S)^{-1} = I + S + S^2 + \cdots = \sum_{k=0}^{\infty} S^k \qquad (7.55)$$

converging in norm to a bounded operator in $L_2(B_0)$. It follows that when $f \in L_2(B_0)$, we have on B_0 (using that $r_{B_0} P 1_{B_0} v = r_{B_0} P v$ since P is a differential operator):

$$r_{B_0} P 1_{B_0} Q (I - S)^{-1} f = r_{B_0} PQ (I - S)^{-1} f = r_{B_0} (I - 1_{B_0} \mathcal{R})(I - S)^{-1} f = f,$$

which shows that $u = 1_{B_0}Q(I - S)^{-1}f$ solves the equation $Pu = f$ on B_0. Since Q is of order $-d$, $r_{B_0}u$ lies in $H^d(B_0)$. $\square$

More powerful conclusions can be obtained for ψdo's on compact manifolds; they will be taken up in Chapter 8.

7.5 Strongly elliptic operators, the Gårding inequality

A polyhomogeneous pseudodifferential operator $P = \mathrm{Op}(p(x,\xi))$ of order d on Ω is said to be *strongly elliptic*, when the principal symbol satisfies

$$\mathrm{Re}\, p^0(x,\xi) \geq c_0(x)|\xi|^d, \text{ for } x \in \Omega,\ |\xi| \geq 1, \tag{7.56}$$

with $c_0(x)$ continuous and positive. It is *uniformly strongly elliptic* when $c_0(x)$ has a positive lower bound (this holds of course on compact subsets of Ω). The definition extends to $(N \times N)$-matrix-formed operators, when we define $\mathrm{Re}\, p^0 = \frac{1}{2}(p^0 + p^{0*})$ and read (7.56) in the matrix sense:

$$(\mathrm{Re}\, p^0(x,\xi)v, v) \geq c_0(x)|\xi|^d|v|^2, \text{ for } x \in \Omega,\ |\xi| \geq 1,\ v \in \mathbb{C}^N. \tag{7.57}$$

When P is uniformly strongly elliptic of order $d > 0$, one can show that $\mathrm{Re}\, P = \frac{1}{2}(P + P^*)$ has a certain lower semiboundedness property called the *Gårding inequality* (it was shown for differential operators by Gårding in [G53]):

$$\mathrm{Re}(Pu, u) \geq c_1\|u\|_{d/2}^2 - c_2\|u\|_0^2 \tag{7.58}$$

holds for $u \in C_0^\infty(\Omega)$, with some $c_1 > 0$, $c_2 \in \mathbb{R}$.

Before giving the proof, we shall establish a useful interpolation property of Sobolev norms.

Theorem 7.22. *Let s and $t \in \mathbb{R}$.*
1° *For any $\theta \in [0, 1]$ one has for all $u \in H^{\max\{s,t\}}(\mathbb{R}^n)$:*

$$\|u\|_{\theta s + (1-\theta)t, \wedge} \leq \|u\|_{s,\wedge}^\theta \|u\|_{t,\wedge}^{1-\theta}. \tag{7.59}$$

2° *Let $s < r < t$. For any $\varepsilon > 0$ there exists $C(\varepsilon) > 0$ such that*

$$\|u\|_{r,\wedge} \leq \varepsilon\|u\|_{t,\wedge} + C(\varepsilon)\|u\|_{s,\wedge} \text{ for } u \in H^t(\mathbb{R}^n). \tag{7.60}$$

Proof. 1°. When $\theta = 0$ or 1, the inequality is trivial, so let $\theta \in\,]0, 1[$. Then we use the Hölder inequality with $p = 1/\theta$, $p' = 1/(1 - \theta)$:

$$\|u\|_{\theta s+(1-\theta)t,\wedge}^2 = \int_{\mathbb{R}^n} \langle\xi\rangle^{2\theta s+2(1-\theta)t}|\hat{u}(\xi)|^2\,d\xi$$

$$= \int_{\mathbb{R}^n} (\langle\xi\rangle^{2s}|\hat{u}(\xi)|^2)^{1/p}(\langle\xi\rangle^{2t}|\hat{u}(\xi)|^2)^{1/p'}\,d\xi$$

$$\leq \left(\int_{\mathbb{R}^n} \langle\xi\rangle^{2s}|\hat{u}(\xi)|^2\,d\xi\right)^{1/p} \left(\int_{\mathbb{R}^n} \langle\xi\rangle^{2t}|\hat{u}(\xi)|^2\,d\xi\right)^{1/p'}$$

$$= \|u\|_{s,\wedge}^{2\theta}\|u\|_{t,\wedge}^{2(1-\theta)}.$$

2°. Taking $\theta = (r-t)/(s-t)$, we have (7.59) with $\theta s+(1-\theta)t = r$. For $\theta \in [0,1]$ and a and $b \geq 0$ there is the general inequality:

$$a^\theta b^{1-\theta} \leq \max\{a,b\} \leq a+b \tag{7.61}$$

(since e.g. $a^\theta \leq b^\theta$ when $0 \leq a \leq b$). We apply this to $b = \varepsilon\|u\|_{t,\wedge}$, $a = \varepsilon^{(-1+\theta)/\theta}\|u\|_{s,\wedge}$, $\theta = (r-t)/(s-t)$, which gives:

$$\|u\|_{r,\wedge} \leq (\varepsilon^{(-1+\theta)/\theta}\|u\|_{s,\wedge})^\theta(\varepsilon\|u\|_{t,\wedge})^{1-\theta}$$

$$\leq \varepsilon^{(-1+\theta)/\theta}\|u\|_{s,\wedge} + \varepsilon\|u\|_{t,\wedge}. \quad \square$$

Theorem 7.23 (THE GÅRDING INEQUALITY). *Let A be a properly supported $(N \times N)$-matrix-formed ψdo on Ω_1 of order $d > 0$, strongly elliptic on Ω_1. Denote $\frac{1}{2}d = d'$. Let Ω be an open subset such that $\overline{\Omega}$ is compact in Ω_1. There exist constants $c_0 > 0$ and $k \in \mathbb{R}$ such that*

$$\mathrm{Re}(Au,u) \geq c_0\|u\|_{d'}^2 - k\|u\|_0^2, \text{ when } u \in C_0^\infty(\Omega)^N. \tag{7.62}$$

Proof. The symbol $\mathrm{Re}\,a^0(x,\xi) = \frac{1}{2}(a^0(x,\xi)+a^0(x,\xi)^*)$ is the principal symbol of $\mathrm{Re}\,A = \frac{1}{2}(A+A^*)$ and is positive definite by assumption. Let

$$p^0(x,\xi) = \sqrt{\mathrm{Re}\,a^0(x,\xi)} \text{ for } |\xi| \geq 1,$$

extended smoothly to $|\xi| \leq 1$, it is elliptic of order d'. (When $N > 1$, one can define the square root as

$$\frac{i}{2\pi}\int_{\mathcal{C}} \lambda^{\frac{1}{2}}(\mathrm{Re}\,a^0(x,\xi)-\lambda)^{-1}\,d\lambda,$$

where $\mathcal{C}$ is a closed curve in $\mathbb{C} \setminus \overline{\mathbb{R}}_-$ encircling the spectrum of $\mathrm{Re}\,a^0(x,\xi)$ in the positive direction.)

Let P be a properly supported ψdo on Ω_1 with symbol $p^0(x,\xi)$; its adjoint P^* is likewise properly supported and has principal symbol $p^0(x,\xi)$. Then P^*P has principal symbol $\mathrm{Re}\,a^0(x,\xi)$, so

$$\mathrm{Re}\,A = P^*P + S,$$

where S is of order $d-1$. For any $s \in \mathbb{R}$, let Λ_s be a properly supported ψdo on Ω_1 equivalent with $\mathrm{Op}(\langle \xi \rangle^s I_N)$ (I_N denotes the $(N \times N)$-unit matrix). Then $\Lambda_{-d'}\Lambda_{d'} \sim I$, and we can rewrite

$$S = S_1 S_2 + \mathcal{R}_1, \text{ where } S_1 = S\Lambda_{-d'}, \; S_2 = \Lambda_{d'},$$

with S_1 of order $d'-1$, S_2 of order d' and $\mathcal{R}_1$ of order $-\infty$, all properly supported.

Since P is elliptic of order d', it has a properly supported parametrix Q of order $-d'$. Because of the properly supportedness there are bounded sets Ω' and Ω'' with $\overline{\Omega} \subset \Omega'$, $\overline{\Omega'} \subset \Omega''$, $\overline{\Omega''} \subset \Omega_1$, such that when $\mathrm{supp}\, u \subset \Omega$, then $Au, Pu, S_1 u, S_1^* u, S_2 u$ are supported in Ω' and $QPu, S_1 S_2 u, \mathcal{R}_1 u, \mathcal{R}_2 u$ are supported in Ω''.

For $u \in C_0^\infty(\Omega)^N$ we have

$$\begin{aligned}
\mathrm{Re}(Au, u) = \tfrac{1}{2}[(Au, u) + (u, Au)] &= ((\mathrm{Re}\, A)u, u) \\
&= (P^* Pu, u) + (S_1 S_2 u, u) + (\mathcal{R}_1 u, u) \\
&= \|Pu\|_0 + (S_2 u, S_1^* u) + (\mathcal{R}_1 u, u).
\end{aligned} \tag{7.63}$$

To handle the first term, we note that by the Sobolev space mapping properties of Q and $\mathcal{R}_2$ we have (similarly to (7.54))

$$\|u\|_{d'}^2 \le (\|QPu\|_{d'} + \|\mathcal{R}_2 u\|_{d'})^2 \le C(\|Pu\|_0 + \|u\|_0)^2 \le 2C(\|Pu\|_0^2 + \|u\|_0^2)$$

with $C > 0$, hence

$$\|Pu\|_0^2 \ge (2C)^{-1}\|u\|_{d'}^2 - \|u\|_0^2. \tag{7.64}$$

The last term in (7.63) satisfies

$$|(\mathcal{R}_1 u, u)| \le \|\mathcal{R}_1 u\|_0 \|u\|_0 \le c\|u\|_0^2. \tag{7.65}$$

For the middle term, we estimate

$$|(S_2 u, S_1^* u)| \le \|S_2 u\|_0 \|S_1^* u\|_0 \le c' \|u\|_{d'} \|u\|_{d'-1} \le \tfrac{1}{2} c'(\varepsilon^2 \|u\|_{d'}^2 + \varepsilon^{-2} \|u\|_{d'-1}^2),$$

any $\varepsilon > 0$. If $d'-1 > 0$, we refine this further by using that by (7.60),

$$\|u\|_{d'-1}^2 \le \varepsilon' \|u\|_{d'}^2 + C'(\varepsilon')\|u\|_0^2,$$

for any $\varepsilon' > 0$. Taking first ε small and then ε' small enough, we can obtain that

$$|(S_2 u, S_1^* u)| \le (4C)^{-1}\|u\|_{d'}^2 + C''\|u\|_0^2. \tag{7.66}$$

Application of (7.64)–(7.66) in (7.63) gives that

$$\mathrm{Re}(Au, u) \ge (4C)^{-1}\|u\|_{d'}^2 - C'''\|u\|_0^2,$$

an inequality of the desired type. $\qquad\qquad\qquad\qquad\qquad\qquad\qquad \square$

The result can be applied to differential operators in the following way (generalizing the applications of the Lax-Milgram lemma given in Section 4.4):

Theorem 7.24. *Let $\Omega \subset \mathbb{R}^n$ be bounded and open, and let $A = \sum_{|\alpha| \leq d} a_\alpha D^\alpha$ be strongly elliptic on a neighborhood Ω_1 of $\overline{\Omega}$ (with $(N \times N)$-matrix-formed C^∞-functions $a_\alpha(x)$ on Ω_1). Then d is even, $d = 2m$, and the realization A_γ of A in $L_2(\Omega)^N$ with domain $D(A_\gamma) = H_0^m(\Omega)^N \cap D(A_{\max})$ is a variational operator (hence has its spectrum and numerical range in an angular set (12.50)).*

Proof. The order d is even, because $\operatorname{Re} a^0(x, \xi)$ and $\operatorname{Re} a^0(x, -\xi)$ are both positive definite. Theorem 7.23 assures that

$$\operatorname{Re}(Au, u) \geq c_0 \|u\|_m^2 - k \|u\|_0^2, \text{ when } u \in C_0^\infty(\Omega)^N, \tag{7.67}$$

with $c_0 > 0$. We can rewrite A in the form

$$Au = \sum_{|\beta|, |\theta| \leq m} D^\beta (b_{\beta\theta} D^\theta u);$$

with suitable matrices $b_{\beta\theta}(x)$ that are C^∞ on Ω_1, using the backwards Leibniz formula (7.42). Then define the sesquilinear form $a(u, v)$ by

$$a(u, v) = \sum_{|\beta|, |\theta| \leq m} (b_{\beta\theta} D^\theta u, D^\beta v)_{L_2(\Omega)^N}, \text{ for } u, v \in H_0^m(\Omega)^N;$$

it is is bounded on $V = H_0^m(\Omega)^N$ since the $b_{\beta\theta}$ are bounded on Ω. Moreover, by distribution theory,

$$(Au, v) = a(u, v) \text{ for } u \in D(A_{\max}), v \in C_0^\infty(\Omega)^N; \tag{7.68}$$

this identity extends by continuity to $v \in V$ (recall that $H_0^m(\Omega)^N$ is the closure of $C_0^\infty(\Omega)^N$ in m-norm). Then by (7.67), $a(u, v)$ is V-coercive (12.39), with $H = L_2(\Omega)^N$.

Applying the Lax-Milgram construction from Section 12.4 to the triple $\{H, V, a\}$, we obtain a variational operator A_γ. In the following, use the notation of Chapter 4. In view of (7.68), A_γ is a closed extension of $A_{C_0^\infty}$, hence of $A_{\min}$. Similarly, the Hilbert space adjoint $A_\gamma{}^*$ extends $A'_{\min}$, by the same construction applied to the triple $\{H, V, a^*\}$, so A_γ is a realization of A. It follows that $D(A_\gamma) \subset H_0^m(\Omega)^N \cap D(A_{\max})$. By (7.68)ff., this inclusion is an identity. $\square$

We find moreover that $D(A_\gamma) \subset H_{\text{loc}}^{2m}(\Omega)^N$, in view of the ellipticity of A and Corollary 7.20.

Note that there were no smoothness assumptions on Ω whatsoever in this theorem. With some smoothness, one can moreover show that the domain is in $H^{2m}(\Omega)^N$. This belongs to the deeper theory of elliptic boundary value problems (for which a pseudodifferential strategy is presented in Chapters 10 and 11), and will be shown for smooth sets at the end of Chapter 11.

A_γ is regarded as the Dirichlet realization of A, since its domain consists of those functions u in the maximal domain that belong to $H_0^m(\Omega)^N$; when Ω is sufficiently smooth, this means that the Dirichlet boundary values $\{\gamma_0 u, \gamma_1 u, \dots, \gamma_{m-1} u\}$ are 0.

One can also show a version of Theorem 7.24 for suitable ψdo's, cf. [G96, Sect. 1.7].

Exercises for Chapter 7

7.1. For an arbitrary $s \in \mathbb{R}$, find the asymptotic expansion of the symbol $\langle\xi\rangle^s$ in homogeneous terms.

7.2. Show, by insertion in the formula (7.16) and suitable reductions, that the ψdo defined from the symbol (7.28) is zero.

7.3. Let $p(x,\xi)$ and $p'(x,\xi)$ be polyhomogeneous pseudodifferential symbols, of order d resp. d'. Consider $p'' \sim p \circ p'$, defined according to (7.41) and Definition 7.15; it is polyhomogeneous of degree $d'' = d + d'$.

(a) Show that $p''_{d''} = p_d\, p'_{d'}$.

(b) Show that

$$p''_{d''-1} = \sum_{j=1}^{n} D_{\xi_j} p_d\, \partial_{x_j} p'_{d'} + p_d\, p'_{d'-1} + p_{d-1}\, p'_{d'}.$$

(c) Find $p''_{d''-2}$.

7.4. Consider the fourth-order operator $a(x)\Delta^2 + b(x)$ on $\mathbb{R}^n$, where a and b are C^∞-functions with $a(x) > 0$ for all x.

(a) Show that A is elliptic. Find a parametrix symbol, where the first three homogeneous terms (of order $-4, -5, -6$) are worked out in detail.

(b) Investigate the special case $a = |x|^2 + 1$, $b = 0$.

7.5. The operator $L_\sigma = -\operatorname{div}\operatorname{grad} + \sigma \operatorname{grad}\operatorname{div} = -\Delta + \sigma \operatorname{grad}\operatorname{div}$, applied to n-vectors, is a case of the Lamé operator. In details

$$L_\sigma u = \begin{pmatrix} -\Delta + \sigma\partial_1^2 & \sigma\partial_1\partial_2 & \cdots & \sigma\partial_1\partial_n \\ \sigma\partial_1\partial_2 & -\Delta + \sigma\partial_2^2 & \cdots & \sigma\partial_2\partial_n \\ \vdots & \vdots & \ddots & \vdots \\ \sigma\partial_1\partial_n & \sigma\partial_2\partial_n & \cdots & -\Delta + \sigma\partial_n^2 \end{pmatrix} \begin{pmatrix} u_1 \\ u_2 \\ \vdots \\ u_n \end{pmatrix};$$

here σ is a real constant. Let $n = 2$ or 3.

(a) For which σ is L_σ elliptic?

(b) For which σ is L_σ strongly elliptic?

Chapter 8
Pseudodifferential operators on manifolds, index of elliptic operators

8.1 Coordinate changes

Pseudodifferential operators will now be defined on compact manifolds. Our presentation here is meant as a useful orientation for the reader, with relatively brief explanations.

In order to define ψdo's on manifolds, we have to investigate how they behave under coordinate changes. Let Ω and $\underline{\Omega}$ be open subsets of $\mathbb{R}^n$ together with a diffeomorphism κ of Ω onto $\underline{\Omega}$. When P is a ψdo defined on Ω, we define $\underline{P}$ on $\underline{\Omega}$ by

$$\underline{P}\underline{u} = P(\underline{u} \circ \kappa) \circ \kappa^{-1}, \text{ when } \underline{u} \in C_0^\infty(\underline{\Omega}), \tag{8.1}$$

the definition extends to larger spaces as explained earlier.

Theorem 8.1. $1°$ *Let* $P = \mathrm{Op}(q(x,y,\xi))$ *be a* ψdo *with* $q(x,y,\xi) \in S_{1,0}^m(\Omega \times \Omega, \mathbb{R}^n)$ *and let* K *be a compact subset of* Ω. *With* κ' *denoting the Jacobian matrix* $(\partial \kappa_i / \partial x_j)$, *let* $M(x,y)$ *be the matrix defined by (8.6) below; it satisfies*

$$\underline{x} - \underline{y} = M(x,y)(x - y), \tag{8.2}$$

and is invertible on $U_\varepsilon = \{(x,y) \mid x,y \in K, |x - y| < \varepsilon\}$ *for a sufficiently small* $\varepsilon > 0$, *with* M, M^{-1} *and their derivatives bounded there. In particular,* $M(x,x) = \kappa'(x)$.

If $q(x,y,\xi)$ *vanishes for* $(x,y) \notin U_\varepsilon$, *then* $\underline{P}$ *is a* ψdo *with a symbol* $\underline{q}(\underline{x}, \underline{y}, \underline{\xi}) \in S_{1,0}^m(\underline{\Omega} \times \underline{\Omega}, \mathbb{R}^n)$ *(vanishing for* $(\underline{x}, \underline{y})$ *outside the image of* U_ε*); it satisfies, with* $\underline{x} = \kappa(x)$, $\underline{y} = \kappa(y)$,

$$\underline{q}(\underline{x}, \underline{y}, \underline{\xi}) = q(x, y, {}^t M(x,y)\underline{\xi}) \cdot |\det {}^t M(x,y)| \, |\det \kappa'(y)^{-1}|. \tag{8.3}$$

In particular, $\underline{q}(\underline{x}, \underline{x}, \underline{\xi}) = q(x, x, {}^t\kappa'(x)\underline{\xi})$. *If* q *is polyhomogeneous, then so is* $\underline{q}$.

If $q(x, y, \xi)$ vanishes for $(x, y) \notin K \times K$, let $\varrho(x, y) = \chi(|x - y|/r)$, then $\underline{P}$ is, for sufficiently small r, the sum of an operator with symbol (8.3) multiplied by $\chi(|x - y|/r)$ and a negligible operator.

For general $q(x, y, \xi)$, let $(\varphi_j)_{j \in \mathbb{N}_0}$ be a locally finite partition of unity on Ω and write $q(x, y, \xi) = \sum_{j,k} \varphi_j(x) q(x, y, \xi) \varphi_k(y)$, so that $P = \sum_{j,k} \varphi_j P \varphi_k$; then $\underline{P} = \sum_{j,k} \underline{\varphi}_j \underline{P} \underline{\varphi}_k$ with $\underline{\varphi}(\kappa(x)) = \varphi(x)$. The terms where $\operatorname{supp} \varphi_j \cap \operatorname{supp} \varphi_k \neq \emptyset$ are treated as above, and the others define a negligible operator that transforms to a negligible operator.

$2°$ *When $P = \operatorname{Op}(p(x, \xi))$, $\underline{P}$ is the sum of a ψdo in $\underline{x}$-form $\operatorname{Op}(p_\kappa(\underline{x}, \underline{\xi}))$ and a negligible operator, where the symbol p_κ has the asymptotic expansion (Hörmander's formula)*

$$
\begin{aligned}
p_\kappa(\kappa(x), \underline{\xi}) &= e^{-i\kappa(x) \cdot \underline{\xi}} P(e^{i\kappa(x) \cdot \underline{\xi}}) \\
&\sim \sum_{\alpha \in \mathbb{N}_0^n} \tfrac{1}{\alpha!} D_\xi^\alpha p(x, {}^t\kappa'(x)\underline{\xi}) \partial_y^\alpha e^{i\mu_x(y) \cdot \underline{\xi}}|_{y=x}
\end{aligned}
\tag{8.4}
$$

in $S_{1,0}^m(\Omega, \mathbb{R}^n)$; here $\mu_x(y) = \kappa(y) - \kappa(x) - \kappa'(x)(y - x)$, and each factor $\varphi_\alpha(x, \underline{\xi}) = \partial_y^\alpha e^{i\mu_x(y) \cdot \underline{\xi}}|_{y=x}$ is in $S^{|\alpha|/2}(\Omega, \mathbb{R}^n)$ and is a polynomial in $\underline{\xi}$ of degree $\leq |\alpha|/2$. In particular, $\varphi_0 = 1$ and $\varphi_\alpha = 0$ for $|\alpha| = 1$, and in the case where p is polyhomogeneous,

$$
p_\kappa^0(\underline{x}, \underline{\xi}) = p^0(x, {}^t\kappa'(x)\underline{\xi}),
\tag{8.5}
$$

on the set where it is homogeneous.

Proof. By Taylor's formula applied to each κ_i, (8.2) holds with

$$
M(x, y) = \int_0^1 \kappa'(x + t(y - x)) \, dt,
\tag{8.6}
$$

a smooth function of x and y where it is defined; the domain includes a neighborhood of the diagonal in $\Omega \times \Omega$. (Note that an M satisfying (8.2) is uniquely determined only if $n = 1$.) In particular, $M(x, x) = \kappa'(x)$, hence is invertible. Since $M(x, y) = M(x, x)[I + M(x, x)^{-1}(M(x, y) - M(x, x))]$, it is seen by a Neumann series construction that $M(x, y)^{-1}$ exists and is bounded (with bounded derivatives) for $(x, y) \in U_\varepsilon$, for a sufficiently small $\varepsilon > 0$.

If $q(x, y, \xi)$ vanishes for $(x, y) \notin U_\varepsilon$, we have for $\underline{u} \in C_0^\infty(\underline{\Omega})$, setting $\xi = {}^t M(x, y)\underline{\xi}$:

$$
\begin{aligned}
(\underline{P}\underline{u})(\kappa(x)) &= \int e^{i(x-y) \cdot \xi} q(x, y, \xi) \underline{u}(\kappa(y)) \, dy d\xi \\
&= \int e^{i(x-y) \cdot {}^t M \underline{\xi}} q(x, y, \xi) \underline{u}(y) |\det \kappa'(y)^{-1}| \, |\det {}^t M(x, y)| \, dy d\underline{\xi} \\
&= \int e^{i(\underline{x}-\underline{y}) \cdot \underline{\xi}} \underline{q}(\underline{x}, \underline{y}, \underline{\xi}) u(\underline{y}) \, d\underline{y} d\underline{\xi},
\end{aligned}
$$

with q defined by (8.3). Clearly, q is a symbol in $S_{1,0}^d$ as asserted. When $x = y$, $\det \kappa'(y)^{-1}$ and $\det {}^t M(x,y)$ cancel out. The formula (8.3) shows moreover that polyhomogeneity is preserved.

If $q(x,y,\xi)$ vanishes for $(x,y) \notin K \times K$, we write

$$P = \mathcal{R} + P_1,$$

where $\mathcal{R} = \mathrm{Op}((1 - \chi(|x-y|/r))q(x,y,\xi))$ is negligible by Lemma 7.9, and P_1 is as above; we can e.g. take $r = \varepsilon/2$. Since $\mathcal{R}$ is an integral operator with C^∞-kernel, so is the transformed operator $\underline{\mathcal{R}}$.

For the general q, one uses that the summation of the terms with $\mathrm{supp}\,\varphi_j \cap \mathrm{supp}\,\varphi_k \neq \emptyset$ is finite locally in (x,y).

If we now consider an operator given in x-form, one can find the x-form p_κ of the symbol of $\underline{P}$ by an application of Theorem 7.13 1°. The formula (8.5) follows easily from this.

As for (8.4), the first formula is the characterization we know from (7.29). The second formula is given in more detail in [H85, Th. 18.1.17]; it was first proved in [H65]. The present method of proof going via (x,y)-forms is slightly different from that of [H65] and is, according to Friedrichs [F68], due to Kuranishi.   □

8.2 Operators on manifolds

The definition of an n-dimensional C^∞-manifold X is explained e.g. in [H63, Sect. 1.8] and [H83, pp. 143–144]. X is a Hausdorff topological space, provided with a family $\mathcal{F}$ of homeomorphisms κ, called coordinate systems, of open sets $U_\kappa \subset X$ onto open sets $V_\kappa \subset \mathbb{R}^n$ (coordinate patches) such that: (i) For any κ_j, κ_k in the family,

$$\kappa_j \kappa_k^{-1} : \kappa_k(U_{\kappa_j} \cap U_{\kappa_k}) \to \kappa_j(U_{\kappa_j} \cap U_{\kappa_k}) \text{ is a diffeomorphism.} \quad (8.7)$$

(This is of course an empty statement unless $U_{\kappa_j} \cap U_{\kappa_k} \neq \emptyset$.) (ii) The sets U_κ cover X. (iii) The family $\mathcal{F}$ is complete, in the sense that when a homeomorphism κ_0 from an open set $U_0 \subset X$ to an open set $V_0 \subset \mathbb{R}^n$ is such that (8.7) holds for $\kappa_j = \kappa_0$, any $\kappa_k \in \mathcal{F}$, then $\kappa_0 \in \mathcal{F}$. A subfamily where the U_κ's cover X is called an *atlas;* it already describes the structure.

Consider just compact manifolds, then a finite atlas $\{\kappa_j : U_j \to V_j \mid j = 1, \ldots, j_0\}$ suffices to describe the structure, and we can assume that the V_j are bounded and mutually disjoint in $\mathbb{R}^n$. We define that a function u on X is C^∞, C^m or $L_{p,\mathrm{loc}}$, when the function $y \mapsto u(\kappa_j^{-1}(y))$ is so on V_j, for each j. Since X is compact, the $L_{p,\mathrm{loc}}$-functions are in fact in $L_p(X)$, which can be provided with a Banach space norm $(\sum_{j=1}^{j_0} \|(\psi_j u) \circ \kappa_j^{-1}\|_{L_p(V_j)}^p)^{\frac{1}{p}}$, defined with the help of a partition of unity as in Lemma 8.4 1° below.

Recall the rule for coordinate changes of distributions in Definition 3.19, which, in the application to test functions, carries a functional determinant factor J along in order to make the rule consistent with coordinate changes in integrals with continuous functions. Namely, when $\kappa : x \mapsto \underline{x}$ is a diffeomorphism from V to $\underline{V}$ in $\mathbb{R}^n$, $\underline{u} = u \circ \kappa^{-1}$ satisfies

$$\langle \underline{u}, \underline{\varphi} \rangle_{\underline{V}} = \langle u, J \cdot \varphi \rangle_V \; \left(= \langle J \cdot u, \varphi \rangle_V \right) \tag{8.8}$$

for $\varphi \in C_0^\infty(V)$, with $J(x) = |\det \kappa'(x)|$, $\underline{\varphi} = \varphi \circ \kappa^{-1}$.

A *distribution density* u on X is defined in [H83, Sect. 6.3] to be a collection of distributions $u_\kappa \in \mathscr{D}'(V_\kappa)$, $\kappa \in \mathcal{F}$, such that the rule (8.8) is respected by the diffeomorphisms $\kappa = \kappa_j \kappa_k^{-1}$ going from $\kappa_k(U_{\kappa_j} \cap U_{\kappa_k})$ to $\kappa_j(U_{\kappa_j} \cap U_{\kappa_k})$, for all $\kappa_j, \kappa_k \in \mathcal{F}$. The value of u on C_0^∞-functions φ on X is then found by linear extension from the cases where φ is supported in a U_κ: When $\varphi \in C_0^\infty(U_\kappa)$, then

$$\langle u, \varphi \rangle = \langle u_\kappa, \varphi \circ \kappa^{-1} \rangle_{V_\kappa}. \tag{8.9}$$

When $\varphi \in C^\infty(X)$, write φ as a finite sum of functions supported in coordinate sets U_κ by use of a partition of unity as in Lemma 8.4 1° below, and apply (8.9) to each term.

In particular, when the structure of the compact manifold X is defined by the atlas $\kappa_j : U_j \to V_j$, $1 \le j \le j_0$, let $(\psi_j)_{1 \le j \le j_0}$ be a partition of unity as in Lemma 8.4 1°; then the distribution density u defined from a system $u_j \in \mathscr{D}'(V_j)$, $1 \le j \le j_0$, is evaluated on test functions $\varphi \in C^\infty(X)$ by

$$\langle u, \varphi \rangle_X = \sum_{j=1}^{j_0} \langle u_j, (\psi_j \varphi) \circ \kappa_j^{-1} \rangle_{V_j}.$$

When u is a distribution density such that the u_j are in $C^m(V_j)$, one says that u is a C^m-density, it carries a multiplication by the functional determinant along in coordinate changes.

So, distribution densities do not quite generalize continuous functions. [H83] defines genuine distributions as a strict generalization of functions with the usual rule for coordinate changes as for functions, without the functional determinant factor. Such distributions can be evaluated, not on C_0^∞-functions, but on C_0^∞-*densities*, by use of local coordinates.

If one provides X with a smooth measure (or volume form) dx compatible with Lebesgue measure in local coordinates, e.g. coming from a Riemannian structure, one can identify the distribution densities with the distributions, giving $\langle u, \varphi \rangle_X$ a meaning for distributions u and C_0^∞-functions φ. This also gives a scalar product and norm in $L_2(X)$. *We shall assume from now on that such a choice has been made*, and denote the distribution space $\mathscr{D}'(X)$. With a notation from [H83, Sect. 6.3], the local representatives are denoted $u_\kappa = u \circ \kappa^{-1}$, $\kappa \in \mathcal{F}$. We refer to the quoted book for an explanation of how the identification between distributions and distribution densities is obtained with the help of a fixed choice of a positive C^∞-density.

There is a more refined presentation of distribution spaces over X in [H71] (and in [H85, p. 92]), where the introduction of densities of order $\frac{1}{2}$ (essentially carrying $J^{\frac{1}{2}}$ along with the measure) makes the situation for distributions and test functions more symmetric under coordinate changes. We shall make do with the old-fashioned explanations given above.

Sobolev spaces $H^s(X)$ can be defined by use of local coordinates and a partition of unity as in Lemma 8.4 1°: $u \in H^s(X)$ when, for each j, $(\psi_j u) \circ \kappa_j^{-1} \in H^s(\mathbb{R}^n)$ (here an extension by zero in $\mathbb{R}^n \setminus V_j$ is understood), and a Hilbert space norm on $H^s(X)$ can be defined by

$$\|u\|_s = (\sum_{j=1}^{j_0} \|(\psi_j u) \circ \kappa_j^{-1}\|_{H^s}^2)^{\frac{1}{2}}. \tag{8.10}$$

This formula depends on many choices and is in no way "canonical", so $H^s(X)$ could be viewed as a "hilbertable" space rather than a Hilbert space (with an expression heard in a lecture by Seeley).

It is not hard to see that $C^\infty(X)$ is dense in $H^s(X)$ for all s. Indeed, when $u \in H^s(X)$, one can approximate each piece $\psi_j u \circ \kappa_j^{-1}$ in H^s-norm by C_0^∞-functions $(v_{jk})_{k \in \mathbb{N}_0}$ on V_j; then $u_k(x) = \sum_j v_{jk}(\kappa_j(x))$ is an approximating sequence for u.

One can, after fixing the norms on $H^s(X)$ for $s \geq 0$, choose the norms in the $H^{-s}(X)$ so that $H^{-s}(X)$ identifies with the dual space of $H^s(X)$ in such a way that the duality is consistent with the L_2-duality.

Theorem 8.2 (RELLICH'S THEOREM). *The injection of $H^s(X)$ into $H^{s'}(X)$ is compact when $s > s'$.*

This can be proved in the following steps: 1) A reduction to compactly supported distributions in each coordinate patch by a partition of unity, 2) an embedding of a compact subset of a coordinate patch into $\mathbb{T}^n$ (the n-dimensional torus), 3) a proof of the property for $\mathbb{T}^n$ by use of Fourier series expansions.

We shall carry this program out below. To begin with, let us explain the Sobolev spaces over the torus.

A basic result in the theory of Fourier series is that the system $\{e^{ik\cdot x} \mid k \in \mathbb{Z}^n\}$ is an orthonormal basis of $L_2(\mathbb{T}^n, d\!\!\!/x)$; here $d\!\!\!/x = (2\pi)^{-n} dx$, and $\mathbb{T}^n$ is identified with $Q = [-\pi, \pi]^n$ glued together at the edges (in other words, the functions on $\mathbb{T}^n$ identify with functions on $\mathbb{R}^n$ that are periodic with period 2π in each coordinate $x_1, \ldots, x_n$). Then the mapping $f \mapsto (c_k(f))_{k \in \mathbb{Z}^n}$, $c_k(f) = (f, e^{ik\cdot x})$, defines an isometry of $L_2(\mathbb{T}^n, d\!\!\!/x)$ onto $\ell_2(\mathbb{Z}^n)$.

There is an easy way to define distributions on the torus. We have the explicit bilinear form $\int_{\mathbb{T}^n} f(x)g(x)\, d\!\!\!/x = \int_Q f(x)g(x)\, d\!\!\!/x$. The test functions are the C^∞-functions on $\mathbb{T}^n$ (C^∞-functions on $\mathbb{R}^n$ with period 2π in each coordinate $x_1, \ldots, x_n$), and we can identify $\mathscr{D}'(\mathbb{T}^n)$ with the dual space, such that an L_2-function f identifies with the distribution acting like $\varphi \mapsto \int_{\mathbb{T}^n} f(x)\varphi(x)\, dx$.

For a function $f \in C^m(\mathbb{T}^n)$ having the Fourier series $\sum_{k \in \mathbb{Z}^n} c_k e^{ik\cdot x}$, an integration by parts shows that

$$D^\alpha f = \sum_{k \in \mathbb{Z}^n} k^\alpha c_k e^{ik\cdot x}.$$

For $u \in C^m(\mathbb{T}^n)$, the m-th Sobolev norm therefore satisfies, in view of (5.2),

$$\|u\|_m = \Big(\sum_{|\alpha| \le m} \|D^\alpha u\|_{L_2}^2 \Big)^{\frac{1}{2}} \begin{cases} \le \|(\langle k \rangle^m c_k(u))_{k \in \mathbb{Z}^n}\|_{\ell_2}, \\ \ge c\|(\langle k \rangle^m c_k(u))_{k \in \mathbb{Z}^n}\|_{\ell_2}. \end{cases}$$

Since $C^\infty(\mathbb{T}^n)$ and hence $C^m(\mathbb{T}^n)$ is dense in $H^m(\mathbb{T}^n)$, we conclude that the m-th Sobolev space ($m \in \mathbb{N}_0$) satisfies

$$\begin{aligned} H^m(\mathbb{T}^n) &= \{\, u \in L_2(\mathbb{T}^n) \mid (k^\alpha c_k(u))_{k \in \mathbb{Z}^n} \in \ell_2 \text{ for } |\alpha| \le m \,\} \\ &= \{\, u \in L_2(\mathbb{T}^n) \mid (\langle k \rangle^m c_k(u))_{k \in \mathbb{Z}^n} \in \ell_2 \,\}. \end{aligned}$$

Denote by $\ell_2^s(\mathbb{Z}^n)$ (or ℓ_2^s) the space of sequences $\underline{a} = (a_k)_{k \in \mathbb{Z}^n}$ for which $\sum_{k \in \mathbb{Z}^n} |\langle k \rangle^s a_k|^2 < \infty$. It is a Hilbert space with scalar product and norm

$$(\underline{a}, \underline{b})_{\ell_2^s} = \sum_{k \in \mathbb{Z}^n} \langle k \rangle^{2s} a_k \overline{b}_k, \quad \|\underline{a}\|_{\ell_2^s} = \Big(\sum_{k \in \mathbb{Z}^n} |\langle k \rangle^s a_k|^2 \Big)^{\frac{1}{2}}; \qquad (8.11)$$

this follows immediately from the fact that multiplication $M_{\langle k \rangle^s} : (a_k) \mapsto (\langle k \rangle^s a_k)$ maps ℓ_2^s isometrically onto the well-known Hilbert space ℓ_2 ($= \ell_2^0$). Then the above calculations show that $H^m(\mathbb{T}^n)$ may be equivalently provided with the scalar product and norm

$$(u, v)_{m,\wedge} = \sum_{k \in \mathbb{Z}^n} \langle k \rangle^{2m} c_k(u) \overline{c_k(v)} = \big((c_k(u))_{k \in \mathbb{Z}^n}, (c_k(v))_{k \in \mathbb{Z}^n} \big)_{\ell_2^m},$$

$$\|u\|_{m,\wedge} = (u, u)_{m,\wedge}^{\frac{1}{2}} = \|(c_k(u))_{k \in \mathbb{Z}^n}\|_{\ell_2^m}.$$

For $s \in \mathbb{R}_+$, this generalizes immediately to define, as subspaces of $L_2(\mathbb{T}^n)$,

$$H^s(\mathbb{T}^n) = \{u \mid (c_k(u))_{k \in \mathbb{Z}^n} \in \ell_2^s(\mathbb{Z}^n)\}, \qquad (8.12)$$

provided with the scalar product and norm

$$(u, v)_{s,\wedge} = \sum_{k \in \mathbb{Z}^n} \langle k \rangle^{2s} c_k(u) \overline{c_k(v)} = \big((c_k(u)), (c_k(v)) \big)_{\ell_2^s},$$

$$\|u\|_{s,\wedge} = (u, u)_{s,\wedge}^{\frac{1}{2}} = \|(c_k(u))\|_{\ell_2^s}. \qquad (8.13)$$

Also for noninteger s, the definition of $H^s(\mathbb{T}^n)$ is consistent with the general definition on compact manifolds given further above; this can be shown e.g. by use of interpolation theory (cf. Lions and Magenes [LM68]: When $s \in {]}0,2{[}$, the space $H^s(X)$ is the domain of $A^{s/2}$, whenever A is a selfadjoint positive operator in $L_2(X)$ with domain $H^2(X)$). Details will not be given here.

We can moreover make a generalization to arbitrary $s \in \mathbb{R}$.

Observe that when f_k is a sequence in $\ell_2^s(\mathbb{Z}^n)$ for some $s \in \mathbb{R}$, then the series $\sum_{k\in\mathbb{Z}^n} f_k e^{ik\cdot x}$ converges in $\mathscr{D}'$ to a distribution f: Take a test function $\varphi \in C^\infty(\mathbb{T}^n)$; its Fourier series $\sum_{k\in\mathbb{Z}^n} a_k e^{ik\cdot x}$ has the coefficient sequence $(a_k)_{k\in\mathbb{Z}^n}$ lying in ℓ_2^r for any r. Let $f^N = \sum_{|k|\le N} f_k e^{ik\cdot x}$; then

$$\langle f^N, \varphi \rangle = \sum_{|k|\le N} f_k \bar{a}_k, \text{ where } \sum_{|k|\le N} |f_k \bar{a}_k| = \sum_{|k|\le N} |\langle k \rangle^s f_k| |\langle k \rangle^{-s} a_k|$$

$$\le \Big(\sum_{|k|\le N} |\langle k \rangle^s f_k|^2 \Big)^{\frac{1}{2}} \Big(\sum_{|k|\le N} |\langle k \rangle^{-s} a_k|^2 \Big)^{\frac{1}{2}} \le \|(f_k)\|_{\ell_2^s} \|(a_k)\|_{\ell_2^{-s}}.$$

Thus $\langle f^N, \varphi \rangle$ converges for each φ when $N \to \infty$, and it follows from the limit theorem (Theorem 3.9) that f^N converges to a distribution f. In particular, $\langle f, e^{-ik\cdot x} \rangle = \lim_{N\to\infty} \langle f^N, e^{-ik\cdot x} \rangle = f_k$ for each k.

So there is a subset of the distributions $f \in \mathscr{D}'(\mathbb{T}^n)$ that can be written as $\sum_{k\in\mathbb{Z}^n} f_k e^{ik\cdot x}$, with $f_k = c_k(f) = \langle f, e^{-ik\cdot x} \rangle$ and $(f_k) \in \ell_2^s$, and we define $H^s(\mathbb{T}^n)$ to consist of these; in other words it is defined by (8.12) and (8.13).

It can now be remarked that $H^s(\mathbb{T}^n)$ and $H^{-s}(\mathbb{T}^n)$ identify with each other's dual spaces with a duality extending the L_2 scalar product. The proof is similar to that of Theorem 6.15: First of all, ℓ_2 identifies with its own dual space by the Riesz representation theorem. By use of the isometry $M_{\langle k \rangle^s}$ this extends to an identification of ℓ_2^{-s} and ℓ_2^s with each other's dual spaces, and this carries over to the duality between H^{-s} and H^s when we carry $(a_k)_{k\in\mathbb{Z}^n}$ over to $\sum_{k\in\mathbb{Z}^n} a_k e^{ik\cdot x}$.

Another observation is that since $\mathbb{T}^n$ is compact, any distribution u has a finite order M. Now when φ is as above we have for $|\alpha| \le M$,

$$\sup |D^\alpha \varphi(x)| = \sup \Big| \sum_{k\in\mathbb{Z}^n} k^\alpha a_k e^{ik\cdot x} \Big| \le \sum_{k\in\mathbb{Z}^n} |\langle k \rangle|^{|\alpha|} a_k|$$

$$\le \sum_{k\in\mathbb{Z}^n} \langle k \rangle^{M+b} |a_k| \langle k \rangle^{-b} \le \|(a_k)\|_{\ell_2^{M+b}} \|(\langle k \rangle^{-b})\|_{\ell_2^0},$$

where $\|(\langle k \rangle^{-b})\|_{\ell_2^0} < \infty$ for $b > \frac{n}{2}$ (cf. (8.15)ff. below). Thus $\|\varphi\|_{C^M} \le c_b \|(a_k)\|_{\ell_2^{M+b}}$ for $b > \frac{n}{2}$. Hence

$$|\langle u, \varphi \rangle| \le C_M \sup\{|D^\alpha \varphi(x)| \mid |\alpha| \le M, x \in \mathbb{T}^n\} \le C' \|\varphi\|_{M+b,\wedge},$$

for all φ, so u defines a continuous functional on $H^{M+b}(\mathbb{T}^n)$. It follows that $u \in H^{-M-b}(\mathbb{T}^n)$. Consequently,

$$\mathscr{D}'(\mathbb{T}^n) = \bigcup_{s \in \mathbb{R}} H^s(\mathbb{T}^n) = \bigcup_{s \in \mathbb{Z}} H^s(\mathbb{T}^n).$$

Let us define Λ_s for $s \in \mathbb{R}$ as the operator

$$\Lambda_s : \sum_{k \in \mathbb{Z}^n} c_k e^{ik \cdot x} \mapsto \sum_{k \in \mathbb{Z}^n} \langle k \rangle^s c_k e^{ik \cdot x} \qquad (8.14)$$

(corresponding to the multiplication operator $M_{\langle k \rangle^s}$ on the coefficient sequence); then clearly

$$\Lambda_s \text{ maps } H^t(\mathbb{T}^n) \text{ isometrically onto } H^{t-s}(\mathbb{T}^n),$$

when the norms $\|u\|_{r,\wedge}$ are used, and Λ_{-s} is the inverse of Λ_s for any s.

Theorem 8.3. 1° *When $s > 0$, Λ_{-s} defines a bounded selfadjoint operator in $L_2(\mathbb{T}^n)$, which is a compact operator in $L_2(\mathbb{T}^n)$.*
The injection of $H^s(\mathbb{T}^n)$ into $L_2(\mathbb{T}^n)$ is compact.
2° *When $s > s'$, the injection of $H^s(\mathbb{T}^n)$ into $H^{s'}(\mathbb{T}^n)$ is compact.*

Proof. Let $s > 0$. Since $\|u\|_{s,\wedge} \geq \|u\|_{0,\wedge}$, Λ_{-s} defines a bounded operator T in $L_2(\mathbb{T}^n)$, and it is clearly symmetric, hence selfadjoint. Moreover, the orthonormal basis $(e^{ik \cdot x})_{k \in \mathbb{Z}^n}$ is a complete system of eigenvectors of T, with eigenvalues $\langle k \rangle^{-s}$. Then T is a compact operator in $L_2(\mathbb{T}^n)$, since $\langle k \rangle^{-s} \to 0$ for $|k| \to \infty$.

It follows that the injection of $H^s(\mathbb{T}^n)$ into $H^0(\mathbb{T}^n)$ is compact. Namely, when u_l is a bounded sequence in $H^s(\mathbb{T}^n)$, we can write $u_l = \Lambda_{-s} f_l = T f_l$, where (f_l) is bounded in $L_2(\mathbb{T}^n)$, so the compactness of T in $L_2(\mathbb{T}^n)$ implies that u_l has a convergent subsequence in $L_2(\mathbb{T}^n)$. This shows 1°.

For more general s and s', one carries the injection $E : H^s(\mathbb{T}^n) \hookrightarrow H^{s'}(\mathbb{T}^n)$ over into the injection $E' : H^{s-s'}(\mathbb{T}^n) \hookrightarrow H^0(\mathbb{T}^n)$ by use of the isometries $\Lambda_{-s'}$ and $\Lambda_{s'}$, setting $E = \Lambda_{-s'} E' \Lambda_{s'}$; it is compact in $H^{s'}(\mathbb{T}^n)$, since E' is so in $H^0(\mathbb{T}^n)$.   □

Proof (of Theorem 8.2). Let u_l be a bounded sequence in $H^s(X)$, then for each j (cf. (8.10)), $(\varphi_j u) \circ \kappa_j^{-1}$ is bounded in $H^s(\mathbb{R}^n)$. The support is in a fixed compact subset of V_j, and we can assume (after a scaling and translation if necessary) that this is a compact subset of Q°, so that the sequence identifies with a bounded sequence in $H^s(\mathbb{T}^n)$. Then Theorem 8.3 gives that there is a subsequence that converges in $H^{s'}(\mathbb{T}^n)$. Taking subsequences in this way successively for $j = 1, \ldots, j_0$, we arrive at a numbering such that the corresponding subsequence of u_l converges in $H^{s'}(X)$.   □

The statement on the compactness of the operator defined from Λ_{-s} for $s > 0$ can be made more precise by reference to *Schatten classes*. A compact selfadjoint operator $T \geq 0$ is said to belong to the Schatten class $\mathcal{C}_p$ (for some $p > 0$), when the eigenvalue sequence $\{\lambda_j\}_{j \in \mathbb{N}_0}$ satisfies $\sum_{j \in \mathbb{N}_0} \lambda_j^p < \infty$.

In particular, the operators in $\mathcal{C}_1$ are the trace-class operators, those in $\mathcal{C}_2$ are the Hilbert-Schmidt operators. For nonselfadjoint T, the Schatten class is defined according to the behavior of the eigenvalues of $(T^*T)^{\frac{1}{2}}$.

We here observe that Λ_{-s}, or the embedding $H^s(\mathbb{T}^n) \hookrightarrow H^0(\mathbb{T}^n)$, is in the Schatten classes $\mathcal{C}_p$ with $p > n/s$, since

$$\sum_{k\in\mathbb{Z}^n} \langle k \rangle^{-sp} < \infty \text{ for } s > n/p \tag{8.15}$$

(which is seen by comparison with $\int_{\mathbb{R}^n} \langle x \rangle^{-sp}\, dx$). In particular, the injection is trace-class for $s > n$, and it is Hilbert-Schmidt for $s > n/2$.

An operator $P : C^\infty(X) \to C^\infty(X)$ is said to be a pseudodifferential operator of order d, when $P_j : C_0^\infty(V_j) \to C^\infty(V_j)$, defined by

$$P_j v = P(v \circ \kappa_j) \circ \kappa_j^{-1}, v \in C_0^\infty(V_j), \tag{8.16}$$

is a ψdo of order d on V_j for each j.

To see that the definition makes good sense and is independent of the choice of a particular atlas, we appeal to Theorem 8.1, which shows that the property of being a ψdo on an open set is preserved under diffeomorphisms. Here the pieces P_j generally have $S_{1,0}^d$ symbols, but if they are polyhomogeneous with respect to one atlas, they are so in all atlases, and we say that P is polyhomogeneous. In the following we restrict our attention to the polyhomogeneous case.

The symbols of the localized pieces P_j of course depend on the choice of atlas. However, there is a remarkable fact, namely, that the principal symbol has an invariant meaning.

To explain this, we need the concept of *vector bundles* over a manifold, that we shall now briefly explain. A *trivial vector bundle* over X with fiber dimension N is simply the manifold $X \times \mathbb{C}^N$; the points are denoted for example $\{x, v\}$ ($x \in X$ and $v \in \mathbb{C}^N$), and for each $x \in X$, the subset $\{x\} \times \mathbb{C}^N$ is called the *fiber* (or fibre) over x. X is then called the base space. On the space $X \times \mathbb{C}^N$, the mapping $\pi : \{x, v\} \mapsto \{x, 0\}$ is a projection. Here we identify $\{x, 0\}$ with x, such that X is the range of the projection. Then, for each $x \in X$, $\pi^{-1}(\{x\}) = \{x\} \times \mathbb{C}^n$, the fiber over x. The *sections* of $X \times \mathbb{C}^N$ are the vector-valued functions $f : X \to \mathbb{C}^N$.

A general C^∞-vector bundle E over X is a C^∞-manifold provided with a projection $\pi : E \to X$ such that $\pi^{-1}(\{x\})$ is an N-dimensional complex vector space (the fiber over x). Again, X is identified with a subset of E. Here we require that E is covered by open sets of the form $\pi^{-1}(U)$ with U open $\subset X$, and there is an associated mapping $\Psi : \pi^{-1}(U) \to V \times \mathbb{C}^N$, such that the restriction of Ψ to U is a coordinate mapping $\kappa : U \to V$, and at each $x \in U$, Ψ maps the fiber $\pi^{-1}(\{x\})$ over x *linearly* onto $\mathbb{C}^N$, the fiber over $\kappa(x)$. Such a mapping Ψ (or rather, a triple $\{\Psi, U, V\}$) is called a *local trivialization*, and the associated mapping $\kappa : U \to V$ is called the base space mapping. When Ψ_1 and Ψ_2 are local trivializations with $U_1 \cap U_2 \neq \emptyset$, the mapping $g_{12} = \Psi_1 \circ \Psi_2^{-1}$ from $\kappa_2(U_1 \cap U_2) \times \mathbb{C}^N$ to $\kappa_1(U_1 \cap U_2) \times \mathbb{C}^N$ is called a transition function (skiftefunktion); it is a smooth family of regular $N \times N$-matrices parametrized by $y \in \kappa_2(U_1 \cap U_2)$. The (continuous, say) sections of E are the continuous functions $f : X \to E$ such that $f(x) \in \pi^{-1}(\{x\})$ (i.e., $f(x)$ lies in the fiber over x). In each local trivialization, they carry over to continuous functions from V to $\mathbb{C}^N$. They are said to be C^k ($k \leq \infty$) when they carry over to C^k functions in the local trivializations. — Each section can be viewed as a subset of E (a "graph").

The zero section sends x into the origin of the fiber over x; one often identifies the zero section with X itself.

One can of course also define real vector bundles, where the fibers are real N-dimensional vector spaces.

We use these concepts in two ways: For one thing, we can let our ψdo's be matrix-valued, and then we can allow the matrix structure to "twist" when one moves around on X, by letting the ψdo's act on sections of vector bundles. Here it is natural to take complex vector bundles. One can provide such vector bundles with a Hermitian structure — this means that one chooses a scalar product in each fiber, varying smoothly along X; then L_2 scalar products can be defined for the sections (not just for functions), and all that was said about function spaces (and distribution spaces) above extends to sections of vector bundles.

The other use we make of vector bundles is crucial even for scalar pseudodifferential operators: There is a special real vector bundle with fiber dimension n associated with X called the *cotangent bundle* $T^*(X)$. It can be described as follows: It has an atlas consisting of open sets $\pi^{-1}(U_j)$, $j = 1, \ldots, j_0$, and local trivializations $\Psi_i : \pi^{-1}(U_j) \to V_j \times \mathbb{R}^n$, such that the associated base space mappings $\kappa_j : U_j \to V_j$ are connected with the transition functions in the following way:

When $x \in U_i \cap U_j$, the linear map $\Psi_j \circ \Psi_i^{-1}$ from $\{\kappa_i(x)\} \times \mathbb{R}^n$ to $\{\kappa_j(x)\} \times \mathbb{R}^n$ equals the inverse transpose of the Jacobian of $\kappa_j \circ \kappa_i^{-1}$ at $\kappa_i(x)$.

(A full discussion can be found in textbooks on differential geometry. One can also find a detailed description in [H83] pages 146–148. The cotangent bundle is the dual bundle of the *tangent bundle* $T(X)$ where the transition functions are the Jacobians of the $\kappa_j \circ \kappa_i^{-1}$.)

This is a way of describing the cotangent bundle that fits directly with our purpose, which is to define the principal symbol of a ψdo as a function on $T^*(X) \setminus 0$ (the cotangent bundle with the zero section removed). Indeed, formula (8.5) shows that when $p^0(x, \xi)$ is given in some coordinate system, one gets the same value after a change to new coordinates $\underline{x} = \kappa(x)$ if one maps $\{x, \xi\}$ to $\{\underline{x}, \underline{\xi}\} = \{\kappa(x), {}^t\kappa'(x)^{-1}\xi\}$. (We here consider the version of p^0 that is homogeneous outside $\xi = 0$.) So it is independent of the choice of local trivializations.

When P is a polyhomogeneous ψdo of order d on X, there is defined a principal symbol in each local coordinate system, and this allows us to define a principal symbol that is a function on the nonzero cotangent bundle $T^*(X) \setminus 0$ (since the value can be found in a consistent way from the value in any local coordinate system). We call it $p^0(x, \xi)$ again, where $x \in X$ and ξ indicates a point in the fiber over x.

One can even let p^0 take its values in a vector bundle E over X (so that p^0 is matrix-formed in local trivializations).

With these preparations, it is meaningful to speak of *elliptic* ψdo's on X. For ψdo's acting on sections of bundles — sending $C^\infty(E)$ into $C^\infty(E')$ where E and E' are vector bundles of fiber dimension N resp. N' — we can even speak of injectively elliptic resp. surjectively elliptic ψdo's (meaning that they are so in local trivializations).

There are some considerations on cut-off functions that are useful when dealing with ψdo's on manifolds.

Lemma 8.4. *Let X be a compact C^∞ manifold.*

1° *To every finite open cover $\{U_1, \ldots, U_J\}$ of X there exists an* **associated** *partition of unity $\{\psi_1, \ldots, \psi_J\}$, that is, a family of nonnegative functions $\psi_j \in C_0^\infty(U_j)$ such that $\sum_{j \le J} \psi_j = 1$.*

2° *There exists a finite family of local coordinates $\kappa_i : U_i \to V_i$, $i = 1, \ldots, I_1$ for which there is a* **subordinate** *partition of unity $\{\varrho_1, \ldots, \varrho_{J_0}\}$ (nonnegative smooth functions having the sum 1), such that any four of the functions $\varrho_j, \varrho_k, \varrho_l, \varrho_m$ have their support in some U_i. (This extends to families of trivializations when a vector bundle is considered.)*

Proof. The statement in 1° is a simple generalization of the well-known statement for compact sets in $\mathbb{R}^n$ (Theorem 2.17): We can choose compact subsets K_j, K_j' of the U_j such that $K_j' \subset K_j^\circ$, $X = \bigcup K_j'$, and we can find smooth functions ζ_j that are 1 on K_j' and have support in K_j°; then take $\psi_j = \zeta_j / \sum_k \zeta_k$.

For 2°, a proof goes as follows: We can assume that X is provided with a Riemannian metric. Consider a system of coordinates $\{\kappa_i : U_i \to V_i\}_{i=1}^{I_0}$; we assume that the patches V_i are disjoint, and note that they need not be connected sets. By the compactness of X, there is a number $\delta > 0$ such that any subset of X with geodesic diameter $\le \delta$ is contained in one of the sets U_i. Now cover X by a finite system of open balls B_j, $j = 1, \ldots, J_0$, of radius $\le \delta/8$. We claim that this system has the following 4-*cluster property:* Any four sets $B_{j_1}, B_{j_2}, B_{j_3}, B_{j_4}$ can be grouped in clusters that are mutually disjoint and where each cluster lies in one of the sets U_i.

The 4-cluster property is seen as follows: Let $j_1, j_2, j_3, j_4 \le J_0$ be given. First adjoin to B_{j_1} those of the B_{j_k}, $k = 2, 3, 4$, that it intersects with; next, adjoin to this union those of the remaining sets that it intersects with, and finally do it once more; this gives the first cluster. If any sets are left, repeat the procedure with these (at most three). Now the procedure is repeated with the remaining sets, and so on; this ends after at most four steps. The clusters are clearly mutually disjoint, and by construction, each cluster has diameter $\leq \delta$, hence lies in a set U_i. (One could similarly obtain covers with an N-cluster property, taking balls of radius $\leq \delta/2N$.)

To the original coordinate mappings we now adjoin the following new ones: Assume that $B_{j_1}, B_{j_2}, B_{j_3}, B_{j_4}$ gave rise to the disjoint clusters $U', U'', \ldots$, where $U' \subset U_{i'}$, $U'' \subset U_{i''}, \ldots$. Then use $\kappa_{i'}$ on U', $\kappa_{i''}$ on $U'', \ldots$ (if necessary followed by linear transformations $\Phi'', \ldots$ to separate the images) to define the mapping $\kappa : U' \cup U'' \cup \cdots \to \kappa_{i'}(U') \cup \Phi'' \kappa_{i''}(U'') \cup \cdots$. This gives a new coordinate mapping, for which $B_{j_1} \cup B_{j_2} \cup B_{j_3} \cup B_{j_4}$ equals the initial set $U' \cup U'' \cup \cdots$. In this way, finitely many new coordinate mappings, say $\{\kappa_i : U_i \to V_i\}_{i=I_0+1}^{I_1}$, are adjoined to the original ones, and we have established a mapping $(j_1, j_2, j_3, j_4) \mapsto i = i(j_1, j_2, j_3, j_4)$ for which

$$B_{j_1} \cup B_{j_2} \cup B_{j_3} \cup B_{j_4} \subset U_{i(j_1,j_2,j_3,j_4)}.$$

Let $\{\varrho_j\}_{j=1}^{J_0}$ be a partition of unity associated with the cover $\{B_j\}_{j=1}^{J_0}$ (here we use $1°$), then it has the desired property with respect to the system $\{\kappa_i : U_i \to V_i\}_{i=1}^{I_1}$.   □

The refined partition of unity in $2°$ is convenient when we consider compositions of operators. (It was used in this way in [S69], but a proof was not included there.) If P and Q are ψdo's on X, write

$$R = PQ = \sum_{j,k,l,m} \varrho_j P \varrho_k \varrho_l Q \varrho_m, \qquad (8.17)$$

then each term $\varrho_j P \varrho_k \varrho_l Q \varrho_m$ has support in a set U_i that carries over to $V_i \subset \mathbb{R}^n$ by κ_i, so that we can use the composition rules for ψdo's on $\mathbb{R}^n$. It follows that R is again a ψdo on X; in the local coordinates it has symbols calculated by the usual rules. In particular, the principal symbol of the composition is found from the principal symbols, carried back to $T^*(X) \setminus 0$ and added up:

$$r^0(x,\xi) = \sum_{j,k,l,m} \varrho_j(x)\varrho_k(x)p^0(x,\xi)\varrho_l(x)\varrho_m(x)q^0(x,\xi) = p^0(x,\xi)q^0(x,\xi).$$
$$(8.18)$$

One shows that the adjoint P^* of a ψdo P (with respect to the chosen scalar product in $L_2(X)$) is again a ψdo, by using a partition of unity as in $2°$ in the consideration of the identities

$$\int_X (Pf)\overline{g}\, dx = \int_X f\overline{(P^*g)}\, dx. \qquad (8.19)$$

The principal symbol follows the rule from Theorem 7.13; in particular, P^* is elliptic when P is so.

We can also use the partitions of unity to show continuity in Sobolev spaces over X:

Theorem 8.5. *Let P be a pseudodifferential operator on X of order d. Then P is continuous from $H^s(X)$ to $H^{s-d}(X)$ for all $s \in \mathbb{R}$.*

Proof. Let ϱ_k be a partition of unity as in Lemma 8.4 $2°$. Then $P = \sum_{j,k \le J_0} \varrho_j P \varrho_k$. For each j,k there is an $i \le I_1$ such that ϱ_j and ϱ_k are supported in U_i. Then when $\varrho_j P \varrho_k$ is carried over to V_i, we find a ψdo with symbol (in (x,y)-form) vanishing for (x,y) outside a compact subset of $V_i \times V_i$; it identifies (by extension by 0) with a symbol in $S_{1,0}^d(\mathbb{R}^{2n}, \mathbb{R}^n)$, and hence defines a continuous operator from $H^s(\mathbb{R}^n)$ to $H^{s-d}(\mathbb{R}^n)$ for any $s \in \mathbb{R}$. It follows that $\varrho_j P \varrho_k$ is continuous from $H^s(X)$ to $H^{s-d}(X)$. Adding the pieces, we find the statement in the theorem. $\qquad\square$

For the construction of a parametrix of a given elliptic operator we use Lemma 8.4 $1°$ combined with some extra cut-off functions:

Theorem 8.6. *Let P be an elliptic ψdo of order d on X. Then there is a ψdo Q on X, elliptic of order $-d$, such that*

$$
\begin{aligned}
\text{(i)} \quad & PQ = I + \mathcal{R}_1, \\
\text{(ii)} \quad & QP = I + \mathcal{R}_2,
\end{aligned}
\tag{8.20}
$$

with $\mathcal{R}_1$ and $\mathcal{R}_2$ of order $-\infty$.

Proof. Along with the ψ_j in $1°$ we can find ζ_j and $\theta_j \in C_0^\infty(U_j)$ such that $\zeta_j(x) = 1$ on a neighborhood of $\operatorname{supp} \psi_j$, $\theta_j(x) = 1$ on a neighborhood of $\operatorname{supp} \zeta_j$. Since P_j (recall (8.16)) is elliptic on V_j, it has a parametrix Q'_j there. Denoting the functions $\psi_j, \zeta_j, \theta_j$ carried over to V_j (i.e., composed with κ_j^{-1}) by $\underline{\psi}_j, \underline{\zeta}_j, \underline{\theta}_j$, we observe that

$$
C'_j = \underline{\theta}_j P_j \underline{\theta}_j \underline{\zeta}_j Q'_j \underline{\psi}_j - \underline{\psi}_j \sim 0.
$$

Namely,

$$
\begin{aligned}
C'_j &= (\underline{\theta}_j P_j \underline{\theta}_j \underline{\zeta}_j Q'_j - I)\underline{\psi}_j = (\underline{\theta}_j P_j \underline{\zeta}_j Q'_j - I)\underline{\psi}_j \\
&= (\underline{\theta}_j P_j Q'_j - I)\underline{\psi}_j + \underline{\theta}_j P_j (\underline{\zeta}_j - 1) Q'_j \underline{\psi}_j \sim (\underline{\theta}_j - 1)\underline{\psi}_j = 0;
\end{aligned}
$$

it is used here that $1 - \zeta_j$ and ψ_j have disjoint supports, and that $\theta_j = 1$ on $\operatorname{supp} \psi_j$. Moreover, C'_j carries over to X as a negligible operator C_j with support properties corresponding to the fact that C'_j has kernel support in $\operatorname{supp} \underline{\theta}_j \times \operatorname{supp} \underline{\psi}_j$. Now let $Qu = \sum_j (\underline{\zeta}_j Q'_j((\psi_j u) \circ \kappa_j^{-1})) \circ \kappa_j$. Then

$$
\begin{aligned}
(PQ - I)u &= \sum_j (P[\underline{\zeta}_j Q'_j((\psi_j u) \circ \kappa_j^{-1}) \circ \kappa_j] - \psi_j u) \\
&= \sum_j (\theta_j P \theta_j [\underline{\zeta}_j Q'_j((\psi_j u) \circ \kappa_j^{-1}) \circ \kappa_j] - \psi_j u) + \mathcal{R}u,
\end{aligned}
$$

with $\mathcal{R}$ negligible, since $1 - \theta_j$ and ζ_j have disjoint supports. The j-th term in the last sum equals $C_j u$, and $\sum_j C_j$ is negligible. This shows (8.20) (i).

There is a similar construction of a left parametrix, satisfying (8.20) (ii), and then each of them is a two-sided parametrix, by the usual argument (as in Corollary 7.19).

In each coordinate system, the principal symbol of Q'_j is $p^0(\underline{x}, \xi)^{-1}$ (for those ξ where it is homogeneous). Then the principal symbol of Q, the sum of the $\overline{\zeta_j} Q'_j \underline{\psi}_j$ carried over to X, is equal to

$$\sum_j \zeta_j(x) p^0(x, \xi)^{-1} \psi_j(x) = (p^0)^{-1}(x, \xi),$$

since $\zeta_j \psi_j = \psi_j$ and $\sum_j \psi_j = 1$. In particular, Q is elliptic of order $-d$. $\square$

There are also one-sided versions, that we state here in full generality:

Theorem 8.7. *Let E and E' be complex vector bundles over X of fiber dimensions N resp. N', and let P be a pseudodifferential operator of order d from the sections of E to the sections of E'.*

1° If $N' \leq N$ and P is surjectively elliptic of order d, then there exists a right parametrix Q, which is a ψdo from the sections of E' to E, injectively elliptic of order $-d$, such that (8.20) (i) holds.

2° If $N' \geq N$ and P is injectively elliptic of order d, then there exists a left parametrix Q, which is a ψdo from the sections of E' to E, surjectively elliptic of order $-d$, such that (8.20) (ii) holds.

3° If $N = N'$ and either 1° or 2° holds for some Q, then P is elliptic of order d, and both (i) and (ii) in (8.20) hold with that Q.

In the case of trivial bundles over X, the proof can be left to the reader (to deduce it from Theorem 7.18 by the method in Theorem 8.6). In the situation of general bundles one should replace the coordinate changes κ_j by the local trivializations Ψ_j, with an appropriate notation for the trivialized sections. (Here the proof can be left to readers who are familiar with working with vector bundles.) There is again a corollary on regularity of solutions of elliptic problems as in Corollary 7.20. For existence and uniqueness questions we can get much better results than in Chapter 7, as will be seen in the following.

8.3 Fredholm theory, the index

Let V_1 and V_2 be vector spaces. A Fredholm operator from V_1 to V_2 is a linear operator T from V_1 to V_2 such that $\ker T$ and $\operatorname{coker} T$ have finite dimension; here $\ker T$ is the nullspace (also denoted $Z(T)$) and $\operatorname{coker} T$ is the quotient space $H_2/R(T)$, where $R(T)$ is the range of T. The dimension of $H_2/R(T)$ is also called $\operatorname{codim} R(T)$.

We shall recall some facts concerning Fredholm operators between Hilbert spaces. More comprehensive treatments can be found in various books, e.g.

[H85, Sect./ 19.1], Conway [C90] or Schechter [S02]. Our presentation is much inspired from the Lund University lecture notes of Hörmander [H89].

Fredholm operators can also be studied in Banach spaces (and even more general spaces) but we shall just need them in Hilbert spaces, where some proofs are simpler to explain.

Let H_1 and H_2 be Hilbert spaces; we denote as usual the space of bounded linear operators from H_1 to H_2 by $\mathbf{B}(H_1, H_2)$. First of all one can observe that when $T \in \mathbf{B}(H_1, H_2)$ is a Fredholm operator, then $R(T)$ is closed in H_2. To see this, note that $T : H_1 \ominus Z(T) \to H_2$ is again bounded and is a Fredholm operator. If $\dim \operatorname{coker} T = n$, we can choose a linear mapping $S : \mathbb{C}^n \to H_2$ that maps $\mathbb{C}^n$ onto a complement of $R(T)$ in H_2; then $T_1 : \{x, y\} \to Tx + Sy$ from $(H_1 \ominus Z(T)) \oplus \mathbb{C}^n$ to H_2 is bijective. T_1 is continuous, hence so is its inverse (by the closed graph theorem). But then $R(T) = T_1((H_1 \ominus Z(T)) \oplus \{0\})$ is closed.

The property of having closed range is often included in the definition of Fredholm operators, but we see that it holds automatically here.

When T is a Fredholm operator, its index is defined by

$$\operatorname{index} T = \dim \ker T - \dim \operatorname{coker} T. \tag{8.21}$$

The following property is fundamental:

Proposition 8.8. *When $T \in \mathbf{B}(H_1, H_2)$ is bijective, and $K \in \mathbf{B}(H_1, H_2)$ is compact, then $T + K$ is a Fredholm operator.*

Proof. Recall that a compact operator maps any bounded sequence into a sequence that has a convergent subsequence. We first show why $Z(T + K)$ is finite dimensional: $Z(T + K)$ is a linear space, and for $x \in Z(T + K)$,

$$Tx = -Kx.$$

Let x_j be a bounded sequence in $Z(T + K)$. By the compactness of K, Kx_j has a convergent subsequence Kx_{j_k}, but then Tx_{j_k} is also convergent, and so is x_{j_k} since T^{-1} is bounded. We have shown that any bounded sequence in $Z(T + K)$ has a convergent subsequence. Then the Hilbert space $Z(T + K)$ must be finite dimensional; for an infinite dimensional Hilbert space has an infinite orthonormal sequence (with no convergent subsequences).

Recall the general property $H_2 = \overline{R(T + K)} \oplus Z(T^* + K^*)$. Since T^* is invertible and K^* is compact, $Z(T^* + K^*)$ has finite dimension. So to see that $R(T + K)$ has finite codimension, we just have to show that it is closed.

Consider $(T + K) : \widetilde{H}_1 \to H_2$ where $\widetilde{H}_1 = H_1 \ominus Z(T + K)$. We claim that for all $x \in \widetilde{H}_1$,

$$\|x\| \le c\|(T + K)x\| \text{ for some } c > 0. \tag{8.22}$$

Because if not, then there exist sequences $x_j \in \widetilde{H}_1$ and c_j such that $\|x_j\| = 1$ and $c_j \to \infty$, with

$$1 = \|x_j\| \geq c_j \|(T + K)x_j\|, \text{ for all } j.$$

But then $\|(T + K)x_j\| \leq 1/c_j \to 0$. Since K is compact and $\|x_j\| = 1$, there is a subsequence x_{j_k} with $Kx_{j_k} \to v$ for some $v \in H_2$. Then $Tx_{j_k} \to -v$, so $x_{j_k} = T^{-1}Tx_{j_k} \to w = -T^{-1}v$; note that $w \in \widetilde{H}_1$ and has norm 1, since this holds for the x_{j_k}. But $(T + K)w = \lim_k(Tx_{j_k} + Kx_{j_k}) = 0$, contradicting the fact that $\widetilde{H}_1 \perp Z(T + K)$. It follows from (8.22) that $R(T + K)$ is closed. $\square$

We shall see further below that the index of $T + K$ in this case is 0. A special case is where $H_1 = H_2$ and $T = I$, in this case the result was proved by Fredholm and Riesz, and is known as the *Fredholm alternative*: "For an operator $I + K$ in a Hilbert space H with K compact, the equation

$$(I + K)u = f \tag{8.23}$$

is *either* uniquely solvable for all $f \in H$, *or* there exist subspaces Z, Z' with the same finite dimension, such that (8.23) is solvable precisely when $f \perp Z'$, and the solution is unique modulo Z ". The general rules for Fredholm operators have been established by Atkinson and others.

Lemma 8.9. *An operator* $T \in \mathbf{B}(H_1, H_2)$ *is Fredholm if and only if there exist* $S_1, S_2 \in \mathbf{B}(H_2, H_1)$, K_1 *compact in* H_1 *and* K_2 *compact in* H_2, *such that*

$$S_1 T = I + K_1, \quad T S_2 = I + K_2. \tag{8.24}$$

Proof. When X is a closed subspace of a Hilbert space H, we denote by pr_X the orthogonal projection onto X, and by i_X the injection of X into H. (When $X \subset X_1 \subset H$, we denote the injection of X into X_1 by $\mathrm{i}_{X \to X_1}$.)

Let T be Fredholm, then T defines a bijective operator $\widetilde{T}$ from $\widetilde{H}_1 = H_1 \ominus Z(T)$ to $\widetilde{H}_2 = R(T) = H_2 \ominus Z(T^*)$. Let $S_2 = \mathrm{i}_{\widetilde{H}_1}(\widetilde{T})^{-1}\mathrm{pr}_{R(T)}$; then

$$T S_2 = T\,\mathrm{i}_{\widetilde{H}_1}(\widetilde{T})^{-1}\mathrm{pr}_{R(T)} = \mathrm{pr}_{R(T)} = I - \mathrm{pr}_{Z(T^*)},$$

so the second equation in (8.24) is achieved, even with a finite rank operator $K_2 = -\mathrm{pr}_{Z(T^*)}$. The adjoint T^* is likewise Fredholm, so the same argument shows the existence of S_3, K_3 such that $T^*S_3 = I + K_3$ with K_3 of finite rank. Taking adjoints, we see that S_3^*, K_3^* can be used as S_1, K_1.

Conversely, assume that (8.24) holds. The equations show that

$$Z(T) \subset Z(I + K_1), \quad R(T) \supset R(I + K_2).$$

The space $Z(I + K_1)$ has finite dimension, and $R(I + K_2)$ has finite codimension, in view of Proposition 8.8. Here $\dim Z(T) \leq \dim Z(I + K_1)$ and $\mathrm{codim}\,R(T) \leq \mathrm{codim}\,R(I + K_2)$, so T is Fredholm. $\square$

Some of the most important properties of Fredholm operators are:

Theorem 8.10. 1° MULTIPLICATIVE PROPERTY OF THE INDEX. *When $T_1 \in$ $\mathbf{B}(H_1, H_2)$ and $T_2 \in \mathbf{B}(H_2, H_3)$ are Fredholm operators, then $T_2T_1 \in \mathbf{B}(H_1, H_3)$ is also Fredholm, and*

$$\text{index}\, T_2T_1 = \text{index}\, T_2 + \text{index}\, T_1. \tag{8.25}$$

2° INVARIANCE OF FREDHOLM PROPERTY AND INDEX UNDER SMALL PERTURBATIONS. *Let $T \in \mathbf{B}(H_1, H_2)$ be a Fredholm operator. There is a constant $c > 0$ such that for all operators $S \in \mathbf{B}(H_1, H_2)$ with norm $< c$, $T + S$ is Fredholm and*

$$\text{index}(T + S) = \text{index}\, T. \tag{8.26}$$

Thus the index is invariant under a homotopy of Fredholm operators.

3° INVARIANCE OF FREDHOLM PROPERTY AND INDEX UNDER COMPACT PERTURBATIONS. *Let $T \in \mathbf{B}(H_1, H_2)$ be a Fredholm operator. Then for any compact operator $S \in \mathbf{B}(H_1, H_2)$, $T + S$ is Fredholm and* (8.26) *holds.*

Proof. For 1°, we give a brief indication of the proof, found with more details in [C90, IX §3]. When T is an operator of the form

$$T = \begin{pmatrix} T' & R \\ 0 & T'' \end{pmatrix} : \begin{matrix} H_1' \\ \oplus \\ H_1'' \end{matrix} \to \begin{matrix} H_2' \\ \oplus \\ H_2'' \end{matrix},$$

with finite dimensional H_1'' and H_2'', and T' bijective, then T is Fredholm with index equal to $\dim H_1'' - \dim H_2''$, for so is $T'' : H_1'' \to H_2''$, by an elementary rule from linear algebra. The given Fredholm operators T_1 and T_2 can be written in this form,

$$T_1 = \begin{pmatrix} T_1' & R_1 \\ 0 & T_1'' \end{pmatrix} : \begin{matrix} H_1' \\ \oplus \\ H_1'' \end{matrix} \to \begin{matrix} H_2' \\ \oplus \\ H_2'' \end{matrix}, \quad T_2 = \begin{pmatrix} T_2' & R_2 \\ 0 & T_2'' \end{pmatrix} : \begin{matrix} H_2' \\ \oplus \\ H_2'' \end{matrix} \to \begin{matrix} H_3' \\ \oplus \\ H_3'' \end{matrix},$$

when we define

$$H_2' = R(T_1) \cap Z(T_2)^\perp, \; H_2'' = H_2 \ominus H_2' = R(T_1)^\perp + Z(T_2),$$
$$H_1' = T_1^{-1}(H_2') \cap Z(T_1)^\perp, \; H_1'' = H_1 \ominus H_1',$$
$$H_3' = T_1(H_2'), \; H_3'' = H_3 \ominus H_3'.$$

Then

$$T_2T_1 = \begin{pmatrix} T_2'T_1' & R_3 \\ 0 & T_2''T_1'' \end{pmatrix} : \begin{matrix} H_1' \\ \oplus \\ H_1'' \end{matrix} \to \begin{matrix} H_3' \\ \oplus \\ H_3'' \end{matrix},$$

with $\text{index}\, T_2T_1 = \dim H_3'' - \dim H_1'' = \dim H_3'' - \dim H_2'' + \dim H_2'' - \dim H_1'' = \text{index}\, T_2 + \text{index}\, T_1$.

For 2°, note that it holds obviously when T is bijective, for then

$$T + S = T(I + T^{-1}S), \quad \text{where } (I + T^{-1}S)^{-1} = \sum_{k \in \mathbb{N}_0} (-T^{-1}S)^k,$$

converging in norm when $\|S\| < \|T^{-1}\|^{-1}$. Here (8.26) holds since both $T + S$ and T have index 0, being bijective.

In the general case we can write, with notation as in Lemma 8.9,

$$T = \begin{pmatrix} \widetilde{T} & 0 \\ 0 & 0 \end{pmatrix} : \begin{matrix} \widetilde{H}_1 \\ \oplus \\ Z \end{matrix} \to \begin{matrix} \widetilde{H}_2 \\ \oplus \\ Z' \end{matrix},$$

where we have set $Z = Z(T)$, $Z' = Z(T^*)$. Then

$$T + S = \begin{pmatrix} \widetilde{T} + S_{11} & S_{12} \\ S_{21} & S_{22} \end{pmatrix} : \begin{matrix} \widetilde{H}_1 \\ \oplus \\ Z \end{matrix} \to \begin{matrix} \widetilde{H}_2 \\ \oplus \\ Z' \end{matrix},$$

and $\|S_{11}\| < c$ if $\|S\| < c$. For c sufficiently small, $\widetilde{T} + S_{11}$ will be bijective. The range of $T + S$ contains all elements of the form $(\widetilde{T} + S_{11})u_1 \oplus S_{21}u_1$, $u_1 \in \widetilde{H}_1$; hence since $\widetilde{T} + S_{11}$ maps $\widetilde{H}_1$ onto $\widetilde{H}_2$, every equivalence class in H_2/Z' is reached, so $\operatorname{codim} R(T + S) \leq \operatorname{codim} R(T)$. The adjoints have a similar structure, so $\dim Z(T + S) \leq \dim Z(T)$. We conclude that $T + S$ is Fredholm, and will now determine its index.

We have that
$$\widetilde{T} + S_{11} = \operatorname{pr}_{\widetilde{H}_2}(T + S) \, \mathrm{i}_{\widetilde{H}_1},$$

where each factor is a Fredholm operator. The product rule 1° gives

$$0 = \operatorname{index}(\widetilde{T} + S_{11}) = \operatorname{index} \operatorname{pr}_{\widetilde{H}_2} + \operatorname{index}(T + S) + \operatorname{index} \mathrm{i}_{\widetilde{H}_1}$$
$$= \dim Z' + \operatorname{index}(T + S) - \dim Z,$$

and we conclude that $\operatorname{index}(T + S) = \dim Z - \dim Z' = \operatorname{index} T$.

For 3°, we know from Proposition 8.8 that $T + S$ is Fredholm if T is bijective. Applying 2° to the family of Fredholm operators $T + \lambda S$, $\lambda \in [0, 1]$, we see that (8.26) holds in this case with indices zero. In particular, $I + S$ has index 0.

For general Fredholm operators T we use Lemma 8.9: With auxiliary operators as in (8.24), we have that

$$S_1(T + S) = I + K_1 + S_1 S = I + K_1',$$
$$(T + S)S_2 = I + K_2 + SS_2 = I + K_2',$$

with K_1' and K_2' compact, hence $T + S$ is Fredholm by the lemma. Here $I + K_1$ has index 0, so $\operatorname{index} S_1 = -\operatorname{index} T$ by the product rule. Moreover, $I + K_1'$ has index zero, so $\operatorname{index}(T + S) = -\operatorname{index} S_1 = \operatorname{index} T$. $\qquad \square$

One sometimes attaches the Fredholm property to unbounded operators too. This can be justified when T is a closed operator, as follows:

Let T be a closed operator from $D(T) \subset H_1$ to H_2. Then T is bounded as an operator from the Hilbert space $D(T)$ provided with the graph norm $\|u\|_{\text{graph}} = (\|u\|_{H_1}^2 + \|Tu\|_{H_2}^2)^{\frac{1}{2}}$, to H_2. It is said to be Fredholm when $\dim \ker T$ and $\dim \operatorname{coker} T$ are finite, and the index is defined as usual by (8.21). $R(T)$ is again found to be closed in H_2, and Theorem 8.10 holds when the graph norm is used on $D(T)$.

We shall now show that elliptic pseudodifferential operators on compact manifolds are Fredholm operators.

Theorem 8.11. *Let X be a compact n-dimensional C^∞ manifold, and let P be an elliptic pseudodifferential operator of order d on X. Then $P : H^s(X) \to H^{s-d}(X)$ is a Fredholm operator for any $s \in \mathbb{R}$.*

Moreover, the kernel is the same finite dimensional subspace V of $C^\infty(X)$ for all $s \in \mathbb{R}$, and there is a finite dimensional subspace W of $C^\infty(X)$ such that the range for all $s \in \mathbb{R}$ consists of the $f \in H^{s-d}(X)$ satisfying $(f, w) = 0$ for $w \in W$. These statements hold also for $s = \infty$, where $H^\infty(X) = C^\infty(X)$, so P is a Fredholm operator in $C^\infty(X)$.

Thus $P : H^s(X) \to H^{s-d}(X)$ has an index

$$\operatorname{index} P = \dim \ker P - \dim \operatorname{coker} P, \qquad (8.27)$$

that is independent of $s \leq \infty$.

When Q is a parametrix of P,

$$\operatorname{index} Q = - \operatorname{index} P. \qquad (8.28)$$

The statements hold also for matrix-formed operators or operators in vector bundles, in the situation of Theorem 8.7 3°.

Proof. To keep the notation simple, we give the proof details in the scalar case, using Theorem 8.6. P has a parametrix Q, continuous in the opposite direction and satisfying (8.20). Since $\mathcal{R}_1$ and $\mathcal{R}_2$ are bounded operators from $H^s(X)$ to $H^{s+1}(X)$, they are compact operators in $H^s(X)$ by Theorem 8.2, for all s. Then by Theorem 8.10 3°, $PQ = I + \mathcal{R}_1$ and $QP = I + \mathcal{R}_2$ are Fredholm operators in $H^s(X)$ with index 0.

We can write (as in [S91]) $C^\infty(X) = H^\infty(X)$ (since $\bigcap_{s \in \mathbb{R}} H^s(X) = C^\infty(X)$ by the Sobolev embedding theorem). Denote by $Z^s(P)$ the nullspace of $P : H^s(X) \to H^{s-d}(X)$, and similarly by $Z^s(QP)$ the nullspace of $QP : H^s(X) \to H^s(X)$. Clearly, $Z^\infty(P) \subset Z^s(P)$, $Z^\infty(QP) \subset Z^s(QP)$, for $s \in \mathbb{R}$. On the other hand, for $u \in H^s(X)$, any s,

$$Pu = 0 \implies QPu = 0 \implies u = -\mathcal{R}_2 u \in C^\infty(X),$$

by (8.20) (ii), so $Z^s(P) \subset Z^s(QP) \subset C^\infty(X)$. Thus $Z^s(P) = Z^\infty(P)$ and $Z^s(QP) = Z^\infty(QP)$ for all s; we denote $Z^\infty(P) = V$. Moreover, since $Z^s(QP)$ has finite dimension, so has $Z^s(P) = V$.

For the consideration of cokernels, we similarly define $R^s(P)$ and $R^s(PQ)$ to be the ranges of $P : H^{s+d}(X) \to H^s(X)$ and $PQ : H^s(X) \to H^s(X)$. Since $PQ = I + \mathcal{R}_1$ is Fredholm in $H^s(X)$, $R^s(PQ)$ has finite codimension in $H^s(X)$ and then so has $R^s(P) \supset R^s(PQ)$. Thus $P : H^{s+d}(X) \to H^s(X)$ is Fredholm, for any s; in particular, $R^s(P)$ is closed in $H^s(X)$.

Since X is compact, we can identify $\mathscr{D}'(X)$ with $\bigcup_{s\in\mathbb{R}} H^s(X)$. The adjoint $P^* : \mathscr{D}'(X) \to \mathscr{D}'(X)$ is then defined such that its restrictions to Sobolev spaces satisfy:

$P^* : H^{-s}(X) \to H^{-s-d}(X)$ is the adjoint of $P : H^{s+d}(X) \to H^s(X)$, when $H^{-s}(X)$ is identified with the dual space of $H^s(X)$ with respect to a duality (u,v) consistent with the chosen scalar product in $L_2(X)$.

Moreover, since $R^s(P)$ is closed,

$$R^s(P) = \{f \in H^s(X) \mid (f,v) = 0 \text{ for } v \in Z^{-s}(P^*)\}.$$

Since P^* is elliptic, $Z^{-s}(P^*) = Z^\infty(P^*) = W$, a finite dimensional subspace of $C^\infty(X)$ independent of $s \in \mathbb{R}$. This shows that

$$R^s(P) = \{f \in H^s(X) \mid (f,v) = 0 \text{ for } v \in W\},$$

for $s \in \mathbb{R}$, and we finally show it for $s = \infty$. Here the inclusion "$\subset$" is obvious, since $R^\infty(P) \subset R^s(P)$ for any s; on the other hand, when $f \in H^\infty(X)$ with $(f,w) = 0$ for $w \in W$, then there is a $u \in H^s(X)$ (with an arbitrarily chosen $s < \infty$) such that $f = Pu$, and then $u = QPu - \mathcal{R}_2 u \in H^\infty(X)$, so that $f \in R^\infty(P)$.

The last statement follows by another application of Theorem 8.10: Since $\mathcal{R}_1$ is compact in $L_2(X)$, $I + \mathcal{R}_1$ has index 0 there by 3°, and hence index $P +$ index $Q = 0$ by 1°. This extends to the operators in the other spaces since the indices are independent of s. $\square$

For the proof in the vector bundle situations, one needs to introduce a Hermitian scalar product in the fibers in order to speak of the adjoint.

It may be observed that to show the Fredholm property of P, one only needs to use for Q a rough parametrix constructed from the principal symbol, such that the $\mathcal{R}_i$ in (8.20) are of order -1. We also observe:

Corollary 8.12. *The index of P depends only on the principal symbol of P.*

Proof. Let P_1 be a polyhomogeneous ψdo with the same principal symbol as P, i.e., $P_1 - P$ is of order $d - 1$. Then

$$P_1 Q = PQ + (P_1 - P)Q = I + \mathcal{R}_1',$$
$$QP_1 = QP + Q(P_1 - P) = I + \mathcal{R}_2',$$

with $\mathcal{R}'_1$ and $\mathcal{R}'_2$ of order -1. For any $s \in \mathbb{R}$ they are compact as operators in $H^s(X)$, so by Lemma 8.9, P_1 is Fredholm from $H^s(X)$ to $H^{s-d}(X)$. The index satisfies

$$\text{index}\, P_1 = -\,\text{index}\, Q = \text{index}\, P, \qquad (8.29)$$

by Theorem 8.10 1° and 3°, and Theorem 8.11. $\qquad\qquad\qquad\qquad\qquad\square$

For Riemannian manifolds X, the index of the elliptic operator associated with the Riemannian metric, and of related elliptic operators defined in suitable vector bundles over X, have been studied intensively. There is a famous theorem of Atiyah and Singer (see e.g. [AS68], [ABP73], [H85, Sect. 19.2] and Gilkey [Gi74], [Gi85]) showing how the analytical index (the one we have defined above) is related to the so-called topological index defined in terms of algebraic topology associated with the manifold. Also other geometric invariants have been studied, such as e.g. the noncommutative residue (Wodzicki [W84]) and the canonical trace (Kontsevich and Vishik [KV95]), by asymptotic analysis of ψdo's.

Exercises for Chapter 8

8.1. Let A be as in Exercise 7.4. Consider for $n = 2$ the following change of coordinates:

$$\underline{x}_1 = x_1 + x_2,$$
$$\underline{x}_2 = x_2.$$

(a) How does A look in the new coordinates?

(b) Show that the principal symbol follows the rule from Section 8.1.

8.2. Consider the biharmonic operator Δ^2 on functions on $\mathbb{T}^n$.

(a) Show that the functions $e^{ik \cdot x}$ are a complete system of mutually orthogonal eigenfunctions for Δ^2 with eigenvalues $|k|^4$.

(b) Show that $I + \Delta^2$ is a bijection from $H^4(\mathbb{T}^n)$ to $L_2(\mathbb{T}^n)$. To which Schatten classes $\mathcal{C}_p$ does $(I + \Delta^2)^{-1}$ belong?

8.3. Let X be a compact C^∞-manifold, provided with an L_2 scalar product. Let Δ_X be a second-order differential operator defined on X, with principal symbol $|\xi|^2$. Let $A = \Delta_X^* \Delta_X$ (where the formal adjoint is defined with respect to the given L_2 scalar product).

(a) Show that Δ_X has exactly one L_2-realization, with domain $H^2(X)$, and that A has exactly one L_2-realization, with domain $H^4(X)$.

(*Hint.* Consider the maximal and minimal realizations, similarly to the definition in Chapter 4.)

(b) Show (using simple functional analysis) that A has index 0.

Part IV
Boundary value problems

Chapter 9
Boundary value problems
in a constant-coefficient case

9.1 Boundary maps for the half-space

This chapter can be read in direct succession to Chapters 1–6 and does not build on Chapters 7 and 8.

We have in Section 4.4 considered various realizations of the Laplacian and similar operators, defined by homogeneous boundary conditions, and shown how their invertibility properties lead to statements about the solvability of homogeneous boundary value problems. It is of great interest to study also nonhomogeneous boundary value problems

$$Au = f \text{ in } \Omega,$$
$$Tu = g \text{ on } \partial\Omega, \tag{9.1}$$

with $g \neq 0$. Here Ω is an open subset of $\mathbb{R}^n$, A is an elliptic partial differential operator and Tu stands for a set of combinations of boundary values and (possibly higher) normal derivatives (T is generally called a *trace operator*). There are several possible techniques for treating such problems. In specific cases where A is the Laplacian, there are integral formulas (depending on the shape of the boundary), but when A has variable coefficients the methods are more qualitative.

When the boundary is smooth (cf. Appendix C), one succesful method can be described as follows:

At a fixed point x_0 of the boundary, consider the problem with constant coefficients having the value they have at that point, and identify the tangent plane with $\mathbb{R}^{n-1}$ and the interior normal direction with the x_n-axis. The discussion of the resulting constant-coefficient problem is essential for the possibility to obtain solutions of (9.1). Drop the lower-order terms in the equation and boundary condition, and perform a Fourier transformation $\mathscr{F}_{x'\to\xi'}$ in the tangential variables (in $\mathbb{R}^{n-1}$), then we have for each ξ' a one-dimensional problem on $\mathbb{R}_+$. This is what is called the "model problem" at the point x_0.

The original problem (9.1) is called *elliptic* precisely when *the model problem is uniquely solvable in $L_2(\mathbb{R}_+)$ for each $\xi' \neq 0$, at all points x_0 of the boundary*. The conditions for ellipticity are sometimes called the "Shapiro-Lopatinskiĭ conditions".

It is possible to construct a good approximation of the solution in the elliptic case by piecing together the solutions of the model problems in some sense. A very efficient way of doing this is by setting up a pseudodifferential calculus of boundary value problems, based on the theory in Chapter 7 but necessarily a good deal more complicated, since we have operators going back and forth between the boundary and the interior of Ω. An introduction to such a theory is given in Chapter 10. In the present chapter, we give "the flavor" of how the Sobolev spaces and solution operators enter into the picture, by studying in detail a simple special case, namely, that of $1 - \Delta$ on $\mathbb{R}^n_+$.

In the same way as the properties of $I - \Delta$ on $\mathbb{R}^n$ gave a motivational introduction to the properties of general elliptic operators on open sets (or manifolds), the properties of the Dirichlet and Neumann problems for $1 - \Delta$ on $\mathbb{R}^n_+$ serve as a motivational introduction to properties of general elliptic boundary value problems. We shall treat this example in full detail. Here we also study the interpretation of the families of extensions of symmetric (and more general) operators described in Chapter 13. At the end we give some comments on variable-coefficient cases.

The Fourier transform was an important tool for making the treatment of $I - \Delta$ on $\mathbb{R}^n$ easy, cf. Example 5.20. When the operator is considered on $\mathbb{R}^n_+$, we shall use the partial Fourier transform (with respect to the $n-1$ first variables) to simplify things, so we start by setting up how that works.

We shall denote the *restriction* operator from $\mathbb{R}^n$ to $\mathbb{R}^n_+$ by r^+, and the restriction operator from $\mathbb{R}^n$ to $\mathbb{R}^n_-$ by r^-, where

$$\mathbb{R}^n_\pm = \{\, x = (x', x_n) \in \mathbb{R}^n \mid x' = (x_1, \ldots, x_{n-1}) \in \mathbb{R}^{n-1},\ x_n \gtrless 0 \,\}.$$

Here we generally use the notation

$$r_\Omega u = u|_\Omega, \text{ and in particular } r^\pm = r_{\mathbb{R}^n_\pm}, \tag{9.2}$$

defined for $u \in \mathscr{D}'(\mathbb{R}^n)$. Moreover we need the *extension by* 0 operators e_Ω, in particular $e^+ = e_{\mathbb{R}^n_+}$ and $e^- = e_{\mathbb{R}^n_-}$, defined for functions f on Ω by

$$e_\Omega f = \begin{cases} f & \text{on } \Omega, \\ 0 & \text{on } \mathbb{R}^n \setminus \Omega; \end{cases} \qquad e^\pm = e_{\overline{\mathbb{R}}^n_\pm}. \tag{9.3}$$

The spaces of smooth functions on $\overline{\mathbb{R}}^n_+$ that we shall use will be the following:

$$C^\infty_{(0)}(\overline{\mathbb{R}}^n_+) = r^+ C^\infty_0(\mathbb{R}^n), \quad \mathscr{S}(\overline{\mathbb{R}}^n_+) = r^+ \mathscr{S}(\mathbb{R}^n), \tag{9.4}$$

where r^+ is defined in (9.2). Clearly, $C_{(0)}^\infty(\overline{\mathbb{R}}_+^n) \subset \mathscr{S}(\overline{\mathbb{R}}_+^n) \subset H^m(\mathbb{R}_+^n)$ for any m. The space $C_{(0)}^\infty(\overline{\mathbb{R}}_+^n)$ was defined in Chapter 4 as the space of functions on $\overline{\mathbb{R}}_+^n$ that are continuous and have continuous derivatives of all orders on $\mathbb{R}_+^n$, and have compact support in $\overline{\mathbb{R}}_+^n$; this is consistent with (9.4), cf. (C.7).

We note that even though r^+ is the restriction to the *open* set $\mathbb{R}_+^n$ (which makes sense for arbitrary distributions in $\mathscr{D}'(\mathbb{R}^n)$), the functions in $C_{(0)}^\infty(\overline{\mathbb{R}}_+^n)$ and $\mathscr{S}(\overline{\mathbb{R}}_+^n)$ are used as C^∞-functions on the closed set $\overline{\mathbb{R}}_+^n$. When we consider (measurable) *functions*, it makes no difference whether we speak of the restriction to $\mathbb{R}_+^n$ or to $\overline{\mathbb{R}}_+^n$, but for *distributions* it would, since there exist nonzero distributions supported in the set $\{\, x \in \mathbb{R}^n \mid x_n = 0 \,\}$.

Let us define the partial Fourier transform (in the x'-variable) for functions $u \in \mathscr{S}(\overline{\mathbb{R}}_+^n)$, by

$$\mathscr{F}_{x' \to \xi'} u = \acute{u}(\xi', x_n) = \int_{\mathbb{R}^{n-1}} e^{-ix' \cdot \xi'} u(x', x_n)\, dx'. \tag{9.5}$$

The partial Fourier transform (9.5) can also be given a sense for arbitrary $u \in H^m(\mathbb{R}_+^n)$, and for suitable distributions, but we shall perform our calculations on rapidly decreasing functions whenever possible, and extend the statements by continuity.

Just as the standard norm $\|u\|_m$ on $H^m(\mathbb{R}^n)$ could be replaced by the norm $\|u\|_{m,\wedge}$ (6.16) involving $\langle \xi \rangle^m$, which is particularly well suited for considerations using the Fourier transform, we can replace the standard norm $\|u\|_m$ on $H^m(\mathbb{R}_+^n)$ by an expression involving powers of $\langle \xi' \rangle$. Define for $u \in \mathscr{S}(\overline{\mathbb{R}}_+^n)$:

$$\|u\|_{m,\prime}^2 = (2\pi)^{1-n} \int_{\mathbb{R}^{n-1}} \int_0^\infty \sum_{j=0}^m \langle \xi' \rangle^{2(m-j)} |D_{x_n}^j \acute{u}(\xi', x_n)|^2\, dx_n d\xi'$$
$$= (2\pi)^{1-n} \sum_{j=0}^m \|\langle \xi' \rangle^{(m-j)} D_{x_n}^j \acute{u}(\xi', x_n)\|_{L_2(\mathbb{R}_+^n)}^2, \tag{9.6}$$

with the associated scalar product

$$(u,v)_{m,\prime} = (2\pi)^{1-n} \sum_{j=0}^m (\langle \xi' \rangle^{(m-j)} D_{x_n}^j \acute{u}(\xi', x_n), \langle \xi' \rangle^{(m-j)} D_{x_n}^j \acute{v}(\xi', x_n))_{L_2}. \tag{9.7}$$

Lemma 9.1. *For $m \in \mathbb{N}_0$, there are inequalities*

$$\|u\|_m^2 \le \|u\|_{m,\prime}^2 \le C_m' \|u\|_m^2, \tag{9.8}$$

valid for all $u \in \mathscr{S}(\overline{\mathbb{R}}_+^n)$, and hence $\|u\|_{m,\prime}$ (with its associated scalar product) extends by continuity to a Hilbert space norm on $H^m(\mathbb{R}_+^n)$ equivalent with the standard norm. Here $C_0' = 1$.

Proof. We have as in (5.2), for $k \in \mathbb{N}$:

$$\sum_{\beta \in \mathbb{N}_0^{n-1}, |\beta| \leq k} (\xi')^{2\beta} \leq \langle \xi' \rangle^{2k} = \sum_{|\beta| \leq k} C_{k,\beta}(\xi')^{2\beta} \leq C_k' \sum_{|\beta| \leq k} (\xi')^{2\beta}, \qquad (9.9)$$

with positive integers $C_{k,\beta}$ ($= \frac{k!}{\beta!(k-|\beta|)!}$) and $C_k' = \max_{\beta \in \mathbb{N}_0^{n-1}, |\beta| \leq k} C_{k,\beta}$. (The prime indicates that we get another constant than in (5.2) since the dimension has been replaced by $n-1$.)

We then calculate as follows, denoting $D' = (D_1, \ldots, D_{n-1})$, and using the Parseval-Plancherel theorem with respect to the x'-variables:

$$\begin{aligned}
\|u\|_m^2 &= \sum_{|\alpha| \leq m} \|D^\alpha u\|_0^2 = \sum_{j=0}^m \sum_{|\beta| \leq m-j} \|(D')^\beta D_{x_n}^j u\|_0^2 \\
&= \int_{\mathbb{R}^{n-1}} \sum_{j=0}^m \sum_{|\beta| \leq m-j} (\xi')^{2\beta} \int_0^\infty |D_{x_n}^j \hat{u}(\xi', x_n)|^2 \, dx_n d\xi' \\
&\leq \int_{\mathbb{R}^{n-1}} \int_0^\infty \sum_{j=0}^m \langle \xi' \rangle^{2(m-j)} |D_{x_n}^j \hat{u}(\xi', x_n)|^2 \, dx_n d\xi' \\
&\equiv \|u\|_{m,'}^2 \leq C_m' \|u\|_m^2,
\end{aligned} \qquad (9.10)$$

using (9.9) in the inequalities.

This shows (9.8), and the rest follows since the subset $C_{(0)}^\infty(\overline{\mathbb{R}}_+^n)$ of $\mathscr{S}(\overline{\mathbb{R}}_+^n)$, hence $\mathscr{S}(\overline{\mathbb{R}}_+^n)$ itself, is dense in $H^m(\mathbb{R}_+^n)$ (Theorem 4.10). $\qquad \square$

The extended norm is likewise denoted $\|u\|_{m,'}$. Note that in view of the formulas (9.9), we can also write

$$\|u\|_{m,'}^2 = \sum_{j=0}^m \sum_{|\beta| \leq m-j} C_{m-j,\beta} \|(D')^\beta D_n^j u\|_0^2, \qquad (9.11)$$

where the norm is defined without reference to the partial Fourier transform.

The following is a sharper version of the trace theorem shown in Theorem 4.24 (and Exercise 4.22).

Theorem 9.2. *Let m be an integer > 0. For $0 \leq j \leq m-1$, the trace mapping*

$$\gamma_j : u(x', x_n) \mapsto D_{x_n}^j u(x', 0) \qquad (9.12)$$

from $C_{(0)}^\infty(\overline{\mathbb{R}}_+^n)$ to $C_0^\infty(\mathbb{R}^{n-1})$ extends by continuity to a continuous linear mapping (also called γ_j) of $H^m(\mathbb{R}_+^n)$ into $H^{m-j-\frac{1}{2}}(\mathbb{R}^{n-1})$.

Proof. As in Theorem 4.24, we shall use an inequality like (4.47), but now in a slightly different and more precise way. For $v(t) \in C_0^\infty(\mathbb{R})$ one has

$$|v(0)|^2 = -\int_0^\infty \frac{d}{dt}[v(t)\overline{v}(t)]\,dt = -\int_0^\infty [v'(t)\overline{v}(t) + v(t)\overline{v'}(t)]\,dt$$

$$\leq 2\|v\|_{L_2(\mathbb{R}_+)}\|\frac{dv}{dt}\|_{L_2(\mathbb{R}_+)} \tag{9.13}$$

$$\leq a^2\|v\|_{L_2(\mathbb{R}_+)}^2 + a^{-2}\|\frac{dv}{dt}\|_{L_2(\mathbb{R}_+)}^2,$$

valid for any $a > 0$. For general functions $u \in C_{(0)}^\infty(\overline{\mathbb{R}}_+^n)$ we apply the partial Fourier transform in x' (cf. (9.5)) and apply (9.13) with respect to x_n, setting $a = \langle\xi'\rangle^{\frac{1}{2}}$ for each ξ' and using the norm (6.16) for $H^{m-j-\frac{1}{2}}(\mathbb{R}^{n-1})$:

$$\|\gamma_j u\|_{H^{m-j-\frac{1}{2}}(\mathbb{R}^{n-1})}^2 = \int_{\mathbb{R}^{n-1}} \langle\xi'\rangle^{2m-2j-1}|D_{x_n}^j \hat{u}(\xi',0)|^2\,d\xi'$$

$$\leq \int_{\mathbb{R}^{n-1}} \langle\xi'\rangle^{2m-2j-1}\int_0^\infty \big(a^2|D_{x_n}^j \hat{u}(\xi',x_n)|^2$$
$$+ a^{-2}|D_{x_n}^{j+1}\hat{u}(\xi',x_n)|^2\big)\,dx_n\,d\xi' \tag{9.14}$$

$$= \int \big[\langle\xi'\rangle^{2(m-j)}|D_{x_n}^j \hat{u}(\xi',x_n)|^2$$
$$+ \langle\xi'\rangle^{2(m-j-1)}|D_{x_n}^{j+1}\hat{u}(\xi',x_n)|^2\big]\,dx_n\,d\xi'$$

$$\leq \|u\|_{m,\prime}^2 \leq C_m'\|u\|_m^2,$$

cf. (9.8). This shows the continuity of the mapping, for the dense subset $C_{(0)}^\infty(\overline{\mathbb{R}}_+^n)$ of $H^m(\mathbb{R}_+^n)$, so the mapping γ_j can be extended to all of $H^m(\mathbb{R}_+^n)$ by closure. $\qquad\square$

The range space in Theorem 9.2 is *optimal*, for one can show that the mappings γ_j are surjective. In fact this holds for the whole *system* of trace operators γ_j with $j = 0, \ldots, m-1$. Let us for each $m > 0$ define the *Cauchy trace operator* $\rho_{(m)}$ associated with the order m by

$$\rho_{(m)} = \begin{pmatrix} \gamma_0 \\ \gamma_1 \\ \vdots \\ \gamma_{m-1} \end{pmatrix} : H^m(\mathbb{R}_+^n) \to \prod_{j=0}^{m-1} H^{m-j-\frac{1}{2}}(\mathbb{R}^{n-1}); \tag{9.15}$$

for $u \in H^m(\mathbb{R}_+^n)$ we call $\rho_{(m)}u$ the *Cauchy data* of u. (The indexation with (m) may be omitted if it is understood from the context.) The space $\prod_{j=0}^{m-1} H^{m-j-\frac{1}{2}}(\mathbb{R}^{n-1})$ is provided with the product norm $\|\varphi\|_{\prod H^{m-j-\frac{1}{2}}} = \big(\|\varphi_0\|_{m-\frac{1}{2}}^2 + \cdots + \|\varphi_{m-1}\|_{\frac{1}{2}}^2\big)^{\frac{1}{2}}$. Then one can show that the mapping $\varrho_{(m)}$ in (9.15) is surjective.

We first give a proof of the result for γ_0, that shows the basic idea and moreover gives some insight into boundary value problems for $I - \Delta$. For

$\varphi \in \mathscr{S}'(\mathbb{R}^{n-1})$ we denote its Fourier transform by $\hat{\varphi}$ (the usual notation, now applied with respect to the variable x').

Theorem 9.3. *Define the Poisson operator K_γ from $\mathscr{S}(\mathbb{R}^{n-1})$ to $\mathscr{S}(\overline{\mathbb{R}}_+^n)$ by*

$$K_\gamma : \varphi(x') \mapsto \mathscr{F}_{\xi' \to x'}^{-1}(e^{-\langle \xi' \rangle x_n} \hat{\varphi}(\xi')). \qquad (9.16)$$

It satisfies
  1° $\gamma_0 K_\gamma \varphi = \varphi$ for $\varphi \in \mathscr{S}(\mathbb{R}^{n-1})$.
  2° $(I - \Delta)K_\gamma \varphi = 0$ for $\varphi \in \mathscr{S}(\mathbb{R}^{n-1})$.
  3° K_γ extends to a continuous mapping (likewise denoted K_γ) from $H^{m-\frac{1}{2}}(\mathbb{R}^{n-1})$ to $H^m(\mathbb{R}_+^n)$ for any $m \in \mathbb{N}_0$; the identity in 1° extends to $\varphi \in H^{\frac{1}{2}}(\mathbb{R}^{n-1})$, and the identity in 2° extends to $\varphi \in H^{-\frac{1}{2}}(\mathbb{R}^{n-1})$.

Proof. It is easily checked that $e^{-\langle \xi' \rangle x_n}$ belong to $\mathscr{S}(\overline{\mathbb{R}}_+^n)$ (one needs to check derivatives of $e^{-\langle \xi' \rangle x_n}$, where (5.7) is useful); then also $e^{-\langle \xi' \rangle x_n} \hat{\varphi}(\xi')$ is in $\mathscr{S}(\overline{\mathbb{R}}_+^n)$, and so is its inverse partial Fourier transform; hence K_γ maps $\mathscr{S}(\mathbb{R}^{n-1})$ into $\mathscr{S}(\overline{\mathbb{R}}_+^n)$. Now

$$\gamma_0 K_\gamma \varphi = \left[\mathscr{F}_{\xi' \to x'}^{-1}(e^{-\langle \xi' \rangle x_n} \hat{\varphi}(\xi')) \right]_{x_n = 0} = \mathscr{F}_{\xi' \to x'}^{-1}(\hat{\varphi}(\xi')) = \varphi$$

shows 1°. By partial Fourier transformation,

$$\mathscr{F}_{x' \to \xi'}((I - \Delta)K_\gamma \varphi) = (1 + |\xi'|^2 - \partial_{x_n}^2)(e^{-\langle \xi' \rangle x_n} \hat{\varphi}(\xi'))$$
$$= (\langle \xi' \rangle^2 - \langle \xi' \rangle^2)e^{-\langle \xi' \rangle x_n} \hat{\varphi}(\xi') = 0;$$

this implies 2°.

For 3°, consider first the case $m = 0$. Here we have

$$\|K_\gamma \varphi\|_0^2 = \int_{\mathbb{R}^{n-1}} \int_0^\infty |K_\gamma \varphi(x)|^2 \, dx_n d\xi'$$
$$= \int_{\mathbb{R}^{n-1}} \int_0^\infty e^{-2\langle \xi' \rangle x_n} |\hat{\varphi}(\xi')|^2 \, dx_n d\xi'$$
$$= \int_{\mathbb{R}^{n-1}} (2\langle \xi' \rangle)^{-1} |\hat{\varphi}(\xi')|^2 \, d\xi' = \tfrac{1}{2}\|\varphi\|_{-\frac{1}{2},\wedge}^2,$$

showing that K_γ is not only continuous from $H^{-\frac{1}{2}}(\mathbb{R}^{n-1})$ to $L_2(\mathbb{R}_+^n)$, but even proportional to an isometry (into the space).

In the cases $m > 0$, we must work a little more, but get simple formulas using the new norms:

$$\|K_\gamma \varphi\|_{m,\prime}^2 = \int_{\mathbb{R}^{n-1}} \int_0^\infty \sum_{j=0}^m \langle \xi' \rangle^{2(m-j)} |D_{x_n}^j e^{-\langle \xi' \rangle x_n} \hat{\varphi}(\xi')|^2 \, dx_n d\xi'$$

$$= \int_{\mathbb{R}^{n-1}} \int_0^\infty \sum_{j=0}^m \langle\xi'\rangle^{2m} e^{-2\langle\xi'\rangle x_n} |\hat{\varphi}(\xi')|^2 \, dx_n d\xi' \tag{9.17}$$

$$= \tfrac{m+1}{2} \int_{\mathbb{R}^{n-1}} \langle\xi'\rangle^{2m-1} |\hat{\varphi}(\xi')|^2 \, d\xi' = \tfrac{m+1}{2} \|\varphi\|^2_{m-\frac{1}{2},\wedge};$$

again the mapping is proportional to an isometry.

Since γ_0 is well-defined as a continuous operator from $H^1(\mathbb{R}^n_+)$ to $H^{\frac{1}{2}}(\mathbb{R}^{n-1})$ and K_γ is continuous in the opposite direction, the identity in 1° extends from the dense subset $\mathscr{S}(\mathbb{R}^{n-1})$ to $H^{\frac{1}{2}}(\mathbb{R}^{n-1})$.

For 2°, let $\varphi \in H^{-\frac{1}{2}}(\mathbb{R}^{n-1})$ and let φ_k be a sequence in $\mathscr{S}(\mathbb{R}^{n-1})$ converging to φ in $H^{-\frac{1}{2}}(\mathbb{R}^{n-1})$. Then $K_\gamma \varphi_k$ converges to $v = K_\gamma \varphi$ in $L_2(\mathbb{R}^n_+)$. Here $(I-\Delta)K_\gamma\varphi_k = 0$ for all k, so in fact $K_\gamma\varphi_k$ converges to v in the graph norm for the maximal realization $A_{\max}$ defined from $I - \Delta$. Then $v \in D(A_{\max})$ with $A_{\max}v = 0$. So $(I-\Delta)v = 0$ in the distribution sense, in other words, $(I-\Delta)K_\gamma\varphi = 0$. This shows that 2° extends to $H^{-\frac{1}{2}}(\mathbb{R}^{n-1})$. $\qquad\square$

Remark 9.4. The mapping K_γ can be extended still further down to "negative Sobolev spaces". One can define $H^s(\mathbb{R}^n_+)$ for all orders (including noninteger and negative values, and consistently with the definitions for $s \in \mathbb{N}_0$) by

$$H^s(\mathbb{R}^n_+) = \{u \in \mathscr{D}'(\mathbb{R}^n_+) \mid u = U|_{\mathbb{R}^n_+} \text{ for some } U \in H^s(\mathbb{R}^n)\},$$

$$\|u\|_{s,\wedge} = \inf_{\text{such } U} \|U\|_{s,\wedge}; \tag{9.18}$$

and there is for any $s \in \mathbb{R}$ a continuous linear extension mapping $p_s :$ $H^s(\mathbb{R}^n_+) \to H^s(\mathbb{R}^n)$ such that $r^+ p_s$ is the identity on $H^s(\mathbb{R}^n_+)$. Then one can show that K_γ extends to map $H^{s-\frac{1}{2}}(\mathbb{R}^{n-1})$ continuously into $H^s(\mathbb{R}^n_+)$ for any $s \in \mathbb{R}$.

On the other hand, the mapping γ_0 *does not* extend to negative Sobolev spaces. More precisely, one can show that γ_0 *makes sense on* $H^s(\mathbb{R}^n_+)$ *if and only if* $s > \frac{1}{2}$. To show the sufficiency, we use the inequality $|v(0)| \le \int_{\mathbb{R}} |\hat{v}(t)| \, dt$, and the calculation, valid for $s > \frac{1}{2}$,

$$\int \langle\xi\rangle^{-2s} \, d\xi_n = \langle\xi'\rangle^{-2s+1} \int (1 + |\xi_n/\langle\xi'\rangle|^2)^{-s} \, d(\xi_n/\langle\xi'\rangle) = c_s \langle\xi'\rangle^{-2s+1}. \tag{9.19}$$

Then we have for $u \in \mathscr{S}(\mathbb{R}^n)$:

$$\|u(x',0)\|^2_{s-\frac{1}{2},\wedge} = \int_{\mathbb{R}^{n-1}} |\hat{u}(\xi',0)|^2 \langle\xi'\rangle^{2s-1} \, d\xi'$$

$$\le c \int_{\mathbb{R}^{n-1}} \langle\xi'\rangle^{2s-1} \Big(\int_{\mathbb{R}} |\hat{u}(\xi)| \, d\xi_n \Big)^2 d\xi'$$

$$\le c' \int_{\mathbb{R}^{n-1}} \langle\xi'\rangle^{2s-1} \Big(\int_{\mathbb{R}} |\hat{u}(\xi)|^2 \langle\xi\rangle^{2s} \, d\xi_n \Big) \Big(\int_{\mathbb{R}} \langle\xi\rangle^{-2s} \, d\xi_n \Big) d\xi'$$

$$= c'' \int_{\mathbb{R}^n} \langle \xi \rangle^{2s} |\hat{u}(\xi)|^2 \, d\xi = c_3 \|u\|_{s,\wedge}^2. \tag{9.20}$$

By density, this defines a bounded mapping $\tilde{\gamma}_0$ from $H^s(\mathbb{R}^n)$ to $H^{s-\frac{1}{2}}(\mathbb{R}^{n-1})$, and it follows that γ_0 extends to a bounded mapping from $H^s(\mathbb{R}^n_+)$ to $H^{s-\frac{1}{2}}(\mathbb{R}^{n-1})$. The necessity of $s > \frac{1}{2}$ is seen in case $n = 1$ as follows: Assume that $\gamma_0 : u \mapsto u(0)$ is bounded from $H^s(\mathbb{R})$ to $\mathbb{C}$ when $u \in \mathscr{S}(\mathbb{R})$. Write

$$u(0) = \int_{\mathbb{R}} \hat{u}(t) \, dt = \int_{\mathbb{R}} \langle t \rangle^{-s} \langle t \rangle^s \hat{u}(t) \, dt = \langle \langle t \rangle^{-s}, g(t) \rangle,$$

where $g(t) = \frac{1}{2\pi} \langle t \rangle^s \hat{u}(t)$ likewise runs through $\mathscr{S}(\mathbb{R})$. The boundedness implies that

$$|u(0)| = |\langle \langle t \rangle^{-s}, g(t) \rangle| \leq c\|u\|_{s,\wedge} = c'\|g\|_{L_2(\mathbb{R})},$$

which can only hold when $\langle t \rangle^{-s} \in L_2$, i.e., $s > \frac{1}{2}$. The general case is treated in [LM68, Th. I 4.3].

Nevertheless, γ_0 does have a sense, for any $s \in \mathbb{R}$, on the *subset* of distributions $u \in H^s(\mathbb{R}^n_+)$ for which $(I - \Delta)u = 0$. We shall show this below for $s = 0$. Variable-coefficient A's are treated in Lions and Magenes [LM68]; a general account is given in Theorem 11.4 later.

To show the surjectiveness of the Cauchy data map $\varrho_{(m)}$, one can construct a right inverse as a linear combination of operators defined as in (9.16) with $x_n^j e^{-\langle \xi' \rangle x_n}$, $j = 0, \ldots, m - 1$, inserted. Here follows another choice that is easy to check (and has the advantage that it could be used for $\mathbb{R}^n_-$ too). It is called a *lifting operator*.

Theorem 9.5. *Let* $\psi \in \mathscr{S}(\mathbb{R})$ *with* $\psi(t) = 1$ *on a neighborhood of* 0*. Define the Poisson operator* $\mathcal{K}_{(m)}$ *from* $\varphi = \{\varphi_0, \ldots, \varphi_{m-1}\} \in \mathscr{S}(\mathbb{R}^{n-1})^m$ *to* $\mathcal{K}_{(m)}\varphi \in \mathscr{S}(\overline{\mathbb{R}}^n_+)$ *by*

$$\mathcal{K}_{(m)}\varphi = \mathcal{K}_0\varphi_0 + \cdots + \mathcal{K}_{m-1}\varphi_{m-1}, \quad \text{where}$$
$$\mathcal{K}_j\varphi_j = \mathscr{F}^{-1}_{\xi' \to x'}\left(\frac{(ix_n)^j}{j!}\psi(\langle \xi' \rangle x_n)\hat{\varphi}_j(\xi')\right), \quad \text{for } j = 0, \ldots, m - 1. \tag{9.21}$$

It satisfies
$1°$ $\varrho_{(m)}\mathcal{K}_{(m)}\varphi = \varphi$ *for* $\varphi \in \mathscr{S}(\mathbb{R}^{n-1})^m$.
$2°$ $\mathcal{K}_{(m)}$ *extends to a continuous mapping (likewise denoted* $\mathcal{K}_{(m)}$*) from* $\Pi_{j=0}^{m-1} H^{m-j-\frac{1}{2}}(\mathbb{R}^{n-1})$ *to* $H^m(\mathbb{R}^n_+)$*, and the identity in* $1°$ *extends to* $\varphi \in \Pi_{j=0}^{m-1} H^{m-j-\frac{1}{2}}(\mathbb{R}^{n-1})$.

Proof. Since $\psi(0) = 1$ and $D_{x_n}^j \psi(0) = 0$ for $j > 0$, and $[D_{x_n}^k(\frac{1}{j!}i^j x_n^j)]|_{x_n=0} = \delta_{kj}$ (the Kronecker delta),

$$\gamma_k \mathcal{K}_j\varphi_j = (D_{x_n}^k \frac{(ix_n)^j}{j!})|_{x_n=0}\varphi_j = \delta_{kj}\varphi_j; \tag{9.22}$$

showing $1°$.

For the estimates in 2°, we observe the formulas that arise from replacing x_n by $t = \langle \xi' \rangle x_n$ in the integrals:

$$\int_0^\infty |D_{x_n}^k (x_n^j \psi(\langle \xi' \rangle x_n))|^2 \, dx_n$$
$$= \langle \xi' \rangle^{2k-2j-1} \int_0^\infty |D_t^k (t^j \psi(t))|^2 \, dt \equiv \langle \xi' \rangle^{2k-2j-1} c_{kj}. \quad (9.23)$$

Then

$$\|\mathcal{K}_j \varphi_j\|_{m,'}^2$$
$$= \int_{\mathbb{R}^{n-1}} \int_0^\infty \sum_{k \le m} \langle \xi' \rangle^{2m-2k} |D_{x_n}^k [\tfrac{(ix_n)^j}{j!} \psi(\langle \xi' \rangle x_n)] \hat{\varphi}_j(\xi')|^2 \, d\xi' dx_n$$
$$= \int_{\mathbb{R}^{n-1}} \sum_{k \le m} \langle \xi' \rangle^{2m-2k} \langle \xi' \rangle^{2k-2j-1} \tfrac{c_{kj}}{(j!)^2} |\hat{\varphi}_j(\xi')|^2 \, d\xi'$$
$$= c_j \|\varphi_j\|_{m-j-\frac{1}{2},\wedge}^2,$$

which shows the desired boundedness estimate. Now the identity in 1° extends by continuity. □

We know from Theorem 4.25 that the space $H_0^1(\mathbb{R}_+^n)$, defined as the closure of $C_0^\infty(\mathbb{R}_+^n)$ in $H^1(\mathbb{R}_+^n)$, is exactly the space of elements u of $H^1(\mathbb{R}_+^n)$ for which $\gamma_0 u = 0$. This fact extends to higher Sobolev spaces. As usual, $H_0^m(\mathbb{R}_+^n)$ is defined as the closure of $C_0^\infty(\mathbb{R}_+^n)$ in $H^m(\mathbb{R}_+^n)$.

Theorem 9.6. *For all $m \in \mathbb{N}$,*

$$H_0^m(\mathbb{R}_+^n) = \{u \in H^m(\mathbb{R}_+^n) \mid \gamma_0 u = \cdots = \gamma_{m-1} u = 0\}. \quad (9.24)$$

This is proved by a variant of the proof of Theorem 4.25, or by the method described after it, and will not be written in detail here. We observe another interesting fact concerning $H_0^m(\mathbb{R}_+^n)$:

Theorem 9.7. *For each $m \in \mathbb{N}_0$, $H_0^m(\mathbb{R}_+^n)$ identifies, by extension by zero on $\mathbb{R}_-^n$, with the subspace of $H^m(\mathbb{R}^n)$ consisting of the functions supported in $\overline{\mathbb{R}}_+^n$.*

Proof (indications). When $u \in H_0^m(\mathbb{R}_+^n)$, it is the limit of a sequence of function $u_k \in C_0^\infty(\mathbb{R}_+^n)$ in the m-norm. Extending the u_k by 0 (to $e^+ u_k$), we see that $e^+ u_k \to e^+ u$ in $H^m(\mathbb{R}^n)$ with $e^+ u$ supported in $\overline{\mathbb{R}}_+^n$. All such functions are reached, for if $v \in H^m(\mathbb{R}^n)$ has support in $\overline{\mathbb{R}}_+^n$, we can approximate it in m-norm by functions in $C_0^\infty(\mathbb{R}_+^n)$ by 1) truncation, 2) translation *into* $\mathbb{R}_+^n$ and 3) mollification. □

For any $s \in \mathbb{R}$, one can define the closed subspace of $H^s(\mathbb{R}^n)$:

$$H_0^s(\overline{\mathbb{R}}_+^n) = \{u \in H^s(\mathbb{R}^n) \mid \operatorname{supp} u \subset \overline{\mathbb{R}}_+^n\}; \qquad (9.25)$$

it identifies with $H_0^m(\mathbb{R}_+^n)$ when $s = m \in \mathbb{N}_0$. For negative s (more precisely, when $s \leq -\frac{1}{2}$), this is *not* a space of distributions on $\mathbb{R}_+^n$. It serves as a dual space: For any s one can show that $H^s(\mathbb{R}_+^n)$ and $H_0^{-s}(\overline{\mathbb{R}}_+^n)$ are dual spaces of one another, with a duality extending the scalar product in $L_2(\mathbb{R}_+^n)$, similarly to Theorem 6.15. (The duality is shown e.g. in [H63, Sect. 2.5].)

9.2 The Dirichlet problem for $I - \Delta$ on the half-space

We now consider the elliptic operator $A = I - \Delta$ on $\mathbb{R}_+^n$. Recall the definition of $A_{\max}$ and $A_{\min}$ from Chapter 4; note that by Theorem 6.24, $D(A_{\min}) = H_0^2(\mathbb{R}_+^n)$.

First we will show that γ_0 can be extended to $D(A_{\max})$. An important ingredient is the following denseness result (adapted from [LM68]):

Theorem 9.8. *The space* $C_{(0)}^\infty(\overline{\mathbb{R}}_+^n)$ *is dense in* $D(A_{\max})$.

Proof. This follows if we show that when ℓ is a continuous antilinear (conjugate linear) functional on $D(A_{\max})$ which vanishes on $C_{(0)}^\infty(\overline{\mathbb{R}}_+^n)$, then $\ell = 0$. So let ℓ be such a functional; it can be written

$$\ell(u) = (f, u)_{L_2(\mathbb{R}_+^n)} + (g, Au)_{L_2(\mathbb{R}_+^n)} \qquad (9.26)$$

for some $f, g \in L_2(\mathbb{R}_+^n)$. We know that $\ell(\varphi) = 0$ for $\varphi \in C_{(0)}^\infty(\overline{\mathbb{R}}_+^n)$. Any such φ is the restriction to $\mathbb{R}_+^n$ of a function $\Phi \in C_0^\infty(\mathbb{R}^n)$, and in terms of such functions we have

$$\ell(r^+\Phi) = (e^+f, \Phi)_{L_2(\mathbb{R}^n)} + (e^+g, (I-\Delta)\Phi)_{L_2(\mathbb{R}^n)} = 0, \text{ all } \Phi \in C_0^\infty(\mathbb{R}^n). \qquad (9.27)$$

Now use that $I - \Delta$ on $\mathbb{R}^n$ has the formal adjoint $I - \Delta$, so the right-hand side equations in (9.27) imply

$$\langle e^+f + (I-\Delta)e^+g, \overline{\Phi}\rangle = 0, \text{ all } \Phi \in C_0^\infty(\mathbb{R}^n),$$

i.e.,

$$e^+f + (I-\Delta)e^+g = 0, \text{ or } (I-\Delta)e^+g = -e^+f, \qquad (9.28)$$

as distributions on $\mathbb{R}^n$. Here we know that e^+g and e^+f are in $L_2(\mathbb{R}^n)$, and Theorem 6.12 then gives that $e^+g \in H^2(\mathbb{R}^n)$. Since it has support in $\overline{\mathbb{R}}_+^n$, it identifies with a function in $H_0^2(\mathbb{R}_+^n)$ by Theorem 9.7, i.e., $g \in H_0^2(\mathbb{R}_+^n)$. Then by Theorem 6.24, g is in $D(A_{\min})$! And (9.28) implies that $Ag = -f$. But then, for any $u \in D(A_{\max})$,

$$\ell(u) = (f, u)_{L_2(\mathbb{R}^n_+)} + (g, Au)_{L_2(\mathbb{R}^n_+)} = -(Ag, u)_{L_2(\mathbb{R}^n_+)} + (g, Au)_{L_2(\mathbb{R}^n_+)} = 0,$$

since $A_{\max}$ and $A_{\min}$ are adjoints. □

In the study of second-order operators, it is customary to omit the factor $-i$ on the normal derivative. So we define the Cauchy data as

$$\varrho u = \begin{pmatrix} \gamma_0 u \\ \nu u \end{pmatrix}, \text{ where } \nu u = \frac{\partial u}{\partial \nu} = \gamma_0(\partial_{x_n} u) = i\gamma_1 u. \tag{9.29}$$

Lemma 9.9. *For u and v in $H^2(\mathbb{R}^n_+)$ one has Green's formula*

$$(Au, v)_{L_2(\mathbb{R}^n_+)} - (u, Av)_{L_2(\mathbb{R}^n_+)} = (\nu u, \gamma_0 v)_{L_2(\mathbb{R}^{n-1})} - (\gamma_0 u, \nu v)_{L_2(\mathbb{R}^{n-1})}. \tag{9.30}$$

Proof. For functions in $C^\infty_{(0)}(\overline{\mathbb{R}}^n_+)$, the formula follows directly from (A.20). (It is easily verified by integration by parts.) Since γ_0 and ν by Theorem 9.2 map $H^2(\mathbb{R}^n_+)$ to spaces continuously injected in $L_2(\mathbb{R}^{n-1})$, the equation extends by continuity to $u, v \in H^2(\mathbb{R}^n_+)$. □

Now we can show:

Theorem 9.10. *Let $A = I - \Delta$ on $\mathbb{R}^n_+$. The Cauchy trace operator $\varrho = \{\gamma_0, \nu\}$, defined on $C^\infty_{(0)}(\overline{\mathbb{R}}^n_+)$, extends by continuity to a continuous mapping from $D(A_{\max})$ to $H^{-\frac{1}{2}}(\mathbb{R}^{n-1}) \times H^{-\frac{3}{2}}(\mathbb{R}^{n-1})$. Here Green's formula (9.30) extends to the formula*

$$(Au, v)_{L_2(\mathbb{R}^n_+)} - (u, Av)_{L_2(\mathbb{R}^n_+)} = \langle \nu u, \overline{\gamma_0 v} \rangle_{H^{-\frac{3}{2}}, H^{\frac{3}{2}}} - \langle \gamma_0 u, \overline{\nu v} \rangle_{H^{-\frac{1}{2}}, H^{\frac{1}{2}}}, \tag{9.31}$$

for $u \in D(A_{\max})$, $v \in H^2(\mathbb{R}^n_+)$.
Moreover, $\gamma_0 K_\gamma = I$ on $H^{-\frac{1}{2}}(\mathbb{R}^{n-1})$.

Proof. Let $u \in D(A_{\max})$. In the following, we write $H^s(\mathbb{R}^{n-1})$ as H^s. We want to define $\varrho u = \{\gamma_0 u, \nu u\}$ as a continuous antilinear functional on $H^{\frac{1}{2}} \times H^{\frac{3}{2}}$, depending continuously (and of course linearly) on $u \in D(A_{\max})$. For a given $\varphi = \{\varphi_0, \varphi_1\} \in H^{\frac{1}{2}} \times H^{\frac{3}{2}}$, we can use Theorem 9.5 to define

$$w_\varphi = K_0 \varphi_1 + i K_1 \varphi_0; \text{ then } \gamma_0 w_\varphi = \varphi_1, \ \nu w_\varphi = -\varphi_0.$$

Now we set

$$\begin{aligned} &\ell_u(\varphi) = (Au, w_\varphi) - (u, Aw_\varphi), \text{ noting that} \\ &|\ell_u(\varphi)| \le C\|u\|_{D(A_{\max})}\|w_\varphi\|_{H^2(\mathbb{R}^n_+)} \le C'\|u\|_{D(A_{\max})}\|\varphi\|_{H^{\frac{1}{2}} \times H^{\frac{3}{2}}}. \end{aligned} \tag{9.32}$$

So, ℓ_u is a continuous antilinear functional on $\varphi \in H^{\frac{1}{2}} \times H^{\frac{3}{2}}$, hence defines an element $\psi = \{\psi_0, \psi_1\} \in H^{-\frac{1}{2}} \times H^{-\frac{3}{2}}$ such that

$$\ell_u(\varphi) = \langle \psi_0, \overline{\varphi_0} \rangle_{H^{-\frac{1}{2}}, H^{\frac{1}{2}}} + \langle \psi_1, \overline{\varphi_1} \rangle_{H^{-\frac{3}{2}}, H^{\frac{3}{2}}}.$$

Moreover, it depends continuously on $u \in D(A_{\max})$, in view of the estimates in (9.32). If u is in $C_{(0)}^{\infty}(\overline{\mathbb{R}}_{+}^{n})$, the defining formula in (9.32) can be rewritten using Green's formula (9.30), which leads to

$$l_u(\varphi) = (Au, w_\varphi) - (u, Aw_\varphi) = (\nu u, \gamma_0 w_\varphi) - (\gamma_0 u, \nu w_\varphi) = (\gamma_0 u, \varphi_0) + (\nu u, \varphi_1),$$

for such u. Since φ_0 and φ_1 run through full Sobolev spaces, it follows that $\psi_0 = \gamma_0 u$, $\psi_1 = \nu u$, when $u \in C_{(0)}^{\infty}(\overline{\mathbb{R}}_{+}^{n})$, so the functional ℓ_u is consistent with $\{\gamma_0 u, \nu u\}$ then. Since $C_{(0)}^{\infty}(\overline{\mathbb{R}}_{+}^{n})$ is dense in $D(A_{\max})$, we have found the unique continuous extension.

(9.31) is now obtained in general by extending (9.30) by continuity from $u \in C_{(0)}^{\infty}(\overline{\mathbb{R}}_{+}^{n})$, $v \in H^2(\mathbb{R}_{+}^{n})$.

For the last statement, let $\eta \in H^{-\frac{1}{2}}(\mathbb{R}^{n-1})$. Note that we have already shown in Theorem 9.3 that K_γ maps $\eta \in H^{-\frac{1}{2}}$ into a function in $L_2(\mathbb{R}_{+}^{n})$ satisfying $(I - \Delta)K_\gamma \eta = 0$. But then $K_\gamma \eta \in D(A_{\max})$, so γ_0 can be applied in the new sense. The formula $\gamma_0 K_\gamma \eta = \eta$ is then obtained by extension by continuity from the case $\eta \in \mathscr{S}(\mathbb{R}^{n-1})$. $\qquad\square$

The definition was made via a choice w_φ of the "lifting" of the vector φ, but the final result shows that the definition is independent of this choice.

Consider now the Dirichlet problem (with a scalar φ)

$$Au = f \text{ in } \mathbb{R}_{+}^{n}, \quad \gamma_0 u = \varphi \text{ on } \mathbb{R}^{n-1}, \tag{9.33}$$

and its two semihomogeneous versions

$$Au = f \text{ in } \mathbb{R}_{+}^{n}, \quad \gamma_0 u = 0 \text{ on } \mathbb{R}^{n-1}, \tag{9.34}$$

$$Az = 0 \text{ in } \mathbb{R}_{+}^{n}, \quad \gamma_0 z = \varphi \text{ on } \mathbb{R}^{n-1}. \tag{9.35}$$

For (9.34), we find by application of the variational theory in Section 12.4 to the triple $(L_2(\mathbb{R}_{+}^{n}), H_0^1(\mathbb{R}_{+}^{n}), a(u, v))$ with $a(u, v) = (u, v)_1$, the variational operator A_γ, which is selfadjoint with lower bound 1. It is seen as in Section 4.4 (cf. Exercise 4.23), that A_γ is a realization of A with domain $D(A_\gamma) = D(A_{\max}) \cap H_0^1(\mathbb{R}_{+}^{n})$; the Dirichlet realization. Thus problem (9.34) has for $f \in L_2(\mathbb{R}_{+}^{n})$ the unique solution $u = A_\gamma^{-1} f$ in $D(A_{\max}) \cap H_0^1(\mathbb{R}_{+}^{n})$.

For (9.35), we have just shown that when $m \in \mathbb{N}_0$, it has a solution in $H^m(\mathbb{R}_{+}^{n})$ for any $\varphi \in H^{m-\frac{1}{2}}(\mathbb{R}^{n-1})$, namely, $K_\gamma \varphi$, by Theorem 9.3 and the last statement in Theorem 9.10. We have not yet investigated the uniqueness of the latter solution; it will be obtained below.

We can improve the information on (9.34) by use of K_γ and the structure of $I - \Delta$: To solve (9.34) for a given $f \in L_2(\mathbb{R}_{+}^{n})$, introduce

$$v = r^+ Q e^+ f, \quad Q = \mathrm{Op}(\langle \xi \rangle^{-2}). \tag{9.36}$$

Since Q maps $L_2(\mathbb{R}^n)$ homeomorphically onto $H^2(\mathbb{R}^n)$ (Section 6.2), $r^+ Q e^+$ is continuous from $L_2(\mathbb{R}_{+}^{n})$ to $H^2(\mathbb{R}_{+}^{n})$. Application of the differential operator

$A = I - \Delta$ to v gives that

$$Av = Ar^+Qe^+f = r^+(I - \Delta)Qe^+f = f,$$

so when u solves (9.34), the function $z = u - v$ must solve (9.35) with $\varphi = -\gamma_0 r^+ Qe^+ f$; here φ lies in $H^{\frac{3}{2}}(\mathbb{R}^{n-1})$. The problem for z has a solution $z = K_\gamma \varphi = -K_\gamma \gamma_0 r^+ Qe^+ f \in H^2(\mathbb{R}^n_+)$. Writing $u = v + z$, we finally obtain a solution of (9.34) of the form

$$u = r^+Qe^+f - K_\gamma\gamma_0 r^+Qe^+f. \tag{9.37}$$

It lies in $H^2(\mathbb{R}^n_+) \subset D(A_{\max})$, and in $H^1_0(\mathbb{R}^n_+)$ since $\gamma_0 u = 0$, so it must equal the unique solution already found. This improves the regularity! We have shown:

Theorem 9.11. *The solution operator A_γ^{-1} of (9.34) equals*

$$A_\gamma^{-1} = r^+Qe^+ - K_\gamma\gamma_0 r^+Qe^+, \quad \text{also denoted } R_\gamma. \tag{9.38}$$

It maps $L_2(\mathbb{R}^n_+)$ continuously into $H^2(\mathbb{R}^n_+)$, hence

$$D(A_\gamma) = H^2(\mathbb{R}^n_+) \cap H^1_0(\mathbb{R}^n_+). \tag{9.39}$$

For the special case we are considering, this gives the optimal regularity statement for the problem (9.34) with $f \in L_2(\mathbb{R}^n_+)$. And not only that; it also gives a *solution formula* (9.38) which is more constructive than the existence-and-uniqueness statement we had from the variational theory.

Remark 9.12. With further efforts it can be shown that r^+Qe^+ maps $H^k(\mathbb{R}^n_+)$ into $H^{k+2}(\mathbb{R}^n_+)$ also for $k \geq 1$; this is not clear from its form, since e^+ does not map $H^k(\mathbb{R}^n_+)$ into $H^k(\mathbb{R}^n)$ for $k \geq 1$, but hinges on the so-called transmission property of Q. Then one can obtain that u defined by (9.37) is in $H^{k+2}(\mathbb{R}^n_+)$, when $f \in H^k(\mathbb{R}^n_+)$: *higher elliptic regularity*. A systematic result of this kind is shown in Chapter 10.

As a corollary, we obtain a uniqueness statement for (9.35) in a regular case:

Corollary 9.13. *For any $\varphi \in H^{\frac{3}{2}}(\mathbb{R}^{n-1})$, $K_\gamma\varphi$ is the unique solution in $H^2(\mathbb{R}^n_+)$ of (9.35).*

Proof. Let $v = K_\gamma\varphi$, then $v \in H^2(\mathbb{R}^n_+)$ with $Av = 0$ and $\gamma_0 v = \varphi$. So a function z solves (9.35) if and only if $w = z - v$ solves

$$Aw = 0, \quad \gamma_0 w = 0.$$

Here if $z \in H^2(\mathbb{R}^n_+)$, also $w \in H^2(\mathbb{R}^n_+)$, so by the uniqueness of solutions of (9.34) in $H^2(\mathbb{R}^n_+)$, w must equal 0. $\square$

The argument can be extended to cover $\varphi \in H^{\frac{1}{2}}(\mathbb{R}^{n-1})$, $z \in H^1(\mathbb{R}^n_+)$, but since we are aiming for $z \in L_2(\mathbb{R}^n_+)$ we go on toward that case (and include the H^1 case afterwards).

Define, for $k \in \mathbb{N}_0$,

$$Z^k(A) = \{z \in H^k(\mathbb{R}^n_+) \mid Az = 0\}, \text{ closed subspace of } H^k(\mathbb{R}^n_+); \qquad (9.40)$$

note that $Z^0(A) = Z(A_{\max})$.

Corollary 9.13 and Theorem 9.3 together imply:

Corollary 9.14. *The mappings*

$$\gamma_0 : Z^2(A) \to H^{\frac{3}{2}}(\mathbb{R}^{n-1}) \ and \ K_\gamma : H^{\frac{3}{2}}(\mathbb{R}^{n-1}) \to Z^2(A)$$

are inverses of one another.

Proof. The mappings are well-defined according to Theorem 9.3, which also shows the identity $\gamma_0 K_\gamma = I$ on $H^{\frac{3}{2}}(\mathbb{R}^{n-1})$; it implies surjectiveness of γ_0 and injectiveness of K_γ. Corollary 9.13 implies that when $z \in Z^2(A)$, $K_\gamma \gamma_0 z = z$, so $K_\gamma : H^{\frac{3}{2}}(\mathbb{R}^{n-1}) \to Z^2(A)$ is surjective. □

Now we can show:

Proposition 9.15. $Z^2(A)$ *is dense in* $Z^0(A)$.

Proof. Let $z \in Z^0(A)$. By Theorem 9.8 there exists a sequence $u_k \in C^\infty_{(0)}(\overline{\mathbb{R}}^n_+)$ such that $u_k \to z$ in $L_2(\mathbb{R}^n_+)$, $Au_k \to 0$. Let $v_k = A^{-1}_\gamma Au_k$, then $v_k \to 0$ in $H^2(\mathbb{R}^n_+)$ by Theorem 9.11, so also $z_k = u_k - v_k \to z$ in $D(A_{\max})$, with $z_k \in H^2(\mathbb{R}^n_+)$. Here $Au_k = Av_k$ by definition, so indeed $z_k \in Z^2(A)$. □

Then we finally obtain:

Theorem 9.16. *The mappings*

$$\gamma_0 : Z^0(A) \to H^{-\frac{1}{2}}(\mathbb{R}^{n-1}) \ and \ K_\gamma : H^{-\frac{1}{2}}(\mathbb{R}^{n-1}) \to Z^0(A)$$

are inverses of one another.

Proof. We already have the identity $\gamma_0 K_\gamma \varphi = \varphi$ for $\varphi \in H^{-\frac{1}{2}}(\mathbb{R}^{n-1})$ from Theorem 9.10, and the other identity $K_\gamma \gamma_0 z = z$ for $z \in Z^0(A)$ now follows from the identity valid on $Z^2(A)$ by extension by continuity, using Proposition 9.15. □

Corollary 9.17. *Let* $m \in \mathbb{N}_0$. *The mappings*

$$\gamma_0 : Z^m(A) \to H^{m-\frac{1}{2}}(\mathbb{R}^{n-1}) \ and \ K_\gamma : H^{m-\frac{1}{2}}(\mathbb{R}^{n-1}) \to Z^m(A) \qquad (9.41)$$

are inverses of one another.

Proof. That the mappings are well-defined follows from Theorem 9.3 and its extension in Theorem 9.10. Then the statements $\gamma_0 K_\gamma \varphi = \varphi$ and $K_\gamma \gamma_0 z = z$ for the relevant spaces follow by restriction from Theorem 9.16. $\square$

Theorem 9.11 and Corollary 9.17 give highly satisfactory results on the solvability of the semihomogeneous problems (9.34) and (9.35). We collect some consequences for the nonhomogeneous problem (9.33):

Theorem 9.18. *The Dirichlet problem* (9.33) *is uniquely solvable in* $D(A_{\max})$ *for* $f \in L_2(\mathbb{R}^n_+)$, $\varphi \in H^{\frac{3}{2}}(\mathbb{R}^{n-1})$; *the solution belongs to* $H^2(\mathbb{R}^n_+)$ *and is defined by the formula*

$$u = r^+ Q e^+ f - K_\gamma \gamma_0 r^+ Q e^+ f + K_\gamma \varphi = R_\gamma f + K_\gamma \varphi; \tag{9.42}$$

cf. (9.38). *In particular, the mapping*

$$\begin{pmatrix} A \\ \gamma_0 \end{pmatrix} : H^2(\mathbb{R}^n_+) \to \begin{matrix} L_2(\mathbb{R}^n_+) \\ \times \\ H^{\frac{3}{2}}(\mathbb{R}^{n-1}) \end{matrix} \quad \text{has inverse} \quad \begin{pmatrix} R_\gamma & K_\gamma \end{pmatrix} : \begin{matrix} L_2(\mathbb{R}^n_+) \\ \times \\ H^{\frac{3}{2}}(\mathbb{R}^{n-1}) \end{matrix} \to H^2(\mathbb{R}^n_+).$$

$$\tag{9.43}$$

The above gives a complete clarification of the basic solvability properties for the Dirichlet problem (9.33) in the very special case of the simple constant-coefficient operator $I - \Delta$ on $\mathbb{R}^n_+$.

The deduction is of general interest. In fact, one can to some extent treat Dirichlet problems for strongly elliptic operators A (those where the real part of the principal symbol is positive) on bounded smooth domains Ω according to the same scheme. For such operators one can again define a variational realization A_γ representing the Dirichlet condition (this is done at the end of Chapter 7). One can add a constant to A to obtain that A_γ is invertible.

By use of localizations as in Appendix C and Section 4.2, one can establish mapping properties of the Cauchy trace operator analogous to (9.15). Moreover, one can establish a lifting operator $\mathcal{K}$ (a Poisson operator) with properties similar to those in Theorem 9.5. The denseness of $C^\infty(\overline{\Omega})$ in the maximal domain, the generalized trace operators and the extended Green's formula can be obtained much as in Theorems 9.8 and 9.10.

It is somewhat harder to extend Theorem 9.3 which defines the Poisson operator K_γ mapping into the nullspace, to the general situation, though; one can easily get a first approximation, but from then on there is as much work in it as in the general treatment of (9.33). The strategy in Lions and Magenes [LM68] is to begin with the regularity theory for the fully nonhomogeneous boundary value problem, next to pass to an adjoint situation in negative Sobolev spaces and finally use interpolation theory. (Their method works for far more general boundary conditions too.) Methods for constructing K_γ more directly can be based on the theory of Boutet de Monvel [B71] (see Chapters 10 and 11).

Note that in the case of $I - \Delta$ we could use the straightforward inverse $Q = \mathrm{Op}(\langle\xi\rangle^{-2})$ on $\mathbb{R}^n$; in general cases it will be replaced by more complicated expressions, pseudodifferential parametrices.

In the calculus of pseudodifferential boundary problems initiated by Boutet de Monvel, operators like r^+Qe^+ are called *truncated pseudodifferential operators*, operators like K_γ and $\mathcal{K}_{(m)}$ are called *Poisson operators* (we have already used this name) and operators like $K_\gamma\gamma_0$ and $K_\gamma\gamma_0 r^+Qe^+$, acting on functions on $\mathbb{R}^n_+$ but not of the form r^+Pe^+ with a ψdo P, are called *singular Green operators*. More on such operators in Chapter 10.

9.3 The Neumann problem for $I - \Delta$ on the half-space

For the special operator $I - \Delta$ it is now also easy to discuss the Neumann problem on $\mathbb{R}^n_+$. First we have, similarly to Theorem 9.3:

Theorem 9.19. *Define the Poisson operator K_ν from $\mathscr{S}(\mathbb{R}^{n-1})$ to $\mathscr{S}(\overline{\mathbb{R}}^n_+)$ by*

$$K_\nu : \varphi(x') \mapsto \mathscr{F}^{-1}_{\xi'\to x'}(-\langle\xi'\rangle^{-1}e^{-\langle\xi'\rangle x_n}\hat{\varphi}(\xi')). \tag{9.44}$$

It satisfies

$1°$ $\nu K_\nu\varphi = \varphi$ *for* $\varphi \in \mathscr{S}(\mathbb{R}^{n-1})$.

$2°$ $(I - \Delta)K_\nu\varphi = 0$ *for* $\varphi \in \mathscr{S}(\mathbb{R}^{n-1})$.

$3°$ K_ν *extends to a continuous mapping (likewise denoted K_ν) from $H^{m-\frac{3}{2}}(\mathbb{R}^{n-1})$ to $H^m(\mathbb{R}^n_+)$ for any $m \in \mathbb{N}_0$; the identity in $1°$ extends to $\varphi \in H^{\frac{1}{2}}(\mathbb{R}^{n-1})$, and the identity in $2°$ extends to $\varphi \in H^{-\frac{3}{2}}(\mathbb{R}^{n-1})$.*

Proof. $1°$ is seen from

$$\nu K_\nu\varphi = \left[\mathscr{F}^{-1}_{\xi'\to x'}(-\partial_{x_n}\langle\xi'\rangle^{-1}e^{-\langle\xi'\rangle x_n}\hat{\varphi}(\xi'))\right]_{x_n=0} = \mathscr{F}^{-1}_{\xi'\to x'}(\hat{\varphi}(\xi')) = \varphi.$$

$2°$ holds since

$$\mathscr{F}_{x'\to\xi'}((I - \Delta)K_\nu\varphi) = (\langle\xi'\rangle^2 - \partial^2_{x_n})(-\langle\xi'\rangle^{-1}e^{-\langle\xi'\rangle x_n}\hat{\varphi}(\xi'))$$
$$= (-\langle\xi'\rangle + \langle\xi'\rangle)e^{-\langle\xi'\rangle x_n}\hat{\varphi}(\xi') = 0.$$

For the continuity statements in $3°$, we calculate as in (9.17), with the modification that the extra factor $-\langle\xi'\rangle^{-1}$ changes the norm on φ to be the norm in $H^{m-\frac{3}{2}}(\mathbb{R}^{n-1})$.

Since ν is well-defined as a continuous operator from $H^2(\mathbb{R}^n_+)$ to $H^{\frac{1}{2}}(\mathbb{R}^{n-1})$ by Theorem 9.2, and K_ν is continuous in the opposite direction, the identity in $1°$ extends by continuity to $H^{\frac{1}{2}}(\mathbb{R}^{n-1})$. The proof of the extension of $2°$ goes in the same way as in Theorem 9.3. $\qquad\square$

Now consider the Neumann problem

$$Au = f \text{ in } \mathbb{R}^n_+, \quad \nu u = \varphi \text{ on } \mathbb{R}^{n-1}, \tag{9.45}$$

and its two semihomogeneous versions

$$Au = f \text{ in } \mathbb{R}^n_+, \quad \nu u = 0 \text{ on } \mathbb{R}^{n-1}, \tag{9.46}$$

$$Az = 0 \text{ in } \mathbb{R}^n_+, \quad \nu z = \varphi \text{ on } \mathbb{R}^{n-1}. \tag{9.47}$$

For (9.46) we find by application of the variational theory in Section 12.4 to the triple $(L_2(\mathbb{R}^n_+), H^1(\mathbb{R}^n_+), a(u,v))$ with $a(u,v) = (u,v)_1$, the variational operator A_ν, which is selfadjoint with lower bound 1. It is seen as in Section 4.4 (cf. Exercise 4.23), that A_ν is a realization of A with domain

$$D(A_\nu) = \{u \in D(A_{\max}) \cap H^1(\mathbb{R}^n_+) \mid (Au, v) = a(u,v) \text{ for all } v \in H^1(\mathbb{R}^n_+)\}; \tag{9.48}$$

the so-called Neumann realization. The smooth elements of $D(A_\nu)$ satisfy the Neumann condition $\nu u = 0$, but the interpretation of the conditions in (9.48) in the general case was not fully clarified (we do so below). However, in this generalized sense, problem (9.46) has for $f \in L_2(\mathbb{R}^n_+)$ the unique solution $u = A_\nu^{-1} f$ in $D(A_\nu)$.

For the problem (9.47), Theorem 9.19 shows that when $m \geq 2$, it has the solution $K_\nu \varphi \in H^m(\mathbb{R}^n_+)$ for any $\varphi \in H^{m-\frac{3}{2}}(\mathbb{R}^{n-1})$. To include the value $m = 0$, we use the statement from Theorem 9.10 that ν extends to a continuous operator from $D(A_{\max})$ to $H^{-\frac{3}{2}}$. The identity $\varphi = \nu K_\nu \varphi$, valid for $\varphi \in H^{\frac{1}{2}}$, then extends by continuity to $H^{-\frac{3}{2}}$. We also know from Theorem 9.19 that $(1 - \Delta)K_\nu \varphi = 0$ for $\varphi \in H^{-\frac{3}{2}}$. Thus $K_\nu \varphi$ is indeed a solution of (9.47) when $\varphi \in H^{-\frac{3}{2}}$.

Now the information on (9.46) can be improved by our knowledge of K_ν. Define again v by (9.36) and subtract it from u in (9.46), then calculations as after (9.36) with γ_0, K_γ replaced by ν, K_ν lead to the formula

$$u = r^+ Q e^+ f - K_\nu \nu r^+ Q e^+ f, \tag{9.49}$$

giving a solution in $H^2(\mathbb{R}^n_+)$ of (9.46). It belongs to the domain of A_ν, since the "halfways Green's formula"

$$(Au, v) - a(u,v) = (\nu u, \gamma_0 v), \tag{9.50}$$

known from (A.20) for smooth functions, extends by continuity (using Theorem 9.2) to $u \in H^2(\mathbb{R}^n_+)$ and $v \in H^1(\mathbb{R}^n_+)$, where νu and $\gamma_0 v$ are in $H^{\frac{1}{2}}(\mathbb{R}^{n-1}) \subset L_2(\mathbb{R}^{n-1})$.

Then u defined by (9.49) must be equal to $A_\nu^{-1} f$, and we conclude, similarly to Theorem 9.11:

Theorem 9.20. *The solution operator A_ν^{-1} of (9.46) equals*

$$A_\nu^{-1} = r^+ Q e^+ - K_\nu \nu r^+ Q e^+, \text{ also denoted } R_\nu. \tag{9.51}$$

It maps $L_2(\mathbb{R}^n_+)$ *continuously into* $H^2(\mathbb{R}^n_+)$, *hence*

$$D(A_\nu) = \{u \in H^2(\mathbb{R}^n_+) \mid \nu u = 0\}. \tag{9.52}$$

Having thus obtained optimal regularity for the Neumann problem (9.46), we can also clear up (9.47) completely. The proofs of Corollaries 9.13 and 9.14 generalize immediately to show:

Corollary 9.21. *For any* $\varphi \in H^{\frac{1}{2}}(\mathbb{R}^{n-1})$, $K_\nu\varphi$ *is the unique solution in* $H^2(\mathbb{R}^n_+)$ *of* (9.47). *The mappings*

$$\nu : Z^2(A) \to H^{\frac{1}{2}}(\mathbb{R}^{n-1}) \text{ and } K_\nu : H^{\frac{1}{2}}(\mathbb{R}^{n-1}) \to Z^2(A)$$

are inverses of one another.

Then we get, using the denseness of $Z^2(A)$ in $Z^0(A)$ shown in Proposition 9.15:

Theorem 9.22. *Let* $m \in \mathbb{N}_0$. *The mappings*

$$\nu : Z^m(A) \to H^{m-\frac{3}{2}}(\mathbb{R}^{n-1}) \text{ and } K_\nu : H^{m-\frac{3}{2}}(\mathbb{R}^{n-1}) \to Z^m(A) \tag{9.53}$$

are inverses of one another.

This is shown just as in the proofs of Theorem 9.16 and Corollary 9.17.

We collect some facts from Theorem 9.20 and Corollary 9.21 in a theorem on the fully nonhomogeneous problem:

Theorem 9.23. *The Neumann problem* (9.45) *is uniquely solvable in* $D(A_{\max})$ *for* $f \in L_2(\mathbb{R}^n_+)$, $\varphi \in H^{\frac{1}{2}}(\mathbb{R}^{n-1})$; *the solution belongs to* $H^2(\mathbb{R}^n_+)$ *and is defined by the formula*

$$u = r^+ Q e^+ f - K_\nu \nu r^+ Q e^+ f + K_\nu \varphi = R_\nu f + K_\nu \varphi; \tag{9.54}$$

cf. (9.51). *In particular, the mapping*

$$\begin{pmatrix} A \\ \nu \end{pmatrix} : H^2(\mathbb{R}^n_+) \to \begin{matrix} L_2(\mathbb{R}^n_+) \\ \times \\ H^{\frac{1}{2}}(\mathbb{R}^{n-1}) \end{matrix} \text{ has inverse } (R_\nu \ K_\nu) : \begin{matrix} L_2(\mathbb{R}^n_+) \\ \times \\ H^{\frac{1}{2}}(\mathbb{R}^{n-1}) \end{matrix} \to H^2(\mathbb{R}^n_+).$$

$$\tag{9.55}$$

9.4 Other realizations of $I - \Delta$

Besides the Dirichlet and the Neumann boundary conditions, it is of interest to study *Neumann-type* conditions:

$$\nu u + B\gamma_0 u = \varphi, \tag{9.56}$$

when B is a differential operator on $\mathbb{R}^{n-1}$ of order 1. This type of condition is called *normal*, since the γ_j with highest j appears with an invertible coefficient.

To keep things simple, we shall here only discuss constant-coefficient operators B. On the other hand, we shall allow the operators B to be pseudodifferential — however, just with x'-independent symbols as in Chapter 6. We take them closely related to the example $\langle \xi' \rangle$, to avoid reference to the general definition of the order of a ψdo at this moment.

The problem

$$\begin{aligned} Au &= f \text{ in } \mathbb{R}^n_+, \\ \nu u + B\gamma_0 u &= \varphi \text{ on } \mathbb{R}^{n-1}, \end{aligned} \tag{9.57}$$

will for some B behave like the Neumann problem (having $H^2(\mathbb{R}^n_+)$-regularity of solutions), for other B not, in particular when complex coefficients or pseudodifferential terms are allowed.

Still more general problems can be considered, for example with a Dirichlet boundary condition on part of the boundary and a Neumann-type condition on the rest of the boundary; then further complications arise.

An interesting question is, how all this fits into the abstract analysis established in Chapter 13, when it is applied with $A_0 = A_{\min}$ and $A_1 = A_{\max}$: How do the concrete boundary conditions fit into the picture? Can all the closed operators $\widetilde{A} \in \mathcal{M}$ be interpreted as representing boundary conditions? We shall deal with this question in the following, assuming that the reader is familiar with Sections 13.1–13.2.

With $A = I - \Delta$ on $\mathbb{R}^n_+$, consider the operators in $H = L_2(\mathbb{R}^n_+)$:

$$A_0 = A_{\min}, \quad A_1 = A_{\max}, \quad A_\gamma = \text{ the Dirichlet realization}; \tag{9.58}$$

then we have a setup as in Section 13.2 with

$$D(A_\gamma) = H^2(\mathbb{R}^n_+) \cap H^1_0(\mathbb{R}^n_+), \quad Z(A_1) = Z(A_{\max}) = Z^0(A), \tag{9.59}$$

and the positive selfadjoint operator A_γ has lower bound 1. We shall use that $Z(A_{\max})$ is isomorphic with $H^{-\frac{1}{2}}(\mathbb{R}^{n-1})$, as established in Theorem 9.16. In view of (9.17) we have in fact an isometry:

$$2^{\frac{1}{2}} K_\gamma : H^{-\frac{1}{2}}(\mathbb{R}^{n-1}) \to Z(A_{\max}) \text{ is a surjective isometry.} \tag{9.60}$$

The inverse of K_γ here acts like γ_0 and will be denoted γ_Z.

In the following, we generally omit the indication "$(\mathbb{R}^{n-1})$" from the boundary Sobolev spaces. Recall from Section 6.3 for the spaces $H^{-\frac{1}{2}}$ and $H^{\frac{1}{2}}$ that although they are Hilbert spaces provided with well-defined norms, and are of course self-dual, we put greater emphasis on their identification as dual spaces of one another with respect to the extension $\langle \varphi, \overline{\psi} \rangle_{H^{-\frac{1}{2}}, H^{\frac{1}{2}}}$ of the scalar product in $H^0 = L_2(\mathbb{R}^{n-1})$ (consistent with the distribution duality):

$$H^{\frac{1}{2}} \subset H^0 \subset H^{-\frac{1}{2}}, \quad H^{-\frac{1}{2}} \simeq (H^{\frac{1}{2}})^*, \quad H^{\frac{1}{2}} \simeq (H^{-\frac{1}{2}})^*. \tag{9.61}$$

We take the same point of view for subspaces of them:

Let X be a closed subspace of $H^{-\frac{1}{2}}$. Then we make very little use of the identification of the dual space X^* (the space of antilinear continuous functionals) with X, but regard X^* as a separate object. Since X is not dense in $H^{-\frac{1}{2}}$ when different from $H^{-\frac{1}{2}}$, there is not a natural inclusion between X^* and $H^{\frac{1}{2}}$ (except what comes from identifying them with their duals). But there is a surjective mapping:

Definition 9.24. For $\psi \in H^{\frac{1}{2}}$ we define the element $\psi|_X$ of X^* as the functional acting like ψ on X:

$$\langle \psi|_X, \overline{\varphi} \rangle_{X^*, X} = \langle \psi, \overline{\varphi} \rangle_{H^{\frac{1}{2}}, H^{-\frac{1}{2}}} \text{ for } \varphi \in X. \tag{9.62}$$

If $\psi \in H^{\frac{1}{2}}$ and $\eta \in X^*$, the identity $\psi|_X = \eta$ may also be expressed as

$$\psi = \eta \text{ on } X. \tag{9.63}$$

The functional $\psi|_X$ is continuous on X since the norm on this space is inherited from $H^{-\frac{1}{2}}$. All elements of X^* are obtained in this way. (For example, if η is in X^*, one can extend it to an element $\tilde{\eta}$ of $H^{\frac{1}{2}}$ by defining it to be zero on $H^{-\frac{1}{2}} \ominus X$. Other choices of complement of X in $H^{-\frac{1}{2}}$ will give other extensions.) The perhaps slightly abusive formulation (9.63) is standard for functions.

The following operator will play an important role:

Definition 9.25. Define the Dirichlet-to-Neumann operator P by

$$P\varphi = \nu K_\gamma \varphi; \tag{9.64}$$

it is continuous from $H^{m-\frac{1}{2}}$ to $H^{m-\frac{3}{2}}$ for all $m \in \mathbb{N}_0$.

For $m \geq 2$ and for $m = 0$, the continuity follows from the continuity of K_γ from $H^{m-\frac{1}{2}}$ to $Z^m(A)$ by Theorem 9.3, and the continuity of $\nu = i\gamma_1$ from $H^m(\mathbb{R}^n_+)$ to $H^{m-\frac{3}{2}}$ for $m \geq 2$ in Theorem 9.2 and from $D(A_{\max})$ to $H^{-\frac{3}{2}}$ for $m = 0$ by Theorem 9.10 ($m = 1$ is easily included). In fact, we can say more about P: From the exact form of K_γ in (9.16) (for smooth functions) we see that P is the pseudodifferential operator (in the sense of Definition 6.1)

$$P = \text{Op}(-\langle \xi' \rangle), \tag{9.65}$$

and it is clearly an isometry of H^s onto H^{s-1} for all $s \in \mathbb{R}$ (cf. (6.17)). Moreover, P is formally selfadjoint and $-P$ is positive with lower bound 1 in $L_2(\mathbb{R}^{n-1})$, since the symbol $-\langle \xi' \rangle$ is real and ≤ -1; cf. Theorem 6.3.

Next, we define a somewhat strange trace operator:

Definition 9.26. The trace operator μ is defined on $D(A_{\max})$ by

$$\mu u = \nu u - P\gamma_0 u. \tag{9.66}$$

At first sight, it ranges in $H^{-\frac{3}{2}}$, since ν and $P\gamma_0$ do so. But this information can be improved:

Proposition 9.27. *The trace operator μ satisfies*

$$\mu u = \nu A_\gamma^{-1} A u, \text{ for } u \in D(A_{\max}), \tag{9.67}$$

hence maps $D(A_{\max})$ continuously into $H^{\frac{1}{2}}$. The following Green's formula holds for all $u, v \in D(A_{\max})$:

$$(Au, v) - (u, Av) = \langle \mu u, \overline{\gamma_0 v} \rangle_{H^{\frac{1}{2}}, H^{-\frac{1}{2}}} - \langle \gamma_0 u, \overline{\mu v} \rangle_{H^{-\frac{1}{2}}, H^{\frac{1}{2}}}. \tag{9.68}$$

Furthermore,

$$\mu z = 0 \text{ for } z \in Z(A_{\max}), \tag{9.69}$$

and

$$(Au, z) = \langle \mu u, \overline{\gamma_0 z} \rangle_{H^{\frac{1}{2}}, H^{-\frac{1}{2}}} \text{ when } u \in D(A_{\max}), z \in Z(A_{\max}). \tag{9.70}$$

Proof. When $u \in D(A_{\max})$, we decompose it in

$$u = u_\gamma + u_\zeta, \quad u_\gamma = A_\gamma^{-1} A u \in D(A_\gamma),$$

where $u_\zeta \in Z(A_{\max})$ (as in Lemma 13.1). Then since $\gamma_0 u_\gamma = 0$,

$$\gamma_0 u = \gamma_0 u_\zeta, \text{ so } P\gamma_0 u = P\gamma_0 u_\zeta = \nu u_\zeta \tag{9.71}$$

by definition of P, and hence

$$\nu u - P\gamma_0 u = \nu u - \nu u_\zeta = \nu u_\gamma = \nu A_\gamma^{-1} A u.$$

This shows (9.67), and the continuity follows from the continuity properties of the three factors.

Consider Green's formula (9.30) for functions $u, v \in \mathscr{S}(\overline{\mathbb{R}}_+^n)$ and subtract from it the identity $(P\gamma_0 u, \gamma_0 v) - (\gamma_0 u, P\gamma_0 v) = 0$, then we get

$$(Au, v) - (u, Av) = (\nu u - P\gamma_0 u, \gamma_0 v) - (\gamma_0 u, \nu v - P\gamma_0 v),$$

for functions in $\mathscr{S}(\overline{\mathbb{R}}_+^n)$; this shows (9.68) for smooth functions. It extends by continuity to $u, v \in D(A_{\max})$ in view of the continuity of μ shown above and the continuity of γ_0 shown in Theorem 9.10.

The equation (9.69) is clear from the fact that $\nu z = P\gamma_0 z$ when $z \in Z(A_{\max})$, and (9.70) follows from (9.68) with $v = z$ since Az and μz vanish. $\square$

Recall from Section 13.2 that any closed realization $\widetilde{A}$ corresponds (in a unique way) to a closed, densely defined operator $T : V \to W$, where V and W are closed subspaces of $Z(A_{\max})$. We carry this over to a situation with spaces over the boundary as follows:

Definition 9.28. Let V and W be closed subspaces of $Z(A_{\max})$, and let $T : V \to W$ be closed, densely defined. The corresponding setup over the boundary is then defined by letting

$$X = \gamma_0 V,\ Y = \gamma_0 W,\ \text{closed subspaces of } H^{-\frac{1}{2}}, \qquad (9.72)$$

and defining $L : X \to Y^*$ by

$$\begin{aligned} D(L) &= \gamma_0 D(T), \\ \langle L\gamma_0 v, \overline{\gamma_0 w}\rangle_{Y^*,Y} &= (Tv, w),\ \text{for all } v \in D(T), w \in W. \end{aligned} \qquad (9.73)$$

Since γ_0 equals the invertible operator γ_Z in all the formulas in this definition, T is determined from L, and when $T : V \to W$ runs through all choices of closed subspaces V, W of $Z(A_{\max})$ and closed densely defined operators from V to W, then $L : X \to Y^*$ runs through all choices of closed subsaces X, Y of $H^{-\frac{1}{2}}$ and closed densely defined operators from X to Y^*. This is obvious if we do provide X and Y with the norm in $H^{-\frac{1}{2}}$, identify them with their duals and use that $2^{-\frac{1}{2}}\gamma_Z$ is an isometry. (So the norm comes in useful here, but in general we focus on properties that are expressed without reference to a choice of norm.)

Note that T and L have similar properties: They are simultaneously injective, or surjective. The nullspace of L is $Z(L) = \gamma_Z Z(T)$. The adjoint of T, $T^* : W \to V$, corresponds to the adjoint L^* of L defined as an operator from Y to X^* with domain $D(L^*) = \gamma_0 D(T^*)$:

$$(v, T^*w) = (Tv, w) = \langle L\gamma_0 v, \overline{\gamma_0 w}\rangle_{Y^*,Y} = \langle \gamma_0 v, \overline{L^*\gamma_0 w}\rangle_{X,X^*},$$
$$v \in D(T), w \in D(T^*).$$

(It is understood that $\langle \varphi, \overline{\psi}\rangle_{X,X^*} = \overline{\langle \psi, \overline{\varphi}\rangle}_{X^*,X}$.) Note that $Y = \overline{D(L^*)}$, the closure in $H^{-\frac{1}{2}}$. When $V = W$, i.e., $X = Y$, T is selfadjoint if and only if L is so. Also lower boundedness is preserved in the correspondence: When $V \subset W$, then $X \subset Y$, and

$$\operatorname{Re}(Tv, v) \geq c\|v\|^2 \text{ for } v \in D(T) \qquad (9.74)$$

holds if and only if

$$\mathrm{Re}\langle L\varphi, \overline{\varphi}\rangle_{Y^*,Y} \geq c' \|\varphi\|_Y^2 \text{ for } \varphi \in D(L), \tag{9.75}$$

for some c' whose *value* depends on the choice of norm in $H^{-\frac{1}{2}}$, but whose *sign* (positive, negative or zero) is the same as that of c, independently of the choice of norm.

We can now interpret the general realizations of A by boundary conditions.

Theorem 9.29. *Consider a closed realization $\widetilde{A}$ of A, corresponding to T : $V \to W$ as in Theorem 13.7 (with $A_1 = A_1' = A_{\max}$), and let $L : X \to Y^*$ be the corresponding operator introduced in Definition 9.28. Then $D(\widetilde{A})$ consists of the functions $u \in D(A_{\max})$ for which*

$$\gamma_0 u \in D(L), \quad \mu u|_Y = L\gamma_0 u. \tag{9.76}$$

In this correspondence, $D(L) = \gamma_0 D(\widetilde{A})$.

Proof. We have from Theorem 13.5 that the elements of $D(\widetilde{A})$ are characterized by the two conditions:

$$u_\zeta \in D(T), \quad (Au, w) = (Tu_\zeta, w) \text{ for all } w \in W. \tag{9.77}$$

We just have to translate this to boundary conditions. In view of the definition of L, we have for any $u \in D(A_{\max})$, since $\gamma_0 u_\gamma = 0$, that

$$u_\zeta \in D(T) \iff \gamma_0 u_\zeta \in D(L) \iff \gamma_0 u \in D(L), \tag{9.78}$$

showing that the first conditions in (9.76) and (9.77) are equivalent. When $w \in W \subset Z(A_{\max})$, we have in view of (9.70):

$$(Au, w) = \langle \mu u, \overline{\gamma_0 w}\rangle_{H^{\frac{1}{2}}, H^{-\frac{1}{2}}},$$

whereas

$$(Tu_\zeta, w) = \langle L\gamma_0 u, \overline{\gamma_0 w}\rangle_{Y^*,Y},$$

in view of (9.71) and (9.73). Then the second conditions in (9.76) and (9.77) are equivalent, in view of Definition 9.24.

The last assertion follows from (9.78), since $D(T) = \mathrm{pr}_\zeta D(\widetilde{A})$; cf. Theorem 13.5. $\qquad\square$

Let us consider some examples.

Example 9.30. For A_γ itself, $V = W = \{0\}$, and T is trivial (zero) then. So $X = Y = \{0\}$ and $L = 0$. The boundary condition is

$$\gamma_0 u = 0, \tag{9.79}$$

as we know very well.

Example 9.31. Consider A_M from Example 13.10, the von Neumann (or Kreĭn) extension; it is the realization with domain

$$D(A_M) = D(A_{\min}) \dotplus Z(A_{\max}) = H_0^2(\mathbb{R}_+^n) \dotplus Z^0(A). \tag{9.80}$$

Here $V = W = Z^0(A)$, and $T = 0$ on $D(T) = Z^0(A)$. So $X = Y = H^{-\frac{1}{2}}$ with $L = 0$ on $D(L) = H^{-\frac{1}{2}}$. The boundary condition is

$$\mu u = 0, \text{ i.e., } \nu u - P\gamma_0 u = 0. \tag{9.81}$$

This is a normal boundary condition (cf. (9.56)ff.), with $B = -P$. The realization, it defines, has *no regularity:* The domain is not in $H^m(\mathbb{R}_+^n)$ for any $m > 0$, since it contains $Z^0(A)$. (By Corollary 9.17, the $Z^m(A)$ are strictly different for different m, since the $H^{m-\frac{1}{2}}$ are so.)

Example 9.32. The Neumann realization is defined by the boundary condition

$$\nu u = 0. \tag{9.82}$$

We know from Section 9.3 that the domain of the hereby defined realization A_ν equals $\{u \in H^2(\mathbb{R}_+^n) \mid \nu u = 0\}$.

Since $\varrho = \{\gamma_0, \nu\}$ is surjective from $H^2(\mathbb{R}_+^n)$ to $H^{\frac{3}{2}} \times H^{\frac{1}{2}}$ (cf. Theorem 9.5), $\gamma_0 u$ runs through $H^{\frac{3}{2}}$ when u runs through the domain, so $D(L) = H^{\frac{3}{2}}$. Moreover, $X = H^{-\frac{1}{2}}$ since $H^{\frac{3}{2}}$ is dense in $H^{-\frac{1}{2}}$. Since A_ν is selfadjoint, also $Y = H^{-\frac{1}{2}}$. Then $D(A_\nu)$ must be characterized by a condition of the form

$$\mu u - L\gamma_0 u = 0, \text{ for } \gamma_0 u \in H^{\frac{3}{2}},$$

so since $\mu u = \nu u - P\gamma_0 u$,

$$\nu u - P\gamma_0 u - L\gamma_0 u = 0, \text{ for } \gamma_0 u \in H^{\frac{3}{2}},$$

when $u \in D(A_\nu)$. Since $\nu u = 0$ there, $L + P = 0$ on $H^{\frac{3}{2}}$. This shows that

$$L \text{ acts like } -P, \quad D(L) = H^{\frac{3}{2}}, \tag{9.83}$$

in the case of A_ν.

Let us now consider more general realizations defined by Neumann-type boundary conditions

$$\nu u + B\gamma_0 u = 0. \tag{9.84}$$

Here we let B be a pseudodifferential operator with symbol

$$b(\xi') = ib_1\xi_1 + \cdots + ib_{n-1}\xi_{n-1} + c\langle\xi'\rangle, \text{ some } c \in \mathbb{R}; \tag{9.85}$$

i.e., it is the sum of a differential operator B_1 and a multiple of P:

$$B = \mathrm{Op}(b) = B_1 - cP, \quad B_1 = b_1 \partial_{x_1} + \cdots + b_{n-1} \partial_{x_{n-1}}. \tag{9.86}$$

(9.84) determines the realization $\widetilde{A}$ with domain

$$D(\widetilde{A}) = \{ u \in D(A_{\max}) \mid \nu u + B\gamma_0 u = 0 \}. \tag{9.87}$$

Since the trace operator $\nu + B\gamma_0$ is continuous from $D(A_{\max})$ to $H^{-\frac{3}{2}}$, the domain is closed.

Clearly,

$$D(\widetilde{A}) \supset \{ u \in H^2(\mathbb{R}^n_+) \mid \nu u + B\gamma_0 u = 0 \} = D(\widetilde{A}) \cap H^2(\mathbb{R}^n_+),$$

but we do not know a priori whether this is an equality; in fact, we shall look for criteria for whether it is so.

Since $\varrho = \{\gamma_0, \nu\}$ is surjective from $H^2(\mathbb{R}^n_+)$ to $H^{\frac{3}{2}} \times H^{\frac{1}{2}}$, $\gamma_0 u$ runs through $H^{\frac{3}{2}}$ when u runs through $D(\widetilde{A}) \cap H^2(\mathbb{R}^n_+)$, so $D(L) \supset H^{\frac{3}{2}}$. Since the latter space is dense in $H^{-\frac{1}{2}}$, $X = H^{-\frac{1}{2}}$.

Observe that in view of Corollary 9.17 and the decomposition of $D(A_{\max})$ into $D(A_\gamma)$ and $Z(A_{\max})$,

$$D(L) = H^{\frac{3}{2}} \iff D(\widetilde{A}) = D(\widetilde{A}) \cap H^2(\mathbb{R}^n_+). \tag{9.88}$$

Concerning the adjoint $\widetilde{A}^*$, we have from the general Green's formula (9.31) that when $v \in D(\widetilde{A}^*)$ and $u \in D(\widetilde{A}) \cap H^2(\mathbb{R}^n_+)$, then

$$
\begin{aligned}
0 = (Av, u) - (v, Au) &= \langle \nu v, \overline{\gamma_0 u} \rangle_{H^{-\frac{3}{2}}, H^{\frac{3}{2}}} - \langle \gamma_0 v, \overline{\nu u} \rangle_{H^{-\frac{1}{2}}, H^{\frac{1}{2}}} \\
&= \langle \nu v, \overline{\gamma_0 u} \rangle_{H^{-\frac{3}{2}}, H^{\frac{3}{2}}} - \langle \gamma_0 v, -\overline{B\gamma_0 u} \rangle_{H^{-\frac{1}{2}}, H^{\frac{1}{2}}} \\
&= \langle \nu v + B^* \gamma_0 v, \overline{\gamma_0 u} \rangle_{H^{-\frac{3}{2}}, H^{\frac{3}{2}}}.
\end{aligned}
$$

This shows that

$$v \in D(\widetilde{A}^*) \implies \nu v + B^* \gamma_0 v = 0,$$

since $\gamma_0 u$ runs freely in $H^{\frac{3}{2}}$. It was used here that $B : H^s \to H^{s-1}$ has the adjoint $B^* : H^{-s+1} \to H^{-s}$, any s, where B^* acts as the ps.d.o. with symbol $\overline{b}(\xi')$. On the other hand, if $v \in H^2(\mathbb{R}^n_+)$ with $\nu v + B^* \gamma_0 v = 0$, then an application of (9.31) with $u \in D(\widetilde{A})$ shows that

$$
\begin{aligned}
(Au, v) - (u, Av) &= \langle \nu u, \overline{\gamma_0 v} \rangle_{H^{-\frac{3}{2}}, H^{\frac{3}{2}}} - \langle \gamma_0 u, \overline{\nu v} \rangle_{H^{-\frac{1}{2}}, H^{\frac{1}{2}}} \\
&= \langle \nu u, \overline{\gamma_0 v} \rangle_{H^{-\frac{3}{2}}, H^{\frac{3}{2}}} - \langle \gamma_0 u, -\overline{B^* \gamma_0 v} \rangle_{H^{-\frac{1}{2}}, H^{\frac{1}{2}}} \\
&= \langle \nu u + B\gamma_0 u, \overline{\gamma_0 v} \rangle_{H^{-\frac{3}{2}}, H^{\frac{3}{2}}} = 0,
\end{aligned}
$$

which implies that $v \in D(\widetilde{A}^*)$. So we conclude that

$$\{v \in H^2(\mathbb{R}^n_+) \mid \nu v + B^* \gamma_0 v = 0\} \subset D(\widetilde{A}^*) \subset \{v \in D(A_{\max}) \mid \nu v + B^* \gamma_0 v = 0\}. \tag{9.89}$$

This gives us the information that $D(L^*)$ contains $H^{\frac{3}{2}}$ in its domain, so since Y equals the closure of $D(L^*)$ in $H^{-\frac{1}{2}}$, $Y = H^{-\frac{1}{2}}$.

It follows that L is an operator from $D(L) \subset H^{-\frac{1}{2}}$ to $Y^* = H^{\frac{1}{2}}$. We can now use that $L\gamma_0 u = \mu u$ on $D(\widetilde{A})$ to find how L acts, when (9.84) holds:

$$0 = L\gamma_0 u - \mu u = L\gamma_0 u - \nu u + P\gamma_0 u = L\gamma_0 u + B\gamma_0 u + P\gamma_0 u,$$

so

$$L = -B - P \text{ on } D(L). \tag{9.90}$$

Similarly,

$$L^* = -B^* - P \text{ on } D(L^*). \tag{9.91}$$

A further precision of L can be made when L is elliptic. Note that

$$L \text{ acts like } -B_1 + cP - P = \text{Op}(l(\xi')),$$
$$l(\xi') = -ib_1\xi_1 - \cdots - ib_{n-1}\xi_{n-1} + (1-c)\langle\xi'\rangle. \tag{9.92}$$

The symbol $\langle\xi'\rangle$ has a series expansion

$$\langle\xi'\rangle = (1 + |\xi'|^2)^{\frac{1}{2}} = |\xi'|(1 + |\xi'|^{-2})^{\frac{1}{2}} = |\xi'| + |\xi'| \sum_{k \geq 1} \binom{\frac{1}{2}}{k} |\xi'|^{-2k},$$

converging for $|\xi'| > 1$. (Here $\binom{\frac{1}{2}}{k} = \frac{1}{2}(\frac{1}{2} - 1) \cdots (\frac{1}{2} - k + 1)/k!$.) With a notation borrowed from Chapter 7, see Definition 7.2 ff. and Definition 7.17, we have that the *principal part* of $l(\xi')$ is

$$l_1(\xi') = -ib_1\xi_1 - \cdots - ib_{n-1}\xi_{n-1} + (1-c)|\xi'|, \tag{9.93}$$

and l is *elliptic* precisely when $l_1(\xi') \neq 0$ for $\xi' \neq 0$. In this case the requirement for Theorem 6.22 with $m = 1$ is satisfied.

Thus in the elliptic case we can conclude from Theorem 6.22 that $\text{Op}(l)\varphi \in H^{\frac{1}{2}}$ implies $\varphi \in H^{\frac{3}{2}}$. Since L acts like $\text{Op}(l)$, has $D(L) \supset H^{\frac{3}{2}}$ and $R(L) \subset H^{\frac{1}{2}}$, we conclude that $D(L) = H^{\frac{3}{2}}$. This shows:

Theorem 9.33. *Let $\widetilde{A}$ be the realization defined by the boundary condition (9.84). If $l(\xi')$ in (9.92) is elliptic, then*

$$D(\widetilde{A}) = \{u \in H^2(\mathbb{R}^n_+) \mid \nu u + B\gamma_0 u = 0\}.$$

In this case also

$$D(\widetilde{A}^*) = \{u \in H^2(\mathbb{R}^n_+) \mid \nu u + B^* \gamma_0 u = 0\}.$$

The last statement follows since the adjoint symbol $\bar{l}(\xi')$ is then likewise elliptic.

Ellipticity clearly holds if

$$c \neq 1, \; b_1, \ldots, b_{n-1} \in \mathbb{R}, \tag{9.94}$$

in particular when $c = 0$ and the b_j are real. So we have as a corollary:

Corollary 9.34. *When B is a differential operator with real coefficients, the boundary condition (9.84) defines a realization with domain in $H^2(\mathbb{R}^n_+)$.*

Nonelliptic examples are found when $c = 1$ — this is so in Example 9.30 — or if some of the coefficients b_j are nonreal, e.g., if one of them equals $\pm i(1-c)$, so that there are points $\xi' \neq 0$ with $l_1(\xi') = 0$. One can also show that ellipticity of $l(\xi')$ is *necessary* for having $D(L) \subset H^{\frac{3}{2}}$, i.e., $D(\widetilde{A}) \subset H^2(\mathbb{R}^n_+)$.

We can also discuss lower boundedness, in view of the results in Section 13.2 and the equivalence of (9.74) and (9.75). The most immediate results are:

Theorem 9.35. *Let $\widetilde{A}$ be the realization determined by the boundary condition (9.84).*

1° *If $\widetilde{A}$ is lower bounded, so is L, with a similar sign of the lower bound.*
2° *If L has positive or zero lower bound, so has $\widetilde{A}$.*

Proof. We use the equivalence of (9.74) and (9.75). Note that we are in a case where we know beforehand that $V = W = Z(A_1)$. The first statement follows from Theorem 13.15. The second statement follows from Theorem 13.17. □

Again this applies easily to the differential case (where $c = 0$ in (9.85)):

Corollary 9.36. *When B is a differential operator with real coefficients, the realization defined by the boundary condition (9.84) is variational with positive lower bound.*

Proof. In this case, $\operatorname{Re} l(\xi') = \langle \xi' \rangle \geq 1$ and $D(L) = H^{\frac{3}{2}}$, so L has lower bound 1, as an operator in $L_2 = H^0$. The same holds for L^*. Moreover,

$$\begin{aligned}
&|\operatorname{Im} l(\xi')| = |b_1\xi_1 + \cdots + b_{n-1}\xi_{n-1}| \leq C|\xi'|, \text{ so} \\
&|\operatorname{Re} l(\xi')| = \langle \xi' \rangle \geq C^{-1}|\operatorname{Im} l(\xi')|,
\end{aligned} \tag{9.95}$$

which imply similar inequalities for L and L^* (as in Theorems 12.13, 6.4 and Exercises 12.35 and 12.36). It follows in view of (9.73) that T and T^* have their numerical ranges in an angular set

$$\mu = \{\, \lambda \in \mathbb{C} \mid |\operatorname{Im} \lambda| \leq C' \operatorname{Re} \lambda, \; \operatorname{Re} \lambda \geq c' \,\}$$

with C' and $c' > 0$. Now Theorem 13.17 applies to show that $\widetilde{A}$ and $\widetilde{A}^*$ have their numerical ranges in a similar set, and the variational property follows from Theorem 12.26. □

We observe in general that if L has lower bound > 0, then it is elliptic, and both $\widetilde{A}$ and $\widetilde{A}^*$ are lower bounded. A closer look at $l_1(\xi')$ will show that then necessarily $c < 1$; moreover, $\widetilde{A}$ and $\widetilde{A}^*$ are variational.

Conditions for the coerciveness inequality

$$\mathrm{Re}(Au, u) \geq c_1 \|u\|_1^2 - c_0 \|u\|_0^2 \text{ for } u \in D(\widetilde{A}), \tag{9.96}$$

can be fully analyzed, and have been done so in the litterature (more on this in [G71] and subsequent works).

It is also possible to analyze the resolvent of $\widetilde{A}$ and its relation to λ-dependent operators over $\mathbb{R}^{n-1}$ by use of the results of Sections 13.3 and 13.4, for this we refer to [BGW08].

Example 9.37. Assume that $n \geq 3$. Consider $\widetilde{A}$ defined by the boundary condition (9.84) with

$$B = \chi(x_1)\partial_{x_1} - P, \tag{9.97}$$

where χ is as defined in Section 2.1 (here we allow a variable coefficient in the differential operator). By (9.90)ff.,

$$L = -\chi(x_1)\partial_{x_1},$$

in this case. It is not elliptic on $\mathbb{R}^{n-1}$. Moreover, L has a large nullspace, containing the functions that are constant in x_1 for $|x_1| \leq 2$. The nullspace is clearly infinite dimensional, and it is not contained in $H^1(\mathbb{R}^{n-1})$ (let alone $H^{\frac{3}{2}}$), since it contains for example all products of L_2-functions of $(x_2, \ldots, x_{n-1})$ with L_2-functions of x_1 that are constant on $[-2, 2]$. We have from Theorem 13.8 that the nullspace of $\widetilde{A}$ equals that of T, and it equals $K_\gamma Z(L)$, cf. (9.62)ff.

So this is an example of a realization with low regularity and a large infinite dimensional nullspace.

9.5 Variable-coefficient cases, higher orders, systems

The case of $I - \Delta$ on the half-space is just a very simple example, having the advantage that precise results can be easily obtained. Let us give some remarks on more general boundary value problems.

Consider a differential operator of a general order $d > 0$ with C^∞ coefficients on an open set $\Omega \subset \mathbb{R}^n$ with smooth boundary:

$$A = \sum_{|\alpha| \leq d} a_\alpha(x) D^\alpha; \tag{9.98}$$

its principal symbol is $a_d(x, \xi) = \sum_{|\alpha|=d} a_\alpha(x)\xi^\alpha$ (cf. (6.5), (6.42)), also called $a^0(x, \xi)$. A is elliptic when $a^0(x, \xi) \neq 0$ for $\xi \neq 0$ (and all x); and A is said to be *strongly elliptic* when

$$\operatorname{Re} a^0(x, \xi) > 0 \text{ for } \xi \neq 0, \text{ all } x. \tag{9.99}$$

It is not hard to see that strongly elliptic operators are necessarily of *even order*, $d = 2m$.

There are a few scalar odd-order elliptic operators, for example the Cauchy-Riemann operator on $\mathbb{R}^2$, of order 1 (cf. Exercise 5.8). But otherwise, odd-order elliptic operators occur most naturally when we consider *systems* — matrix-formed operators — where the $a_\alpha(x)$ are matrix functions (this is considered for pseudodifferential operators in Section 7.4). Ellipticity here means that the principal symbol $a^0(x, \xi)$ is an invertible matrix for all $\xi \neq 0$, all x. Then it must be a square matrix, and the ellipticity means that the determinant of $a^0(x, \xi)$ is nonzero for $\xi \neq 0$.

Examples of first-order systems that are of current interest are Dirac operators, used in Physics and Geometry.

The strongly elliptic *systems* are those for which

$$a^0(x, \xi) + a^0(x, \xi)^* \text{ is positive definite when } \xi \neq 0. \tag{9.100}$$

Here the order must be even, $d = 2m$.

In the strongly elliptic case where Ω is bounded and (9.100) holds for all $x \in \overline{\Omega}$, there is (as shown at the end of Chapter 7) a variational realization A_γ of the Dirichlet problem, representing the boundary condition $\gamma u = 0$, where

$$\gamma u = \{\gamma_0 u, \gamma_1 u, \ldots, \gamma_{m-1} u\}. \tag{9.101}$$

The domain is $D(A_\gamma) = D(A_{\max}) \cap H_0^m(\Omega)$. It is a deeper result to show that in fact

$$D(A_\gamma) = H^{2m}(\Omega) \cap H_0^m(\Omega), \tag{9.102}$$

and that A_γ is a Fredholm operator (cf. Section 8.3) from its domain to $L_2(\Omega)$. Moreover, one can consider the fully nonhomogeneous problem

$$Au = f \text{ in } \Omega, \quad \gamma u = \varphi \text{ on } \partial\Omega, \tag{9.103}$$

showing that it defines a Fredholm operator

$$\begin{pmatrix} A \\ \gamma \end{pmatrix} : H^{2m}(\Omega) \to \begin{matrix} L_2(\Omega) \\ \times \\ \Pi_{j=0}^{m-1} H^{2m-j-\frac{1}{2}}(\partial\Omega) \end{matrix}. \tag{9.104}$$

In the case of bijectiveness, the semihomogeneous problem

$$Au = 0 \text{ in } \Omega, \quad \gamma u = \varphi \text{ on } \partial\Omega, \tag{9.105}$$

has a solution operator K_γ (a so-called Poisson operator) which is bijective:

$$K_\gamma : \Pi_{j=0}^{m-1} H^{s-j-\frac{1}{2}}(\partial\Omega) \xrightarrow{\sim} \{u \in H^s(\Omega) \mid Au = 0\}, \qquad (9.106)$$

for all $s \in \mathbb{R}$, with inverse γ (defined in a generalized sense for low s). When the system in (9.104) is merely Fredholm, (9.106) holds modulo finite dimensional subspaces.

A classical method to show the regularity (9.102) of the solutions to (9.103) with $\varphi = 0$ relies on approximating the derivatives with difference quotients, where the inequalities resulting from the ellipticity hypothesis can be used. The method is due to Nirenberg [N55], and is also described e.g. in Agmon [A65] (including some other variational cases), and in Lions and Magenes [LM68] for general normal boundary problems with nonhomogeneous boundary condition. The difference quotient method allows keeping track of how much smoothness of the coefficients $a_\alpha(x)$ is needed for a specific result. It enters also in modern textbooks such as e.g. Evans [E98].

Another important reference in this connection is the paper of Agmon, Douglis and Nirenberg [ADN64], treating general systems by reduction to local coordinates and complex analysis.

In the case of C^∞ coefficients and domain, there is a modern strategy where the problems are solved in a pseudodifferential framework. Here we need other operators than just ψdo's on open sets of $\mathbb{R}^n$, namely, trace operators T going from functions on Ω to functions on $\partial\Omega$, Poisson operators K going from functions on $\partial\Omega$ to functions on Ω, and composed operators KT, called singular Green operators. Such a theory has been introduced by Boutet de Monvel [B71] (and further developed e.g. in [G84], [G90], [G96]). The use of this framework makes the solvability discussion more operational; one does not just obtain regularity of solutions, but one constructs in a concrete way the operator that maps the data into the solution (obtaining general versions of (9.38), (9.51)).

We give an introduction to the theory of psedodifferential boundary operators (ψdbo's) in Chapter 10, building on the theory of ψdo's on open sets explained in Chapters 7 and 8. In Chapter 11 we introduce the Calderón projector ([C63], [S66], [H66]), which is an efficient tool for the discussion of the most general boundary value problems for elliptic differential operators.

We shall not here go into the various theories for nonsmooth boundary value problems that have been developed through the times.

Exercises for Chapter 9

9.1. An analysis similar to that in Section 9.2 can be set up for the Dirichlet problem for $-\Delta$ on the circle, continuing the notation of Exercise 4.5; you are asked to investigate this. The Sobolev spaces over the boundary can here be defined in terms of the coefficient series (c_k) of Fourier expansions as in Section 8.2, H^s being provided with the norm $(\sum_{k \in \mathbb{Z}} \langle k \rangle^{2s} |c_k|^2)^{\frac{1}{2}}$.

9.2. Consider $l(\xi')$ defined in (9.92). Show that if $\operatorname{Re} l(\xi') > 0$ for all ξ', then $c < 1$. Show that L is then variational.

9.3. Let $B = c_0 + ic_1 \partial_{x_1} + \cdots + ic_{n-1} \partial_{x_{n-1}}$, with real numbers $c_0, \ldots, c_{n-1}$. Show that if $c_1^2 + \cdots + c_{n-1}^2 < 1$, then the realization $\widetilde{A}$ defined by (9.84) with this B is selfadjoint and has $D(\widetilde{A}) \subset H^2(\mathbb{R}_+^n)$.

9.4. Let $A = (1 - \Delta)^2$ on $\mathbb{R}_+^n$, and consider the Dirichlet problem

$$Au = f \text{ in } \mathbb{R}_+^n, \quad \gamma_0 u = \varphi_0 \text{ on } \mathbb{R}^{n-1}, \quad \gamma_1 u = \varphi_1 \text{ on } \mathbb{R}^{n-1}. \tag{9.107}$$

(a) Show that for any $\xi' \in \mathbb{R}^{n-1}$, the equation

$$(\langle \xi' \rangle^2 - \partial_{x_n}^2)^2 u(x_n) = 0 \text{ on } \mathbb{R}_+$$

has the following bounded solutions on $\mathbb{R}_+$:

$$c_0 e^{-\langle \xi' \rangle x_n} + c_1 x_n e^{-\langle \xi' \rangle x_n},$$

where $c_0, c_1 \in \mathbb{C}$.

(b) Show that one can find a linear transformation

$$C(\xi') : (\hat{\varphi}_0(\xi'), \hat{\varphi}_1(\xi')) \mapsto (c_0(\xi'), c_1(\xi'))$$

for each ξ', such that the operator K defined by

$$K : (\varphi_0(x'), \varphi_1(x')) \mapsto \mathscr{F}_{\xi' \to x'}^{-1}(c_0(\xi') e^{-\langle \xi' \rangle x_n} + c_1(\xi') x_n e^{-\langle \xi' \rangle x_n}),$$

goes from $\mathscr{S}(\mathbb{R}^{n-1}) \times \mathscr{S}(\mathbb{R}^{n-1})$ to $\mathscr{S}(\overline{\mathbb{R}}_+^n)$ and solves (9.107) with $f = 0$, φ_0, φ_1 given in $\mathscr{S}(\mathbb{R}^{n-1})$. Determine $C(\xi')$.

(c) Show that K is continuous from $H^{m-\frac{1}{2}}(\mathbb{R}^{n-1}) \times H^{m-\frac{3}{2}}(\mathbb{R}^{n-1})$ to $H^m(\mathbb{R}_+^n)$ for all $m \in \mathbb{N}_0$, and deduce from this a more general theorem on the solution of the problem (9.107) with $f = 0$.

9.5. For the boundary value problem considered in Exercise 9.4, set up an analysis (possibly with some parts only sketched) that leads to a theorem analogous to Theorem 9.18 in the present situation. One can use here that a variational construction with

$$H = L_2(\mathbb{R}_+^n), \quad V = H_0^2(\mathbb{R}_+^n), \quad a(u, v) = ((1 - \Delta)u, (1 - \Delta)v)_{L_2(\mathbb{R}_+^n)}$$

gives an invertible realization A_γ of $A = (I - \Delta)^2$ with $D(A_\gamma) = D(A_{\max}) \cap H_0^2(\mathbb{R}_+^n)$.

Chapter 10
Pseudodifferential boundary operators

10.1 The real formulation

In this chapter we present some essential ingredients from the calculus of pseudodifferential boundary operators (ψdbo's, Green operators) introduced by Boutet de Monvel [B66], [B71], with further developments e.g. from [G84], [G90] and [G96]; see also Rempel and Schulze [RS82]. The basic notions are defined relative to $\mathbb{R}^n$ and the subset $\overline{\mathbb{R}}^n_+ \subset \mathbb{R}^n$; then they are carried over to manifold situations by use of local coordinate systems.

In the case of a differential operator A, the analysis of the relevant boundary conditions is usually based on the polynomial structure of the symbol of A; in particular the roots in ξ_n of the principal symbol polynomial $a^0(x', 0, \xi', \xi_n)$ (in the situation where the domain is $\mathbb{R}^n_+$) play a role. When pseudodifferential operators P are considered, the principal symbol p^0 is generally not a polynomial. It may be a rational function (this happens naturally when one makes reductions in a system of differential operators), in which case one can consider the roots and poles with respect to ξ_n. But then, even when P is elliptic, there is much less control over how these behave than when a^0 is a polynomial; roots and poles may cancel each other or reappear, as the coordinate ξ' varies. For a workable theory, a more universal point of view is needed.

Vishik and Eskin (see [VE67] and [E81]) based a theory on a factorization of a symbol in two factors with different domains of holomorphy in ξ_n. This works well in the scalar case but can be problematic in the case of matrix-formed operators (since the factorization here is generally only piecewise continuous in ξ'). They mainly consider ψdo's of a general kind, with less restrictions on the behavior in ξ than our standard symbol spaces require.

The calculus introduced by Boutet de Monvel takes a special class of ψdo's; one of the advantages of that theory is that it replaces the factorization by a *projection* procedure that works equally well for scalar and matrix-formed operators (depends smoothly on ξ'). This is linked in a natural way with the projections $e^{\pm}r^{\pm}$ of $L_2(\mathbb{R})$ onto $e^{\pm}L_2(\mathbb{R}_{\pm})$ (cf. (9.3)). The description that now follows is given in relation to the latter projections, and the Fourier-transformed version (used in symbol calculations) will be taken up in Section 10.3.

Pseudodifferential operators satisfying the transmission condition

When P is a ψdo on $\mathbb{R}^n$, its truncation (or "restriction") to the subset $\Omega = \mathbb{R}^n_+$ is defined by

$$P_+ u = r^+ P e^+ u, \text{ also denoted } P_{\mathbb{R}^n_+} u \text{ or } P_\Omega u, \tag{10.1}$$

where, as in (9.2) and (9.3), r^+ restricts $\mathscr{D}'(\mathbb{R}^n)$ to $\mathscr{D}'(\mathbb{R}^n_+)$, and e^+ extends locally integrable functions on $\mathbb{R}^n_+$ by zero on $\mathbb{R}^n_-$. We underline that r^+ restricts to the interior of $\overline{\mathbb{R}}^n_+$ so that singularities supported at $x_n = 0$ disappear. As usual, $C^0(\overline{\mathbb{R}}^n_+)$ identifies with a subspace of $\mathscr{D}'(\mathbb{R}^n_+)$.

When P is properly supported and of order d, the operator P_+ is continuous from $L_{2,\mathrm{comp}}(\mathbb{R}^n_+)$ to $H^{-d}_{\mathrm{comp}}(\overline{\mathbb{R}}^n_+)$ (defined in (10.148) below), but in general does not map $H^m_{\mathrm{comp}}(\overline{\mathbb{R}}^n_+)$ into $H^{m-d}_{\mathrm{comp}}(\overline{\mathbb{R}}^n_+)$ for $m > 0$; the discontinuity of e^+u at $x_n = 0$ causes a singularity. Boutet de Monvel singled out a class of ψdo's where one does get these mapping properties for P_+, namely, the ψdo's having the transmission property.

P is said to have *the transmission property* with respect to $\mathbb{R}^n_+$ when $P_{\mathbb{R}^n_+}$ "preserves C^∞ up to the boundary", i.e., $P_{\mathbb{R}^n_+}$ maps $C^\infty_{(0)}(\overline{\mathbb{R}}^n_+)$ into $C^\infty(\overline{\mathbb{R}}^n_+)$. Boutet de Monvel showed in [B66] a necessary and sufficient condition for a polyhomogeneous symbol $p(x,\xi) \sim \sum_{l\in\mathbb{N}_0} p_{d-l}(x,\xi)$ to define an operator $\mathrm{OP}(p)$ with the transmission property w.r.t. $\mathbb{R}^n_+$. It states that the homogeneous terms p_{d-l} have the symmetry property

$$D_x^\beta D_\xi^\alpha p_{d-l}(x',0,0,-\xi_n) = e^{i\pi(d-l-|\alpha|)} D_x^\beta D_\xi^\alpha p_{d-l}(x',0,0,\xi_n), \tag{10.2}$$

for $|\xi_n| \geq 1$ and all indices α, β, l.

We write "OP" instead of "Op" from now on, to allow generalizations to OPT, OPK and OPG definitions below. (The minus on ξ_n should be placed in the left-hand side as in (10.2). Only when d is integer can it equally well be placed in the right-hand side, as done in many texts, e.g. in [G96, (1.2.7)].)

Example 10.1. A parametrix symbol $q(x,\xi)$ for an elliptic *differential operator* P of order d certainly has the transmission property, since its symbol terms are rational functions of ξ. In fact, it has the stronger property

$$q_{-d-l}(x,-\xi) = (-1)^{d-l} q_{-d-l}(x,\xi) \text{ for } |\xi| \geq 1, \text{ all } l, \tag{10.3}$$

guaranteeing the transmission property for whichever direction taken as x_n. Polyhomogeneous symbols having the property (10.3) are in some texts said to have *even-even* alternating parity (the even-order symbols are even), or just to be even-even, for short. The opposite parity

$$q_{-d-l}(x, -\xi) = (-1)^{d-l+1} q_{-d-l}(x, \xi) \text{ for } |\xi| \geq 1, \text{ all } l, \tag{10.4}$$

is then called *even-odd* alternating parity. An example with the latter property is $q(\xi) = \langle \xi \rangle$, the symbol of the square root of $1 - \Delta$. Note that the symbol $\langle \xi \rangle^s$ has the transmission property (and is even-even) if and only if s is an even integer.

When d is *integer* and the equations (10.2) hold, then they also hold for the symbol $p(x, -\xi)$; this implies that also $P_{\mathbb{R}^n}$ preserves C^∞ up to the boundary in $\overline{\mathbb{R}}^n_-$.

When d is not integer, (10.2) implies a "wrong" kind of symmetry for $p(x, -\xi)$, and P will not in general preserve C^∞ in $\overline{\mathbb{R}}^n_-$, but maps the functions in $C^\infty_{(0)}(\overline{\mathbb{R}}^n_-)$ into functions with a specific singular behavior at $x_n = 0$.

There is a complete discussion of necessary and sufficient conditions for the transmission property, extending the analysis to $S^d_{\varrho, \delta}$ symbols with $\varrho > \delta$, in Grubb and Hörmander [GH90]. Besides in [B66], [B71], studies related to the transmission property are found in [VE67], [E81] and in [H85, Sect. 18.2]. There is also an introductory explanation in [G91].

It is shown in [B71] that the properties (10.2) in the case $d \in \mathbb{Z}$ may be rewritten as an expansion property of $p(x, \xi)$ and its derivatives at $x_n = 0$:

$$D^\beta_x D^\alpha_\xi p(x', 0, \xi) \sim \sum_{-\infty < l \leq d - |\alpha|} s_{l,\alpha,\beta}(x', \xi') \xi^l_n, \tag{10.5}$$

where the $s_{l,\alpha,\beta}(x', \xi')$ are *polynomials* in ξ' of degree $d - |\alpha| - l$, for all α and $\beta \in \mathbb{N}^n_0$. We denote $s_{l,0,0} = s_l$. More precisely, let us introduce (with X representing x', y' or (x', y')):

Definition 10.2. A symbol $p(X, x_n, y_n, \xi)$ of order $d \in \mathbb{Z}$ satisfies the transmission condition (at $x_n = y_n = 0$), when there exist symbols $s_l(X, \xi')$, polynomial in ξ' of order $d - l$, such that for all indices,

$$|D^\beta_X D^\alpha_\xi [\xi^m_n p(X, 0, 0, \xi) - \sum_{-m \leq l \leq d - |\alpha|} s_l(X, \xi') \xi^{l+m}_n]|$$
$$\leq c(X) \langle \xi' \rangle^{d+1+m-|\alpha|} \langle \xi \rangle^{-1}, \tag{10.6}$$

with continuous functions $c(X)$, and there are similar expansions of $\partial^j_{x_n} \partial^{j'}_{y_n} p$.

For polyhomogeneous symbols of integer order, (10.2) holds if and only if Definition 10.2 is satisfied by p (and by each term p_{d-j} in its symbol). But (10.5) and (10.6) have the advantage that they make sense for $S^d_{1,0}$ symbols also and guarantee their C^∞ preserving properties, as noted in [B71]. For such symbols it is a sufficient condition, which is why we call it the transmission *condition*.

In the following, we formulate the principles for symbols in *x*-form (cf. Remark 7.4). Formula (10.5) means that $p(x', 0, \xi)$ (and in a similar way the derivatives $D_x^\beta D_\xi^\alpha p(x',0,\xi))$ has an expansion in integer powers of ξ_n such that at each (x',ξ') one has for any $m \in \mathbb{N}_0$ that

$$\xi_n^m p(\xi_n) - \sum_{-m < l \le d} s_l \xi_n^{l+m} \tag{10.7}$$

has the same limit (namely, s_{-m}) for $\xi_n \to +\infty$ as for $\xi_n \to -\infty$. The C^∞-functions of ξ_n with this property form the important space $\mathcal{H}$ that we study in Section 10.2 below. In the present section we do not want to go into details with the complex analysis involved in studying $\mathcal{H}$; instead we shall explain the point of view one gets by studying the inverse Fourier transforms in ξ_n (the "real" point of view).

Note in particular that the definition implies

$$p(x',0,\xi) = \sum_{0 \le l \le d} s_l(x',\xi')\xi_n^l + p'(x',\xi), \tag{10.8}$$

where p' is $O(\langle \xi \rangle^{-1} \langle \xi' \rangle^{d+1})$, the sum over l is *polynomial* in ξ of order d and the top coefficient s_d is a *function of* x',

$$s_d(x',\xi') = s_d(x'). \tag{10.9}$$

Thus, *if p is constant in x_n*, P is the sum of a differential operator and an operator that preserves L_2 with respect to x_n. More generally, p could be much less well-behaved for $x_n \ne 0$, but here a Taylor expansion in x_n gives a number of good terms $\frac{1}{j!}x_n^j \partial_{x_n}^j p(x',0,\xi)$ where $\partial_{x_n}^j p$ behaves similarly to (10.8), plus a term with a factor x_n^M that can make it harmless when M is large.

The partial inverse Fourier transform $\tilde{p}(x',0,\xi',z_n) = \mathscr{F}^{-1}_{\xi_n \to z_n} p(x',0,\xi)$ is always rapidly decreasing for $z_n \to \pm\infty$, since $D_{\xi_n}^k(\xi_n^m p)$ is integrable in ξ_n when $k \ge m+d+2$. But in case of general symbols, it has a singularity at $z_n = 0$. However, as shown in [B66] and [GH90], symbols of integer order d satisfy the transmission condition *if and only if*, for any $|\xi'| \ge 1$, any α, β, the functions

$$\tilde{p}_{\alpha,\beta}(x',0,\xi',z_n) = \mathscr{F}^{-1}_{\xi_n \to z_n} D_x^\beta D_\xi^\alpha p(x',0,\xi), \text{ considered for } (x',z_n) \in \mathbb{R}^n_\pm,$$

$$\text{extend to } C^\infty\text{-functions of } (x',z_n) \in \overline{\mathbb{R}}^n_\pm. \tag{10.10}$$

The limits for $z_n \to 0+$ and $z_n \to 0-$ are in general different, and the condition does not exclude singularities supported in $\{z_n = 0\}$. (See also Exercises 10.2 and 10.3.)

Let us show that the transmission condition implies (10.10). In view of (10.8), $\tilde{p}$ is the sum of a distribution $\sum_{0 \le l \le d} s_l(x',\xi') D_{z_n}^l \delta(z_n)$ supported in $\{z_n = 0\}$ and a *function* $\tilde{p}'(x',\xi',z_n)$ in $\overline{L}_2$ with respect to $z_n \in \mathbb{R}$. Here

$$\|\tilde{p}'(x',\xi',z_n)\|_{L_{2,z_n}(\mathbb{R})} = (2\pi)^{-\frac{1}{2}}\|p'(x',\xi)\|_{L_{2,\xi_n}(\mathbb{R})}$$
$$\leq c(x')\langle\xi'\rangle^{d+1}\|\langle\xi\rangle^{-1}\| \leq c'(x')\langle\xi'\rangle^{d+\frac{1}{2}}$$

(recall (9.19)); so in particular, the functions $r^{\pm}_{z_n}\tilde{p}'$ ($= r^{\pm}_{z_n}\tilde{p}$) satisfy such estimates in $L_2(\mathbb{R}_{\pm})$-norm, respectively:

$$\|r^+\tilde{p}(x',\xi',z_n)\|_{L_{2,z_n}(\mathbb{R}_+)} = \|r^+\tilde{p}'(x',\xi',z_n)\|_{L_{2,z_n}(\mathbb{R}_+)} \leq c(x')\langle\xi'\rangle^{d+\frac{1}{2}},$$
$$\|r^-\tilde{p}(x',\xi',z_n)\|_{L_{2,z_n}(\mathbb{R}_-)} = \|r^-\tilde{p}'(x',\xi',z_n)\|_{L_{2,z_n}(\mathbb{R}_-)} \leq c(x')\langle\xi'\rangle^{d+\frac{1}{2}},$$

with $c(x')$ continuous. For any k, k', α', β', the derived distribution

$$z_n^k D_{z_n}^{k'} D_{x'}^{\beta'} D_{\xi'}^{\alpha'} \tilde{p}(x',0,\xi',z_n) = \mathscr{F}^{-1}_{\xi_n \to z_n} D_{\xi_n}^k \xi_n^{k'} D_{\xi'}^{\alpha'} D_{x'}^{\beta'} p(x',0,\xi)$$

comes from a symbol satisfying Definition 10.2 with d replaced by $d - k + k' - |\alpha'|$, hence we likewise find

$$\|z_n^k D_{z_n}^{k'} D_{x'}^{\beta'} D_{\xi'}^{\alpha'} r^{\pm}\tilde{p}'\|_{L_{2,z_n}(\mathbb{R}_{\pm})} = \|z_n^k D_{z_n}^{k'} D_{x'}^{\beta'} D_{\xi'}^{\alpha'} r^{\pm}\tilde{p}\|_{L_{2,z_n}(\mathbb{R}_{\pm})}$$
$$\leq c(x')\langle\xi'\rangle^{d+\frac{1}{2}-k+k'-|\alpha'|}, \qquad (10.11)$$

using again that $\tilde{p} - \tilde{p}'$ is supported in $\{z_n = 0\}$. Similar arguments apply to $\partial_{x_n}^j p(x',0,\xi)$. Since all derivatives have bounded L_2-norms, they have bounded sup-norms (as in Theorem 4.18), locally uniformly in (x',ξ'), so it follows that the functions extend to C^∞-functions of $(x',z_n,\xi') \in \overline{\mathbb{R}}^n_+ \times \mathbb{R}^{n-1}$ resp. $\overline{\mathbb{R}}^n_- \times \mathbb{R}^{n-1}$.

Function spaces defined by estimates as in (10.11) will now be introduced systematically. For this we first introduce some abbreviations: We use the notation $\mathbb{R}^2_{++} = \mathbb{R}_+ \times \mathbb{R}_+$, $\overline{\mathbb{R}}^2_{++} = \overline{\mathbb{R}}_+ \times \overline{\mathbb{R}}_+$, and $\mathscr{S}(\overline{\mathbb{R}}^2_{++}) = r^+_{x_n} r^+_{y_n} \mathscr{S}(\mathbb{R}^2)$ (where r^+_z indicates restriction to $\{z > 0\}$); recall also (9.4). Here we write for short:

$$\mathscr{S}(\overline{\mathbb{R}}_+) = \mathscr{S}_+, \quad \mathscr{S}(\overline{\mathbb{R}}^2_{++}) = \mathscr{S}_{++}. \qquad (10.12)$$

Definition 10.3. Let $d \in \mathbb{R}$ and let Ξ be open $\subset \mathbb{R}^{n'}$.

$1°$ The space $S_{1,0}^d(\Xi, \mathbb{R}^{n-1}, \mathscr{S}_+)$ consists of the functions $\tilde{f}(X,x_n,\xi') \in C^\infty(\Xi \times \overline{\mathbb{R}}_+ \times \mathbb{R}^{n-1})$, lying in $\mathscr{S}_+$ with respect to x_n, such that for all α, β, k, k',

$$\|x_n^k D_{x_n}^{k'} D_X^\beta D_{\xi'}^\alpha \tilde{f}(X,x_n,\xi')\|_{L_2(\mathbb{R}_+)} \leq c(X)\langle\xi'\rangle^{d+\frac{1}{2}-k+k'-|\alpha|}, \qquad (10.13)$$

with continuous functions $c(X)$. Moreover, $\tilde{f}$ is said to have the asymptotic expansion $\tilde{f} \sim \sum_{l \in \mathbb{N}_0} \tilde{f}_{d-l}$ in $S_{1,0}^d(\Xi, \mathbb{R}^{n-1}, \mathscr{S}_+)$, when there is a sequence of functions $\tilde{f}_{d-l}$ lying in $S_{1,0}^{d-l}(\Xi, \mathbb{R}^{n-1}, \mathscr{S}_+)$ such that $\tilde{f} - \sum_{l < M} \tilde{f}_{d-l} \in S_{1,0}^{d-M}(\Xi, \mathbb{R}^{n-1}, \mathscr{S}_+)$ for any $M \in \mathbb{N}_0$.

$2°$ The subspace $S^d(\Xi, \mathbb{R}^{n-1}, \mathscr{S}_+)$ of polyhomogeneous elements consists of the functions $\tilde{f} \in S_{1,0}^d(\Xi, \mathbb{R}^{n-1}, \mathscr{S}_+)$ that have asymptotic expansions $\tilde{f} \sim \sum_{l \in \mathbb{N}_0} \tilde{f}_{d-l}$ where the functions $\tilde{f}_{d-l}$ have the quasi-homogeneity property

$$\tilde{f}_{d-l}(X, \tfrac{1}{\lambda}x_n, \lambda\xi') = \lambda^{d+1-l}\,\tilde{f}_{d-l}(X, x_n, \xi') \text{ for } \lambda \geq 1 \text{ and } |\xi'| \geq 1. \quad (10.14)$$

$3°$ The space $S^d_{1,0}(\Xi, \mathbb{R}^{n-1}, \mathscr{S}_{++})$ consists of the functions $\tilde{g}(X, x_n, y_n, \xi') \in C^\infty(\Xi \times \overline{\mathbb{R}}^2_{++} \times \mathbb{R}^{n-1})$, lying in $\mathscr{S}_{++}$ with respect to (x_n, y_n), such that for all $\alpha, \beta, k, k', m, m'$,

$$\|x_n^k D_{x_n}^{k'} y_n^m D_{y_n}^{m'} D_X^\beta D_{\xi'}^\alpha \tilde{g}(x', x_n, y_n, \xi')\|_{L_2(\mathbb{R}^2_{++})}$$
$$\leq c(X)\langle\xi'\rangle^{d+1-k+k'-m+m'-|\alpha|}. \quad (10.15)$$

Here $\tilde{g}$ is said to have the asymptotic expansion $\tilde{g} \sim \sum_{l\in\mathbb{N}_0} \tilde{g}_{d-l}$ in $S^d_{1,0}(\Xi, \mathbb{R}^{n-1}, \mathscr{S}_{++})$, when there are functions $\tilde{g}_{d-l} \in S^{d-l}_{1,0}(\Xi, \mathbb{R}^{n-1}, \mathscr{S}_{++})$ such that $\tilde{g} - \sum_{l<M} \tilde{g}_{d-l} \in S^{d-M}_{1,0}(\Xi, \mathbb{R}^{n-1}, \mathscr{S}_{++})$ for any $M \in \mathbb{N}_0$.

$4°$ The subspace $S^d(\Xi, \mathbb{R}^{n-1}, \mathscr{S}_{++})$ of polyhomogeneous elements consists of the functions $\tilde{g} \in S^d_{1,0}(\Xi, \mathbb{R}^{n-1}, \mathscr{S}_{++})$ that have asymptotic expansions $\tilde{g} \sim \sum_{l\in\mathbb{N}_0} \tilde{g}_{d-l}$ where the functions $\tilde{g}_{d-l}$ have the quasi-homogeneity property

$$\tilde{g}_{d-l}(X, \tfrac{1}{\lambda}x_n, \tfrac{1}{\lambda}y_n, \lambda\xi') = \lambda^{d+2-l}\,\tilde{g}_{d-l}(X, x_n, y_n, \xi') \text{ for } \lambda \geq 1 \text{ and } |\xi'| \geq 1. \quad (10.16)$$

What we showed in (10.11) is that when p satisfies Definition 10.2, then $r^+\tilde{p}$ belongs to $S^d_{1,0}(\mathbb{R}^{n-1}, \mathbb{R}^{n-1}, \mathscr{S}_+)$. This is taken up again in Theorem 10.21 below. When p is polyhomogeneous, $r^+\tilde{p}$ is in $S^d(\mathbb{R}^{n-1}, \mathbb{R}^{n-1}, \mathscr{S}_+)$. Similar statements hold for $(r^-\tilde{p})|_{x_n=-z_n}$.

Functions with the properties in Definition 10.3 are called *symbol-kernels*. Just like for pseudodifferential symbols, one can turn a series of symbol-kernels of decreasing orders into an asymptotic series for a suitable symbol-kernel of the highest order (cf. Lemma 7.3). $3°$ and $4°$ will be used in the description of singular Green operators below.

The properties of the functions could equivalently be formulated in terms of sup-norms: $\tilde{f}$ is in $S^d_{1,0}(\Xi, \mathbb{R}^{n-1}, \mathscr{S}_+)$ if and only if

$$\sup_{x_n>0} |x_n^k D_{x_n}^{k'} D_X^\beta D_{\xi'}^\alpha \tilde{f}(x', x_n, \xi')| \leq c(X)\langle\xi'\rangle^{d+1-k+k'-|\alpha|} \quad (10.17)$$

(one can use estimates as in the proof of Theorem 4.18 to go from L_2-norms to sup-norms, and insert factors like $(1+ix_n\langle\xi'\rangle)(1+ix_n\langle\xi'\rangle)^{-1}$ for the other direction). We use the L_2-norms for convenience in Fourier transformation.

The quasi-homogeneity properties correspond to homogeneity of the Fourier-transformed functions

$$f(X, \xi', \xi_n) = \mathscr{F}_{x_n\to\xi_n}e^+\tilde{f}(X, x_n, \xi'),$$
$$g(X, \xi', \xi_n, \eta_n) = \mathscr{F}_{x_n\to\xi_n}\overline{\mathscr{F}}_{y_n\to\eta_n}e^+_{x_n}e^+_{y_n}\tilde{g}(X, x_n, y_n, \xi');$$

namely,

$$f_{d-l}(X, \lambda\xi) = \lambda^{d-l}f_{d-l}(X, \xi) \quad \text{for } \lambda \geq 1, |\xi'| \geq 1,$$

$$g_{d-l}(X, \lambda\xi', \lambda\xi_n, \lambda\eta_n) = \lambda^{d-l}g_{d-l}(X, \xi', \xi_n, \eta_n), \ \lambda \geq 1, |\xi'| \geq 1;$$

as is easily checked from the definition of Fourier transformation.

In the following, all ψdo symbols on $\mathbb{R}^n$ will tacitly be assumed to be of integer order satisfying the transmission condition. We denote by $p(x', \xi', D_n)$ or $\mathrm{OP}_n(p(x, \xi))$ the operator on $\mathbb{R}$ where the ψdo definition (7.2) is applied with respect to (x_n, ξ_n) only, and by $\mathrm{OP}'(p(x, \xi))$ the operator on $\mathbb{R}^{n-1}$ where (7.2) is applied with respect to (x', ξ') only.

Systems (Green operators)

We shall now introduce the other ingredients in the Boutet de Monvel calculus. Let P be $N' \times N$-matrix formed. Along with P_+, which operates on $\mathbb{R}^n_+$, we shall consider operators going to and from the boundary $\mathbb{R}^{n-1}$, forming together with P a system

$$\mathcal{A} = \begin{pmatrix} P_+ + G & K \\ T & S \end{pmatrix} : \begin{matrix} C^\infty_{(0)}(\overline{\mathbb{R}}^n_+)^N \\ \times \\ C^\infty_0(\mathbb{R}^{n-1})^M \end{matrix} \rightarrow \begin{matrix} C^\infty(\overline{\mathbb{R}}^n_+)^{N'} \\ \times \\ C^\infty(\mathbb{R}^{n-1})^{M'} \end{matrix}. \tag{10.18}$$

Here T is a so-called *trace operator*, going from $\mathbb{R}^n_+$ to $\mathbb{R}^{n-1}$; K is a so-called *Poisson operator* (called a potential operator or coboundary operator in some other texts), going from $\mathbb{R}^{n-1}$ to $\mathbb{R}^n_+$; S is a *pseudodifferential operator on $\mathbb{R}^{n-1}$*; and G is an operator on $\mathbb{R}^n_+$ called a *singular Green operator*, a non-pseudodifferential term that has to be included in order to have adequate composition rules. The full system $\mathcal{A}$ was called a *Green operator* by Boutet de Monvel, or a *pseudodifferential boundary operator* (that we can write ψdbo). Since some authors later used the name "Green operator" for a generalization of singular Green operators, we shall mainly write ψdbo's. We shall usually take $N = N'$, whereas the dimensions M and M' can have all values, including zero.

When P is a differential operator, it is classical to study systems of the form

$$\mathcal{A} = \begin{pmatrix} P_+ \\ T \end{pmatrix}; \tag{10.19}$$

here $M = 0$ and $M' > 0$. When this $\mathcal{A}$ has an inverse in the calculus, the inverse will be of the form

$$\mathcal{A}^{-1} = \begin{pmatrix} Q_+ + G & K \end{pmatrix}; \tag{10.20}$$

where $M > 0$ and $M' = 0$. Simple examples are found in Chapter 9, with $P = 1 - \Delta$, $T = \gamma_0$ or γ_1, the inverses described in Theorems 9.18 and 9.23.

The entries in (10.18) will now be explained.

Trace operators

The trace operators include the usual differential trace operators $\gamma_j : u \mapsto (D^j_{x_n} u)|_{x_n=0}$ composed with ψdo's on $\mathbb{R}^{n-1}$, and moreover some integral operator types, governed by the fact that

$$T = \gamma_0 P_+ \tag{10.21}$$

should be a trace operator whenever P is a ψdo satisfying the transmission condition. Here, it is found for the right-hand side that when $P = \mathrm{OP}(p(x,\xi))$, then $\gamma_0 P_+ u = \gamma_0 r^+ \mathrm{OP}(p(x',0,\xi))e^+ u$, where we can insert (10.8). This gives a sum of differential trace operators plus a term where the ψdo symbol is p'; here $r^\pm \tilde{p}'$ satisfy estimates as in (10.11). The term is (using the notation $\mathscr{F}_{x' \to \xi'} u = \acute{u}$, cf. (9.5))

$$\gamma_0 \mathrm{OP}(p')_+ u(x) = \gamma_0 \int_{\mathbb{R}^{2n}} e^{i(x-y)\cdot\xi} p'(x',0,\xi)e^+ u(y)dy\,d\xi$$

$$= \gamma_0 \int_{\mathbb{R}^{n-1}} e^{ix'\cdot\xi'} \int_{\mathbb{R}} \tilde{p}'(x',0,\xi',x_n - y_n)e^+ \acute{u}(\xi',y_n)dy_n\,d\xi' \tag{10.22}$$

$$= \int_{\mathbb{R}^{n-1}} e^{ix'\cdot\xi'} \int_0^\infty \tilde{p}'(x',0,\xi',-y_n)\acute{u}(\xi',y_n)dy_n\,d\xi'.$$

As noted above, $\tilde{p}(x',0,\xi',-y_n)|_{y_n>0} \in S_{1,0}^d(\mathbb{R}^{n-1},\mathbb{R}^{n-1},\mathscr{S}_+)$.

The general definition goes as follows:

Definition 10.4. A trace operator of order d ($\in \mathbb{R}$) and class r ($\in \mathbb{N}_0$) is an operator of the form

$$Tu = \sum_{0 \le j \le r-1} S_j \gamma_j + T', \tag{10.23}$$

where γ_j denotes the standard trace operator $(\gamma_j u)(x') = D_{x_n}^j u(x',0)$, the S_j are ψdo's in $\mathbb{R}^{n-1}$ of order $d - j$, and T' is an operator of the form

$$(T'u)(x') = \int_{\mathbb{R}^{n-1}} e^{ix'\cdot\xi'} \int_0^\infty \tilde{t}'(x',x_n,\xi')\acute{u}(\xi',x_n)dx_n\,d\xi', \tag{10.24}$$

with $\tilde{t}' \in S_{1,0}^d(\mathbb{R}^{n-1},\mathbb{R}^{n-1},\mathscr{S}_+)$. $\tilde{t}'$ is called the *symbol-kernel* of T'. See also (10.25), (10.26), (10.27).

When $u \in \mathscr{S}(\overline{\mathbb{R}}_+^n)$, $\|\acute{u}(\xi',x_n)\|_{L_2(\mathbb{R}_+)}$ is $O(\langle\xi'\rangle^{-N})$ for all N, so $T'u$ is C^∞ in x'.

Observe the meaning of the *class* number r; it counts the number of standard trace operators γ_j that enter into T. *Class zero* means that there are no such terms; we shall see that T is then well-defined on $L_2(\mathbb{R}_+^n)$.

Symbol-kernels depending on (x',y') instead of x' can also be allowed; then the defining formula (10.24) should be written

$$(T'u)(x') = \int_{\mathbb{R}^{2(n-1)}} e^{i(x'-y')\cdot\xi'} \int_0^\infty \tilde{t}'(x',y',y_n,\xi')u(y',y_n)dy\,d\xi', \tag{10.25}$$

and an interpretation via oscillatory integrals is understood. To keep down the volume of the formulas, we mostly write x'-forms in the following, leaving some generalizations to (x',y')-form to the reader (also for the other operator types to come). Symbol-kernels in (x', y')-form are reduced to x'-form or y'-form by the formulas in Theorem 7.13 1°, applied in the primed variables.

Continuity properties of the operators will be systematically investigated in Section 10.5.

One can show (by use of the Seeley extension [S64]) that for any T' with symbol-kernel in $S_{1,0}^d(\mathbb{R}^{n-1}, \mathbb{R}^{n-1}, \mathscr{S}_+)$, there exists a ψdo Q of $S_{1,0}$ type, satisfying the transmission condition at $x_n = 0$, such that $T' = \gamma_0 Q_+$.

The subclass of *polyhomogeneous* trace operators are those where the S_j are polyhomogeneous, and $\tilde{t}'$ lies in the subspace $S^d(\mathbb{R}^{n-1}, \mathbb{R}^{n-1}, \mathscr{S}_+)$. Note that, in contrast with polyhomogeneous ψdo's, the homogeneity in ξ of the symbol terms is only required for $|\xi'| \geq 1$, not $|\xi| \geq 1$.

The function (distribution when $r > 0$)

$$\tilde{t}(x', x_n, \xi') = \sum_{0 \leq j < r} s_j(x', \xi') D_{x_n}^j \delta(x_n) + \tilde{t}'(x', x_n, \xi'), \tag{10.26}$$

understood as extended by 0 on $\mathbb{R}_-$ if needed, is called the *symbol-kernel* of T; its conjugate Fourier transform

$$t(x', \xi) = \overline{\mathscr{F}}_{x_n \to \xi_n} e^+ \tilde{t}(x, \xi') = \sum_{0 \leq j < r} s_j(x', \xi') \xi_n^j + t'(x', \xi) \tag{10.27}$$

(where $t'(x', \xi) = \overline{\mathscr{F}}_{x_n \to \xi_n} e^+ \tilde{t}'(x, \xi')$) is the *symbol* of T.

In the polyhomogeneous case, we often denote t_d and $\tilde{t}_d$ by t^0 resp. $\tilde{t}^0$, the *principal* symbol and symbol-kernel.

Application of the operator definition with respect to only the x_n-variable gives the *boundary symbol operator* $t(x', \xi', D_n)$ (resp. *principal* boundary symbol operator $t^0(x', \xi, D_n)$) from $\mathscr{S}(\overline{\mathbb{R}}_+)$ to $\mathbb{C}$,

$$t(x', \xi', D_n)u = \sum_{0 \leq j < r} s_j(x', \xi') \gamma_j u + \int_0^\infty \tilde{t}'(x', x_n, \xi') u(x_n) dx_n; \tag{10.28}$$

it is also denoted $\mathrm{OPT}_n(t)$ or $\mathrm{OPT}_n(\tilde{t})$. We can then write

$$Tu = \mathrm{OP}'(t(x', \xi', D_n))u = \mathrm{OP}' \, \mathrm{OPT}_n(t(x', \xi))u, \text{ also denoted}$$
$$\mathrm{OPT}(t(x', \xi))u \text{ or } \mathrm{OPT}(\tilde{t}(x, \xi'))u.$$

Poisson operators

We use the same symbol-kernel spaces from Definition 10.3 1° and 2° to define Poisson operators, but now doing a multiplication instead of taking a scalar product in the x_n-variable.

Definition 10.5. A Poisson operator of order d is an operator defined by a formula

$$(Kv)(x', x_n) = \int_{\mathbb{R}^{n-1}} e^{ix' \cdot \xi'} \tilde{k}(x', x_n, \xi') \hat{v}(\xi') \, d\xi' \tag{10.29}$$

where the *symbol-kernel* $\tilde{k}$ belongs to $S_{1,0}^{d-1}(\mathbb{R}^{n-1}, \mathbb{R}^{n-1}, \mathscr{S}_+)$. See also (10.30), (10.31).

Again, symbol-kernels depending on (x', y') can be allowed:

$$(Kv)(x', x_n) = \int_{\mathbb{R}^{2(n-1)}} e^{i(x'-y')\cdot\xi'} \tilde{k}(x', y', x_n, \xi') v(y') \, dy' d\xi'. \qquad (10.30)$$

The *symbol* corresponding to $\tilde{k}(x, \xi')$ is

$$k(x', \xi) = \mathscr{F}_{x_n \to \xi_n} e^+ \tilde{k}(x, \xi'). \qquad (10.31)$$

In the polyhomogeneous case, it has an expansion in homogeneous terms in ξ (for $|\xi'| \geq 1$) of degree $d - 1 - l$. In this case we often denote $\tilde{k}_{d-1} = \tilde{k}^0$ and $k_{d-1} = k^0$, the *principal* symbol-kernel or symbol. Again, one can view K defined in (10.29) (also denoted OPK(k) or OPK$(\tilde{k})$) as an operator $K = \text{OP}'(k(x', \xi', D_n))$, where $k(x', \xi', D_n)$ is the *boundary symbol operator* from $\mathbb{C}$ to $\mathscr{S}(\overline{\mathbb{R}}_+)$:

$$k(x', \xi', D_n)a = \tilde{k}(x', x_n, \xi') \cdot a \text{ for } a \in \mathbb{C}, \qquad (10.32)$$

also denoted OPK$_n(k)$ or OPK$_n(\tilde{k})$.

Remark 10.6. The above order convention, introduced originally in [B71], may seem a bit strange: Polyhomogeneous Poisson operators of order d have principal symbols homogeneous of degree $d - 1$. But the convention will fit the purpose that the composition of two operators of order d resp. d' will be of order $d + d'$ (valid, e.g., for the ψdo TK on $\mathbb{R}^{n-1}$).

The trace operators T' of class 0 (and order d) have as adjoints precisely the Poisson operators (of order $d + 1$), and vice versa. This is obvious on the boundary-symbol-operator level:

$$t(D_n) : u \in L_2(\mathbb{R}_+) \mapsto (u, \overline{\tilde{f}}) \in \mathbb{C} \text{ and } k(D_n) : v \in \mathbb{C} \mapsto v \cdot \tilde{f} \in L_2(\mathbb{R}_+)$$

are adjoints of one another. On the full operator level it is easy to show for symbols depending on (x', y') instead of x'; here if T' has symbol-kernel $\tilde{f}(x', y', x_n, \xi')$, the Poisson operator T'^* has symbol-kernel $\overline{\tilde{f}}(y', x', x_n, \xi')$. Details will be given in Theorem 10.29 later.

Trace operators of class $r > 0$ do not have adjoints within the calculus.

Example 10.7. The operator K_γ introduced in Theorem 9.3 is the Poisson operator with symbol-kernel $\tilde{k}(x_n, \xi') = e^{-\langle\xi'\rangle x_n}$; it is of order 0. Its symbol is

$$k(\xi', \xi_n) = \frac{1}{\langle\xi'\rangle + i\xi_n}, \qquad (10.33)$$

cf. Exercise 5.3. Inserting the expansion (7.11) of $\langle\xi'\rangle$ in (10.33) one can expand in homogeneous terms of falling degree (beginning with degree -1), showing that the symbol and symbol-kernel are polyhomogeneous of degree -1.

The adjoint of K_γ is the trace operator T of class 0 with symbol-kernel $\tilde{t}(x_n, \xi') = e^{-\langle \xi' \rangle x_n}$ and symbol $t(\xi', \xi_n) = \frac{1}{\langle \xi' \rangle - i\xi_n}$; it is of degree and order -1. Furthemore, a calculation shows that for $Q = \text{OP}(\langle \xi \rangle^{-2})$ as in (9.36)ff., $\gamma_0 Q_+$ is the trace operator with symbol-kernel $\frac{1}{2\langle \xi' \rangle} e^{-\langle \xi' \rangle x_n}$.

Poisson operators also arise from the following situation: Let $v(x') \in \mathscr{S}(\mathbb{R}^{n-1})$, and consider the distribution $v(x') \otimes \delta(x_n)$ (the product of $v(x')$ and $\delta(x_n)$). When P is a ψdo satisfying the transmission condition, one can show that $r^+ P(v \otimes \delta)$ makes sense as a function in $C^\infty(\overline{\mathbb{R}}^n_+)$, and the mapping $K : v \mapsto r^+ P(v \otimes \delta)$ is a Poisson operator. See Theorem 10.25 later.

Singular Green operators

We now get to the most unfamiliar element G of $\mathcal{A}$ in (10.18). A singular Green operator (s.g.o.) G arises, for instance, when we compose a Poisson operator K with a trace operator T as $G = KT$; this operator acts in $\mathbb{R}^n_+$ but is not a P_+. Another situation where s.g.o.s enter is when we compose two ψdo's P_+ and Q_+ (satisfying the transmission condition); then the " leftover operator"

$$L(P,Q) \equiv (PQ)_+ - P_+ Q_+ = r^+ PQe^+ - r^+ Pe^+ r^+ Qe^+$$
$$= r^+ P(I - e^+ r^+)Qe^+ \qquad (10.34)$$

is an operator acting in $\mathbb{R}^n_+$, that is not a ψdo (more about $L(P,Q)$ in Section 10.4). It turns out that these cases are covered by operators of the following form (they are in fact convergent series of products of Poisson and trace operators, cf. (10.107) later):

Definition 10.8. A singular Green operator G of order d ($\in \mathbb{R}$) and class r ($\in \mathbb{N}_0$) is an operator

$$G = \sum_{0 \leq j \leq r-1} K_j \gamma_j + G', \qquad (10.35)$$

where the K_j are Poisson operators of order $d - j$, the γ_j are standard trace operators and G' is an operator of the form

$$(G'u)(x) = \int_{\mathbb{R}^{n-1}} e^{ix' \cdot \xi'} \int_0^\infty \tilde{g}'(x', x_n, y_n, \xi') \hat{u}(\xi', y_n) dy_n d\xi', \qquad (10.36)$$

where $\tilde{g}'$, the *symbol-kernel* of G', is in $S_{1,0}^{d-1}(\mathbb{R}^{n-1}, \mathbb{R}^{n-1}, \mathscr{S}_{++})$, cf. Definition 10.3 3°. There is a corresponding *symbol* g', defined by

$$g'(x', \xi', \xi_n, \eta_n) = \mathscr{F}_{x_n \to \xi_n} \overline{\mathscr{F}}_{y_n \to \xi_n} e^+_{x_n} e^+_{y_n} \tilde{g}'(x', x_n, y_n, \xi'). \qquad (10.37)$$

The symbol-kernel and symbol of G itself are

$$\tilde{g}(x', x_n, y_n, \xi') = \sum_{0 \leq j < r} \tilde{k}_j(x', x_n, \xi') D_{y_n}^j \delta(y_n) + \tilde{g}'(x', x_n, y_n, \xi'),$$

$$g(x', \xi', \xi_n, \eta_n) = \sum_{0 \leq j < r} k_j(x', \xi) \eta_n^j + g'(x', \xi', \xi_n, \eta_n). \tag{10.38}$$

In the polyhomogeneous case, the *principal* symbol-kernel and symbol are $\tilde{g}^0 = \tilde{g}_{d-1}$ resp. $g^0 = g_{d-1}$. (Both for singular Green symbols and for trace symbols, the definition can be refined further to allow the notion of negative class $r < 0$, see [G96, Sect. 2.8].) In some recent works, it has been practical to replace the enumeration $d - 1 - l$ by $d - l$, but we here stick to the notation of [G96].

We define the *boundary symbol operator* $g(x', \xi', D_n)$ from $\tilde{g}$ by

$$g(x', \xi', D_n)u(x_n) = \sum_{0 \leq j < r} \tilde{k}_j(x', x_n, \xi')\gamma_j u + \int_0^\infty \tilde{g}'(x', x_n, y_n, \xi')u(y_n)dy_n,$$

$$\tag{10.39}$$

also called $\mathrm{OPG}_n(g)$ or $\mathrm{OPG}_n(\tilde{g})$; then G (also called $\mathrm{OPG}(g)$ or $\mathrm{OPG}(\tilde{g})$) can be viewed as $G = \mathrm{OP}'(g(x', \xi', D_n)) = \mathrm{OP}' \, \mathrm{OPG}_n(g)$.

Example 10.9. As a simple example of a singular Green symbol-kernel, let us take $\tilde{g}(x_n, y_n, \xi') = e^{-\langle \xi' \rangle(x_n + y_n)}$. Its symbol is

$$g(\xi', \xi_n, \eta_n) = \frac{1}{(\langle \xi' \rangle + i\xi_n)(\langle \xi' \rangle - i\eta_n)},$$

it is of degree -2 and order -1.

In view of the last remark in Example 10.7, $-\frac{1}{2\langle \xi' \rangle} \tilde{g}$ is the symbol-kernel of the last term $-K_\gamma \gamma_0 Q_+$ in (9.38); it is the singular Green operator part of the solution operator for the Dirichlet problem considered there.

Singular Green operators of class 0 have *adjoints* of the same kind: Allowing symbols depending on (x', y') we have that

$$G = \mathrm{OPG}(\tilde{g}(x', y', x_n, y_n, \xi')) \quad \text{implies}$$

$$G^* = \mathrm{OPG}(\overline{\tilde{g}}(y', x', y_n, x_n, \xi')), \tag{10.40}$$

when G is of class 0. More on this in Remark 10.35. Singular Green operators of class $r > 0$ do not have adjoints within the calculus.

Negligible operators

To the above operators defined by Fourier integral formulas, one adds the *negligible operators of each type*, defined as operators of the form (10.23), (10.29), (10.35) with S_j, T', K, K_j and G' replaced by integral operators with C^∞-kernels (up to the boundary) over the respective domains:

$$S_j \gamma_j u = \int_{\mathbb{R}^{n-1}} \mathcal{K}_{S_j}(x', y')(\gamma_j u)(y')\, dy', \quad T'u = \int_{\mathbb{R}^n} \mathcal{K}_{T'}(x', y)u(y)\, dy,$$

$$Kv = \int_{\mathbb{R}^{n-1}} \mathcal{K}_K(x, y')v(y')\, dy', \quad G'u = \int_{\mathbb{R}^n} \mathcal{K}_{G'}(x, y)u(y)\, dy, \qquad (10.41)$$

with $\mathcal{K}_{S_j} \in C^\infty(\mathbb{R}^{2n-2})$, $\mathcal{K}_{T'} \in C^\infty(\mathbb{R}^{n-1} \times \overline{\mathbb{R}}_+^n)$, $\mathcal{K}_K \in C^\infty(\overline{\mathbb{R}}_+^n \times \mathbb{R}^{n-1})$, $\mathcal{K}_{G'} \in C^\infty(\overline{\mathbb{R}}_+^n \times \overline{\mathbb{R}}_+^n)$. They are of class r when they contain trace operators γ_j for $j \le r - 1$. It will be seen from the mapping properties that we show in detail in Section 10.5 that the negligible operators include the operators defined above with symbol-kernel of order $-\infty$ in (x', y')-form. However, the kernels of these are decreasing in x_n and y_n, and we observe that more has been included now: $r^+ P e^+$ is included as a negligible G when P is negligible.

We shall show the mapping property indicated in (10.18) as well as mapping properties in Sobolev spaces in Section 10.5 below.

The various operator classes defined above are *invariant* under coordinate changes in $\overline{\mathbb{R}}_+^n$ preserving the boundary $\{x_n = 0\}$; this holds both for the polyhomogeneous classes and the $S_{1,0}$ classes. This is stated in [B71], with an indication of how to conclude the invariance for $S_{1,0}$ Poisson operators once it is shown for $S_{1,0}$ ψdo's satisfying the transmission condition. [RS82] proves the invariance under coordinate changes in x' alone, where the rules for ψdo's in x' apply. A complete proof, with formulas for the symbols of the transformed operators, is found in [G96, Sect. 2.4 and Th. 2.2.13], covering also parameter-dependent symbols.

Thanks to the invariance, one can also define the operators as acting in vector bundles over manifolds, by use of local coordinates. The book [G96] allows noncompact manifolds, but we at present just consider the compact case: X is an n-dimensional compact C^∞-manifold with boundary $\partial X = X'$, smoothly embedded in a neighboring n-dimensional manifold $\widetilde{X}$, and $\widetilde{E}$, $E = \widetilde{E}|_X$, $\widetilde{E}'$ and $E' = \widetilde{E}'|_X$, resp. F and F', are vector bundles of dimension N and N', resp. M and M', over $\widetilde{X}$ and X, resp. X' (described by local coordinates and trivializations). P is a ψdo in $\widetilde{E}$ over $\widetilde{X}$ satisfying the transmission condition at X' and the ψdbo's considered in connection with P are of the form

$$\mathcal{A} = \begin{pmatrix} P_+ + G & K \\ T & S \end{pmatrix} : \begin{matrix} C^\infty(E) & & C^\infty(E') \\ \times & \rightarrow & \times \\ C^\infty(F) & & C^\infty(F') \end{matrix} ; \qquad (10.42)$$

here $P_+ = r_X \circ P e_{X^\circ}$ (defined similarly to (10.1) for $X^\circ \subset \widetilde{X}$). All the operators are assumed to be such that they in local trivializations act in the way we have described above for $\mathbb{R}_+^n$. The terms T, K and S (and sometimes even $P_+ + G$) are often given as block matrices with different orders for different entries (fitting together in a suitable way); see the remarks on multi-order systems around (11.13).

Having defined the ingredients in $\mathcal{A}$ in (10.18), we shall now look at composition rules. When $\mathcal{A}'$ is another such system, going from $C_{(0)}^{\infty}(\overline{\mathbb{R}}_{+}^{n})^{N'} \times C_{0}^{\infty}(\mathbb{R}^{n-1})^{M'}$ to $C^{\infty}(\overline{\mathbb{R}}_{+}^{n})^{N''} \times C^{\infty}(\mathbb{R}^{n-1})^{M''}$, and one of the operators is properly supported (with a suitable generalization of Definition 7.6), the composition may be written

$$\mathcal{A}'' = \begin{pmatrix} P_{+} + G & K \\ T & S \end{pmatrix} \begin{pmatrix} P_{+}' + G' & K' \\ T' & S' \end{pmatrix} = \begin{pmatrix} P_{+}'' - L + G''' & K'' \\ T'' & S'' \end{pmatrix}. \qquad (10.43)$$

The point is to show that $\mathcal{A}''$ again has the structure of a ψdbo, which really amounts to showing 14 different composition rules:

$$\begin{aligned} &\text{(i)} && P'' = PP' \text{ is a } \psi\text{do with transm. cond.,} && (10.44) \\ &\text{(ii)} && L(P, P') = (PP')_{+} - P_{+}P_{+}' \text{ is an s.g.o.,} \\ &\text{(iii)} && G''' = P_{+}G' + GP_{+}' + GG' + KT' \text{ is an s.g.o.,} \\ &\text{(iv.)} && T'' = TP_{+}' + TG' + ST' \quad \text{is a trace operator,} \\ &\text{(v)} && K'' = P_{+}K' + GK' + KS' \quad \text{is a Poisson operator,} \\ &\text{(vi)} && S'' = TK' + SS' \text{ is a } \psi\text{do on } \mathbb{R}^{n-1}. \end{aligned}$$

We observe right away that the first rule (i) is easy to check from Definition 10.2, in view of the standard composition rule Theorem 7.13 for ψdo's (recall that we only consider P's of integer order). For the other rules, there will be some information in Proposition 10.10 below and a full treatment in Section 10.4. In this sense, the ψdbo's $\mathcal{A}$ form an "algebra".

It is also of interest to see whether $\mathcal{A}$ has an adjoint within the ψdbo calculus. In view of the preceding information, this occurs for general K, and for G and T when they are of class 0. For P_{+}, the adjoint $(P_{+})^{*}$ equals $(P^{*})_{+}$ when P is of order ≤ 0. More on adjoints in Theorem 10.29 and Remark 10.35.

There is a technical device worth mentioning here, which can transform the system $\mathcal{A}$ into one that does have an adjoint, namely, that there exists a family of ψdo's Λ_{-}^{r} of orders $r \in \mathbb{Z}$, such that $(\Lambda_{-}^{r})_{+}$ maps $H^{s}(\mathbb{R}_{+}^{n})$ homeomorphically onto $H^{s-r}(\mathbb{R}_{+}^{n})$ for all s. They are called order-reducing operators. (See Exercise 10.11.)

Now some details on compositions. The new operators are introduced in such a way that they have a special definition with respect to the x_{n}-variable (defined by OPT_{n}, OPK_{n}, OPG_{n}), whereas the definition with respect to the x'-variable is the standard ψdo definition OP'. In compositions, the new thing to deal with is therefore just what happens in the x_{n}-direction, whereas the rules in the x'-direction are as in Chapter 7. So let us now study x_{n}-compositions (denoted $\circ_{n}$ for the symbols).

From the real formulation given above we have easily:

Proposition 10.10. *Consider a ψdo symbol $p(x',0,\xi',\xi_n)$ and trace, Poisson and singular Green symbol-kernels $\tilde{t}(x',x_n,\xi')$, $\tilde{k}(x',x_n,\xi')$, $\tilde{g}(x',x_n,y_n,\xi')$ of order d, and $\tilde{t}'(x',x_n,\xi')$, $\tilde{k}'(x',x_n,\xi')$, $\tilde{g}'(x',x_n,y_n,\xi')$ of order d', all of class 0. Let $d'' = d + d'$. We have the rules:*

(i) *If p is $O(\langle\xi_n\rangle^{-1})$, then $\gamma_0\operatorname{OP}_n(p)_+ = \operatorname{OPT}_n(\tilde{t}'')$, where $\tilde{t}''(x',x_n,\xi') = \tilde{p}(x',0,\xi',-x_n)|_{x_n>0}$, a trace symbol-kernel of order d and class 0.*

(ii) *$\gamma_0\operatorname{OPK}_n(\tilde{k}) = s''$, where $s''(x',\xi') = \tilde{k}(x',0,\xi')$, a ψdo symbol of order d.*

(iii) *$\gamma_0\operatorname{OPG}_n(\tilde{g}) = \operatorname{OPT}_n(\tilde{t}'')$, where $\tilde{t}''(x',y_n,\xi') = \tilde{g}(x',0,y_n,\xi')$, a trace symbol-kernel of order d and class 0.*

(iv) *$\operatorname{OPK}_n(\tilde{k})\operatorname{OPT}_n(\tilde{t}') = \operatorname{OPG}_n(\tilde{g}'')$, where*

$$\tilde{g}''(x',x_n,y_n,\xi') = \tilde{k}(x',x_n,\xi')\tilde{t}'(x',y_n,\xi'),$$

an s.g.o. symbol-kernel of order d'' and class 0.

(v) *$\operatorname{OPT}_n(\tilde{t})\operatorname{OPK}_n(\tilde{k}') = s''$, a ψdo symbol of order d'' and defined by*

$$s''(x',\xi') = \int_0^\infty \tilde{t}(x',x_n,\xi')\tilde{k}'(x',x_n,\xi')\,dx_n.$$

(vi) *$\operatorname{OPG}_n(\tilde{g})\operatorname{OPK}_n(\tilde{k}') = \operatorname{OPK}_n(\tilde{k}'')$, of order d'' and defined by*

$$\tilde{k}''(x',x_n,\xi') = \int_0^\infty \tilde{g}(x',x_n,y_n,\xi')\tilde{k}'(x',y_n,\xi')\,dy_n.$$

(vii) *$\operatorname{OPT}_n(\tilde{t})\operatorname{OPG}_n(\tilde{g}') = \operatorname{OPT}_n(\tilde{t}'')$, of order d'' and class 0, defined by*

$$\tilde{t}''(x',y_n,\xi') = \int_0^\infty \tilde{t}(x',x_n,\xi')\tilde{g}'(x',x_n,y_n,\xi')\,dx_n.$$

(viii) *$\operatorname{OPG}_n(\tilde{g})\operatorname{OPG}_n(\tilde{g}') = \operatorname{OPG}_n(\tilde{g}'')$, of order d'' and class 0, defined by*

$$\tilde{g}''(x',x_n,y_n,\xi') = \int_0^\infty \tilde{g}(x',x_n,z_n,\xi')\tilde{g}''(x',z_n,y_n,\xi')\,dz_n.$$

(ix) *One has for all $l,m \in \mathbb{N}_0$, with resulting operators of order $d - l + m$:*

$$x_n^l D_{x_n}^m \operatorname{OPK}_n(\tilde{k}(x',x_n,\xi'))\varphi = \operatorname{OPK}_n(x_n^l D_{x_n}^m \tilde{k}(x',x_n,\xi'))\varphi,$$
$$x_n^l D_{x_n}^m \operatorname{OPG}_n(\tilde{g}(x',x_n,y_n,\xi'))u = \operatorname{OPG}_n(x_n^l D_{x_n}^m \tilde{g}(x',x_n,y_n,\xi'))u.$$

Proof. The first rule follows from (10.22), in view of the estimates (10.11). For the second rule, we have in view of (10.32) that for $a \in \mathbb{C}$,

$$\gamma_0\operatorname{OPK}_n(k)a = \gamma_0\tilde{k}(x',x_n,\xi')a = s''(x',\xi')a$$

with $s''(x',\xi') = \tilde{k}(x',0,\xi')$. The third rule follows similarly. The other rules are likewise verified immediately from the defining formulas. The symbol-kernel estimates for (v)–(viii) are shown using the Cauchy-Schwarz inequality; the detailed proofs can be left as an exercise for the reader (Exercise 10.4). The rules in (ix) follow from the definitions of the symbol-kernel spaces. □

As an important corollary we observe that the "singular" ingredients in $\mathcal{A}$ — the operators G, T and K — are negligible at a distance from the boundary:

Proposition 10.11. *Let* $\zeta \in C^\infty(\overline{\mathbb{R}}^n_+)$ *be such that* $\zeta(x) = 0$ *for* $x_n \leq \varepsilon$, *some* $\varepsilon > 0$ *(e.g.,* $\zeta(x) = 1 - \chi(x_n/\varepsilon)$ *on* $\overline{\mathbb{R}}^n_+$*). Let* G, T *and* K *be operators as in* (10.18) *of order* $d \in \mathbb{R}$, *and class* $r \geq 0$ *when relevant. Then*
  1° ζG *is negligible of class* r, $G\zeta$ *is negligible of class* 0.
  2° ζK *is negligible,* $T\zeta$ *is negligible of class* 0.

Proof. For any $N \in \mathbb{N}_0$, $\zeta_N(x) = \zeta(x)/x_n^N$ is in $C^\infty(\overline{\mathbb{R}}^n_+)$, supported in $\{x_n \geq \varepsilon\}$. Then
$$\zeta K = \zeta_N x_n^N K, \quad \zeta G = \zeta_N x_n^N G,$$
where it is seen from Proposition 10.10 (ix) that $x_n^N K$ and $x_n^N G$ are of order $d - N$ (the class of G remains the same), hence so are $\zeta_N x_n^N K$ and $\zeta_N x_n^N G$. Since N can be arbitrarily large, the orders are $-\infty$, so the operators are negligible.

For $T = \sum_{j<r} S_j \gamma_j + T'$, $T\zeta$ equals $T'\zeta$, of class 0. This is the adjoint of the Poisson operator $\bar{\zeta} T'^*$, to which the preceding result applies; so it is negligible. There is a similar proof for $G\zeta$.   □

The rules in Proposition 10.10 are quite straightforward and unsophisticated. However, when we get to compositions of $\mathrm{OP}_n(p)_+$ with the general boundary operators, the real formulation requires convolutions of $\tilde{p}'$, in a suitably truncated version, with the other symbol-kernels. Here the usual ψdo experience tells us that it should be an advantage to work with *symbols*, after Fourier transformation from x_n to ξ_n, where convolutions are replaced by products. But then the cut-off operator $e^+ r^+$ must be replaced by its effect in the ξ_n-variable, and here the transmission condition will be important.

There is a little piece of function theory that takes care of all this, which we shall now explain.

10.2 Fourier transform and Laguerre expansion of $\mathscr{S}_+$

The treatment of ψdo symbols satisfying the transmission condition (10.5) takes place in the following spaces of functions of a real variable t (playing the role of ξ_n).

Definition 10.12. For each integer $d \in \mathbb{Z}$, the space $\mathcal{H}_d$ is defined as the space of C^∞-functions $f(t)$ on $\mathbb{R}$ with the asymptotic property: There exist complex numbers $s_d, s_{d-1}, \ldots$ such that for all indices k, l and $N \in \mathbb{N}_0$,

$$\partial_t^l[t^k f(t) - \sum_{d-N \leq j \leq d} s_j t^{j+k}] \text{ is } O(|t|^{d-N-1+k-l}) \text{ for } |t| \to \infty. \quad (10.45)$$

Clearly, the s_j are uniquely determined from f. We denote

$$\mathcal{H} = \bigcup_{d \in \mathbb{Z}} \mathcal{H}_d \quad (10.46)$$

and observe also the decomposition in a direct sum

$$\mathcal{H} = \mathcal{H}_{-1} \dotplus \mathbb{C}[t], \quad (10.47)$$

where $\mathbb{C}[t]$ is the space of polynomials in t. The corresponding projection of $\mathcal{H}$ onto $\mathcal{H}_{-1}$ is denoted h_{-1}, and $(I - h_{-1})f$ is called the polynomial part of f. Occasionally, we also use the projector h_0 of $\mathcal{H}$ onto $\mathcal{H}_0$, which removes $\sum_{1 \leq j \leq d} s_j t^j$ from f. The spaces $\mathcal{H}_d$ have Fréchet topologies (defined by families of seminorms in relation to (10.45)), and $\mathcal{H}$ is an inductive limit of such spaces.

Lemma 10.13. *Let $\sigma > 0$ and let $d \in \mathbb{Z}$. Let $f(t) \in C^\infty(\mathbb{R})$, and define*

$$\tau = t^{-1}, \; k(\tau) = \tau^d f(\tau^{-1}) \; \text{for } \tau \in \mathbb{R} \setminus \{0\};$$

$$z = \frac{\sigma - it}{\sigma + it} \quad (\text{hence } t = \frac{\sigma}{i}\frac{1-z}{1+z}, \; 1+z = \frac{2\sigma}{\sigma+it}), \quad (10.48)$$

$$g(z) = (1+z)^d f(\frac{\sigma}{i}\frac{1-z}{1+z}) \; \text{for } z \in S^1 = \{z \in \mathbb{C} \mid |z| = 1\}, \; z \neq -1.$$

The following statements (i)–(iii) *are equivalent:*

(i) $f \in \mathcal{H}_d$.
(ii) k *extends to a function in* $C^\infty(\mathbb{R})$.
(iii) g *extends to a function in* $C^\infty(S^1)$.

Proof. Consider first the case $d = 0$. Assume that f satisfies the conditions (10.45) (which then also hold with $\partial_t^l t^k$ replaced by $t^k \partial_t^l$). Since $f(t) - s_0$ is $O(t^{-1})$ for $t \to \pm\infty$, $k(\tau)$ extends to a continuous function k on $\mathbb{R}$ (it is the point $\tau = 0$ that needs checking). Now

$$\partial_\tau k(\tau) = -t^2 \partial_t f(t)\big|_{t=\tau^{-1}}, \quad (10.49)$$

so since $-t^2 \partial_t(f(t) - s_0 - s_{-1}t^{-1}) = -t^2 \partial_t f(t) - s_{-1}$ is $O(t^{-1})$ for $t \to \pm\infty$, $\partial_\tau k(\tau) - s_{-1}$ is $O(\tau)$ for $\tau \to 0\pm$. Similarly, for each m,

$$(-t^2 \partial_t)^m (f(t) - \sum_{0 \leq l \leq m} s_{-l} t^{-l}) = (-t^2 \partial_t)^m f(t) - m! s_{-m} \quad (10.50)$$

is $O(t^{-1})$ for $t \to \pm\infty$, showing that $\partial_\tau^m k(\tau) - m! s_{-m}$ is $O(\tau)$ for $\tau \to 0\pm$. Thus (i) implies (ii) with

$$\partial_\tau^m k(0) = m! s_{-m} \; \text{for } m \in \mathbb{N}_0. \quad (10.51)$$

Conversely, the formulas (10.49), (10.50) allow us to conclude from (ii) to (i) with the coefficients s_{-m} determined successively from (10.51) (the particular family of estimates $(-t^2\partial_t)^m(f(t) - \sum_{-m \leq j \leq 0} s_j t^j) = O(t^{-1})$, $m \in \mathbb{N}_0$, implies the full family (10.45)).

For the transition between (ii) and (iii) we now just observe that

$$1 + z = 1 + \frac{\sigma - i/\tau}{\sigma + i/\tau} = \frac{2\sigma\tau}{\sigma\tau + i};$$

so smoothness of k at $\tau = 0$ is equivalent with smoothness of g at $z = -1$.

When $d < 0$, one reduces to the case $d = 0$ by replacing $f(t)$ by $f^*(t) = t^{-d}f(t)$, corresponding to the same $k(\tau)$ and to $g^*(z) = (i/\sigma(1 - z))^d g(z)$; and when $d > 0$, one makes a slight generalization to the above proof. □

Note that the coefficients s_j in (10.45) are proportional to the Taylor coefficients of $k(\tau)$ at $\tau = 0$; in this sense, they are "Taylor coefficients of f at ∞".

The function $k(\tau)$ here is just an auxiliary function, whereas $g(z)$ has a particular interest, in an analysis of $\mathcal{H}_{-1}$ that we shall now describe. Let $f(t) \in \mathcal{H}_{-1}$, let $g(z) = (1 + z)^{-1}f(-i\sigma(1 - z)/(1 + z))$ (by (10.48) with $d = -1$), and consider its Fourier series expansion (for $z = e^{i\theta}$), with the convention

$$g(z) = (2\sigma)^{-\frac{1}{2}} \sum_{k \in \mathbb{Z}} b_k z^k, \quad \text{decomposed in}$$

$$g^+(z) = (2\sigma)^{-\frac{1}{2}} \sum_{k \geq 0} b_k z^k \quad \text{and} \quad g^-(z) = (2\sigma)^{-\frac{1}{2}} \sum_{k < 0} b_k z^k. \tag{10.52}$$

This decomposition gives rise to a decomposition of $\mathcal{H}_{-1}$,

$$\mathcal{H}_{-1} = \mathcal{H}_{-1}^+ \dotplus \mathcal{H}_{-1}^-, \tag{10.53}$$

where f is decomposed in the sum of $f^\pm(t) \in \mathcal{H}_{-1}^\pm$ corresponding to the functions $g^\pm(z)$ as in (10.48). We shall analyze the spaces $\mathcal{H}_{-1}^\pm$, showing in particular that they are the Fourier transforms of the spaces $e^\pm \mathscr{S}(\overline{\mathbb{R}}_\pm)$. This will be done in an elementary way based on orthogonal expansions.

We know from the theory of trigonometric series that the function g on the circle $\{|z| = 1\}$ is C^∞ if and only if the sequence $(b_k)_{k \in \mathbb{Z}}$ is *rapidly decreasing* for $|k| \to \infty$, i.e., the sequences $(k^N b_k)_{k \in \mathbb{Z}}$ are bounded, for all N. Equivalently, the series $\sum_{k \in \mathbb{Z}} |(1 + |k|)^N b_k|^2$ are convergent for all N. The space of rapidly decreasing sequences (b_k) will be denoted by $\mathfrak{s}(\mathbb{Z})$, and we shall use

$$\|(b_k)_{k \in \mathbb{Z}}\|_{\ell_2^N} = (\sum_{k \in \mathbb{Z}} |(1 + |k|)^N b_k|^2)^{1/2}, \tag{10.54}$$

as a norm on the Hilbert space $\ell_2^N(\mathbb{Z})$; it is equivalent with the norm (8.11). (One can replace $\mathbb{Z}$ by $\mathbb{N}_0$ or other index sets.) So $\mathfrak{s}(\mathbb{Z}) = \bigcap_{N \geq 0} \ell_2^N(\mathbb{Z})$.

For $f(t) = (1 + z)g(z)$, the expansion (10.52) of g gives an expansion of f in terms of the functions

$$\widehat{\varphi}_k(t,\sigma) = (2\sigma)^{\frac{1}{2}} \frac{(\sigma - it)^k}{(\sigma + it)^{k+1}}, \quad \text{corresponding to } (2\sigma)^{-\frac{1}{2}}(1 + z)z^k, \quad (10.55)$$

cf. (10.48). They are easily checked to be *orthogonal in* $L_2(\mathbb{R})$ (with norms $(2\pi)^{\frac{1}{2}}$); and the completeness of the trigonometric system $(z^k)_{k\in\mathbb{Z}}$ implies the completeness of the system $(\widehat{\varphi}_k)_{k\in\mathbb{Z}}$.

Now the inverse Fourier transform carries the $\widehat{\varphi}_k(t,\sigma)$ over to the functions $\varphi_k(x,\sigma)$ defined by

$$\varphi_k(x,\sigma) = \begin{cases} (2\sigma)^{\frac{1}{2}}(\sigma - \partial_x)^k(x^k e^{-x\sigma})/k! & \text{for } k \geq 0,\, x \geq 0, \\ 0 & \text{for } k \geq 0,\, x < 0; \end{cases} \quad (10.56)$$

$$\varphi_k(x,\sigma) = \varphi_{-k-1}(-x,\sigma) \text{ for } k < 0;$$

they are a variant of the *Laguerre functions*. By the Parseval-Plancherel theorem, the functions $(\varphi_k)_{k\in\mathbb{Z}}$ form a complete orthonormal system in $L_2(\mathbb{R})$, and hence the functions with $k \geq 0$, resp. $k < 0$, span $L_2(\mathbb{R}_+)$, resp. $L_2(\mathbb{R}_-)$. (We often write φ_k for $r^+\varphi_k$ when $k \geq 0$, and φ_k for $r^-\varphi_k$ when $k < 0$.)

One can check that the φ_k with $k \geq 0$ are the eigenfunctions of the (variant of a) Laguerre operator

$$\mathcal{L}_{\sigma,+} = \sigma^{-1}(\sigma + \partial_x)x(\sigma - \partial_x) = -\sigma^{-1}\partial_x x \partial_x + \sigma x + 1 \quad (10.57)$$

in $L_2(\mathbb{R}_+)$, with simple eigenvalues $2(k + 1)$; and the φ_k with $k < 0$ are similarly the eigenfunctions for $\mathcal{L}_{\sigma,-}$ defined by the same expression on $\mathbb{R}_-$.

A property of expansions in the Laguerre system that is of particular interest here is the fact that rapidly decreasing coefficient series correspond to functions in $\mathscr{S}(\overline{\mathbb{R}}_+)$.

Lemma 10.14. *Let* $u \in L_2(\mathbb{R}_+)$*, expanded in the Laguerre system* $(\varphi_k)_{k\in\mathbb{N}_0}$*, by*

$$u(x) = \sum_{k\in\mathbb{N}_0} b_k \varphi_k(x,\sigma).$$

Then $u \in \mathscr{S}(\overline{\mathbb{R}}_+)$ *if and only if* $(b_k)_{k\in\mathbb{N}_0}$ *is rapidly decreasing. More precisely, one has the identity*

$$\|u\|_{L_2(\mathbb{R}_+)} = \|(b_k(u))_{k\in\mathbb{N}_0}\|_{\ell_2^0}, \quad (10.58)$$

and there are estimates of $(b_k)_{k\in\mathbb{N}_0}$ *in terms of* u:

$$\|(b_k(u))_{k\in\mathbb{N}_0}\|_{\ell_2^N} = 2^{-N}\|(b_k(\mathcal{L}_{\sigma,+}^N u))\|_{\ell_2}$$
$$= 2^{-N}\|\mathcal{L}_{\sigma,+}^N u\|_{L_2} \leq c_N \max_{j+l \leq N} \sigma^{j-l}\|x^{j+l}\partial_x^{2l}u\|_{L_2}, \quad (10.59)$$

and estimates of u in terms of $(b_k)_{k \in \mathbb{N}_0}$ (with any $\varepsilon > 0$):

$$\|x^j \partial_x^l u\|_{L_2} \le c_\varepsilon \, \sigma^{-j+l} \|(b_k(u))\|_{\ell_2^{j+(1+\varepsilon)l}}. \tag{10.60}$$

Proof. The identity (10.58) follows from the orthonormality and completeness of the system φ_k in $L_2(\mathbb{R}_+)$. (10.59) then follows easily from the eigenvalue property of the φ_k:

$$
\begin{aligned}
\|(b_k)_{k \in \mathbb{N}_0}\|_{\ell_2^N}^2 &= \sum_k |(1+k)^N b_k|^2 = 2^{-2N} \|\sum_k b_k \, \mathcal{L}_{\sigma,+}^N \varphi_k\|_{L_2}^2 \\
&= 2^{-2N} \|\mathcal{L}_{\sigma,+}^N u\|_{L_2}^2 = 2^{-2N} \|(-\sigma^{-1} \partial_x x \partial_x + \sigma x + 1)^N u\|_{L_2}^2 \\
&\le c_N \max_{j+l \le N} \sigma^{2(j-l)} \|x^{j+l} \partial_x^{2l} u\|_{L_2}^2,
\end{aligned}
$$

since $\partial_x(xu) = x \partial_x u + u$. For the estimates (10.60) one calculates the expansion coefficients of $xu(x)$ and $\partial_x u(x)$ in terms of those of u; details are found in [G96, Lemma 2.2.1]. We refer to the proof given there, and shall just quote the formulas that give the general idea:

$$
\begin{aligned}
x \varphi_k(x, \sigma) &= \tfrac{1}{2\sigma}(k \varphi_{k-1} + (2k+1)\varphi_k + (k+1)\varphi_{k+1}) \\
\partial_x \varphi_k(x, \sigma) &= -\sigma \varphi_k + 2\sigma \sum_{0 \le j < k} (-1)^{k-1-j} \varphi_j + (-1)^k (2\sigma)^{\frac{1}{2}} \delta,
\end{aligned}
\tag{10.61}
$$

for $k \ge 0$ (easily proved using the $\hat{\varphi}_k$). $\qquad\square$

For certain combinations of x^j and ∂_x^l there are better estimates than (10.60), see [G96, Lemma 2.2.1].

We now return to the decomposition (10.52), which has the counterpart for $f(t)$, in view of (10.55),

$$f(t) = f^+(t) + f^-(t), \text{ where}$$
$$f^+(t) = \sum_{k \ge 0} b_k \hat{\varphi}_k(t, \sigma) \text{ and } f^-(t) = \sum_{k < 0} b_k \hat{\varphi}_k(t, \sigma); \tag{10.62}$$

here the sequences $(b_k)_{k \ge 0}$ and $(b_k)_{k < 0}$ can be arbitrary rapidly decreasing sequences. The hereby defined decomposition of the space $\mathcal{H}_{-1}$ is denoted $\mathcal{H}_{-1}^+ \dotplus \mathcal{H}_{-1}^-$ as already stated in (10.53); we shall denote the corresponding projections h_{-1}^+ resp. h_{-1}^-. Note that they are *orthogonal projections* with respect to $L_2(\mathbb{R})$-norm (since the $\hat{\varphi}_k$ are mutually orthogonal), so that

$$\|f^+\|_{L_2} \le \|f\|_{L_2}; \quad \|f^-\|_{L_2} \le \|f\|_{L_2} \text{ for } f \in \mathcal{H}_{-1}. \tag{10.63}$$

Observe that

$$\overline{\hat{\varphi}}_k = \hat{\varphi}_{-k-1} \text{ for all } k, \tag{10.64}$$

which implies that

$$f \in \mathcal{H}_{-1}^+ \iff \overline{f} \in \mathcal{H}_{-1}^-. \tag{10.65}$$

By Lemma 10.14 we now see that $\mathcal{H}^+_{-1}$ *is precisely the space of Fourier transforms of functions* $e^+u(x)$, *where* $u \in \mathscr{S}(\overline{\mathbb{R}}_+)$,

$$\mathcal{H}^+_{-1} = \mathscr{F}(e^+\mathscr{S}(\overline{\mathbb{R}}_+)). \tag{10.66}$$

Similarly (cf. (10.65))

$$\mathcal{H}^-_{-1} = \mathscr{F}(e^-\mathscr{S}(\overline{\mathbb{R}}_-)). \tag{10.67}$$

So, when $f = f^+ + f^-$ is such that $f^\pm = \mathscr{F}(e^\pm u^\pm)$, then the projections $h^\pm_{-1}$ in $\mathcal{H}_{-1}$ correspond to the projections $e^\pm r^\pm$ applied to $u = e^+u^+ + e^-u^- \in e^+\mathscr{S}(\overline{\mathbb{R}}_+)\dot{+}e^-\mathscr{S}(\overline{\mathbb{R}}_-)$.

The above analysis is concerned with $\mathcal{H}_{-1}$; for the complete description of $\mathcal{H}$ one has to adjoin $\mathbb{C}[t]$ (cf. (10.47)), and it is customary to define (with a slight asymmetry)

$$\begin{aligned} \mathcal{H}^+ &= \mathcal{H}^+_{-1}, \\ \mathcal{H}^- &= \mathcal{H}^-_{-1} \dot{+} \mathbb{C}[t]. \end{aligned} \tag{10.68}$$

Then $\mathcal{H} = \mathcal{H}^+\dot{+}\mathcal{H}^-$, and the corresponding projections are denoted h^+ and h^-, extending h^+_{-1} resp. h^-_{-1}.

Note that $\mathbb{C}[t]$ is the space of Fourier transforms of the "polynomials" $\sum c_k \delta^{(k)}$ where $\delta^{(k)} = D^k_x\delta$; we call the latter space $\mathbb{C}[\delta']$. Then the analysis can be summed up in the following statement.

Theorem 10.15. 1° *The space* $\mathcal{H} = \bigcup_{d \in \mathbb{Z}} \mathcal{H}_d$ *admits a decomposition in a direct sum*

$$\mathcal{H} = \mathcal{H}^+ \dot{+} \mathcal{H}^-, \tag{10.69}$$

with projections denoted h^+ *and* h^-; *moreover*

$$\mathcal{H}^- = \mathcal{H}^-_{-1} \dot{+} \mathbb{C}[t], \quad \mathcal{H}^+ = \mathcal{H}^+_{-1}.$$

The decompositions are defined in such a way that the space $\mathcal{H}$ *by inverse Fourier transformation is mapped onto the space*

$$\dot{\mathscr{S}}(\mathbb{R}) \equiv e^+\mathscr{S}(\overline{\mathbb{R}}_+) \dot{+} e^-\mathscr{S}(\overline{\mathbb{R}}_-) \dot{+} \mathbb{C}[\delta'], \text{ where} \tag{10.70}$$

$$\mathscr{F}^{-1}\mathcal{H}^\pm_{-1} = e^\pm\mathscr{S}(\overline{\mathbb{R}}_+), \quad \mathscr{F}^{-1}\mathcal{H}^- = e^-\mathscr{S}(\overline{\mathbb{R}}_-) \dot{+} \mathbb{C}[\delta']. \tag{10.71}$$

The projectors h^+ *and* h^- *carry over to the projectors* e^+r^+ *and* $I - e^+r^+$ *by* $\mathscr{F}^{-1}$; *here* $(I - e^+r^+)v = e^-r^-v$ *when* $v \in e^+\mathscr{S}(\overline{\mathbb{R}}_+) \dot{+} e^-\mathscr{S}(\overline{\mathbb{R}}_-)$.

2° *The spaces* $\mathscr{S}(\overline{\mathbb{R}}_+)$ *and* $\mathcal{H}^+$ *can be described as the spaces of functions*

$$u(x) = \sum_{k \in \mathbb{N}_0} b_k\varphi_k(x,\sigma) \text{ resp. } f(t) = \sum_{k \in \mathbb{N}_0} b_k\hat{\varphi}_k(t,\sigma), \tag{10.72}$$

expanded in the orthonormal Laguerre system $(\varphi_k)_{k \in \mathbb{N}_0}$ *on* $\mathbb{R}_+$ *(cf. (10.56)), resp. the Fourier-transformed Laguerre system* $(\hat{\varphi}_k)_{k \in \mathbb{N}_0}$ *in* $L_2(\mathbb{R})$ *(cf. (10.55)), with rapidly decreasing coefficient series* $(b_k)_{k\in\mathbb{N}_0}$. *There are similar statements for* $\mathscr{S}(\overline{\mathbb{R}}_-)$ *and* $\mathcal{H}_{-1}^-$ *using* $(\varphi_k)_{k<0}$ *on* $\mathbb{R}_-$, *resp.* $(\hat{\varphi}_k)_{k<0}$ *in* $L_2(\mathbb{R})$ *(the latter system is the same as* $(\overline{\hat{\varphi}}_l)_{l\geq 0}$).

3° *The general element* $f \in \mathcal{H}_d$ *has an expansion with uniquely determined coefficients*

$$f(t) = \sum_{0\leq j\leq d} s_j t^j + \sum_{k\in\mathbb{Z}} b_k \hat{\varphi}_k(t,\sigma), \qquad (10.73)$$

where the last sum equals $h_{-1}f$, *and*

$$h^+ f(t) = \sum_{k\geq 0} b_k \hat{\varphi}_k(t,\sigma),$$

$$h^- f(t) = \sum_{0\leq j\leq d} s_j t^j + \sum_{k<0} b_k \hat{\varphi}_k(t,\sigma) = \sum_{0\leq j\leq d} s_j t^j + \sum_{l\geq 0} b_l \overline{\hat{\varphi}}_l(t,\sigma);$$

here

$$\sum_{k\in\mathbb{Z}} |b_k|^2 = (2\pi)^{-1} \|h_{-1}f\|_{L_2(\mathbb{R})}^2.$$

For later reference we define

$$\mathcal{H}_d^+ = \mathcal{H}^+ \cap \mathcal{H}_d, \quad \mathcal{H}_d^- = \mathcal{H}^- \cap \mathcal{H}_d, \quad \text{any } d \in \mathbb{Z}. \qquad (10.74)$$

Remark 10.16. There is another decomposition related to (10.73) that is sometimes useful. Define the functions, for $k \in \mathbb{Z}$,

$$\hat{\psi}_k(t,\sigma) = \frac{(\sigma - it)^k}{(\sigma + it)^k} \quad \left[= (2\sigma)^{-\frac{1}{2}}(\sigma + it)\hat{\varphi}_k(t,\sigma) \right], \qquad (10.75)$$

and note that

$$\hat{\varphi}_k(t,\sigma) = (2\sigma)^{\frac{1}{2}} \frac{(\sigma - it)^k}{(\sigma + it)^{k+1}} \frac{(\sigma - it + \sigma + it)}{2\sigma} \qquad (10.76)$$

$$= (2\sigma)^{-\frac{1}{2}}(\hat{\psi}_{k+1}(t,\sigma) + \hat{\psi}_k(t,\sigma)).$$

Inserting this in (10.73), we find the expansion

$$f(t) = \sum_{1\leq j\leq d} s_j t^j + \sum_{k\in\mathbb{Z}} a_k \hat{\psi}_k(t,\sigma), \qquad (10.77)$$

where the s_j for $j \geq 1$ are the same as in (10.73) and the other coefficients are determined by the formulas

$$a_0 = (2\sigma)^{-\frac{1}{2}}(b_0 + b_{-1}) + s_0, \quad a_k = (2\sigma)^{-\frac{1}{2}}(b_k + b_{k-1}).$$

The system $\hat{\psi}_k$ is a complete orthogonal system in the weighted L_2-space over $\mathbb{R}$ with weight $(\sigma^2 + t^2)^{-1}$. Their inverse Fourier transforms are the distributions (cf. (10.61))

$$\psi_k(x,\sigma) = (2\sigma)^{\frac{1}{2}} \sum_{0 \leq j \leq k-1} (-1)^{k-1-j}\varphi_j + (-1)^k\delta \text{ for } k \geq 0,$$

$$\psi_k(x,\sigma) = \psi_{-k}(-x,\sigma) \qquad\qquad\qquad \text{for } k \leq 0. \qquad (10.78)$$

Remark 10.17. Since (b_k) is rapidly decreasing, $g^+(z)$ in (10.52) extends to a C^∞-function on the closed unit disk $\{|z| \leq 1\}$ that is holomorphic on the open disk $\{|z| < 1\}$. It follows (cf. (10.48)) that $f^+(t) \in \mathcal{H}^+$ has a C^∞ extension to $\overline{\mathbb{C}}_-$ which is holomorphic in $\mathbb{C}_-$; cf. (A.3). One can moreover show that $f^+(t)$ satisfies estimates (10.45) on $\overline{\mathbb{C}}_-$ with $d = -1$. The complex conjugates in $\mathcal{H}^-_{-1}$ have the analogous behavior with respect to $\mathbb{C}_+$, and for $d \geq -1$, the functions in $\mathcal{H}^-_d$ satisfy estimates (10.45) on $\overline{\mathbb{C}}_+$.

Also the Paley-Wiener theorem, linking the holomorphy of h^+f in $\mathbb{C}_-$ with the fact that $\operatorname{supp}\mathscr{F}^{-1}f \subset \overline{\mathbb{R}}_+$, could be used. Moreover, h^+f and h^-f can be defined from f by Cauchy integrals; this point of view has a prominent role in [B71].

Note that the (Fréchet) topology on the space $\mathcal{H}^+$ can be defined by either of the following three systems of norms (where $f = \mathscr{F}e^+u$, with expansions (10.72)):

$$\|(b_k)_{k\in\mathbb{N}_0}\|_{\ell_2^N}, \quad N \in \mathbb{N}_0,$$

$$\|x^j D_x^m u(x)\|_{L_2(\mathbb{R}_+)} \qquad\qquad\qquad (10.79)$$

$$= (2\pi)^{-\frac{1}{2}}\|h^+(D_t^j[t^m f(t)])\|_{L_2(\mathbb{R})}, \quad j, m \in \mathbb{N}_0.$$

The equivalence of the first and second norm systems was shown in Lemma 10.14, and for the second and third norm we use the Parseval-Plancherel theorem. The application of h^+ here just removes polynomial terms; it corresponds before Fourier transformation to removing those singularities supported in $\{0\}$ that arise from differentiating e^+u.

The mapping that assigns the coefficients s_j in (10.45) to a function $f \in \mathcal{H}$ will now be investigated. As mentioned before, they are linked with the Taylor coefficients of $k(\tau)$ at $\tau = 0$; cf. (10.48). When we consider $\mathscr{F}^{-1}f \in \mathscr{S}'(\mathbb{R})$, we see that the coefficients s_j with $j \geq 0$ appear here as coefficients in a "polynomial" $\sum s_j\delta^{(j)}$. For the coefficients with $j < 0$, the most important step is the analysis of the case where $f \in \mathcal{H}^+$:

Let $f \in \mathcal{H}^+$, having the asymptotic expansion

$$f(t) \sim s_{-1}t^{-1} + s_{-2}t^{-2} + \cdots, \qquad\qquad (10.80)$$

and let $u \in \mathscr{S}_+$ be such that $\mathscr{F}e^+u = f$. We shall see that the s_j are linked with the boundary values of u. The Lagrange (or Green) formula (4.53)

$$(D_x^N u, v)_{\mathbb{R}_+} - (u, D_x^N v)_{\mathbb{R}_+} = i \sum_{0 \leq k \leq N-1} \gamma_{N-k-1} u \cdot \overline{\gamma_k v}, \; u, v \in \mathscr{S}_+ \quad (10.81)$$

(recall that $\gamma_j u = D_x^j u(0)$), can be written in distribution form when we insert $v = r^+ \varphi$ for some $\varphi \in \mathscr{S}(\mathbb{R})$ and observe that

$$(D_x^N u, v)_{\mathbb{R}_+} = (e^+ D_x^N u, \varphi)_{\mathbb{R}} = \langle e^+ D_x^N u, \overline{\varphi} \rangle_{\mathbb{R}},$$
$$(u, D_x^N v)_{\mathbb{R}_+} = \langle e^+ u, \overline{D_x^N \varphi} \rangle_{\mathbb{R}} = \langle D_x^N e^+ u, \overline{\varphi} \rangle_{\mathbb{R}},$$
$$\gamma_j u \cdot \overline{\gamma_k v} = \langle (\gamma_j u) \delta, \overline{D_x^k \varphi} \rangle_{\mathbb{R}} = \langle (\gamma_j u) D_x^k \delta, \overline{\varphi} \rangle_{\mathbb{R}}.$$

Then (10.81) becomes

$$D_x^N e^+ u = -i \sum_{0 \leq k \leq N-1} (\gamma_{N-k-1} u) D_x^k \delta + e^+ D_x^N u, \text{ for any } N \in \mathbb{N}_0. \quad (10.82)$$

This gives by Fourier transformation

$$t^N f(t) = -i \sum_{0 \leq k < N} (\gamma_{N-k-1} u) t^k + g_N(t), \quad (10.83)$$

where $g_N(t) \in \mathcal{H}^+ \subset \mathcal{H}_{-1}$. Comparing (10.83) with (10.80), we conclude from the uniqueness of the coefficients that

$$\gamma_j u = i s_{-1-j} \text{ for all } j \geq 0. \quad (10.84)$$

Another interpretation of the coefficients can be given by use of the so-called *plus-integral* $\int^+ f(t)\, dt$. It is defined on $\mathcal{H}$ by linear extension of the two cases:

$$\int^+ f(t)\, dt = \begin{cases} \int_{\mathbb{R}} f(t)\, dt \text{ when } f \in L_1(\mathbb{R}) \text{ (i.e., } f \in \mathcal{H}_{-2}), \\ \int_C f(t)\, dt \text{ when } f \text{ is meromorphic in } \mathbb{C}_+, \end{cases} \quad (10.85)$$

C denoting a contour around the poles in $\mathbb{C}_+$. This covers all $f \in \mathcal{H}$, since, when $f \in \mathcal{H}_{-1}$ with the expansion (10.45), $f - s_{-1}(t+i)^{-1} \in \mathcal{H}_{-2}$. In particular, in view of Remark 10.17,

$$\int^+ f(t)\, dt = 0 \text{ if } f \in \mathcal{H}^- \text{ or } f \in \mathcal{H}^+ \cap \mathcal{H}_{-2}; \quad (10.86)$$

for in the latter case, the integration can be carried over to a contour in $\mathbb{C}_-$.

Then when $f \in \mathcal{H}^+$ and has the expansion (10.80), we can write, for any $k \in \mathbb{N}_0$,

$$t^k f(t) = \sum_{-k \leq j \leq -1} s_j t^{j+k} + s_{-1-k} \frac{1}{t-i} + g_k(t) \text{ with } g_k(t) \in \mathcal{H}^+ \cap \mathcal{H}_{-2},$$

and conclude by the residue theorem that $\frac{1}{2\pi i}\int^+ t^k f(t)\,dt = s_{-1-k}$. Recalling (10.84), we have obtained:

Lemma 10.18. *When* $f = \mathscr{F}e^+u$ *($u \in \mathscr{S}_+$), with the expansion* (10.80), *then*

$$\frac{1}{2\pi}\int^+ t^k f(t)\,dt = is_{-1-k} = \gamma_k u. \tag{10.87}$$

The coefficient s_{-1-k} can be *estimated* by use of the standard trace estimates

$$
\begin{aligned}
|s_{-1-k}|^2 = |\gamma_k u|^2 &= -\int_0^\infty \partial_x[u^{(k)}\overline{u}^{(k)}]\,dx \\
&\le 2\|D_x^k u\|_{L_2(\mathbb{R}_+)}\|D_x^{k+1}u\|_{L_2(\mathbb{R}_+)} \\
&= \tfrac{1}{\pi}\|h^+(t^k f)\|_{L_2(\mathbb{R})}\|h^+(t^{k+1}f)\|_{L_2(\mathbb{R})}, \text{ when } f \in \mathcal{H}^+.
\end{aligned} \tag{10.88}
$$

In the general case where $f \in \mathcal{H}_{-1}$, one can obtain similar results by applying the above to the components $f^+ = h^+f$ and $f^- = h^-f$.

10.3 The complex formulation

The considerations in Section 10.2 are relevant not only for the ψdo's satisfying the transmission property, but also for the appropriate definition of symbol spaces for trace operators, Poisson operators and singular Green operators. The symbols were mentioned in Section 10.1 along with the symbol-kernels (as their Fourier transforms in the normal variable(s)), but without a systematic definition of symbol spaces. We shall give this now using the notions from Section 10.2.

Definition 10.19. Let $d \in \mathbb{R}$, let $r \in \mathbb{N}_0$, let Ξ be open $\subset \mathbb{R}^{n'}$ and let $\mathcal{K}$ be one of the spaces $\mathcal{H}_{r-1}^-$ or $\mathcal{H}^+$.

1° The space $S_{1,0}^d(\Xi,\mathbb{R}^{n-1},\mathcal{K})$ consists of the functions $f(X,\xi',\xi_n) \in C^\infty(\Xi \times \mathbb{R}^n)$, lying in $\mathcal{K}$ with respect to ξ_n, such that when f is written in the form

$$f(X,\xi',\xi_n) = \sum_{0\le j\le r-1} s_j(X,\xi')\xi_n^j + f'(X,\xi',\xi_n) \tag{10.89}$$

with $f' = h_{-1}f$, then $s_j(X,\xi') \in S_{1,0}^{d-j}(\Xi,\mathbb{R}^{n-1})$ and f' satisfies

$$\|D_X^\beta D_{\xi'}^\alpha h_{-1}(D_{\xi_n}^k \xi_n^{k'} f')\|_{L_{2,\xi_n}} \le c(X)\langle\xi'\rangle^{d+\frac{1}{2}-k+k'-|\alpha|-j}, \text{ all } \xi', \tag{10.90}$$

for all indices $\beta \in \mathbb{N}_0^{n'}$, $\alpha \in \mathbb{N}_0^{n-1}$, k and $k' \in \mathbb{N}_0$, with a continuous function $c(X)$ depending on the indices.

2° The space $S^d(\Xi, \mathbb{R}^{n-1}, \mathcal{K})$ of polyhomogeneous symbols (in $S^d_{1,0}$) consists of those symbols $f \in S^d_{1,0}(\Xi, \mathbb{R}^{n-1}, \mathcal{K})$ that furthermore have asymptotic expansions

$$f \sim \sum_{l \in \mathbb{N}_0} f_{d-l}, \tag{10.91}$$

where $f - \sum_{l<M} f_{d-l} \in S^{d-M}_{1,0}(\Xi, \mathbb{R}^{n-1}, \mathcal{K})$ for any $M \in \mathbb{N}_0$, and the symbols f_{d-l} are homogeneous of degree $d - l$ in ξ on the set where $|\xi'| \geq 1$:

$$f_{d-l}(X, t\xi) = t^{d-l} f_{d-l}(X, \xi) \text{ for } |\xi'| \geq 1, \ t \geq 1. \tag{10.92}$$

In the decomposition (10.89), the sum over j is empty when $r \leq 0$. Note that f' is in the space with r replaced by 0; it is often called "the part of f of class 0". The first term f_d is also denoted f^0.

Definition 10.20. Let $d \in \mathbb{R}$ and $r \in \mathbb{N}_0$.

1° $S^d_{1,0}(\Xi, \mathbb{R}^{n-1}, \mathcal{H}^-_{r-1})$ and $S^d(\Xi, \mathbb{R}^{n-1}, \mathcal{H}^-_{r-1})$ are called, respectively, the space of $S_{1,0}$ or polyhomogeneous **trace symbols** of degree d, class r and order d.

When $r = 0$, the inverse co-Fourier transform gives (after application of $\overline{\mathscr{F}}^{-1}_{\xi_n \to x_n}$ and restriction to $\mathbb{R}_+$) the spaces $S^d_{1,0}(\Xi, \mathbb{R}^{n-1}, \mathscr{S}_+)$ resp. $S^d(\Xi, \mathbb{R}^{n-1}, \mathscr{S}_+)$ of $S_{1,0}$ resp. polyhomogeneous **trace symbol-kernels** of degree d, class 0 and order d, defined in Definition 10.3.

2° $S^d_{1,0}(\Xi, \mathbb{R}^{n-1}, \mathcal{H}^+)$ and $S^d(\Xi, \mathbb{R}^{n-1}, \mathcal{H}^+)$ are called, respectively, the space of $S_{1,0}$ or polyhomogeneous **Poisson symbols** of degree d and **order** $d + 1$.

The inverse Fourier transform gives (after application of $\mathscr{F}^{-1}_{\xi_n \to x_n}$ and restriction to $\mathbb{R}_+$) the spaces $S^d_{1,0}(\Xi, \mathbb{R}^{n-1}, \mathscr{S}_+)$ resp. $S^d(\Xi, \mathbb{R}^{n-1}, \mathscr{S}_+)$, here called the $S_{1,0}$ resp. polyhomogeneous **Poisson symbol-kernels** of degree d and order $d + 1$.

For the order convention, cf. Remark 10.6. The fact that the inverse Fourier-transformed spaces match the ones introduced in Definition 10.3 follows from Theorem 10.15 and the Parseval-Plancherel theorem. The application of h_{-1} in (10.90) serves to remove polynomials (that may arise when $k' > k$); this corresponds to the fact that we are in (10.13), (10.14) considering functions of $x_n \in \mathbb{R}_+$ (extendable to smooth functions on $\overline{\mathbb{R}}_+$), disregarding δ-derivatives that would arise from the discontinuity at 0 if one takes derivatives on $\mathbb{R}$.

For the ψdo symbols satisfying the transmission condition, the $h^\pm$ projections map into these spaces:

Theorem 10.21. *When $p(X, x_n, y_n, \xi)$ satisfies Definition 10.2, then, with $r = \max\{d + 1, 0\}$,*

$$h^+p(X, 0, 0, \xi) \in S_{1,0}^d(\Xi, \mathbb{R}^{n-1}, \mathcal{H}^+),$$
$$h^-p(X, 0, 0, \xi) \in S_{1,0}^d(\Xi, \mathbb{R}^{n-1}, \mathcal{H}_{r-1}^-), \qquad (10.93)$$
$$r^{\pm}\tilde{p}(X, 0, 0, \xi', \pm z_n) \in S_{1,0}^d(\Xi, \mathbb{R}^{n-1}, \mathscr{S}_+).$$

Proof. From Definition 10.2 follows that $h_{-1}[D_X^\beta D_\xi^\alpha(\xi_n^m p(X, 0, 0, \xi))]$ satisfies estimates

$$\|h_{-1}[D_X^\beta D_\xi^\alpha(\xi_n^m p(X, 0, 0, \xi))]\|_{L_{2,\xi_n}(\mathbb{R})} \le c(X)\langle \xi' \rangle^{d+\frac{1}{2}+m-|\alpha|};$$

this implies the estimates for h^+p required in the first line of (10.93) since $\|h^+p\| \le \|h_{-1}p\|$ (cf. (10.63)). Similar estimates are valid for h_{-1}^-p. By Theorem 10.15 this translates to the estimates for $r^\pm\tilde{p}$ in the last line. Since h^-p is the sum of h_{-1}^-p and the polynomial part (as in (10.8)), the assertion in the second line follows easily. $\qquad\qquad\square$

Example 10.22. With $\sigma = \langle \xi' \rangle$, the nonnormalized Fourier-transformed Laguerre functions

$$(2\sigma)^{-\frac{1}{2}}\widehat{\varphi}_l(\xi_n, \sigma) = \frac{(\sigma - i\xi_n)^l}{(\sigma + i\xi_n)^{l+1}}$$

(cf. (10.55)) with $l \ge 0$ lie in $S^{-1}(\mathbb{R}^{n-1}, \mathbb{R}^{n-1}, \mathcal{H}^+)$, so they are Poisson symbols of order 0, degree -1. Their conjugates $(2\sigma)^{-\frac{1}{2}}\overline{\widehat{\varphi}}_l(\xi_n, \sigma)$ lie in $S^{-1}(\mathbb{R}^{n-1}, \mathbb{R}^{n-1}, \mathcal{H}_{-1}^-)$, so they are trace symbols of order and degree -1 and class 0.

In order to describe the symbol spaces for singular Green operators we need to describe the space of Fourier transforms of functions in $\mathscr{S}_{++}$ extended by 0 to $\mathbb{R}^2$.

The tensor product of $\mathscr{S}(\overline{\mathbb{R}}_+)$ with itself is the linear space of functions on $\overline{\mathbb{R}}_{++}^2$ spanned by the products $a(x_n)b(y_n)$, $a, b \in \mathscr{S}(\overline{\mathbb{R}}_+)$. One can define completions of such tensor product spaces in several topologies, but it is known e.g. from Treves [T67] that the space $\mathscr{S}(\overline{\mathbb{R}}_+)$ is *nuclear*, and hence that the various completions coincide and identify with $\mathscr{S}(\overline{\mathbb{R}}_{++}^2)$, the restriction of $\mathscr{S}(\mathbb{R}^2)$ to $\overline{\mathbb{R}}_{++}^2$. We simply write

$$\mathscr{S}(\overline{\mathbb{R}}_+)\hat{\otimes}\mathscr{S}(\overline{\mathbb{R}}_+) = \mathscr{S}(\overline{\mathbb{R}}_{++}^2).$$

The (nuclear Fréchet) topology on $\mathscr{S}(\overline{\mathbb{R}}_{++}^2)$ is described e.g. by the system of seminorms

$$\|x_n^k D_{x_n}^{k'} y_n^m D_{y_n}^{m'} \tilde{g}(x_n, y_n)\|_{L(\mathbb{R}_{++}^2)} \text{ for } k, k', m, m' \in \mathbb{N}_0. \qquad (10.94)$$

By Fourier transformation in x_n and co-Fourier transformation in y_n (i.e., by sesqui-Fourier transformation) of $e_{x_n}^+ e_{y_n}^+ \mathscr{S}(\overline{\mathbb{R}}_{++}^2)$, we obtain (in view of (10.65)–(10.67)) the completed tensor product

$$\mathscr{F}_{x_n \to \xi_n} \overline{\mathscr{F}}_{y_n \to \eta_n} (e_{x_n}^+ e_{y_n}^+ \mathscr{S}(\overline{\mathbb{R}}_{++}^2))$$

$$= \mathscr{F}_{x_n \to \xi_n} \overline{\mathscr{F}}_{y_n \to \eta_n} (e^+ \mathscr{S}(\overline{\mathbb{R}}_+) \hat{\otimes} e^+ \mathscr{S}(\overline{\mathbb{R}}_+)) = \mathcal{H}^+ \hat{\otimes} \mathcal{H}^-_{-1}; \quad (10.95)$$

here the sesqui-Fourier transform acts as a *homeomorphism*. In particular, the (semi)norm (10.94) on $\tilde{g}(x_n, y_n)$ carries over to the (semi)norm

$$\tfrac{1}{2\pi} \| h^+_{\xi_n} h^-_{-1,\eta_n} (D^k_{\xi_n} \xi_n^{k'} D^m_{\eta_n} \eta_n^{m'} g(\xi_n, \eta_n)) \|_{L_2(\mathbb{R}^2)} \text{ for } g \in \mathcal{H}^+ \hat{\otimes} \mathcal{H}^-_{-1}, \quad (10.96)$$

where

$$g(\xi_n, \eta_n) = \int_{\mathbb{R}^2_{++}} e^{-ix_n \xi_n + iy_n \eta_n} \tilde{g}(x_n, y_n) \, dx_n dy_n; \quad (10.97)$$

the system of seminorms (10.96) defines the (nuclear Fréchet) topology on $\mathcal{H}^+ \hat{\otimes} \mathcal{H}^-_{-1}$. (Again, the projections h^+ and h^-_{-1} serve to remove polynomials, which would correspond to δ-derivatives at 0 in the (x_n, y_n)-formulation.)

By use of the direct sum decomposition

$$\mathcal{H}^-_{r-1} = \mathcal{H}^-_{-1} \dotplus \mathbb{C}_{r-1}[t], \text{ for } r \geq 0,$$

where $\mathbb{C}_{r-1}[t]$ denotes the (r-dimensional) space of polynomials of degree $< r$, we can likewise define $\mathcal{H}^+ \hat{\otimes} \mathcal{H}^-_{r-1}$, with elements

$$g(\xi_n, \eta_n) = \sum_{0 \leq j \leq r-1} s_j \eta_n^j + g'(\xi_n, \eta_n),$$

where the s_j are constants and $g' \in \mathcal{H}^+ \hat{\otimes} \mathcal{H}^-_{-1}$. Then we can formulate the appropriate definition:

Definition 10.23. Let $d \in R$, let $r \in \mathbb{N}_0$ and let $\varXi$ be open $\subset \mathbb{R}^{n'}$.

1° The space $S^d_{1,0}(\varXi, \mathbb{R}^{n-1}, \mathcal{H}^+ \hat{\otimes} \mathcal{H}^-_{r-1})$ consists of the functions $g(X, \xi', \xi_n, \eta_n) \in C^\infty(\varXi \times \mathbb{R}^{n+1})$, lying in $\mathcal{H}^+ \hat{\otimes} \mathcal{H}^-_{r-1}$ with respect to (ξ_n, η_n), such that when g is written in the form

$$g(X, \xi', \xi_n, \eta_n) = \sum_{0 \leq j \leq r-1} k_j(X, \xi', \xi_n) \eta_n^j + g'(X, \xi', \xi_n, \eta_n) \quad (10.98)$$

with $g' = h_{-1,\eta_n} g$, then $k_j \in S^{d-j}_{1,0}(\varXi, \mathbb{R}^{n-1}, \mathcal{H}^+)$ for each j, and g' satisfies the estimates

$$\| D^\beta_X D^\alpha_\xi h^+_{-1,\xi_n} h^-_{-1,\eta_n} (D^k_{\xi_n} \xi_n^{k'} D^m_{\eta_n} \eta_n^{m'} g') \|_{L_2(\mathbb{R}^2)}$$

$$\leq c(X) \langle \xi' \rangle^{d+1-k+k'-m+m'-|\alpha|-j}, \quad (10.99)$$

for all $\beta \in \mathbb{N}_0^{n'}, \alpha \in \mathbb{N}_0^{n-1}, k, k', m$ and $m' \in \mathbb{N}_0$.

2° The polyhomogeneous subspace $S^d(\varXi, \mathbb{R}^{n-1}, \mathcal{H}^+ \hat{\otimes} \mathcal{H}^-_{r-1})$ consists of those symbols $g \in S^d_{1,0}(\varXi, \mathbb{R}^{n-1}, \mathcal{H}^+ \hat{\otimes} \mathcal{H}^-_{r-1})$ that furthermore have asymptotic expansions

$$g \sim \sum_{l \in \mathbb{N}_0} g_{d-l},$$

where $g - \sum_{l<M} g_{d-l} \in S_{1,0}^{d-M}(\Xi, \mathbb{R}^{n-1}, \mathcal{H}^+ \hat{\otimes} \mathcal{H}_{r-1}^-)$ for any $M \in \mathbb{N}_0$, and the symbols g_{d-l} are homogeneous of degree $d - l$ in (ξ', ξ_n, η_n):

$$g_{d-l}(X, t\xi', t\xi_n, t\eta_n) = t^{d-l} g_{d-l}(X, \xi', \xi_n, \eta_n) \text{ for } t \text{ and } |\xi'| \geq 1. \quad (10.100)$$

3° The spaces are called, respectively, the space of $S_{1,0}$ or polyhomogeneous **singular Green symbols** of degree d and class r, and of **order** $d + 1$.

When $r = 0$, the inverse Fourier transform from ξ_n to x_n together with the inverse co-Fourier transform from η_n to y_n gives (after restriction to $\mathbb{R}_{++}^2$) the spaces $S_{1,0}^d(\Xi, \mathbb{R}^{n-1}, \mathscr{S}_{++})$ resp. $S^d(\Xi, \mathbb{R}^{n-1}, \mathscr{S}_{++})$ of $S_{1,0}$ resp. polyhomogeneous **singular Green symbol-kernels** of degree d, class 0 and order $d + 1$; they are defined as in Definition 10.3 3° and 4°.

Also for singular Green symbols, the Laguerre expansions are highly relevant. Functions in $\mathscr{S}_{++}$ can be expanded in the orthonormal double sequence $(\varphi_l(x_n, \sigma)\varphi_m(y_n, \sigma))_{l,m \in \mathbb{N}_0}$ formed of the Laguerre functions; it is a complete orthonormal basis for $L_2(\mathbb{R}_{++}^2)$, and its elements lie in $\mathscr{S}_{++}$. Here, when $\tilde{g}$ and g are expanded in double Laguerre series:

$$\tilde{g}(x_n, y_n) = \sum_{l,m \in \mathbb{N}_0} c_{lm} \varphi_l(x_n, \sigma) \varphi_m(y_n, \sigma),$$
$$g(\xi_n, \eta_n) = \sum_{l,m \in \mathbb{N}_0} c_{lm} \hat{\varphi}_l(\xi_n, \sigma) \overline{\hat{\varphi}}_m(\eta_n, \sigma), \quad (10.101)$$

one has that

$$\tilde{g} \in \mathscr{S}_{++} \Leftrightarrow g \in \mathcal{H}^+ \hat{\otimes} \mathcal{H}_{-1}^- \Leftrightarrow (c_{lm})_{l,m \in \mathbb{N}_0} \in \mathfrak{s}(\mathbb{N}_0 \times \mathbb{N}_0), \quad (10.102)$$

where $\mathfrak{s}(\mathbb{N}_0 \times \mathbb{N}_0)$ is the space of rapidly decreasing sequences indexed by $(l, m) \in \mathbb{N}_0 \times \mathbb{N}_0$. The system of (semi)norms where N and N' run through $\mathbb{N}_0$,

$$\|(c_{lm})_{l,m \in \mathbb{N}_0}\|_{\ell_2^{N,N'}} \equiv \left(\sum_{l,m \in \mathbb{N}_0} |(1+l)^N (1+m)^{N'} c_{lm}|^2 \right)^{\frac{1}{2}}, \quad (10.103)$$

defines the topology, equivalently with the systems (10.94) and (10.96).

For the functions in $S_{1,0}^{d-1}(\Xi, \mathbb{R}^{n-1}, \mathscr{S}_{++})$ it is natural to take $\sigma = \langle\xi'\rangle$, then when

$$\tilde{g}(X, x_n, y_n, \xi') = \sum_{l,m \in \mathbb{N}_0} c_{lm}(X, \xi') \varphi_l(x_n, \langle\xi'\rangle) \varphi_m(y_n, \langle\xi'\rangle), \quad (10.104)$$

the $c_{lm}(X, \xi')$ are in $S_{1,0}^d(\Xi, \mathbb{R}^{n-1})$ (ψdo symbols), and the topology of the space $S_{1,0}^{d-1}(\Xi, \mathbb{R}^{n-1}, \mathscr{S}_{++})$ is equivalently defined by the system of seminorms

$$\sup_{\xi'\in\mathbb{R}^{n-1}, X\in K} \langle\xi'\rangle^{-d+|\alpha|}\Big(\sum_{l,m\in\mathbb{N}_0}|(1+l)^N(1+m)^{N'}D_X^\beta D_{\xi'}^\alpha c_{lm}(X,\xi')|^2\Big)^{\frac{1}{2}},$$

$$(10.105)$$

where $\alpha\in\mathbb{N}_0^{n-1}$, $\beta\in\mathbb{N}_0^{n'}$, N and $N'\in\mathbb{N}_0$ and K runs through compact subsets of Ξ. A more refined choice of σ is to take $\sigma=[\xi']$, where

$$[\xi']=|\xi'| \text{ for } |\xi'|\geq 1, \quad [\xi'] \text{ is } C^\infty \text{ and } >0;\qquad (10.106)$$

the modified length. This can be useful in the consideration of polyhomogeneous symbols.

Note that when for example $X=x'$, and $\tilde{g}$ is written in the form (10.104) with $\sigma=\langle\xi'\rangle$, then

$$\tilde{g}(x',x_n,y_n,\xi')=\sum_{m\in\mathbb{N}_0}\tilde{k}_m(x',x_n,\xi')\tilde{t}_m(y_n,\xi'), \text{ where}\qquad (10.107)$$

$$\tilde{k}_m=\sum_{l\in\mathbb{N}_0}c_{lm}(x',\xi')(2\sigma)^{-\frac{1}{2}}\varphi_l(x_n,\sigma),\quad \tilde{t}_m=(2\sigma)^{\frac{1}{2}}\varphi_m(y_n,\sigma);$$

cf. Example 10.22. This shows in view of Proposition 10.10 (iv) that any s.g.o. of order d and class 0 can be written as a series

$$G=\sum_{m\in\mathbb{N}_0}K_mT_m\qquad (10.108)$$

of compositions of Poisson and trace operators of orders d resp. 0. The series converges rapidly in the symbol space seminorms.

For brevity, we shall often omit the indications $\Xi,\mathbb{R}^{n-1}$ from the notation for the various symbol and symbol-kernel spaces, when they are understood from the context.

We now describe how the operators are defined from the symbols. As noted in Section 10.1, the definition with respect to the x'-variable or (x',y')-variable is the standard ψdo definition, so it suffices to describe the definition w.r.t. x_n. Here we have, consistently with (10.24), (10.28), (10.29), (10.36), (10.39), when the plus-integral (10.85) $\int^+$ is used to include the polynomial parts of the trace and s.g.o. symbols in the integrals:

$$\mathrm{OPK}_n(k)v=\tilde{k}(X,x_n,\xi')\cdot v=\int e^{ix_n\xi_n}k(X,\xi)v\,d\!\!\!/\xi_n$$

$$=[k(X,\xi',D_n)v](x_n),\qquad (10.109)$$

$$\mathrm{OPT}_n(t)u=\sum_{0\leq j<r}s_j(X,\xi')\gamma_ju+\int_0^\infty \tilde{t}'(X,y_n,\xi')u(y_n)\,dy_n$$

$$= \int^+ t(X,\xi)\widehat{e^+u}(\xi_n)\,d\!\!\!/\xi_n = t(X,\xi',D_n)u,$$

$$\mathrm{OPG}_n(g)u = \sum_{0 \le j < r} \tilde{k}_j(X,x_n,\xi')\gamma_j u + \int_0^\infty \tilde{g}'(X,x_n,y_n,\xi')u(y_n)dy_n$$

$$= \int e^{ix_n\xi_n} \int^+ g(X,\xi,\eta_n)\widehat{e^+u}(\eta_n)d\eta_n\,d\!\!\!/\xi_n$$

$$= [g(X,\xi',D_n)u](x_n),$$

for $v \in \mathbb{C}$, $u \in \mathscr{S}_+$. The effect of $p(X,\xi)$ with truncation is

$$\mathrm{OP}_n(p)_+ u = r^+ \int e^{ix_n\xi_n}p(X,\xi)\widehat{e^+u}(\xi_n)\,d\!\!\!/\xi_n$$

$$= r^+ \mathscr{F}^{-1}h^+(p\,\widehat{e^+u}) = p(X,\xi',D_n)_+ u. \tag{10.110}$$

All these operators are called boundary symbol operators. The full operator definitions are obtained by combining the above with OP'.

10.4 Composition rules

As noted earlier, the new elements in the proof of the statements (10.44) lie in the compositions with respect to the x_n-variable. We shall write $a \circ_n b = c$, when c is the symbol (in one of our symbol spaces) arising from composing the boundary symbol operators with symbol a resp. b. The rules for (iii)–(vi) in (10.44) are treated in the following theorem.

Theorem 10.24. *Let d and $d' \in \mathbb{R}$, and let r and $r' \in \mathbb{N}_0$. Let*

(i) $p(X,\xi) \in S_{1,0}^d(\Xi,\mathbb{R}^n)$, *with transm. cond.*,

(ii) $g(X,\xi',\xi_n,\eta_n) \in S_{1,0}^{d-1}(\Xi,\mathbb{R}^{n-1},\mathcal{H}^+ \hat{\otimes} \mathcal{H}_{r-1}^-)$,

(iii) $t(X,\xi) \in S_{1,0}^d(\Xi,\mathbb{R}^{n-1},\mathcal{H}_{r-1}^-)$, (10.111)

(iv) $k(X,\xi) \in S_{1,0}^{d-1}(\Xi,\mathbb{R}^{n-1},\mathcal{H}^+)$,

(v) $s(X,\xi') \in S_{1,0}^d(\Xi,\mathbb{R}^{n-1})$,

and let p', g', t', k' and s' be given similarly with symbols in the spaces with d and r replaced by d' and r'. In the formulas where p (resp. p') occur, we assume that d (resp. d') is integer. Define

$$d'' = d + d', \quad r'' = \max\{r + d', 0\}. \tag{10.112}$$

Then the $\circ_n$-compositions give rise to ψdbo's whose symbols are determined by the following formulas (where $S_{1,0}^d(\Xi,\mathbb{R}^{n-1},\mathcal{K})$ is written as $S_{1,0}^d(\mathcal{K})$):

$1°\quad p_+ \circ_n k' = h_{\xi_n}^+ [p(X,\xi)k'(X,\xi)] \in S_{1,0}^{d''-1}(\mathcal{H}^+),$ (10.113)

$2°\quad g \circ_n k' = \int^+ g(X,\xi,\eta_n)k'(X,\xi',\eta_n)d\eta_n \in S_{1,0}^{d''-1}(\mathcal{H}^+),$

$3°\quad k \circ_n s' = k(X,\xi)s'(X,\xi) \in S_{1,0}^{d''-1}(\mathcal{H}^+),$

$4°\quad t \circ_n p'_+ = h_{\xi_n}^- [t(X,\xi)p'(X,\xi)] \in S_{1,0}^{d''}(\mathcal{H}_{r''-1}^-),$

$5°\quad t \circ_n g' = \int^+ t(X,\xi)g'(X,\xi,\eta_n)\,d\xi_n \in S_{1,0}^{d''}(\mathcal{H}_{r'-1}^-),$

$6°\quad s \circ_n t' = s(X,\xi')t'(X,\xi) \in S_{1,0}^{d''}(\mathcal{H}_{r'-1}^-),$

$7°\quad t \circ_n k' = \int^+ t(X,\xi)k'(X,\xi)\,d\xi_n \in S_{1,0}^{d''}(\Xi,\mathbb{R}^{n-1}),$

$8°\quad s \circ_n s' = s(X,\xi')s'(X,\xi') \in S_{1,0}^{d''}(\Xi,\mathbb{R}^{n-1}),$

$9°\quad p_+ \circ_n g' = h_{\xi_n}^+ [p(X,\xi)g'(X,\xi,\eta_n)] \in S_{1,0}^{d''-1}(\mathcal{H}^+ \hat{\otimes} \mathcal{H}_{r'-1}^-),$

$10°\quad g \circ_n p'_+ = h_{\eta_n}^- [g(X,\xi,\eta_n)p'(X,\xi',\eta_n)]$
$$\in S_{1,0}^{d''-1}(\mathcal{H}^+ \hat{\otimes} \mathcal{H}_{r''-1}^-),$$

$11°\quad g \circ_n g' = \int^+ g(X,\xi',\xi_n,\zeta_n)g'(X,\xi,\zeta_n,\eta_n)d\zeta_n$
$$\in S_{1,0}^{d''-1}(\mathcal{H}^+ \hat{\otimes} \mathcal{H}_{r'-1}^-),$$

$12°\quad k \circ_n t' = k(X,\xi',\xi_n)t'(X,\xi',\eta_n) \in S_{1,0}^{d''-1}(\mathcal{H}^+ \hat{\otimes} \mathcal{H}_{r'-1}^-).$

Here composition of polyhomogeneous symbols gives polyhomogeneous symbols.

Proof. The formulas follow rather naturally from the calculus described in Section 10.2. Consider $1°$. Here, for $v \in \mathbb{C}$, $k'v$ is as in (10.109), so

$$p_+(D_n) \circ_n k'(D_n)v = r^+ \int e^{ix_n\xi_n}p(X,\xi)\mathscr{F}[e^+\tilde{k}(X,x_n,\xi')]v\,d\xi_n$$

$$= r^+ \int e^{ix_n\xi_n}p(X,\xi)k(X,\xi)v\,d\xi_n$$

$$= r^+\mathscr{F}^{-1}[p(X,\xi)k(X,\xi)]v = \mathrm{OPK}_n(h^+(pk))v,$$

since h^+ corresponds to e^+r^+ by Fourier transformation (its effect is the restriction to $\mathbb{R}_+$ in the real formulation). There is a slight abuse of notation when we do not always mention the extension by 0 on $\mathbb{R}_-$.

For $4°$ one has that for $u \in \mathscr{S}(\overline{\mathbb{R}}_+)$ with $\mathscr{F}e^+u = f$, the operator $t(D_n) \circ_n p'(D_n)_+$ is defined by

$$t(D_n) \circ_n p'(D_n)_+u = t(D_n)\mathscr{F}^{-1}[h^+(p'f)]$$

$$= \int^+ t \cdot h^+(p'f) \, d\xi_n$$

$$= \int^+ t\, p'f \, d\xi_n \quad \left(\text{since } \int^+ t\, h^-(p'f) \, d\xi_n = 0\right)$$

$$= \int^+ h^-(tp')f \, d\xi_n \quad \left(\text{since } \int^+ h^+(tp')f \, d\xi_n = 0\right)$$

$$= \mathrm{OPT}_n(h^-(tp'))u;$$

here we have used (10.109), (10.110) and (10.86).

The rules 9° and 10° follow the same pattern, except that there is an extra integration (in η_n resp. ξ_n) to carry along.

In rules 3°, 6° and 8°, the effect of s or s' is purely multiplicative.

Rule 7° is the Fourier-transformed version of Proposition 10.10 (v) if $r = 0$. When $r > 0$, we must also deal with compositions $s_j \gamma_j k(D_n)v$; here

$$\gamma_j \tilde{k}(X, x_n, \xi') = \int^+ \xi_n^j k(X, \xi', \xi_n) \, d\xi_n$$

in view of Lemma 10.18, so these terms contribute in the stated way. The rules 2°, 5° and 11° are elaborations of this observation, based on Proposition 10.10 (vi)–(viii) and carrying other variables along. 12° is an obvious extension of Proposition 10.10 (iv).

This shows the formulas, and the estimates required for the indicated spaces are easily checked. □

In preparation for the treatment of the leftover operator $L(P, P')$, we introduce a new rule. When $v \in \mathscr{S}(\mathbb{R}^{n-1})$, it can be multiplied by the distribution $\delta(x_n)$ to define a temperate distribution on $\mathbb{R}^n$; it is traditionally denoted $v(x') \otimes \delta(x_n)$, and acts as follows:

$$\langle v(x') \otimes \delta(x_n), \varphi(x) \rangle_{\mathbb{R}^n} = \langle v(x'), \varphi(x', 0) \rangle_{\mathbb{R}^{n-1}}, \text{ when } \varphi \in \mathscr{S}(\mathbb{R}^n). \quad (10.114)$$

We shall show that when P is a ψdo of order d (satisfying the transmission condition, as always), then the operator K defined by

$$(Kv)(x') = r^+ P(v(x') \otimes \delta(x_n)) \quad (10.115)$$

is a Poisson operator of order $d + 1$.

Theorem 10.25. *Let* $P = \mathrm{OP}(p(x, y, \xi))$ *be of order* d, *and define* K *by* (10.115). *Then* K *is a Poisson operator of order* $d + 1$. *The symbol-kernel* $\tilde{k}$ *and symbol* k *satisfy (with* $\tilde{p} = \mathscr{F}_{\xi_n \to z}^{-1} p$*):*

$$\begin{aligned}\tilde{k}(x', y', x_n, \xi') &= r^+ \tilde{p}(x', 0, y', 0, \xi', z)|_{z=x_n}, \\ k(x', y', \xi) &= h^+ p(x', 0, y', 0, \xi), \text{ if } p \text{ is independent of } x_n;\end{aligned} \quad (10.116)$$

$$\tilde{k}(x',y',x_n,\xi') \sim r^+ \sum_{j\in\mathbb{N}_0} \tfrac{1}{j!} x_n^j \partial_{x_n}^j \tilde{p}(x',0,y',0,\xi',z)|_{z=x_n},$$

$$k(x',y',\xi) \sim \sum_{j\in\mathbb{N}_0} \tfrac{1}{j!} h^+ \big(\overline{D}_{\xi_n}^j \partial_{x_n}^j p(x',0,y',0,\xi) \big) \quad \text{in general.}$$

If P is a differential operator, $K = 0$.

Proof. The last statement is obvious since $P(v(x') \otimes \delta(x_n))$ is supported in $\{x_n = 0\}$ when P is a differential operator.

To show the formula, let first p be independent of x_n. Then

$$r^+ \mathrm{OP}(p)(v \otimes \delta) = r^+ \int_{\mathbb{R}^{2n}} e^{i(x-y)\cdot\xi} p(x',y',y_n,\xi) v(y') \delta(y_n)\, dy d\xi$$

$$= r^+ \int_{\mathbb{R}^{2n-1}} e^{i(x'-y')\cdot\xi' + i x_n \xi_n} p(x',y',0,\xi) v(y')\, dy' d\xi$$

$$= \int_{\mathbb{R}^{2n-2}} e^{i(x'-y')\cdot\xi'} r^+ \tilde{p}(x',y',0,\xi',x_n) v(y')\, dy' d\xi'$$

$$= \mathrm{OPK}(\tilde{k}(x',y',x_n,\xi'))v,$$

where $\tilde{k} = r^+ \tilde{p} = r^+ \mathscr{F}_{\xi_n \to x_n}^{-1} p = r^+ \mathscr{F}_{\xi_n \to x_n}^{-1}(h^+ p)$; the corresponding symbol is $k = h^+ p$. In view of Theorem 10.21, $\tilde{k}(x',y',x_n,\xi')$ is a Poisson symbol-kernel of degree d (order $d+1$). (The application of $\delta(y_n)$ in the first line can be justified by writing it as the limit of an approximate identity.)

When p depends on x_n, it is natural to do a Taylor expansion of p in x_n and apply the preceding result to each term; this gives the third formula in (10.116), except that we have not yet accounted for the justification of the asymptotic series. But this is easy on the symbol level: We can use Theorem 7.13 1° with respect to the variables (x_n, y_n) to replace p by an equivalent symbol p' depending on y_n instead of (x_n, y_n); it has the form

$$p'(x',y',y_n,\xi) \sim \sum_{j\in\mathbb{N}_0} \tfrac{1}{j!} \overline{D}_{\xi_n}^j \partial_{x_n}^j p(x',x_n,y',y_n,\xi)\big|_{x_n=y_n},$$

and is likewise of order d and satisfies the transmission condition at $y_n = 0$. Then we apply the second formula in (10.116) to this symbol, obtaining the fourth formula, and the third formula follows by inverse Fourier transformation in ξ_n.   □

We finally treat rule (ii) in (10.44) for the leftover operator $L(P,P')$. Assume that P and P' are given by symbols $p(X,\xi)$, $p'(X,\xi)$, where $X = x'$, y' or (x',y'), so they are independent of x_n (or y_n). The decomposition of $L(P,P')$ that we shall now explain, was first introduced in [G84]. (In that paper it was useful in obtaining the first complete treatment of $L(P,P')$ in the general case with x_n-dependent symbols.)

According to (10.18) we can write

$$P = \sum_{0 \le l \le d} S_l D_n^l + Q, \quad P' = \sum_{0 \le l \le d'} S_l' D_n^l + Q', \tag{10.117}$$

where the S_l and S_l' are differential operators on $\mathbb{R}^{n-1}$ (with symbols polynomial in ξ') and Q and Q' have symbols that are $O(\langle \xi_n \rangle^{-1})$. One finds the following rules: Since differential operators are local,

$$L(\sum_{0 \le l \le d} S_l D_n^l, P') = 0. \tag{10.118}$$

By Green's formula (10.82),

$$L(P, \sum_{0 \le l \le d'} S_l' D_n^l) u = r^+ \sum_{l \le d'} P S_l' (D_n^l e^+ u - e^+ D_n^l u)$$
$$= -i r^+ P \sum_{1 \le l \le d'} S_l' \sum_{k < l} (\gamma_{l-k-1} u \otimes D_n^k \delta) = \sum_{m \le d'-1} K_m \gamma_m u, \tag{10.119}$$

where K_m is the Poisson operator of order $d + d' - m$ (as in Theorem 10.25),

$$K_m v = -i \sum_{l=m+1}^{d'} r^+ P S_l' D_{x_n}^{l-1-m} (v(x') \otimes \delta(x_n)); \tag{10.120}$$

so (10.119) defines an s.g.o. of class d'. For the analysis of the remaining term we introduce the reflection operator J,

$$J : u(x', x_n) \mapsto u(x', -x_n), \tag{10.121}$$

it sends spaces over $\mathbb{R}_+^n$ into spaces over $\mathbb{R}_-^n$. Then we can write, for $u \in C_{(0)}^\infty(\overline{\mathbb{R}}_+^n)$:

$$L(Q, Q') u = r^+ Q Q' e^+ u - r^+ Q e^+ r^+ Q' e^+ u = r^+ Q(I - e^+ r^+) Q' e^+ u$$
$$= r^+ Q e^- r^- Q' e^+ u = (r^+ Q e^- J)(J r^- Q' e^+) u$$
$$= G^+(Q) G^-(Q') u, \text{ with} \tag{10.122}$$
$$G^+(Q) = r^+ Q e^- J = r^+ P e^- J = G^+(P),$$
$$G^-(Q') = J r^- Q' e^+ = J r^- P' e^+ = G^-(P') = [G^+(P'^*)]^*.$$

We shall show that the latter are singular Green operators of class 0 and orders d resp. d'. Similar formulas hold on the boundary symbol level. Altogether,

$$L(P, P') = \sum_{0 \le m < d'} K_m \gamma_m + G^+(Q) G^-(Q'). \tag{10.123}$$

The resulting symbols will now be analyzed on the one-dimensional level.

Theorem 10.26. *Let $d \in \mathbb{Z}$ and $d' \in \mathbb{N}_0$, let $p(X, \xi) \in S_{1,0}^d(\Xi, \mathbb{R}^n)$ and let $s_l'(X, \xi')$ be polynomial in ξ' of degree $d' - l$ for $0 \le l \le d'$.*

$1°$ One has for the singular Green symbol resulting from the formation of $L(p(X, 0, \xi', D_n), \sum_{0 \le l \le d'} s_l'(X, 0, \xi') D_n^l)$,

$$L(p, \sum s'_l \xi^l_n) = \sum_{0 \le m < d'} k_m(X, \xi) \eta^m_n$$
$$\in S_{1,0}^{d+d'-1}(\Xi, \mathbb{R}^{n-1}, \mathcal{H}^+ \hat{\otimes} \mathcal{H}^-_{d'-1}), \quad (10.124)$$

where the k_m are Poisson symbols

$$k_m(X, \xi) = -ih^+ \left(\sum_{l=m+1}^{d'} p(X, \xi) s'_l(X, \xi') \xi^{l-1-m}_n \right)$$
$$\in S_{1,0}^{d+d'-1-m}(\Xi, \mathbb{R}^{n-1}, \mathcal{H}^+). \quad (10.125)$$

2° For the operators introduced in (10.122) one has

$$g^+(p) = r^+ p(X, 0, \xi', D_n) e^- J, \quad g^-(p) = J r^- p(X, 0, \xi', D_n) e^+, \quad (10.126)$$

are singular Green operators with symbol-kernels (where $\tilde{p} = \mathcal{F}^{-1}_{\xi_n \to z_n} p$), resp. symbols:

$$\tilde{g}^+(p)(X, x_n, y_n, \xi') = r^+ \tilde{p}(X, \xi', z_n)|_{z_n = x_n + y_n} \quad and$$
$$\tilde{g}^-(p)(X, x_n, y_n, \xi') = r^- \tilde{p}(X, \xi', z_n)|_{z_n = -x_n - y_n}$$
$$\in S_{1,0}^{d-1}(\Xi, \mathbb{R}^{n-1}, \mathscr{S}_{++}), \quad (10.127)$$
$$g^+(p)(X, \xi, \eta_n) = \mathcal{F}_{x_n \to \xi_n} \overline{\mathcal{F}}_{y_n \to \eta_n} e^+_{x_n} e^+_{y_n} \tilde{g}^+(p) \quad and$$
$$g^-(p)(X, \xi, \eta_n) = \mathcal{F}_{x_n \to \xi_n} \overline{\mathcal{F}}_{y_n \to \eta_n} e^+_{x_n} e^+_{y_n} \tilde{g}^-(p)$$
$$\in S_{1,0}^{d-1}(\Xi, \mathbb{R}^{n-1}, \mathcal{H}^+ \hat{\otimes} \mathcal{H}^-_{-1}),$$

of order d and class 0.

3° If $p(X, \xi', D_n)$ is a differential operator, these symbols vanish.

Proof. 1°. The statement on k_m follows from Theorem 10.25 1°. The resulting symbol (10.124) is an s.g.o. symbol of the stated type in view of Definition 10.23. Note that only $h^+ p$ enters (cf. also (10.118)).

2°. Denote $h^\pm p = p^\pm$, and omit the variable X. We have, reading the integrals as Fourier transforms and using that the distribution kernel of p equals $\tilde{p}(x_n - y_n)$ where $\tilde{p} \in \mathscr{S}(\mathbb{R})$,

$$r^+ p(\xi', D_n) e^- J u = r^+ \int e^{ix_n \xi_n} p(\xi', \xi_n) \int_{-\infty}^0 e^{-iy_n \xi_n} u(-y_n) \, dy_n d\xi_n$$
$$= r^+ \int e^{ix_n \xi_n} p(\xi', \xi_n) \int_0^\infty e^{iy_n \xi_n} u(y_n) \, dy_n d\xi_n$$
$$= r^+ \int_0^\infty \tilde{p}(\xi', x_n + y_n) u(y_n) \, dy_n = r^+ \int_0^\infty \tilde{p}^+(\xi', x_n + y_n) u(y_n) \, dy_n,$$

since $\tilde{p}(\xi', z_n) = \tilde{p}^+(\xi', z_n)$ for $z_n > 0$. Thus $g^+(p)$ is the integral operator on $\mathbb{R}_+$ with kernel

$$\tilde{g}^+(p)(x_n, y_n, \xi') = r^+ \tilde{p}(\xi', x_n + y_n) = \tilde{p}^+(\xi', x_n + y_n).$$

We have from Theorem 10.21 that $\tilde{p}^+ \in S^d_{1,0}(\Xi, \mathbb{R}^{n-1}, \mathscr{S}_+)$. The kernel is estimated by use of coordinate changes $z = x_n - y_n$, $w = x_n + y_n$:

$$\|\tilde{g}^+(p)(x_n, y_n, \xi')\|^2_{L_{2,x_n,y_n}(\mathbb{R}^2_{++})} = \int_0^\infty \int_0^\infty |\tilde{p}^+(\xi', x_n + y_n)|^2 dx_n dy_n$$

$$= \tfrac{1}{2} \int_0^\infty dw \int_{-w}^w |\tilde{p}^+(\xi', w)|^2 dz dw = \int_0^\infty w |\tilde{p}^+(\xi', w)|^2 dw$$

$$\leq \|w\tilde{p}^+(w, \xi)\| \|\tilde{p}^+(\xi', w)\| \leq c\langle\xi'\rangle^{d-\frac{1}{2}} \langle\xi'\rangle^{d+\frac{1}{2}} = c\langle\xi'\rangle^{2d}.$$

This is the basic estimate, which easily generalizes to derivatives in X and ξ', and to lower-order parts. For the symbol-kernels resulting from application of $x_n^k D_{x_n}^{k'} y_n^m D_{y_n}^{m'}$, we observe that for $w = x_n + y_n$,

$$x_n^k y_n^m \leq w^{k+m} \quad \text{when} \quad x_n, y_n \geq 0,$$

and

$$D_{x_n}^{k'} D_{y_n}^{m'} \tilde{p}(\xi', x_n + y_n) = D_w^{k'+m'} \tilde{p}(\xi', w),$$

so that

$$\|x_n^k D_{x_n}^{k'} y_n^m D_{y_n}^{m'} \tilde{g}^+(p)\|^2_{L_{2,x_n,y_n}(\mathbb{R}^2_{++})}$$

$$\leq \int_0^\infty w^{2k+2m+1} |D_w^{k'+m'} \tilde{p}^+(\xi', w)|^2 dw$$

$$= \int_0^\infty w |w^{k+m} D_w^{k'+m'} \tilde{p}^+(\xi', w)|^2 dw \leq c\langle\xi'\rangle^{2(d-k-m+k'+m')}.$$

It is altogether seen that $\tilde{g}^+(p)$ satisfies all the estimates required of a symbol-kernel in $S^d_{1,0}(\Xi, \mathbb{R}^{n-1}, \mathscr{S}_{++})$. The proof for $\tilde{g}^-(p)$ is similar, based on the identities

$$Jr^- pe^+ u = Jr^- \int_{\mathbb{R}} e^{ix_n\xi_n} p(\xi', \xi_n) \int_0^\infty e^{-iy_n\xi_n} u(y_n)\, dy_n d\xi_n$$

$$= r^+ \int_{\mathbb{R}} e^{-ix_n\xi_n} p(\xi', \xi_n) \int_0^\infty e^{-iy_n\xi_n} u(y_n)\, dy_n d\xi_n$$

$$= r^+ \int_0^\infty \tilde{p}(\xi', -x_n - y_n) u(y_n)\, dy_n.$$

For 3°, recall (10.118). □

From the above analysis we can conclude:

Corollary 10.27. *Let p and p' be as in Theorem 10.24 and write*

$$p' = \sum_{0 \leq j \leq d'} s'_j(X, \xi')\xi_n^j + h_{-1}p'.$$

Then $L(p, p') = (p(D_n)p'(D_n))_+ - p(D_n)_+ \circ_n p'(D_n)_+$ *is a singular Green boundary symbol operator of order* $d+d'$ *and class* $\max\{d', 0\}$. *More precisely, the symbol is defined by*

$$L(p, p')(X, \xi, \eta_n) = \sum_{0 \le m < d'} k_m(X, \xi)\eta_n^m + g^+(p) \circ_n g^-(p'), \qquad (10.128)$$

where k_m, $g^+(p)$ *and* $g^-(p')$ *are as defined in Theorem* 10.26. *In particular,* $L(p, p')$ *depends only on* $h^+ p$ *and* $h^- p'$.

The full composition rules, where the action in x' is also taken into account, look as follows when the ψdo symbols are independent of the normal variable:

Theorem 10.28. *Let symbols be given as in Theorem* 10.24, *with* $\Xi = \mathbb{R}^{n-1}$ *with points* x'. *Let* $a(x', \xi', D_n)$ *stand for the boundary symbol operator, and let* A *stand for the full operator* $\mathrm{OP}'(a(x', \xi', D_n))$ *defined from one of the symbols* $a = p, t, k, g$ *or* s, *and let* $a'(x', \xi', D_n)$ *and* A' *be similarly defined from the primed symbols. (One can also consider symbols in* (x', y')-*form, in which case one can begin by reducing them to* x'-*form, by Theorem* 7.13 1° *applied in the tangential variables.)*
    1° *Consider one of the compositions*

$$A'' = AA' = \mathrm{OP}'(a)\,\mathrm{OP}'(a')$$

listed in (10.44) (iii)–(vi), *with one factor properly supported w.r.t.* (x', y'). *Here* $A'' \sim \mathrm{OP}'(a'')$, *where* a'' *has the asymptotic expansion*

$$a''(x', \xi', D_n) \sim \sum_{\alpha \in \mathbb{N}_0^{n-1}} \tfrac{1}{\alpha!} D_{\xi'}^\alpha a(x', \xi', D_n) \circ_n \partial_{x'}^\alpha a'(x', \xi', D_n), \qquad (10.129)$$

each term being determined by the appropriate composition rule in Theorem 10.24. *The expansion holds in the space of operators and symbols of order* $d + d'$, *and class* r' *resp.* $\max\{r + d', 0\}$ *(in the relevant cases).*
    2° *Consider the singular Green operator*

$$G = L(P, P') \equiv (PP')_+ - P_+ P'_+$$

derived from P *and* P', *with one of the operators properly supported w.r.t.* (x', y'). *Here* $G = \mathrm{OP}'(g)$, *where*

$$g(x', \xi', D_n) \sim \sum_{\alpha \in \mathbb{N}_0^{n-1}} \tfrac{1}{\alpha!} L(D_{\xi'}^\alpha p(x, \xi', D_n), \partial_{x'}^\alpha p'(x, \xi', D_n)), \qquad (10.130)$$

with the terms defined by Corollary 10.27. *The expansion holds in the space of operators and symbols of order* $d + d'$ *and class* $\max\{d', 0\}$.

Proof. 1° If a' were in y'-form, the resulting operator would simply have the boundary symbol operator in (x', y')-form

$$a(x', \xi', D_n) \circ_n a'(y', \xi', D_n).$$

This can be reduced to x'-form as in the proof of Theorem 7.13 1°. The procedure of replacing a' by its y'-form and reducing the resulting product to x'-form gives altogether the formula (10.129). The proof of 2° is similar.

□

When the symbol of P (or P') moreover depends on x_n, one considers each term in its Taylor expansion at $x_n = 0$ separately:

$$p(x', x_n, \xi) \text{ has the expansion } \sum_{j \geq 0} \tfrac{1}{j!} x_n^j \partial_{x_n}^j p(x', 0, \xi); \tag{10.131}$$

the $\partial_{x_n}^j p$ satisfy the transmission condition. In the compositions in (10.44) (ii)–(v), a factor x_n^j decreases the order by j steps (cf. Proposition 10.10 (ix)), so it is indeed possible to collect the resulting symbols as expansions in homogeneous terms of decreasing order. The resulting formulas can e.g. be found in [G96], Section 2.7 (disregard the parameter).

10.5 Continuity

For the proof of continuity properties in Sobolev spaces it will be practical to introduce spaces of different order in the normal and the tangential direction, as in [H63]:

$$H^{s,t}(\mathbb{R}^n) = \{u \in \mathscr{S}'(\mathbb{R}^n) \mid \langle \xi \rangle^s \langle \xi' \rangle^t \hat{u}(\xi) \in L_2(\mathbb{R}^n)\} \text{ with norm}$$
$$\|u\|_{s,t} = (2\pi)^{-\frac{n}{2}} \|\langle \xi \rangle^s \langle \xi' \rangle^t \hat{u}(\xi)\|_{L_2(\mathbb{R}^n)}, \tag{10.132}$$

for $s, t \in \mathbb{R}$. Note that

$$\|u\|_{s,t} \leq \|u\|_{s',t'} \text{ when } s \leq s', \, s + t \leq s' + t'. \tag{10.133}$$

From these spaces we can define the related Hilbert spaces over the half-space:

$$H^{s,t}(\mathbb{R}_+^n) = \{u \in \mathscr{D}'(\mathbb{R}_+^n) \mid u = U|_{\mathbb{R}_+^n} \text{ for some } U \in H^{s,t}(\mathbb{R}^n)\} \text{ with norm}$$
$$\|u\|_{s,t} = \inf_{\text{such } U} \|U\|_{s,t},$$
$$H_0^{s,t}(\overline{\mathbb{R}}_+^n) = \{u \in H^{s,t}(\mathbb{R}^n) \mid \operatorname{supp} u \subset \overline{\mathbb{R}}_+^n\}, \tag{10.134}$$

as mentioned for the case $t = 0$ in (9.18), (9.25). It is shown in [H63, Sect. 2.5] that for all s, t, $H^{s,t}(\mathbb{R}_+^n)$ and $H_0^{-s,-t}(\overline{\mathbb{R}}_+^n)$ are dual spaces of one another, with a duality extending the scalar product in $L_2(\mathbb{R}_+^n)$. $C_{(0)}^\infty(\overline{\mathbb{R}}_+^n)$ is dense in each space $H^{s,t}(\mathbb{R}_+^n)$, and $C_0^\infty(\mathbb{R}_+^n)$ is dense in each space $H_0^{s,t}(\overline{\mathbb{R}}_+^n)$. For $s = m \in \mathbb{N}_0$, one has a more elementary characterization of $H^{m,t}(\mathbb{R}_+^n)$ and

$H_0^{m,t}(\overline{\mathbb{R}}_+^n)$, as the closures of $C_{(0)}^\infty(\overline{\mathbb{R}}_+^n)$ resp. $C_0^\infty(\mathbb{R}_+^n)$ with respect to the norm

$$\|u\|_{m,t,\prime} = \Big((2\pi)^{1-n} \sum_{j=0}^m \|\langle\xi'\rangle^{(t+m-j)} D_{x_n}^j \acute{u}(\xi',x_n)\|_{L_2(\mathbb{R}_+^n)}^2\Big)^{\frac{1}{2}}$$

$$= \Big(\sum_{j=0}^m \|D_{x_n}^j u(x',x_n)\|_{0,t+m-j}^2\Big)^{\frac{1}{2}} \simeq \Big(\sum_{|\alpha|\leq m} \|D^\alpha u\|_{0,t}^2\Big)^{\frac{1}{2}},$$

(10.135)

which is just a slight generalization of (9.6). We shall give proof details for this case, and refer to e.g. [G96] for the documentation that the various properties extend to suitable cases with noninteger first exponent.

For $m = 0$ it is easy to see (using the method of Theorem 6.15 in the tangential variables x') that

$$H^{0,t}(\mathbb{R}_+^n) \text{ and } H^{0,-t}(\mathbb{R}_+^n) \text{ are dual spaces of one another.} \qquad (10.136)$$

In the following, we leave out the indications $\wedge$ and $\prime$ in the norms.

Theorem 10.29. 1° *Let K be a Poisson operator of order $d \in \mathbb{R}$, with symbol-kernel $\tilde{k}(x',y',x_n,\xi') \in S_{1,0}^{d-1}(\mathbb{R}^{2(n-1)}, \mathbb{R}^{n-1}, \mathscr{S}_+)$, compactly supported with respect to x' and y'. For all $m \in \mathbb{Z}$, all $t \in \mathbb{R}$, there are estimates for $v \in \mathscr{S}(\mathbb{R}^{n-1})$:*

$$\|Kv\|_{m,t-d+\frac{1}{2}} \leq c\|v\|_{m+t}; \text{ in particular } \|Kv\|_m \leq c\|v\|_{m+d-\frac{1}{2}}. \qquad (10.137)$$

2° *Let T be a trace operator of class 0 and order d, with symbol-kernel $\tilde{t}(x',y',x_n,\xi') \in S_{1,0}^d(\mathbb{R}^{2(n-1)}, \mathbb{R}^{n-1}, \mathscr{S}_+)$, compactly supported with respect to x' and y'. Then T is the adjoint of the Poisson operator K with symbol-kernel*

$$\tilde{k}(x',y',x_n,\xi') = \overline{\tilde{t}}(y',x',x_n,\xi'); \qquad (10.138)$$

in the sense that

$$(Tu,v)_{L_2(\mathbb{R}^{n-1})} = (u,Kv)_{L_2(\mathbb{R}_+^n)} \text{ for } u \in \mathscr{S}(\overline{\mathbb{R}}_+^n), v \in \mathscr{S}(\mathbb{R}^{n-1}). \qquad (10.139)$$

All trace operators of order d and class 0 (with symbol-kernel compactly supported in x',y') are obtained from Poisson operators of order $d + 1$ (with symbol-kernel compactly supported in x',y') in this way, and vice versa.

For all $m \in \mathbb{N}_0$, all $t \in \mathbb{R}$, there are estimates for $u \in \mathscr{S}(\overline{\mathbb{R}}_+^n)$:

$$\|Tu\|_{m+t-d-\frac{1}{2}} \leq c\|u\|_{0,m+t} \leq c\|u\|_{m,t}, \text{ in particular } \|Tu\|_{m-d-\frac{1}{2}} \leq c\|u\|_m.$$

(10.140)

3° *By extension by continuity, the operators define mappings from the full spaces, satisfying the estimates in (10.137)–(10.140). The estimates (10.137) are also valid with m replaced by any $s \in \mathbb{R}$, and the estimates (10.140) are valid with m replaced by any $s \in \overline{\mathbb{R}}_+$. In 2°,*

$T : L_2(\mathbb{R}^n_+) \to H^{-d-\frac{1}{2}}(\mathbb{R}^{n-1})$ and $K : H^{d+\frac{1}{2}}(\mathbb{R}^{n-1}) \to L_2(\mathbb{R}^n_+)$ are adjoints.
$$(10.141)$$

Proof. We have for $v \in \mathscr{S}(\mathbb{R}^{n-1})$, $u \in \mathscr{S}(\overline{\mathbb{R}}^n_+)$:

$$(Kv, u)_{L_2(\mathbb{R}^n_+)} = \int e^{i(x'-y')\cdot\xi'} \tilde{k}(x', y', x_n, \xi') v(y') \bar{u}(x', x_n) \, dy' d\xi' dx' dx_n$$
$$(10.142)$$

(oscillatory integral w.r.t. ξ'). Here we can insert the Fourier transforms of v w.r.t. y' and u w.r.t. x', and reinterpret the integrations against the exponential functions to Fourier transforms w.r.t. x' and y' of $\tilde{k}$, just as in the proof of Theorem 7.5, obtaining the expression

$$\int \tilde{k}(\widehat{\theta' - \xi'}, \widehat{\xi' - \eta'}, x_n, \xi') \hat{v}(\eta') \overline{\hat{u}}(\theta', x_n) \, d\xi' d\eta' d\theta' dx_n.$$

The $\tilde{k}$ factor is estimated with respect to the primed variables similarly as in the proof of Theorem 7.5, when we consider it as a function on $\mathbb{R}^{3(n-1)}$ valued in $L_{2,x_n}(\mathbb{R}_+)$. This gives

$$\|\tilde{k}(\widehat{\theta'}, \widehat{\eta'}, x_n, \xi')\|_{L_2(\mathbb{R}_+)} \le M_N \langle \xi' \rangle^{d-\frac{1}{2}} \langle \theta' \rangle^{-2N} \langle \eta' \rangle^{-2N},$$

where N can be taken arbitrarily large. Hence, using the Cauchy-Schwarz inequality w.r.t. x_n,

$$|(Kv, u)| \le$$
$$c \int \langle \theta' - \xi' \rangle^{-2N} \langle \xi' - \eta' \rangle^{-2N} \langle \xi' \rangle^{d-\frac{1}{2}} |\hat{v}(\eta')| \|\hat{u}(\theta', x_n)\|_{L_2(\mathbb{R}_+)} \, d\xi' d\eta' d\theta'.$$

Finally, using the Peetre inequality as in Theorem 7.5 and the Cauchy-Schwarz inequality for the integrals over $\mathbb{R}^{3(n-1)}$, followed by integrating out superfluous variables, we arrive at

$$|(Kv, u)| \le c' \|v\|_s \|u\|_{0,d-\frac{1}{2}-s}.$$

Since this holds for all u, we conclude in view of (10.136) that

$$\|Kv\|_{0,s-d+\frac{1}{2}} \le c' \|v\|_s.$$

This shows (10.137) with $m = 0$. For $m \le 0$, it follows immediately in view of (10.133):

$$\|Kv\|_{m,s-d+\frac{1}{2}} \le \|Kv\|_{0,m+s-d+\frac{1}{2}} \le c' \|v\|_{m+s}, \text{ for } m \le 0.$$

To show it for $m > 0$, note that $D^\alpha K$ is a Poisson operator of order $d + |\alpha|$, so

$$\|Kv\|_{m,s-d+\frac{1}{2}} \leq c\big(\sum_{|\alpha|\leq m}\|D^\alpha Kv\|_{0,s-d+\frac{1}{2}}\big)^{\frac{1}{2}}$$

$$\leq c'\big(\sum_{|\alpha|\leq m}\|v\|_{s+|\alpha|}\big)^{\frac{1}{2}} \leq c''\|v\|_{s+m}.$$

This implies (10.137).

For $2°$, the formula (10.139) for T and K with symbol-kernels as in (10.138) follows by changing the order of integration in (10.142). Then we have by the preceding proof:

$$|(Tu,v)| = |(u,Kv)| \leq c\|u\|_{0,d+\frac{1}{2}-s}\|v\|_s,$$

for any s, hence with $t = d + \frac{1}{2} - s$,

$$\|Tu\|_{t-d-\frac{1}{2}} \leq c\|u\|_{0,t}.$$

Replacing t by $m + t$ with $m \geq 0$, we find moreover

$$\|Tu\|_{m+t-d-\frac{1}{2}} \leq c\|u\|_{0,m+t} \leq c\|u\|_{m,t},$$

in view of (10.133). This implies (10.140).

$3°$ The extension by continuity is obvious. The validity for general s, as indicated, can be deduced from the preceding statements by interpolation; this is a technique explained e.g. in [LM68] and we refrain from giving details here.   □

Corollary 10.30. *When* $T = \sum_{0\leq j<r} S_j\gamma_j + T'$ *is a trace operator of class* r *and order* d*, with symbol compactly supported with respect to* x' *and* y'*, and* $s > r - \frac{1}{2}$*,* $s \geq 0$*,* $t \in \mathbb{R}$*, then* T *defines a continuous mapping:*

$$T: H^{s,t}(\mathbb{R}^n_+) \to H^{s+t-d-\frac{1}{2}}(\mathbb{R}^{n-1}). \tag{10.143}$$

Proof. For the part T' of class 0 this follows from Theorem 10.29. For the terms $S_j\gamma_j$, it follows by a straightforward generalization of the estimate for γ_0 shown in Remark 9.4: For $u \in \mathscr{S}(\mathbb{R}^n)$, $s > j + \frac{1}{2}$,

$$\|D_n^j u(x',0)\|^2_{s+t-j-\frac{1}{2}} = \int_{\mathbb{R}^{-1}} |D_n^j \hat{u}(\xi',0)|^2 \langle\xi'\rangle^{2s+2t-2j-1}\,d\xi'$$

$$\leq c\int_{\mathbb{R}^{n-1}} \langle\xi'\rangle^{2s+2t-2j-1}\Big(\int_{\mathbb{R}} |\xi_n^j \hat{u}(\xi)|\,d\xi_n\Big)^2 d\xi'$$

$$\leq c\int_{\mathbb{R}^{n-1}} \langle\xi'\rangle^{2s+2t-2j-1}\Big(\int_{\mathbb{R}} |\hat{u}(\xi)|^2\langle\xi\rangle^{2s}\,d\xi_n\Big)\Big(\int_{\mathbb{R}} \langle\xi\rangle^{2j-2s}\,d\xi_n\Big) d\xi'$$

$$= c'\int_{\mathbb{R}^n} \langle\xi'\rangle^{2s+2t-2j-1+2j-2s+1}\langle\xi\rangle^{2s}|\hat{u}(\xi)|^2\,d\xi = c'\|u\|^2_{s,t};$$

here we used (9.19). This implies the boundedness of the mapping γ_j on the dense subset $\mathscr{S}(\overline{\mathbb{R}}^n_+)$, and it extends by continuity to a bounded mapping from $H^{s,t}(\mathbb{R}^n_+)$ to $H^{s+t-j-\frac{1}{2}}(\mathbb{R}^{n-1})$. This is then combined with the fact that S_j has order $d-j$. $\square$

The operators γ_j do not have adjoints within the calculus, hence neither do trace operators of class $r > 0$.

We can also treat s.g.o.s.

Theorem 10.31. *Let* $G = \sum_{0 \le j < r} K_j \gamma_j + G'$ *be a singular Green operator of class r and order d, with symbol compactly supported with respect to x' and y'. When $s > r - \frac{1}{2}$, $s \ge 0$, $t \in \mathbb{R}$, there are estimates for $u \in \mathscr{S}(\overline{\mathbb{R}}^n_+)$:*

$$\|Gu\|_{s-d,t} \le c\|u\|_{s,t},$$

and hence G defines continuous mappings:

$$G : H^{s,t}(\mathbb{R}^n_+) \to H^{s-d,t}(\mathbb{R}^n_+), \ \text{in particular } G : H^s(\mathbb{R}^n_+) \to H^{s-d}(\mathbb{R}^n_+).$$
$$(10.144)$$

It is also continuous from $H^{0,t}(\mathbb{R}^n_+) \to H^{0,t-d}(\mathbb{R}^n_+)$, all $t \in \mathbb{R}$.

Proof. For the sum over j this is a consequence of the mapping properties shown in Theorem 10.29 and Corollary 10.30. For the part G' of class 0, one can either appeal to the fact that it is a rapidly convergent series of compositions of Poisson and trace operators of class 0 (one should then account for how the norms depend on the enumeration), or one can work out a proof similarly as in Theorem 10.29: For $u, v \in \mathscr{S}(\overline{\mathbb{R}}^n_+)$,

$$(G'u,v)_{L_2(\mathbb{R}^n_+)} \qquad\qquad\qquad\qquad\qquad\qquad\qquad\qquad\qquad (10.145)$$

$$= \int e^{i(x'-y')\cdot\xi'} \tilde{g}'(x',y',x_n,y_n,\xi')u(y',y_n)\bar{v}(x',x_n)\,dy'd\xi'dy_ndx'dx_n$$

$$= \int \tilde{g}'(\widehat{\theta'-\xi'},\widehat{\xi'-\eta'},x_n,y_n,\xi')\acute{u}(\eta',y_n)\bar{\acute{v}}(\theta',x_n)\,d\xi'd\eta'd\theta'dx_ndy_n.$$

One finds from the symbol-kernel estimates satisfied by $\tilde{g}'$ that

$$\|\tilde{g}'(\widehat{\theta'},\widehat{\eta'},x_n,y_n,\xi')\|_{L_{2,x_n,y_n}(\mathbb{R}^2_{++})} \le M_N \langle\xi'\rangle^d \langle\theta'\rangle^{-2N} \langle\eta'\rangle^{-2N},$$

any N. Applications of the inequality

$$\int_{\mathbb{R}^2_{++}} |g(x_n,y_n)\varphi(x_n)\psi(y_n)|\,dx_ndy_n \le \|g\|_{L_2(\mathbb{R}^2_{++})}\|\varphi\|_{L_2(\mathbb{R}_+)}\|\psi\|_{L_2(\mathbb{R}_+)}$$

give, with norms in $L_2(\mathbb{R}_+)$,

$$|(G'u,v)| \le$$

$$c\int \langle\theta'-\xi'\rangle^{-2N} \langle\xi'-\eta'\rangle^{-2N} \langle\xi'\rangle^d \|\acute{u}(\eta',y_n)\| \|\acute{v}(\theta',x_n)\|\,d\xi'd\eta'd\theta',$$

from which one finds (as in Theorems 7.5 and 10.29)

$$|(G'u, v)| \leq c' \|u\|_{0,s} \|v\|_{0,d-s},$$

allowing the conclusion

$$\|G'u\|_{0,s-d} \leq c' \|u\|_{0,s}, \text{ any } s \in \mathbb{R}; \qquad (10.146)$$

which implies the last statement in the theorem. Let $s \geq 0$. We can write $s - d = m + s'$ with $m \in \mathbb{N}_0$ and $s' < 0$, and apply (10.146) to the operators $D^\alpha G'$ of order $d + |\alpha|$ and class 0. This gives in view of (10.133):

$$\|G'u\|_{s-d,t} = \|G'u\|_{m+s',t} \leq \|G'u\|_{m,s'+t} \leq c\big(\textstyle\sum_{|\alpha|\leq m} \|D^\alpha G'u\|_{0,s'+t}^2\big)^{\frac{1}{2}}$$

$$\leq c'\big(\textstyle\sum_{|\alpha|\leq m} \|u\|_{0,s'+d+|\alpha|+t}^2\big)^{\frac{1}{2}} \leq c'' \|u\|_{0,s'+d+m+t} = c'' \|u\|_{0,s} \leq c'' \|u\|_{s,t},$$

showing the continuity in (10.144). $\qquad\qquad\qquad\qquad\qquad\qquad \square$

For ψdo's satisfying the transmission condition, we have:

Theorem 10.32. *Let* $P = \mathrm{OP}(p(x', y', \xi))$ *where* p *is a* ψ*do symbol of order* $d \in \mathbb{Z}$ *satisfying the transmission condition and compactly supported in* (x', y')*. Then for all* $m \in \mathbb{N}_0$*,* $t \in \mathbb{R}$*, there are estimates for* $u \in \mathscr{S}(\overline{\mathbb{R}}_+^n)$*:*

$$\|P_+u\|_{m-d,t} \leq c \|u\|_{m,t}, \text{ in particular } \|P_+u\|_{m-d} \leq c \|u\|_m, \qquad (10.147)$$

extending by continuity to $u \in H^{m,t}(\mathbb{R}_+^n)$ *resp.* $H^m(\mathbb{R}_+^n)$*.*

Proof. As noted in (10.8), we can write p as the sum of a symbol $\sum_{0\leq l\leq d} s_l(x', y', \xi')\xi_n^l$ where the s_l are polynomials in ξ', and a symbol p' which satisfies

$$|D_{x'}^\beta D_\xi^\alpha p'(x', \xi)| \leq c \langle\xi'\rangle^{d-|\alpha|+1} \langle\xi\rangle^{-1}.$$

In the following we only need the slightly weaker estimates

$$|D_{x'}^\beta D_\xi^\alpha p'(x', \xi)| \leq c \langle\xi'\rangle^{d-|\alpha|};$$

they would allow us to include in p' the term that is constant in ξ_n, but this makes no difference for the following. The first part of p reduces to x'-form by a finite and exact version of (7.38); this gives a differential operator of order d, which is known to satisfy (10.147) (by considerations as in Chapter 6). For the other part $P' = \mathrm{OP}(p')$, we can apply the argumentation in the proof of Theorem 7.5 with respect to the tangential variables only; this gives for $w \in \mathscr{S}(\mathbb{R}^n)$, any $r \in \mathbb{R}$:

$$\|P'w\|_{0,r-d} \leq c \|w\|_{0,r},$$

which extends to $w \in H^{0,r}(\mathbb{R}^n)$, and hence implies for u in $H^{0,r}(\mathbb{R}_+^n)$ (identifiable with a closed subspace of $H^{0,r}(\mathbb{R}^n)$ by extension by zero),

$$\|P'_+ u\|_{0,r-d} \le c\|u\|_{0,r}.$$

If $m \le d$, this implies (cf. (10.133))

$$\|P'_+ u\|_{m-d,t} \le \|P'_+ u\|_{0,m-d+t} \le c\|u\|_{0,m+t} \le c\|u\|_{m,t},$$

showing the desired estimate for P' and hence also for P, when $m \le d$.

Now let $m > d$ and let $m' = m - d$. Note that $D^\alpha P_+ = (D^\alpha P)_+$, where $D^\alpha P$ is a ψdo of order $d + |\alpha|$ satisfying the transmission condition. Applying the preceding considerations to $(D^\alpha P)_+$ for $|\alpha| \le m'$, we find

$$\|P_+ u\|_{m-d,t} = \Big(\sum_{|\alpha| \le m'} \|(D^\alpha P)_+ u\|_{0,t}^2 \Big)^{\frac{1}{2}} \le c\Big(\sum_{|\alpha| \le m'} \|u\|_{d+|\alpha|,t}^2 \Big)^{\frac{1}{2}} \le c'\|u\|_{m,t}.$$

$\square$

Remark 10.33. For ψdo's with symbols depending on the normal variable, the result is obtained after an extra reduction. Consider for example a symbol in (x', y', y_n)-form, $p(x', y', y_n, \xi)$. Take a Taylor expansion

$$p(x', y', y_n, \xi) = \sum_{0 \le k < K} \tfrac{1}{k!} \partial_{y_n}^k p(x', y', 0, \xi) y_n^k + p_{(K)}(x', y', y_n, \xi) y_n^K.$$

The terms in the sum define operators that map functions in $H^m(\mathbb{R}^n_+)$ with support in a fixed bounded set continuously into $H^{m-d}(\mathbb{R}^n_+)$, by Theorem 10.32. For a given m, we can take $K \ge m$ so that the functions $y_n^K u$ extend by zero to functions in $H^m(\mathbb{R}^n)$; then $\mathrm{OP}(p_{(K)})_+$ maps them continuously into $H^{m-d}(\mathbb{R}^n_+)$.

The result extends to noninteger $s \ge 0$ (in the place of m) by interpolation. There is moreover an extension down to $s > -\frac{1}{2}$; this hinges on the fact that $H^{s,t}(\mathbb{R}^n_+)$ and $H^{s,t}_0(\overline{\mathbb{R}}^n_+)$ coincide for $-\frac{1}{2} < s < \frac{1}{2}$ (one can show that multiplication by $1_{x_n>0}$ is bounded in $H^{s,t}(\mathbb{R}^n)$ for $|s| < \frac{1}{2}$). The mapping properties in (10.143) and (10.144) likewise extend to $s > -\frac{1}{2}$, when $r = 0$.

We define "loc" and "comp" versions of the Sobolev spaces in the usual way:

$$H^s_{\mathrm{loc}}(\overline{\mathbb{R}}^n_+) = \{u \in \mathscr{D}'(\mathbb{R}^n_+) \mid \varphi u \in H^s(\mathbb{R}^n_+) \text{ for any } \varphi \in C^\infty_{(0)}(\overline{\mathbb{R}}^n_+)\},$$

$$H^s_{\mathrm{comp}}(\overline{\mathbb{R}}^n_+) = \{u \in H^s(\mathbb{R}^n_+) \mid \operatorname{supp} u \text{ is compact in } \overline{\mathbb{R}}^n_+\}; \tag{10.148}$$

the former is a Fréchet space and the latter an inductive limit of such spaces. Note that by the Sobolev embedding theorem,

$$\bigcap_{s \ge 0} H^s_{\mathrm{loc}}(\overline{\mathbb{R}}^n_+) = C^\infty(\overline{\mathbb{R}}^n_+), \quad \bigcap_{s \ge 0} H^s_{\mathrm{comp}}(\overline{\mathbb{R}}^n_+) = C^\infty_{(0)}(\overline{\mathbb{R}}^n_+).$$

Then one can obtain as a corollary of the preceding statements:

Theorem 10.34. *When $\mathcal{A}$ is a ψdbo (a Green operator) as in* (10.18) *with entries of order d and class r, it defines continuous mappings from* $H^s_{\mathrm{comp}}(\overline{\mathbb{R}}^n_+)^N \times H^{s-\frac{1}{2}}_{\mathrm{comp}}(\mathbb{R}^{n-1})^M$ *to* $H^{s-d}_{\mathrm{loc}}(\overline{\mathbb{R}}^n_+)^{N'} \times H^{s-d-\frac{1}{2}}_{\mathrm{loc}}(\mathbb{R}^{n-1})^{M'}$ *for $s >$* $r - \frac{1}{2}$; *it also maps continuously as indicated in* (10.18).

Remark 10.35. A few more observations on adjoints: It was mentioned in (10.40) that the adjoint of a singular Green operator in (x', y')-form $G = \mathrm{OPG}(\tilde{g}(x', y', x_n, y_n, \xi'))$ of order d and class 0 is a singular Green operator G_1 of the same kind, with symbol-kernel $\tilde{g}_1 = \overline{\tilde{g}}(y', x', y_n, x_n, \xi')$. This is seen (when the symbol-kernel is compactly supported in x', y') by changing the order of integration in the first line of (10.145). Here G and G_1 are contained in each other's adjoints as mappings in $\mathscr{S}(\overline{\mathbb{R}}^n_+)$, extending to the duality between $G : H^{0,t}(\mathbb{R}^n_+) \to H^{0,t-d}(\mathbb{R}^n_+)$ and $G_1 : H^{0,d-t}(\mathbb{R}^n_+) \to H^{0,-t}(\mathbb{R}^n_+)$, any $t \in \mathbb{R}$. Relations between symbols in x'-form or y'-form are found by further reductions using Theorem 7.13 1° in the primed variables.

For ψdo's P, we note that when P is of order ≤ 0 and properly supported, P_+ and $(P^*)_+$ are adjoints, since

$$(P_+ u, v)_{L_2(\mathbb{R}^n_+)} = (Pe^+ u, e^+ v)_{L_2(\mathbb{R}^n)} = (e^+ u, P^* e^+ v)_{L_2(\mathbb{R}^n)}$$
$$= (u, (P^*)_+ v)_{L_2(\mathbb{R}^n_+)}, \tag{10.149}$$

for $u, v \in C_0^\infty(\mathbb{R}^n_+)$, extending to $L_2(\mathbb{R}^n_+)$ by approximation.

10.6 Elliptic ψdbo's

For the study of invertible elements in our "algebra" of operators, we need to define the concept of *ellipticity*. This really consists of two conditions, namely, invertibility of the principal interior symbol and invertibility of the principal boundary symbol operator.

Definition 10.36. Let $\mathcal{A}$ be a ψdbo (10.18) — a Green operator — of order d and class r, with symbols $p(x, \xi)$, $g(x', \xi, \eta_n)$, $t(x', \xi)$, $k(x', \xi)$ and $s(x', \xi')$.

1° The **principal interior symbol** is the symbol $p^0(x, \xi)$. The **principal boundary symbol operator** is the operator

$$\mathfrak{a}^0(x', \xi', D_n) = \begin{pmatrix} p^0(x', 0, \xi', D_n)_+ + g^0(x', \xi', D_n) & k^0(x', \xi', D_n) \\ t^0(x', \xi', D_n) & s^0(x', \xi') \end{pmatrix},$$

going from $\mathscr{S}(\overline{\mathbb{R}}_+)^N \times \mathbb{C}^M$ to $\mathscr{S}(\overline{\mathbb{R}}_+)^{N'} \times \mathbb{C}^{M'}$. The interior symbol, resp. boundary symbol operator, are defined similarly from the full symbols.

2° $\mathcal{A}$ is said to be **elliptic**, when p^0 is bijective for all $|\xi| \geq 1$ and all x, and $\mathfrak{a}^0$ is bijective for all $|\xi'| \geq 1$ and all x'. (In particular, $N = N'$ then.)

When P alone is known to be elliptic, it can be shown that $p^0(x', \xi', D_n)_+$ is a Fredholm operator in $\mathscr{S}^N_+$ (and between suitable Sobolev spaces over

$\overline{\mathbb{R}}_+$, the nullspace and range complement being the same as for $\mathscr{S}_+^N$), with an index depending continuously on (x', ξ'). A necessary condition for the ellipticity of the full system $\mathcal{A}$ is then that $M' - M = \text{index } p^0$.

The following theorem holds:

Theorem 10.37. *When $\mathcal{A}$ is elliptic, the inverse $\mathfrak{b}^0$ of the principal boundary symbol operator $\mathfrak{a}^0$ belongs to the calculus; it is a boundary symbol operator of order $-d$ and class $r' = \max\{r - d, 0\}$.*

Proof (indications). The proof of this theorem in the differential operator case, where the ψdbo is as in (10.19) (with $M = 0$), is not so hard, since one can find the inverse constructively by analysis of the solutions in $\mathscr{S}_+^N$ of the equation $p^0(x', 0, \xi', D_n)u = 0$ in terms of the polynomial $\det p^0(x', 0, \xi', \xi_n)$ in ξ_n (cf. Example 10.39 below); it is found that the symbol ingredients in the inverse of the principal symbol are rational functions of ξ_n. In the general pseudodifferential case, another technique is needed. Here one can reduce, by composition with suitable invertible operators, to the situation where $P = I$ and the other operators are of order and class 0. Then the boundary symbol operator is similar to an operator of the form $I + g(x', \xi', D_n)$. Now we can use the Laguerre expansions to write $g(D_n)$ as

$$g(x', \xi', D_n) = \sum_{l,m \in \mathbb{N}_0} c_{lm}(x', \xi')k_l(\xi', D_n)t_m(\xi', D_n);$$

$$k_l = \text{OPK}_n(\hat{\varphi}_l(\xi_n, \langle \xi' \rangle)), \quad t_m = \text{OPT}_n(\overline{\hat{\varphi}}_m(\xi_n, \langle \xi' \rangle)),$$

cf. (10.104). When $I + g(x', \xi', D_n)$ is invertible, we can for large enough M replace g (for x' in a compact set, all $|\xi'| \geq 1$) by

$$g_M(x', \xi', D_n) = \sum_{l,m \leq M} c_{lm}(x', \xi')k_l t_m;$$

such that $I + g_M$ is still invertible (since the c_{lm} are rapidly decreasing). This operator is, for each (x', ξ'), invertible on the finite dimensional space V_M spanned by the orthonormal system $(\varphi_l)_{0 \leq l \leq M}$, where it acts (with respect to the basis) as the matrix $I + (c_{lm}(x', \xi'))_{l \leq M, m \leq M}$. This is an elliptic ψdo symbol in the standard sense, so the inverse is again a ψdo symbol. One can trace this inverse back to an explicit representation of $(\mathfrak{a}^0)^{-1}$, showing that it belongs to the calculus. (Further details are given e.g. in [G96], or in a more elementary form in the first edition from 1986.) $\square$

This allows a parametrix construction:

Theorem 10.38. *When $\mathcal{A}$ is elliptic, there exists a ψdbo $\mathcal{B}$ (with principal boundary symbol operator $\mathfrak{b}^0$ as in Theorem 10.37), which is a **parametrix** of $\mathcal{A}$, in the sense that*

$$\mathcal{AB} - I \text{ and } \mathcal{BA} - I \text{ are negligible.} \tag{10.150}$$

Proof (indications). A first approximation to $\mathcal{B}$ is the operator $\mathcal{B}'$ with boundary symbol $\mathfrak{b}^0$ and interior symbol q, where $q(x,\xi)$ is a parametrix symbol for p. Then $\mathcal{A}\mathcal{B}'$ equals the identity plus a Green operator with interior symbol ~ 0 and with boundary symbol of order -1. Proceeding as in the proof of Theorem 7.18, one can improve $\mathcal{B}'$ to a parametrix $\mathcal{B}$ (still of order $-d$ and class $r' = \max\{r - d, 0\}$), such that $\mathcal{A}\mathcal{B} - I$ is of order $-\infty$. There is an equivalent left parametrix. $\square$

One-sided ellipticity (surjective or injective) can also be considered (some cases of this are discussed in Section 11.2).

Example 10.39. Let $a^0(x', \xi)$ be the principal symbol at $x_n = 0$ of an elliptic partial differential operator A on $\mathbb{R}^n$ of order d, possibly $N \times N$-matrix-formed. Along with A there is given a trace operator $T = \{T_0, T_1, \ldots T_{d-1}\}$, where T_j is $M_j \times N$-matrix-formed with $0 \leq M_j \leq d-1$ (the cases $M_j = 0$ give void boundary conditions, and are just included for notational convenience), each T_j of the form $\sum_{0 \leq k \leq j} S_{jk}\gamma_k$, with differential operators S_{jk} on $\mathbb{R}^{n-1}$ of order $j - k$ having principal symbols $s_{jk}^0(x', \xi')$. We consider the boundary value problem

$$Au = f \text{ on } \mathbb{R}_+^n, \quad Tu = \varphi \text{ on } \mathbb{R}^{n-1},$$

for given vector functions f and φ, with u sought in a suitable Sobolev space in terms of the spaces where f and φ are given. The problem defined by the principal boundary symbol operator

$$a^0(x', \xi', D_n)u(x_n) = f(x_n) \text{ on } \mathbb{R}_+,$$
$$t_j^0 u \equiv \sum_{k \leq j} s_{jk}^0(x', \xi')\gamma_k u = \varphi_j \text{ at } x_n = 0 \quad (0 \leq j \leq d - 1), \qquad (10.151)$$

is considered for all (x', ξ') with $\xi' \neq 0$, usually called the *model problem.*

The problem is easily reduced to a semihomogeneous problem by use of the inverse symbol $q^0(x', \xi', \xi_n) = (a^0)^{-1}$, defined for $\xi' \neq 0$. In fact, $a_+^0 q_+^0 = I$ on $\mathbb{R}_+$, since a^0 is a differential operator (then $L(a^0, q^0) = 0$, cf. Theorem 10.26). Thus when we set $z(x_n) = u(x_n) - (q_+^0 f)(x_n)$, we get the problem for z,

$$a^0(x', \xi', D_n)z(x_n) = 0 \text{ on } \mathbb{R}_+, \quad t_j^0 z = \psi_j \text{ at } x_n = 0 \quad (0 \leq j \leq d - 1),$$
$$\tag{10.152}$$

where $\psi_j = \varphi_j - t_j^0 q_+^0 f$.

For each $(x', \xi') \in \mathbb{R}^{2(n-1)}$, the symbol defines a polynomial in τ,

$$a^0(x', \xi', \tau) = \sum_{0 \leq l \leq d} s_l(x', \xi')\tau^l; \qquad (10.153)$$

the s_l being polynomials in ξ' of degree $d - l$. By the ellipticity, the coefficient of τ^d, $s_d = s_d(x')$, is a bijective matrix function (check by inserting $(\xi', \xi_n) = (0, 1)$). So $\det a^0(x', \xi', \tau)$ has Nd roots in $\mathbb{C}$, counted with multiplicity. When $\xi' \in \mathbb{R}^{n-1} \setminus \{0\}$, the roots lie in $\mathbb{C} \setminus \mathbb{R}$, for otherwise a root $\tau_0 \in \mathbb{R}$ would give

$\det a^0(x',\xi',\tau_0) = 0$ contradicting the ellipticity. Say $m_+(x',\xi')$ roots lie in $\mathbb{C}_+$ (cf. (A.3)), $m_-(x',\xi') = Nd - m_+(x',\xi')$ roots lie in $\mathbb{C}_-$.

When $n \geq 3$, each $\xi' \neq 0$ can be connected to $-\xi'$ by a curve running in $\mathbb{R}^{n-1} \setminus \{0\}$. The number $m_+(x',\xi')$ must be constant along such a curve (the roots in $\mathbb{C}_+$ can be encircled by a large closed curve in $\mathbb{C}_+$, and the polynomial coefficients depend continuously on ξ', so the collected algebraic multiplicity of these roots must be continuous in ξ', by the theorem of Rouché). Then $m_+(x',\xi') = m_+(x',-\xi') = m_-(x',\xi')$. This must then equal $\frac{1}{2}Nd$, constant in (x',ξ') (in particular, Nd must be even). Symbols with the property $m_+(x',\xi') \equiv \frac{1}{2}Nd$ are called *properly elliptic*. Strongly elliptic symbols are properly elliptic.

It is a standard result in ODE that the solutions of $a^0(x',\xi',D_n)z = 0$ are linear combinations of terms $f_j(x_n)e^{i\tau_j x_n}$ where f_j is a (vector valued) polynomial in x_n and τ_j is a root. For the solutions lying in $\mathscr{S}_+^N$, only the terms where $\tau_j \in \mathbb{C}_+$ are nontrivial. More precisely, the space of null-solutions in $\mathscr{S}_+^N$ is spanned by exponential polynomials $f_j(x_n)e^{i\tau_j x_n}$ with $\tau_j \in \mathbb{C}_+$, and has dimension $m_+(x',\xi')$. (Similarly, the space of null-solutions in $\mathscr{S}(\overline{\mathbb{R}}_-)^N$ is spanned by exponential polynomials $f_j(x_n)e^{i\tau_j x_n}$ with $\tau_j \in \mathbb{C}_-$, and has dimension $m_-(x',\xi')$.) When a^0 is properly elliptic, then it is necessary for ellipticity of $\{a^0,t^0\}$ that the dimension $M = M_0 + \cdots + M_{d-1}$ (the number of boundary conditions) equals $\frac{1}{2}Nd$.

The construction of solutions to the model problem is explained in various ways e.g. in [ADN64], [LM68]. A more global construction goes by use of the so-called Calderón projector, which we take up in Chapter 11.

From here on, the construction of a calculus on manifolds with boundary goes very much like in the theory for ψdo's on manifolds without boundary.

The various types of operators can be defined on a compact manifold X with boundary $\partial X = X'$ by use of local coordinates, as already indicated in Section 10.1 around (10.42). In view of Proposition 10.11, the singular ingredients only need special care in a neighborhood of X'; here one can define the local coordinate systems in terms of a covering of the boundary with open sets with an associated normal coordinate that matches on overlapping sets. One can then deduce the following mapping property from Theorem 10.34:

$$\mathcal{A} = \begin{pmatrix} P_+ + G & K \\ T & S \end{pmatrix} : \begin{matrix} H^s(X)^N \\ \times \\ H^{s-\frac{1}{2}}(X')^M \end{matrix} \to \begin{matrix} H^{s-d}(X)^{N'} \\ \times \\ H^{s-d-\frac{1}{2}}(X')^{M'} \end{matrix}, \quad s > r+\tfrac{1}{2}, \quad (10.154)$$

for matrix-formed operators of order d and class r; there is a similar result for operators between sections of vector bundles.

In the elliptic case there is a parametrix $\mathcal{B}$ (of class $r' = \max\{d - r, 0\}$) going in the opposite direction. One uses the parametrix explained in Theorem 10.38 in coordinate patches intersecting the boundary, and in coordinate patches at a distance from the boundary, one simply uses $\left(\begin{smallmatrix} q & 0 \\ 0 & 0 \end{smallmatrix}\right)$ as symbol of $\mathcal{B}$. The local pieces are put together as in the proof of Theorem 8.6. Then

(10.150) holds with remainder operators $\mathcal{R}_1$ and $\mathcal{R}_2$ negligible of class r' resp. r, hence compact in Sobolev spaces of sufficiently high order. From this, one can show regularity of solutions of problems $\mathcal{A}\{u, \varphi\} = \{f, \psi\}$, and deduce that the elliptic ψdbo $\mathcal{A}$ is a Fredholm operator (in spaces (10.154)). There is a vast literature on the index, beginning with Atiyah and Bott [AB64], see also e.g. [B71], [RS82], [H85], [Gi85], [G96]. The noncommutative residue for ψdbo's was introduced in Fedosov, Golse, Leichtnam and Schrohe [FGLS96]; this and the canonical trace are futher studied e.g. in [G08] and its references.

In the treatment of differential operator problems, an advantage of the present theory in comparison with classical methods of "*a priori* estimates" (as in [ADN64], [LM68]) is that the full solution operator for a given boundary value problem is found in a constructive way.

The theory has been extended to L_p-based Sobolev-type spaces ($1 < p < \infty$) in [G90]. A full presentation of the treatment of elliptic systems is given there, with optimal mapping properties (including operators $P_+ + G$ of negative class). Nonsmooth x-dependence is treated in Abels [A05].

There exist other general theories for boundary value problems. To mention a few: Schulze and coauthors have dealt with ψdo's not satisfying the transmission condition in many works, see e.g. [S98] and its references, which also deal with ψdo's on manifolds with singularities. Melrose and coauthors have treated singular situations by different techniques, see e.g. [M93] and its references.

Exercises for Chapter 10

10.1. Show that if the symbol $p(x, \xi)$ is for each x a rational function of ξ for $|\xi| \geq 1$, homogeneous of degree d, then p satisfies the transmission condition.

10.2. *This and the next exercise illustrate the transmission condition in an elementary way.*

Let $p(\xi_n)$ be an x_n-independent symbol in $S_{1,0}^{-1}(\mathbb{R}, \mathbb{R})$ (with variables denoted (x_n, ξ_n)), defining the operator $p(D_n) = \mathrm{OP}_n(p) = \mathscr{F}^{-1}p\mathscr{F}$. Let $\tilde{p}(x_n) = \mathscr{F}^{-1}p(\xi_n)$.

(a) Show that $p \in L_2(\mathbb{R})$ and $\tilde{p} \in L_2(\mathbb{R})$, and moreover that $D_{\xi_n}p \in L_1(\mathbb{R})$ and hence $x_n\tilde{p} \in C_\bullet^0(\mathbb{R})$, the space of continuous functions going to zero for $|x_n| \to \infty$.

(b) Show that $D_{x_n}^k(x_n^j\tilde{p}) = \mathscr{F}^{-1}(\xi_n^k(-D_{\xi_n})^j p) \in C_\bullet^0(\mathbb{R})$, when $j - k \geq 1$.

(c) Show that $\tilde{p}(x_n)$ is C^∞ for $x_n \neq 0$ and is rapidly decreasing for $x_n \to \pm\infty$.

10.3. With p as in Exercise 10.2, consider the operator $p(D_n)_+$, also called p_+,

$$p_+u = r^+p(D_n)e^+u = r^+\mathscr{F}^{-1}(p(\xi_n)f(\xi_n)), \quad f = \mathscr{F}(e^+u),$$

for $u \in \mathscr{S}(\overline{\mathbb{R}}_+)$.

(a) Show that $pf \in L_1(\mathbb{R})$ and hence $p_+u \in C_\bullet^0(\overline{\mathbb{R}}_+)$.

(b) Show, by successive applications of the formula

$$x_n p_+ u = r^+\mathscr{F}^{-1}[-D_{\xi_n}(p(\xi_n)f(\xi_n))] = -\operatorname{OP}_n(D_{\xi_n}p)_+u - \operatorname{OP}_n(p)_+(x_nu),$$

that $x_n p_+ u \in C_\bullet^0(\overline{\mathbb{R}}_+)$, ..., $x_n^N p_+ u \in C_\bullet^0(\overline{\mathbb{R}}_+)$ for all N, hence $(p_+u)(x_n)$ is $O(x_n^{-N})$ for $x_n \to \infty$, any N.

(c) Show, by use of the formula for differentiation of e^+u,

$$D_{x_n}e^+u = e^+D_{x_n}u - iu(0)\delta,$$

that

$$\begin{aligned}
D_{x_n}p_+u &= r^+D_{x_n}p(D_n)e^+u \\
&= r^+p(D_n)e^+D_{x_n}u - i(r^+p(D_n)\delta)\cdot u(0) \\
&= p_+D_{x_n}u - ir^+\tilde{p}(x_n)\cdot u(0).
\end{aligned}$$

Use this to see that $D_{x_n}p_+u \in C^0(\overline{\mathbb{R}}_+)$ for general $u \in \mathscr{S}(\overline{\mathbb{R}}_+)$ if and only if $r^+\tilde{p}(x_n) \in C^0(\overline{\mathbb{R}}_+)$.

(d) Show (by repeated applications of (c))

$$D_{x_n}^j p_+u = p_+D_{x_n}^j u - i\sum_{0 \le l \le j-1} r^+D_{x_n}^l \tilde{p}(x_n)\cdot D_{x_n}^{j-1-l}u(0),$$

and conclude: *In order for p_+u to be in $C^\infty(\overline{\mathbb{R}}_+)$ for all $u \in \mathscr{S}(\overline{\mathbb{R}}_+)$, it is necessary and sufficient that $r^+\tilde{p} \in C^\infty(\overline{\mathbb{R}}_+)$.*

10.4.   Verify the asserted symbol and symbol-kernel estimates for (v)–(viii) of Proposition 10.10.

10.5.   Let σ be a complex constant with $\operatorname{Re}\sigma > 0$.

(a) Show that

$$h^+\frac{1}{\sigma+i\xi_n} = \frac{1}{\sigma+i\xi_n}, \quad h^+\frac{\xi_n}{\sigma+i\xi_n} = \frac{i\sigma}{\sigma+i\xi_n},$$
$$h^+\frac{\xi_n^2}{\sigma+i\xi_n} = \frac{(i\sigma)^2}{\sigma+i\xi_n},$$

and find a formula valid for all powers ξ_n^k in the numerator, $k \in \mathbb{N}_0$.

(b) Show that

$$h^-\frac{1}{\sigma-i\xi_n} = \frac{1}{\sigma-i\xi_n}, \quad h^-\frac{\xi_n}{\sigma-i\xi_n} = i - \frac{i\sigma}{\sigma-i\xi_n},$$
$$h^-\frac{\xi_n^2}{\sigma-i\xi_n} = i\xi_n + \sigma - \frac{\sigma^2}{\sigma-i\xi_n},$$

and find a formula valid for all powers ξ_n^k in the numerator.
(*Hint*. Recall the formula $x^k - y^k = (x - y)(x^{k-1} + x^{k-2}y + \cdots + y^{k-1})$.)
(c) Find, for $k \in \mathbb{N}_0$,

$$h^+ \frac{\xi_n^k}{\sigma - i\xi_n} \text{ and } h^- \frac{\xi_n^k}{\sigma + i\xi_n}.$$

10.6. (a) With σ as in Exercise 10.5, let $a(\xi_n) = \sigma^2 + \xi_n^2$ and let $q(\xi_n) = a^{-1} = 1/(\sigma^2 + \xi_n^2)$. Find

$$h^+(q), \quad h^+(\xi_n q), \quad h^+(\xi_n^2 q),$$
$$h^-(q), \quad h^-(\xi_n q), \quad h^-(\xi_n^2 q),$$
$$\int^+ q(\xi_n)\, d\xi_n, \quad \int^+ \xi_n q(\xi_n)\, d\xi_n, \text{ and } \int^+ \xi_n^2 q(\xi_n)\, d\xi_n.$$

(b) Let $A = -\Delta + m$ on $\mathbb{R}^n$ for some $m > 0$, and let $Q = A^{-1}$. For Q_+ considered on $\mathbb{R}^n_+$, find the trace operators $T_0 = \gamma_0 Q_+$, $T_1 = \gamma_1 Q_+$ and the Poisson operator $K : v(x') \to r^+ Q(v(x') \otimes \delta(x_n))$.

10.7. Let $P = \text{OP}(p(x, y', \xi))$ be of order $d \le -1$ on $\mathbb{R}^n$, where p satisfies the transmission condition at $x_n = 0$, is compactly supported with respect to x' and y', and is independent of y_n.

(a) Show that $T = \gamma_0 P_+$ is a trace operator of class 0 and order d; find its symbol and symbol-kernel.

(b) Show that $K : v(x') \mapsto r^+ P^*(v(x') \otimes \delta(x_n))$ is a Poisson operator of order $d + 1$; find its symbol and symbol-kernel.

(c) Show that T and K are adjoints, e.g., as $T : L_2(\mathbb{R}^n_+) \to H^{-d-\frac{1}{2}}(\mathbb{R}^{n-1})$ and $K : H^{d+\frac{1}{2}}(\mathbb{R}^{n-1}) \to L_2(\mathbb{R}^n_+)$.

(d) Show that when P is a differential operator, then the Poisson operator $K : v(x') \mapsto r^+ P^*(v(x') \otimes \delta(x_n))$ is zero.

(e) Does the conclusion of (a)–(c) extend to operators of order ≥ 0?
(*Hint*. The answer is negative. Try a simple example.)

10.8. Consider a symbol $p(x', \xi) \in S_{0,1}^d(\mathbb{R}^n, \mathbb{R}^n)$, independent of x_n and satisfying the transmission condition.

(a) Show that when the symbol of p is written in the form

$$p = \sum_{1 \le j \le d} s_j(x', \xi')\xi_n^j + p'', \quad p'' = \sum_{k \in \mathbb{Z}} a_k(x', \xi')\hat{\psi}_k(\xi_n, \sigma), \quad \sigma = [\xi']$$

(cf. (10.8), (10.106) and Remark 10.16), then the bounded part $p'' = \sum_{k \in \mathbb{Z}} a_k \hat{\psi}_k$ defines a "discrete convolution operator" when functions $v \in \mathscr{S}(\mathbb{R})$ are expanded in the Laguerre orthonormal system $\{\varphi_k(x_n, \sigma)\}_{k \in \mathbb{Z}}$, $v = \sum_{m \in \mathbb{Z}} v_m \varphi_m$:

$$p''(x',\xi',D_n)v = \mathscr{F}^{-1}\sum_{k\in\mathbb{Z}} a_k\hat{\psi}_k \sum_{m\in\mathbb{Z}} v_m\hat{\varphi}_m$$

$$= \mathscr{F}^{-1}\sum_{k,m\in\mathbb{Z}} a_k v_m\hat{\varphi}_{k+m} = \sum_{l,m\in\mathbb{Z}} a_{l-m}v_m\varphi_l.$$

(*Hint.* Observe that $\hat{\psi}_k\hat{\varphi}_m = \hat{\varphi}_{k+m}$.)

(b) Show that the truncation to $\mathbb{R}_+$, $p''_+ = p''(x',\xi',D_n)_+$ is then a *Toeplitz operator* with respect to the Laguerre orthonormal system $\{\varphi_k(x_n,\sigma)\}_{k\in\mathbb{N}_0}$ on $\mathbb{R}_+$, namely, the following infinite matrix with diagonals consisting of identical entries:

$$\begin{pmatrix} a_0 & a_1 & a_2 & \cdots \\ a_{-1} & a_0 & a_1 & \\ a_{-2} & a_{-1} & a_0 & \\ \vdots & & & \ddots \end{pmatrix};$$

here $a_l \to 0$ rapidly for $l \to \pm\infty$.

10.9. Continuing in the notation of Exercise 10.8, show that also the differential operator part can be viewed as a Toeplitz operator.
(*Hint.* For ∂_{x_n}, this is seen from (10.61):

$$r^+\partial_{x_n}u = \sum_{m\in\mathbb{N}_0}\left(-\sigma u_m\varphi_m + 2\sigma\sum_{0\le j<m}(-1)^{m-1-j}u_m\varphi_j\right)$$

$$= \sum_{l,m\in\mathbb{N}_0} c_{l-m}u_m\varphi_l, \text{ with}$$

$$c_j = 2\sigma(-1)^{j-1} \text{ for } j > 0, \quad c_0 = -\sigma, \quad c_j = 0 \text{ for } j < 0.$$

Note that these coefficients are large, not rapidly decreasing for $j \to \infty$.)

10.10. Notation as in Exercise 10.8.
(a) Show that the s.g.o. $g^+(p(x',\xi',D_n))$ (cf. (10.126)) is a *Hankel operator* in the Laguerre system, namely, the following infinite matrix whose second-diagonals consist of identical entries:

$$\begin{pmatrix} a_1 & a_2 & a_3 & \cdots \\ a_2 & a_3 & a_4 & \\ a_3 & a_4 & a_5 & \\ \vdots & & & \ddots \end{pmatrix}.$$

(*Hint.* Note that $\varphi_k = J\varphi_{-k-1}$, cf. (10.56).)
(b) Find the corresponding representation for $g^-(p(D_n))$.

10.11 (ORDER-REDUCING OPERATORS). Define, for $r \in \mathbb{Z}$,

$$\chi^r_+(\xi) = (\langle\xi'\rangle + i\xi_n)^r, \quad \chi^r_-(\xi) = (\langle\xi'\rangle - i\xi_n)^r,$$

and denote the operators with these symbols

$$\Xi_+^r = \mathrm{OP}(((\langle\xi'\rangle + i\xi_n)^r), \quad \Xi_-^r = \mathrm{OP}(((\langle\xi'\rangle - i\xi_n)^r);$$

they are pseudodifferential operators in the general sense of Chapter 6, but do not satisfy all requirements for having symbols in $S_{1,0}^r$-spaces (since, for large $|\alpha|$, $D^\alpha\langle\xi'\rangle^r$ is not $O(\langle\xi\rangle^{r-|\alpha|})$, but only $O(\langle\xi'\rangle^{r-|\alpha|})$). However, we can still define $(\Xi_\pm^r)_+$ and $G^\pm(\Xi_\pm^r)$, and leftover operators as in (10.123) (with (10.115) and (10.122)).

(a) Show that $G^\pm(\Xi_\pm^r) = 0$ when $r \geq 0$.

(b) Show that when $r < 0$, $G^+(\Xi_-^r) = 0$ and $G^-(\Xi_+^r) = 0$.

(c) Show that
$$(\Xi_-^r)_+(\Xi_-^t)_+ = (\Xi_-^{r+t})_+$$

for all $r, t \in \mathbb{Z}$; in particular,

$$(\Xi_-^r)_+(\Xi_-^{-r})_+ = I \text{ on } \mathscr{S}(\overline{\mathbb{R}}_+^n),$$

for all $r \in \mathbb{Z}$.

(d) Show that $(\Xi_-^r)_+$ maps $H^s(\mathbb{R}_+^n)$ continuously into $H^{s-r}(\mathbb{R}_+^n)$ for all $s \geq 0$, and conclude (by use of (c)) that the mapping properties extend to

$$(\Xi_-^r)_+ : H^s(\mathbb{R}_+^n) \overset{\sim}{\to} H^{s-r}(\mathbb{R}_+^n), \text{ for all } s \in \mathbb{Z},$$

with inverse $(\Xi_-^{-r})_+$.

(e) Show (by taking adjoints) that the maps $(\Xi_+^r)_+$ extend to maps with the properties
$$(\Xi_+^r)_+ : H_0^s(\mathbb{R}_+^n) \overset{\sim}{\to} H_0^{s-r}(\mathbb{R}_+^n), \text{ for all } s \in \mathbb{Z},$$

with inverse $(\Xi_+^{-r})_+$.

(f) Extend the results to include the statements

$$(\Xi_-^r)_+ : H^{s,t}(\mathbb{R}_+^n) \overset{\sim}{\to} H^{s-r,t}(\mathbb{R}_+^n),$$
$$(\Xi_+^r)_+ : H_0^{s,t}(\mathbb{R}_+^n) \overset{\sim}{\to} H_0^{s-r,t}(\mathbb{R}_+^n),$$

for all $s \in \mathbb{Z}, t \in \mathbb{R}$.

(*Comment.* The mapping properties extend to $s \in \mathbb{R} \setminus \mathbb{Z}$ by interpolation. There is a more refined choice of symbols:

$$\lambda_\pm^r(\xi) = \big(\langle\xi'\rangle\varphi(c\xi_n/\langle\xi'\rangle) \pm i\xi_n\big)^r,$$

where $\varphi \in \mathscr{S}(\mathbb{R})$ with $\varphi(0) = 1$ and $\operatorname{supp} \mathscr{F}^{-1}\varphi \subset \,]-\infty, 0]$, and c is a small positive constant. They work in the same way and have the advantage that $\Lambda_\pm^r = \mathrm{OP}(\lambda_\pm^r)$ belong to the ψdbo calculus, cf. [G90] for details. Both the $\Xi_\pm^r$ and the $\Lambda_\pm^r$ can be used to define similar operators on manifolds with boundary.)

Chapter 11
Pseudodifferential methods for boundary value problems

11.1 The Calderón projector

As an illustration of the usefulness of the systematic ψdbo calculus, we shall briefly explain the definition and application of the Calderón projector C^+ for an elliptic differential operator $A : C^\infty(X, E_1) \to C^\infty(X, E_2)$ of order d, as introduced by Calderón [C63], Seeley [S66], [S69], see also Hörmander [H66], Boutet de Monvel [B66], Grubb [G71], [G77]. Much of this chapter is written in a compact style; it has been included in order to make the material available in textbook form.

We begin, however, with a very simple example.

Example 11.1. Let $\alpha > 0$ and consider the differential operator $Au = -u'' + \alpha^2 u$ on $\mathbb{R}$, $\mathbb{R}_+$ and $\mathbb{R}_-$. The solutions in $\mathscr{S}'(\mathbb{R})$ of $-u'' + \alpha^2 u = 0$ on $\mathbb{R}$ are spanned by $e^{\alpha x}$ and $e^{-\alpha x}$; the only solution in $L_2(\mathbb{R})$ is 0. The nonhomogeneous equation $-u'' + \alpha^2 u = f$ is uniquely solved in $L_2(\mathbb{R})$ by $u(x) = \mathscr{F}^{-1}((\alpha^2 + \xi^2)^{-1}\hat{f}(\xi)) \in H^2(\mathbb{R})$.

The initial values or Cauchy data are defined for general $u \in H^2_{\text{loc}}(\mathbb{R})$ as the pair $\varrho u = \left(\begin{smallmatrix} u(0) \\ u'(0) \end{smallmatrix} \right)$. When precision is needed, we use the notation ϱ^+, ϱ^- or $\tilde{\varrho}$ for the mapping $u \mapsto \left(\begin{smallmatrix} u(0) \\ u'(0) \end{smallmatrix} \right)$ when $u \in H^2(\mathbb{R}_+)$, $u \in H^2(\mathbb{R}_-)$, resp. $u \in H^2_{\text{loc}}(\mathbb{R})$. Let

$$Z_\pm = \{u \in L_2(\mathbb{R}_\pm) \mid Au = 0 \text{ on } \mathbb{R}_\pm\}; \text{ clearly,}$$

$$Z_+ = \text{span}\{r^+ e^{-\alpha x}\} \subset \mathscr{S}(\overline{\mathbb{R}}_+), \quad Z_- = \text{span}\{r^- e^{\alpha x}\} \subset \mathscr{S}(\overline{\mathbb{R}}_-).$$

Define

$$N_+ = \varrho^+ Z_+, \ N_- = \varrho^- Z_-; \text{ clearly,}$$

$$N_+ = \{\left(\begin{smallmatrix} \varphi \\ -\alpha\varphi \end{smallmatrix} \right) \mid \varphi \in \mathbb{C}\}, \quad N_- = \{\left(\begin{smallmatrix} \varphi \\ \alpha\varphi \end{smallmatrix} \right) \mid \varphi \in \mathbb{C}\}.$$

Here N_+ and N_- are linearly independent, so $\mathbb{C}^2 = N_+ \dotplus N_-$. The Calderón projector C^+ is now simply *the projection of* $\mathbb{C}^2$ *onto* N_+ *along* N_-. A straightforward calculation shows

$$C^+ = \begin{pmatrix} \frac{1}{2} & -\frac{1}{2\alpha} \\ -\frac{\alpha}{2} & \frac{1}{2} \end{pmatrix}.$$

The map $C^- = I - C^+$ projects $\mathbb{C}^2$ onto N_- along N_+. There are linear maps $K^\pm : \mathbb{C}^2 \to Z_\pm$ that act as inverses to $\varrho^\pm : Z_\pm \to N_\pm$ and vanish on $N_\mp$, respectively.

The maps $C^\pm$ and $K^\pm$ are easily found explicitly in this case; the efforts in higher-dimensional cases go toward determining them in a constructive way. They serve in the discussion of ellipticity and solvability of boundary value problems.

It should be noted that the general theory would take $\varrho u = \{u(0), Du(0)\}$ with a factor $-i$, this is convenient in Green's formulas in higher-order cases.

We consider a smooth compact n-dimensional manifold X with boundary, provided with two N-dimensional hermitian vector bundles E_1 and E_2. (A reader who is not used to working with vector bundles should just think of $(N \times N)$-matrices of operators.) X can be assumed to be smoothly embedded in an n-dimensional boundaryless manifold $\widetilde{X}$ such that $X' = \partial X$ is an $(n-1)$-dimensional hypersurface in $\widetilde{X}$, and E_1 and E_2 are restrictions to X of N-dimensional hermitian bundles $\widetilde{E}_1$ and $\widetilde{E}_2$ over $\widetilde{X}$. Denote $X^\circ = X_+$, $\widetilde{X} \setminus X = X_-$, and write $\widetilde{E}_i|_{\overline{X}_\pm} = E_{i,\pm}$, $\widetilde{E}_i|_{X'} = E_i'$. We can assume that near the boundary, the manifold and bundles are described as a product situation, with a chosen normal coordinate x_n. More precisely, there is a neighborhood U of X' in $\widetilde{X}$ where the points are represented as $x = (x', x_n)$, $x' \in X'$ and $x_n \in]-1,1[$, such that $x_n = 0$ on the boundary, $x_n > 0$ in $U_+ = X_+ \cap U$ and $x_n < 0$ in $U_- = X_- \cap U$. Moreover, the bundles $\widetilde{E}_i$ are over U simply the liftings of E_i' from X' to $X' \times]-1,1[$. Then $D_n = -i\partial_{x_n}$ has a meaning, over U, on the sections of the bundles $\widetilde{E}_1$, $\widetilde{E}_2$. The coordinate systems on U are taken to be of the form $\kappa_j : U_j' \times]-1,1[\to V_j' \times]-1,1[$, where $\kappa_j(x', x_n) = (\kappa_j'(x'), x_n)$, defined from coordinate systems $\kappa_j' : U_j' \to V_j'$ for X'; trivializations act similarly. The volume form dx on $\widetilde{X}$ is chosen such that $dx = dx' dx_n$ on U, for a volume form dx' on X' and the Lebesgue measure dx_n on $\mathbb{R}$. In these coordinates, D_n is formally selfadjoint.

If A extends to an elliptic operator (also denoted A) from $C^\infty(\widetilde{E}_1)$ to $C^\infty(\widetilde{E}_2)$, we let Q denote a parametrix of A on $\widetilde{X}$; then the formulas of Theorem 8.6 hold with $P = A$. The use of Calderón projectors is simplest if $\widetilde{X}, \widetilde{E}_1, \widetilde{E}_2$ and A can be chosen so that $\widetilde{X}$ is compact and A is *invertible* on $\widetilde{X}$; then Q stands for the inverse, and $\mathcal{R}_1$ and $\mathcal{R}_2$ are zero. *We assume this in the following.* (There can be topological obstructions to a compact choice of $\widetilde{X}$ where A is elliptic; then one can take for $\widetilde{X}$ a neighborhood of X and carry an analysis through modulo negligible operators. Or, if A is elliptic on a compact extension $\widetilde{X}$ but not invertible there, one can get results modulo suitable finite-rank operators. For lack of space we do not treat such cases here.)

As usual, we define $\gamma_0 u = u(x', 0)$ and $\gamma_j u = \gamma_0(D_n^j u)$, and when A is of order d, we also consider the Cauchy data map $\varrho = \{\gamma_0, \dots, \gamma_{d-1}\}$. Each of these maps can be regarded as a map either from sections over $\overline{X}_+$, or from sections over $\overline{X}_-$, or from sections over $\widetilde{X}$, to sections over X'; to distinguish between the three versions, we use notations such as ϱ^+, ϱ^- resp. $\widetilde{\varrho}$ (so what we would originally call ϱ is now ϱ^+). This makes no difference when they act on C^∞ sections, but it is important when various generalizations to function spaces are considered. The operator D_n is the same in all three versions.

When $F = F_0 \oplus \cdots \oplus F_{d-1}$ are vector bundles over X', we denote

$$\begin{aligned}
\mathcal{H}^s(F) &= \textstyle\prod_{0 \le j < d} H^{s-j-\frac{1}{2}}(F_j), \\
\widetilde{\mathcal{H}}^s(F) &= \textstyle\prod_{0 \le j < d} H^{s+j+\frac{1}{2}}(F_j) = (\mathcal{H}^{-s}(F))^*,
\end{aligned} \tag{11.1}$$

(for the last equation we recall the fact that $H^t(F_j)$ and $H^{-t}(F_j)$ are dual spaces). Denoting $\bigoplus_{0 \le j < d} E_i' = E_i'^d$, we can then formulate the well-known mapping properties of $\varrho^\pm$ and $\widetilde{\varrho}$ as follows:

$$\begin{aligned}
\varrho^\pm &: H^s(E_{i,\pm}) \to \mathcal{H}^s(E_i'^d), \\
\widetilde{\varrho} &: H^s(\widetilde{E}_i) \to \mathcal{H}^s(E_i'^d), \quad \text{for } s > d - \tfrac{1}{2}.
\end{aligned} \tag{11.2}$$

The "two-sided" trace map $\widetilde{\gamma}_0 : H^s(\widetilde{E}_i) \to H^{s-\frac{1}{2}}(E_i')$ has an adjoint $\widetilde{\gamma}_0^* : H^{\frac{1}{2}-s}(E_i') \to H^{-s}(\widetilde{E}_i)$ for $s > \frac{1}{2}$. It can also be written

$$\widetilde{\gamma}_0^* v = v(x') \otimes \delta(x_n), \tag{11.3}$$

since

$$\langle v(x') \otimes \delta(x_n), \bar{\varphi}(x) \rangle = \langle v(x'), \bar{\varphi}(x', 0) \rangle = \langle v, \overline{\widetilde{\gamma}_0 \varphi} \rangle,$$

cf. (10.114). Note that $\widetilde{\gamma}_0^*$ ranges in distributions on $\widetilde{X}$ supported in X'. Similarly, the two-sided Cauchy data map $\widetilde{\varrho} : H^s(\widetilde{E}_i) \to \mathcal{H}^s(E_i'^d)$ has the adjoint

$$\widetilde{\varrho}^* : \widetilde{\mathcal{H}}^{-s}(E_i'^d) \to H^{-s}(\widetilde{E}_i), \tag{11.4}$$

for $s > d - \frac{1}{2}$; here since $\widetilde{\varrho} = \widetilde{\gamma}_0 \left(1 \ D_n \ \dots \ D_n^{d-1}\right)^t$,

$$\widetilde{\varrho}^* = \left(1 \ D_n \ \dots \ D_n^{d-1}\right) \widetilde{\gamma}_0^*. \tag{11.5}$$

As usual, we use the notation $P_\pm$ for the truncation of a ψdo P on $\widetilde{X}$ to $X_\pm$ (respectively):

$$P_\pm = r^\pm P e^\pm, \quad \text{when } P \text{ is a } \psi\text{do on } \widetilde{X};$$

here $r^{\pm}$ means restriction to $X_{\pm}$ and $e^{\pm}$ means extension by zero on $X \setminus X_{\pm}$.

Lemma 11.2. *The boundary maps $\varrho^{\pm}$ and $\tilde{\varrho}$ in (11.2) are surjective for $s = d$. In fact, there exists a bounded lifting operator*

$$\tilde{\mathcal{K}}_{(d)} = \left(\tilde{\mathcal{K}}_0 \cdots \tilde{\mathcal{K}}_{d-1} \right) : \mathcal{H}^d(E_i'^d) \to H^d(\tilde{E}_i) \tag{11.6}$$

such that $\mathcal{K}_{(d)}^{\pm} = r^{\pm}\tilde{\mathcal{K}}_{(d)}$ are Poisson operators $(\mathcal{K}_{(d)}^{\pm} = (\mathcal{K}_0^{\pm} \cdots \mathcal{K}_{d-1}^{\pm})$ with $\mathcal{K}_j^{\pm}$ of order $-j$), and

$$\varrho^{\pm}\mathcal{K}_{(d)}^{\pm} = I, \ \tilde{\varrho}\tilde{\mathcal{K}}_{(d)} = I, \ on \ \mathcal{H}^d(E_i'^d). \tag{11.7}$$

Proof. This follows by localization from Theorem 9.5, which treats the Euclidean case where X_+ is replaced by $\mathbb{R}_+^n$. Note first that each operator $\mathcal{K}_j : \mathscr{S}(\mathbb{R}^{n-1}) \to \mathscr{S}(\overline{\mathbb{R}}_+^n)$ defined in Theorem 9.5 (let us call it $\mathcal{K}_j^+$ now) is indeed a Poisson operator of order $-j$, with symbol-kernel $\frac{1}{j!}(ix_n)^j \psi(\langle\xi'\rangle x_n)$. The appropriate estimates (as in Definitions 10.20, 10.3) follow from (9.23), which treats $D_{x_n}^k \frac{1}{j!}(ix_n)^j \psi(\langle\xi'\rangle x_n)$ directly and is easily generalized to expressions with powers of x_n and $D_{\xi'}$ in front. There are similar estimates for this function considered for $x_n \in \mathbb{R}_-$, so it defines a Poisson operator $\mathcal{K}_j^-$ of order $-j$ from $\mathbb{R}^{n-1}$ to $\mathbb{R}_-^n$. Moreover, $\mathcal{K}_j^+$ and $\mathcal{K}_j^-$ can be regarded as $r^+\tilde{\mathcal{K}}_j$ resp. $r^-\tilde{\mathcal{K}}_j$, where $\tilde{\mathcal{K}}_j : \mathscr{S}(\mathbb{R}^{n-1}) \to \mathscr{S}(\mathbb{R}^n)$ is defined by the formula in (9.21) used for all $x_n \in \mathbb{R}$. In the estimates worked out after (9.23), one can replace the integration in $x_n \in \mathbb{R}_+$ by an integration in $x_n \in \mathbb{R}$, showing that $\tilde{\mathcal{K}}_j$ is bounded from $H^{m-j-\frac{1}{2}}(\mathbb{R}^{n-1})$ to $H^m(\mathbb{R}^n)$.

Set $m = d$, then the vectors

$$\tilde{\mathcal{K}}_{(d)} = \left(\tilde{\mathcal{K}}_0 \cdots \tilde{\mathcal{K}}_{d-1} \right), \quad \mathcal{K}_{(d)}^{\pm} = (\mathcal{K}_0^{\pm} \cdots \mathcal{K}_{d-1}^{\pm}),$$

are right inverses of $\tilde{\varrho}_{(d)}$, $\varrho_{(d)}^{\pm}$ in the Euclidean case.

In the manifold situation, we get the right inverses by using the Euclidean construction in local coordinates, after applying a partition of unity. When φ is given in $\mathcal{H}^d(E_i'^d)$, consider a localized piece, say $\underline{\varphi}_l$. It has compact support in $\mathbb{R}^{n-1}$, say M, so if $\tilde{\mathcal{K}}_{(d)}\underline{\varphi}_l$ is multiplied by a function $\underline{\zeta}_l \in C_0^{\infty}(\mathbb{R}^n)$ which is 1 on a neighborhood of M, the identity $\tilde{\varrho}_{(d)}\underline{\zeta}_l\tilde{\mathcal{K}}_{(d)}\underline{\varphi}_l = \underline{\varphi}_l$ still holds. The operators $\underline{\zeta}_l\tilde{\mathcal{K}}_{(d)}$ (where $\underline{\zeta}_l$ is taken with support close to M) are carried over to the manifold situation and added together, giving the desired operator. □

Proposition 11.3. *Let A be a differential operator of order d from $\tilde{E}_1$ to $\tilde{E}_2$, written as*

$$A = \sum_{l=0}^{d} S_l(x', x_n, D') D_n^l \tag{11.8}$$

on U, with differential operators S_l of order $d - l$ acting in X' for each $x_n \in\,] - 1, 1[$. The following Green's formula holds for $u \in H^d(E_{1,+})$ and $v \in H^d(E_{2,+})$:

$$(Au, v)_{X_+} - (u, A^*v)_{X_+} = (\mathfrak{A}\varrho^+ u, \varrho^+ v)_{X'}, \tag{11.9}$$

where $\mathfrak{A}$ is a (uniquely determined) matrix

$$\mathfrak{A} = (\mathfrak{A}_{jk})_{j,k=0,\dots,d-1}$$

of differential operators $\mathfrak{A}_{jk}$ (from E_1' to E_2') of orders $d - j - k - 1$, with

$$\mathfrak{A}_{jk}(x', D') = iS_{j+k+1}(x', 0, D') + \text{lower-order terms} \tag{11.10}$$

(zero if $j + k + 1 > d$). Here $\mathfrak{A}$ maps $\mathcal{H}^s(E_1'^d)$ continuously into $\widetilde{\mathcal{H}}^{s-d}(E_2'^d)$ for all $s \in \mathbb{R}$, bijectively if S_d is bijective at $x_n = 0$.   □

Proof. We show the formula for smooth u and v; then it extends by continuity to H^d spaces. By definition of A^*, $(Au, v)_{X_+} - (u, A^*v)_{X_+} = 0$ if u or v has compact support in X_+, so the only nontrivial contribution comes from cases where u and v are supported in U. For such sections we have the usual formula (by integration by parts)

$$(D_n u, v)_{U_+} - (u, D_n v)_{U_+} = i(\gamma_0^+ u, \gamma_0^+ v)_{X'},$$

and its iterated version (as in (4.53))

$$(D_n^l u, v)_{U_+} - (u, D_n^l v)_{U_+} = i \sum_{0 \le k \le l-1} (\gamma_{l-1-k}^+ u, \gamma_k^+ v)_{X'}.$$

Then

$$\begin{aligned}
(Au, v)_{X_+} - (u, A^*v)_{X_+} &= (\sum_{l \le d} S_l D_n^l u, v)_{X_+} - (u, \sum_{l \le d} D_n^l S_l^* v)_{X_+} \\
&= \sum_{l \le d} [(D_n^l u, S_l^* v)_{X_+} - (u, D_n^l S_l^* v)_{X_+}] \\
&= i \sum_{l \le d} \sum_{k \le l-1} (\gamma_{l-1-k}^+ u, \gamma_k^+ (S_l^* v))_{X'}.
\end{aligned} \tag{11.11}$$

Since $D_n^k(S_l^* v) = S_l^* D_n^k v + \sum_{j<k} S_{lj}' D_n^j v$ with differential operators S_{lj}' on X' of order $d - l$, the last expression can be further reorganized to have the form given in the right-hand side of (11.9). The asserted continuity holds since $\mathfrak{A}_{jk}$ is of order $d - j - k - 1$.

$\mathfrak{A}$ is uniquely determined since ϱ is *surjective* from the smooth sections over $\overline{X}_+$ to the d-tuples of smooth sections over X'.

Observe that $\mathfrak{A}$ has a skew-triangular character

$$\mathfrak{A} = \mathfrak{A}^0 + \mathfrak{A}' = i \begin{pmatrix} S_1^0 & \cdots & S_{d-1}^0 & S_d^0 \\ S_2^0 & \cdots & S_d^0 & 0 \\ \vdots & & \vdots & \vdots \\ S_d^0 & \cdots & 0 & 0 \end{pmatrix} + \begin{pmatrix} \text{lower} & & \cdots & 0 \\ \text{order} & \cdots & 0 & 0 \\ \vdots & & & \vdots \\ 0 & & \cdots & 0 \end{pmatrix}. \tag{11.12}$$

The terms in the second diagonal of $\mathfrak{A}$ and $\mathfrak{A}^0$ equal $iS_d(x', 0, D') = is_d(x')$; it is a zero-order differential operator and hence a multiplication operator (a vector bundle morphism from E_1' to E_2', trivializing to multiplication by an x'-dependent matrix). The operator $\mathfrak{A}$ (and also $\mathfrak{A}^0$) is then *invertible* if and only if $s_d(x')$ is a *bijective* morphism (the matrix is regular). □

When A is elliptic, $s_d(x')$ is invertible at each point, so $\mathfrak{A}$ is bijective. For more general differential operators, bijectiveness of $s_d(x')$ means that X' is *noncharacteristic* for A.

We see from the example of $\mathfrak{A}$ that it is interesting to allow matrix-formed operators that are *multi-order systems*, with different orders of the various entries, fitting together in a convenient way. When $S = (S_{jk})_{j_0 \le j \le j_1, k_0 \le k \le k_1}$, we say that

S has multi-order $(t_j, s_k)_{j_0 \le j \le j_1, k_0 \le k \le k_1}$, when each S_{jk} has order $t_j - s_k$. (11.13)

The principal part then consists of the $(t_j - s_k)$-order parts of the S_{jk}. Ellipticity means that the matrix of principal symbols defined accordingly, is invertible for $|\xi| \ge 1$; it is sometimes called Douglas-Nirenberg ellipticity (also Volevich should be mentioned in this context). The operator $\mathfrak{A}$ in (11.12) is of multi-order $(d - j - 1, k)_{j,k=0,\ldots,d-1}$, elliptic when $s_d(x')$ is bijective for all x'.

In relation to the given elliptic operator A we define the spaces

$$\begin{aligned} Z_\pm^s &= \{\, z \in H^s(E_{1,\pm}) \mid Az = 0 \text{ on } X_\pm \,\}, \\ N_\pm^s &= \varrho^\pm Z_\pm^s \subset \mathcal{H}^s(E_1'^d). \end{aligned} \tag{11.14}$$

Although the trace operators $\varrho^\pm$ are defined on $H^s(E_{1,\pm})$ for $s > d - \frac{1}{2}$ only, they can be given a sense on $Z_\pm^s$ (respectively) for all $s \in \mathbb{R}$.

Consider for example ϱ^+. Accounts of the extension to Z_+^s are found in several places: For one thing, there is Seeley's own deduction in [S66] where the Calderón projector was originally worked out with full proofs (many of which are reproduced here); this particular point is shown at the end of the paper, also for L_p Sobolev spaces, $1 < p < \infty$. Another account is by Lions and Magenes in the paper [LM63], where it is shown that when A is elliptic, $C^\infty(\overline{X}_+)$ is dense in $D_A^s = \{\, u \in H^s(X_+) \mid Au \in L_2(X_+) \,\}$ for all $s \le 0$, and this allows an extension of Green's formula to such u; we have included a particular case of the proof in Theorems 9.8, 9.10. (The paper [LM63] covers also L_p-related spaces; more general results in an L_2 framework are found in the book [LM68].) Third, the result follows from Theorems 4.3.1 and 2.5.6 in [H63] for the localized situation in $\mathbb{R}^n_+$, using the spaces $H^{s,t}(\mathbb{R}^n_+)$ (cf. (10.132), (10.134)). The idea is to use the equation $Au = 0$ to conclude that $u \in Z^s_+$ implies $u \in H^{s+k,-k}(\mathbb{R}^n_+)$ for all k ("partial hypoellipticity at the boundary"); then for sufficiently large k, the boundary value $\gamma^+_j u$ makes sense. For completeness, we indicate a proof of the third type.

Theorem 11.4. *The map ϱ^+ extends to a continuous map from Z^s_+ to $\mathcal{H}^s(E^{\prime d}_1)$ for all $s \in \mathbb{R}$.*

In fact, when $u \in Z^s_+$, then in local coordinates at the boundary, u is in $H^{s+k,-k}(\mathbb{R}^n_+)$ for any $k \in \mathbb{N}_0$, where $\gamma^+_j : H^{s+k,-k}(\mathbb{R}^n_+) \to H^{s-j-\frac{1}{2}}(\mathbb{R}^{n-1})$ for $s + k - j > \frac{1}{2}$.

Similarly, ϱ^- extends to a continuous map from Z^s_- to $\mathcal{H}^s(E^{\prime d}_1)$ for all $s \in \mathbb{R}$.

Proof (indications). One ingredient in the proof is the fact that for any $r \in \mathbb{Z}$, the operator $(\Xi^r_-)_+ = \mathrm{OP}((\langle \xi' \rangle - i\xi_n)^r)_+$ maps $H^{s,t}(\mathbb{R}^n_+)$ homeomorphically onto $H^{s-r,t}(\mathbb{R}^n_+)$, with inverse $(\Xi^{-r}_-)_+$, for all $s,t \in \mathbb{R}$. This can be inferred from [H63] and appears in Seeley's proof [S66], and in joint works of Vishik and Eskin as well as Eskin's book [E81]. See Exercise 10.11, where the reader is guided through a proof.

Consider the localized situation. Let $A = D^d_n + \sum_{0 \le l \le d-1} S_l(x, D')D^l_n$, where the S_l are differential operators with respect to x' of order $d - l$ with bounded smooth coefficients. (An elliptic operator is reduced to this form by dividing out the coefficient of D^d_n.) If $u \in H^{s,t}(\mathbb{R}^n_+)$ solves $Au = 0$, then since $S_l D^l_n u \in H^{s-l,t-d+l}(\mathbb{R}^n_+) \subset H^{s-d+1,t-1}(\mathbb{R}^n_+)$ (recall (10.133)),

$$D^d_n u = - \sum_{0 \le l \le d-1} S_l D^l_n \in H^{s-d+1,t-1}(\mathbb{R}^n_+).$$

Then also $(\Xi^d_-)_+ u = \sum_{0 \le l \le d} c_l \mathrm{OP}'(\langle \xi' \rangle^{d-l})D^l_n u$ lies in $H^{s-d+1,t-1}(\mathbb{R}^n_+)$, and it follows by application of $(\Xi^{-d}_-)_+$ that $u \in H^{s+1,t-1}(\mathbb{R}^n_+)$. This gives the induction step in a proof that when $u \in H^{s,0}(\mathbb{R}^n_+)$ solves $Au = 0$, then $u \in H^{s+k,-k}(\mathbb{R}^n_+)$ for all k. The asserted mapping property of γ^+_j was shown in Corollary 10.30. $\square$

Since $(Au, v)_{\widetilde{X}} - (u, A^*v)_{\widetilde{X}} = 0$, we have in addition to (11.9):

$$(Au, v)_{X_-} - (u, A^*v)_{X_-} = -(\mathfrak{A}\varrho^- u, \varrho^- v)_{X'}, \tag{11.15}$$

for $u \in H^d(E_{1,-})$ and $v \in H^d(E_{2,-})$.

The central step in the Calderón construction is to introduce the operators

$$K = Q\widetilde{\varrho}^*\mathfrak{A}, \quad K^\pm = \mp r^\pm K = \mp r^\pm Q\widetilde{\varrho}^*\mathfrak{A}. \tag{11.16}$$

Since $\mathfrak{A} : \mathcal{H}^s(E^{\prime d}_1) \to \widetilde{\mathcal{H}}^{s-d}(E^{\prime d}_2)$ for $s \in \mathbb{R}$, $\widetilde{\varrho}^* : \widetilde{\mathcal{H}}^{s-d}(E^{\prime d}_2) \to H^{s-d}(\widetilde{E}_2)$ for $s < \frac{1}{2}$ (cf. (11.4)), and $Q : H^{s-d}(\widetilde{E}_2) \to H^s(\widetilde{E}_1)$ for $s \in \mathbb{R}$, we have that

$$K : \mathcal{H}^s(E_1'^d) \to H^s(\widetilde{E}_1), \quad \text{when } s < \tfrac{1}{2};$$
$$K^\pm : \mathcal{H}^s(E_1'^d) \to H^s(E_{1,\pm}), \quad \text{when } s < \tfrac{1}{2}. \tag{11.17}$$

Observe moreover that the operators $K^\pm$ in fact map into $Z_\pm^s$, since

$$AK\varphi = \widetilde{\varrho}^* \mathfrak{A}\varphi \quad \text{is supported in } X'. \tag{11.18}$$

The operators $K^\pm$ are *Poisson operators* in view of Theorem 10.25, so by Theorem 10.29, the mapping properties in the second line of (11.17) extend to all $s \in \mathbb{R}$. This is not true for the operator K itself that gives an important singularity at $x_n = 0$.

Proposition 11.5. *The Poisson operators $K^\pm$ defined in (11.17) satisfy, for any $s \in \mathbb{R}$:*

$$K^+ \varrho^+ z = z, \text{ for } z \in Z_+^s, \quad K^- \varrho^- z = z, \text{ for } z \in Z_-^s. \tag{11.19}$$

It follows that for all $s \in \mathbb{R}$:

$$\varrho^+ K^+ \varphi = \varphi, \text{ for } \varphi \in N_+^s, \quad \varrho^- K^- \varphi = \varphi, \text{ for } \varphi \in N_-^s. \tag{11.20}$$

Proof. We begin by showing that

$$K^+ \varrho^+ z = z, \text{ for } z \in Z_+^d, \quad K^- \varrho^- z = z, \text{ for } z \in Z_-^d. \tag{11.21}$$

The first statement is seen as follows: Let $z \in Z_+^d$, let $w \in L_2(E_{1,+})$ and let $v = Q^* e^+ w$; it lies in $H^d(\widetilde{E}_2)$. Note that $r^+ A^* v = w$. Then by Green's formula (11.9), since $Az = 0$ on X_+,

$$-(z,w)_{X_+} = (Az, r^+ v)_{X_+} - (z, r^+ A^* v)_{X_+} = (\mathfrak{A}\varrho^+ z, \varrho^+ v)_{X'}$$
$$= (\mathfrak{A}\varrho^+ z, \widetilde{\varrho} v)_{X'} = \langle \widetilde{\varrho}^* \mathfrak{A}\varrho^+ z, \overline{Q^* e^+ w} \rangle_{\widetilde{X}} = (Q\widetilde{\varrho}^* \mathfrak{A}\varrho^+ z, e^+ w)_{\widetilde{X}}$$
$$= -(K^+ \varrho^+ z, w)_{X_+}.$$

Since w was arbitrary, it follows that $z = K^+ \varrho^+ z$. The second statement in (11.21) is shown similarly, using (11.15).

The identities in (11.19) then also hold for $s \geq d$, by the continuity properties of the maps. Our next task is to show them for $s < d$. We here take recourse to the description in Theorem 11.4 of the elements of the nullspace in local coordinates. Let $z \in Z_+^s$ for some $s < d$. Let $\{\eta_1, \ldots, \eta_{J_0}\}$ be a partition of unity subordinate to an atlas of local trivializations $\Psi_i, U_i, V_i, i = 1, \ldots, I_0$, for E_1, as in Lemma 8.4 2°. Decompose z as $z = \sum_{j \leq J_0} z_j$, $z_j = \eta_j z$. We use the notation $\underline{z}_j$ for the localized version of z_j for a choice of trivialization Ψ_i where η_j is supported in U_i (and we likewise indicate localized operators by underlining). By Theorem 11.4, each $\underline{z}_j$ is in $H^{s+k,-k}(\mathbb{R}_+^n)^N$ for $k \in \mathbb{N}_0$, where $\underline{\varrho}^+$ is well-defined if $s + k \geq d$, mapping into $\mathcal{H}^s(\mathbb{R}^{n-1}, (\mathbb{C}^N)^d)$. Take k so large that $s + k \geq d$. Since $C_{(0)}^\infty(\overline{\mathbb{R}}_+^n)$ is dense in $H^{s+k,-k}(\mathbb{R}_+^n)$, we can find a sequence $u_{j,l}$ of C^∞-functions supported in U_i such that $\underline{u}_{j,l} \to \underline{z}_j$ in $H^{s+k,-k}(\mathbb{R}^n_+)^N$ for $l \to \infty$. Then $\underline{A}\,\underline{u}_{j,l} \to \underline{A}\,\underline{z}_j$ in $H^{s+k-d,-k}(\mathbb{R}^n_+)^N$. Let $v_{j,l} = Q_+Au_{j,l}$; then $Av_{j,l} = Au_{j,l}$, so $z_{j,l} = u_{j,l} - v_{j,l}$ has $Az_{j,l} = 0$. We set $z_{(l)} = \sum_{j \le J_0} z_{j,l}$; it lies in Z^∞_+.

Moreover, we can write each $v_{j,l}$ as

$$v_{j,l} = \sum_{m \le J_0} v_{j,m,l}, \quad v_{j,m,l} = \eta_m v_{j,l}.$$

Consider a term $v_{j,m,l}$; here when U_i contains the supports of both η_m and η_j, $v_{j,m,l}$ localizes to $\underline{v}_{j,m,l} = \underline{\eta}_m \underline{Q}_+ \underline{A}\,\underline{u}_{j,l}$. It converges to $\underline{v}_{j,m,0} = \underline{\eta}_m \underline{Q}_+ \underline{A}\,\underline{z}_j$ in $H^{s+k,-k}$ for $l \to \infty$, so in this sense,

$$\sum_{j \le J_0} v_{j,l} = \sum_{j,m \le J_0} v_{j,m,l} \to Q_+Az = 0, \text{ for } l \to \infty. \tag{11.22}$$

Now $z_{(l)} \to z$ in the sense that the localized pieces $\underline{z}_{j,l} = \underline{u}_{j,l} - \sum_{m \le J_0} \underline{v}_{j,m,l}$ have $\underline{u}_{j,l} \to \underline{z}_j$ in $H^{s+k,-k}$, and $\underline{v}_{j,m,l} \to \underline{v}_{j,m,0}$ in $H^{s+k,-k}$, with $\sum_{j,m \le J_0} v_{j,m,0} = 0$.

Since $z_{(l)} \in Z^\infty_+$, (11.21) applies to show that

$$K^+ \varrho^+ z_{(l)} = z_{(l)}. \tag{11.23}$$

For the localized pieces we have for $l \to \infty$ (after insertion of an extra partition of unity (η_r) to localize the action of K^+):

$$\underline{\eta}_r \underline{K}^+ \underline{\varrho}^+ (\underline{\eta}_m \underline{u}_{j,l} - \underline{v}_{j,m,l}) \to \underline{\eta}_r \underline{K}^+ \underline{\varrho}^+ \underline{\eta}_m \underline{z}_j,$$

$$\underline{\eta}_m \underline{u}_{j,l} - \underline{v}_{j,m,l} \to \underline{\eta}_m \underline{z}_j - \underline{v}_{m,j,0},$$

in $H^{s+k,-k}$, using that $\underline{\varrho}^+$ maps $H^{s+k,-k}$ to $\mathcal{H}^s$ and $\underline{\eta}_r \underline{K}^+$ maps $\mathcal{H}^s$ to $H^{s+k,-k}$. The formulas hold in trivializations Ψ_i where η_r, η_m, η_j are supported in U_i. Adding the pieces carried back to X and using (11.23) before passing to the limit, we find the desired identity $K^+\varrho^+z = z$.

There is a similar proof for the minus-case.

Finally, (11.21) follows immediately, since $\varphi \in N^s_+$ means that $\varphi = \varrho^+z$ for some $z \in Z^s_+$, and

$$\varrho^+ K^+ \varphi = \varrho^+ K^+ \varrho^+ z = \varrho^+ z = \varphi,$$

by the just proved identity. There is a similar proof for the minus-case. $\square$

Definition 11.6. The Calderón projectors $C^\pm$ associated with A are defined by

$$C^\pm = \varrho^\pm K^\pm. \tag{11.24}$$

They are ψdo's in $\bigoplus_{0 \le j < d} E_1'$, by the composition rule Theorem 10.24 7°
and Theorem 10.28, with the continuity property

$$C^{\pm} : \mathcal{H}^s(E_1'^d) \to \mathcal{H}^s(E_1'^d), \quad \text{all } s \in \mathbb{R}. \tag{11.25}$$

Moreover,

$$C^{\pm} \text{ maps } \mathcal{H}^s(E_1'^d) \text{ into } N_{\pm}^s, \tag{11.26}$$

respectively, since $K^{\pm}$ maps into $Z_{\pm}^s$, respectively. The projection property
will now be shown.

Proposition 11.7. 1° *The ψdo's $C^{\pm}$ defined in Definition 11.6 are projections in $\mathcal{H}^s(E_1'^d)$ for all $s \in \mathbb{R}$,*

$$(C^+)^2 = C^+, \quad (C^-)^2 = C^-.$$

2° *Moreover, C^+ and C^- are complementing projections,*

$$C^+ + C^- = I, \quad C^+ C^- = 0 = C^- C^+.$$

Proof. The projection property follows, since

$$(C^+)^2 = \varrho^+ K^+ \varrho^+ K^+ = \varrho^+ K^+ = C^+$$

in view of (11.19); the identity $(C^+)^2 = C^+$ thus holds for smooth sections and extends by continuity to general distribution sections. The identity
$(C^-)^2 = C^-$ follows similarly from (11.19). This shows 1°.

To show 2°, let $\varphi \in \mathcal{H}^s(E_1'^d)$, and let $z^{\pm} = K^{\pm}\varphi$. For each $\psi \in C^{\infty}(E_1'^d)$,
choose a section $v \in C^{\infty}(\widetilde{E})$ with $\widetilde{\varrho}v = \psi$. Then we have, using the Green's
formulas (11.9), (11.15) "backwards", together with the fact that $z^{\pm} \in Z_{\pm}^d$,

$$(\mathfrak{A}(C^+ + C^-)\varphi, \psi)_{X'} = (\mathfrak{A}\varrho^+ z^+, \varrho^+ r^+ v) + (\mathfrak{A}\varrho^- z^-, \varrho^- r^- v)$$
$$= (Az^+, r^+ v)_{X_+} - (z^+, r^+ A^* v)_{X_+} - (Az^-, r^- v)_{X_-} + (z^-, r^- A^* v)_{X_-}$$
$$= -(K^+\varphi, r^+ A^* v)_{X_+} + (K^-\varphi, r^- A^* v)_{X_-}$$
$$= (Q\widetilde{\varrho}^* \mathfrak{A}\varphi, A^* v)_{\widetilde{X}} = \langle \widetilde{\varrho}^* \mathfrak{A}\varphi, \overline{v}\rangle_{\widetilde{X}} = (\mathfrak{A}\varphi, \psi)_{X'}.$$

Since φ and ψ were arbitrary and $\mathfrak{A}$ is invertible, it follows that $C^+ + C^- = I$ holds on $\mathcal{H}^d(E_1'^d)$, and the validity on $\mathcal{H}^s(E_1'^d)$ for general s follows by
extension by continuity.

It follows moreover that $C^+ C^- = C^+(I - C^+) = C^+ - (C^+)^2 = 0$. $\quad\square$

We have hereby established the essential ingredients in the main theorem:

Theorem 11.8. *Assume that A has the inverse Q on $\widetilde{X}$. Define the spaces
$Z_{\pm}^s$ and $N_{\pm}^s$ by (11.14). Then the spaces $N_{\pm}^s$ are complementing closed subspaces of $\mathcal{H}^s(E_1'^d)$;*

$$\mathcal{H}^s(E_1'^d) = N_+^s \dotplus N_-^s, \quad \text{for any } s \in \mathbb{R}. \tag{11.27}$$

When we define

$$K^{\pm} = \mp r^{\pm} Q \tilde{\varrho}^{*} \mathfrak{A}, \quad C^{\pm} = \varrho^{\pm} K^{\pm} = \mp \varrho^{\pm} r^{\pm} Q \tilde{\varrho}^{*} \mathfrak{A}, \tag{11.28}$$

the Poisson operators $K^{\pm} : \mathcal{H}^{s}(E_1^{\prime d}) \to H^{s}(E_{1,\pm})$ have range equal to $Z_{\pm}^{s}$; moreover, they act as homeomorphisms

$$K^{\pm} : N_{\pm}^{s} \xrightarrow{\sim} Z_{\pm}^{s}, \quad \text{with inverse } \varrho^{\pm}, \tag{11.29}$$

respectively. This gives us a parametrization of the nullspace $Z_{\pm}^{s}$ by its Cauchy data.

The ψdo's $C^{\pm}$ (the Calderón projectors for A) are the projections of $\mathcal{H}^{s}(E_1^{\prime d})$ onto $N_{\pm}^{s}$ along $N_{\mp}^{s}$, respectively. In particular,

$$C^{+} + C^{-} = I, \quad (C^{+})^{2} = C^{+}, \quad (C^{-})^{2} = C^{-}, \quad C^{+}C^{-} = 0. \tag{11.30}$$

Proof. It remains to account for surjectiveness and homeomorphism properties.

The surjectiveness of $K^{+} : \mathcal{H}^{s}(E_1^{\prime d}) \to Z_{+}^{s}$ follows from the identity $z = K^{+}\varrho^{+}z$ for $z \in Z_{+}^{s}$ shown in Proposition 11.5.

It follows that the $C^{\pm}$ are surjective onto the spaces $N_{\pm}^{s}$. These are complementing closed subspaces of $\mathcal{H}^{s}(E_1^{\prime d})$, since they are the range spaces for the complementing projections.

Now the $K^{\pm}$ in (11.29) are injective, since e.g. $K^{+}\varphi = 0$ for a $\varphi \in N_{+}^{s}$ implies $C^{+}\varphi = 0$, hence $\varphi = 0$ since C^{+} acts like the identity on N_{+}^{s}. They are also surjective, since $z \in Z_{+}^{s}$ can be written as $z = K^{+}\varphi$ with $\varphi = \varrho^{+}z \in N_{+}^{s}$, by Proposition 11.5 and the definition of N_{+}^{s}. So indeed (11.29) holds with plus; there is a similar proof with minus. $\square$

When Q is merely a parametrix of A, one can still define operators $K^{\pm}$ by formulas as in (11.28) supplied with smoothing terms, setting

$$C^{+} = \varrho^{+} K^{+} = -\varrho^{+} r^{+} Q \tilde{\varrho}^{*} \mathfrak{A} + \mathcal{T} \tag{11.31}$$

and $C^{-} = I - C^{+}$ (with a ψdo $\mathcal{T}$ of order $-\infty$); then they have the listed mapping properties only modulo smoothing operators. Such a construction is worked out in [G77] for general multi-order operators A, with applications. Seeley gives in [S69] an *optimal* construction, where K^{+} maps $\mathcal{H}^{s}$ injectively onto a subspace of Z_{+}^{s} with complement Z_0, and where $C^{+} = \varrho^{+}K^{+}$ *is a projection of $\mathcal{H}^{s}$ onto N_{+}^{s}*. The book of Booss-Bavnbek and Wojciechowski [BW93] goes through a proof of Theorem 11.8 for first-order operators A.

The *principal symbols* of $C^{\pm}$ and $K^{\pm}$ are found by carrying out the analogous construction in the one-dimensional model case (in the style of Example 11.1). For fixed x', fixed $\xi' \neq 0$, consider the model operator

$$a^{0}(x', \xi', D_n) = \text{OP}_n(a^{0}(x', 0, \xi', \xi_n)),$$

acting on N-vector functions on $\mathbb{R}$. Recall the notation $\mathscr{S}(\overline{\mathbb{R}}_\pm) = \mathscr{S}_\pm$. The solutions of $a^0 u = 0$ that are bounded on $\mathbb{R}_+$ resp. $\mathbb{R}_-$ form the vector spaces

$$Z_\pm(x', \xi') = \{u(x_n) \in \mathscr{S}_\pm^N \mid a^0(x', \xi', D_n)u = 0 \text{ on } \mathbb{R}_\pm\}. \tag{11.32}$$

They were recalled in Example 10.39: The bounded solutions on $\mathbb{R}_+$ are spanned by exponential polynomials $f_j(x_n)e^{i\tau_j x_n}$, where τ_j is a root in $\mathbb{C}_+$ of the polynomial $\det a^0(x', \xi', \tau)$ (so the solutions are rapidly decreasing for $x_n \to \infty$). The dimension of $Z_+(x', \xi')$ equals $m_+(x', \xi')$, the number of roots in $\mathbb{C}_+$ counted with multiplicity. Similarly, $Z_-(x', \xi')$ is spanned by exponential polynomials with $\tau_j \in \mathbb{C}_-$ and has dimension $m_-(x', \xi')$. Note that requiring the solutions to be in $H^s(\mathbb{R}_+)^N$ for some $s \in \mathbb{R}$ gives the same result; only the functions $f_j(x_n)e^{i\tau_j x_n}$ with $\tau_j \in \mathbb{C}_+$ can satisfy this. Hence the nullspaces coincide, for all s, with the space defined in (11.32),

$$Z_\pm(x', \xi') = \{u(x_n) \in H^s(\mathbb{R}_\pm)^N \mid a^0(x', \xi', D_n)u = 0 \text{ on } \mathbb{R}_\pm\}. \tag{11.33}$$

So we need not refer to a Sobolev scale of nullspaces in the model case.

It is standard ODE knowledge that the null solutions are in a 1–1 correspondence with their Cauchy data, so if we define

$$N_+(x', \xi') = \varrho^+ Z_+(x', \xi'), \quad N_-(x', \xi') = \varrho^- Z_-(x', \xi'), \tag{11.34}$$

we have immediately that $N_\pm(x', \xi')$ have dimension $m_\pm(x', \xi')$. Moreover, $N_+(x', \xi')$ and $N_-(x', \xi')$ must be complementing subspaces of $\mathbb{C}^{Nd}$, for their intersection is zero (a linear combination of the occurring exponential polynomials cannot vanish both for $x_n \to \infty$ and for $x_n \to -\infty$ without being 0), and the sum of their dimensions is Nd.

So, just from the knowledge of the solution structure of ODE, we have the *existence* of homeomorphisms $K^\pm(x', \xi') : N_\pm(x', \xi') \to Z_\pm(x', \xi')$ inverse to $\varrho^\pm : Z_\pm(x', \xi') \to N_\pm(x', \xi')$, and complementing projections $C^\pm(x', \xi')$ of $\mathbb{C}^{Nd}$ onto $N_\pm(x', \xi')$, respectively. The $K_\pm(x', \xi')$ are extended linearly to map $N_\mp(x', \xi')$ to 0, respectively.

The Calderón construction gives us useful formulas for these operators. There is a matrix $\mathfrak{a}^0(x', \xi')$ such that

$$(a^0(D_n)u, v)_{\mathbb{R}_+} - (u, a^0(D_n)^* v)_{\mathbb{R}_+} = (\mathfrak{a}^0 \varrho^+ u, \varrho^+ v)_{\mathbb{C}^{Nd}}, \tag{11.35}$$

for $u, v \in H^d(\mathbb{R}_+)^N$, it is the principal symbol of $\mathfrak{A}$ from Proposition 11.3. (It is a $d \times d$-matrix of $N \times N$-matrices.) The inverse of $a^0(x', \xi', D_n)$ is defined from the principal symbol of Q, $q^0 = (a^0)^{-1}$, by

$$a^0(x', \xi', D_n)^{-1} = q^0(x', \xi', D_n) = \text{OP}_n(q^0(x', 0, \xi)).$$

We can consider the adjoint of $\tilde{\varrho} : H^s(\mathbb{R})^N \to \mathbb{C}^{Nd}$ $(s > \frac{1}{2})$, writing

$$\tilde{\varrho}^* v = \begin{pmatrix} 1 & D_n & \cdots & D_n^{d-1} \end{pmatrix} \tilde{\gamma}_0^* v; \quad \tilde{\gamma}_0^* v = v\delta(x_n) \text{ for } v \in \mathbb{C}^{Nd}. \tag{11.36}$$

Then when we define

$$K(x', \xi') = q^0(x', \xi', D_n)\widetilde{\varrho}^* \mathfrak{a}^0(x', \xi'),$$
$$K^\pm(x', \xi') = \mp r^\pm K(x', \xi') = \mp r^\pm q^0(x', \xi', D_n)\widetilde{\varrho}^* \mathfrak{a}^0(x', \xi'), \quad (11.37)$$

the $K^\pm(x', \xi')$ are verified to be the desired operators (that we already have defined), with $C^\pm(x', \xi') = \varrho^\pm K^\pm(x', \xi')$, by simple variants of the proofs of Propositions 11.5 and 11.7 and Theorem 11.8.

The reader is encouraged to check the details. In these calculations, doing the $\circ_n$ compositions starting with the originally given principal symbols, we arrive at the *principal symbols* of the operators that were first defined in the full calculus. Thus the operators $K^\pm(x', \xi')$ and $C^\pm(x', \xi')$ are the principal boundary symbol operators for the operators $K^\pm$, $C^\pm$ in the PDE situation, usually denoted $k^{\pm,0}(x', \xi', D_n)$ resp. $c^{\pm,0}(x', \xi')$.

For later reference, we sum up the results in a proposition.

Proposition 11.9. *Let $\xi' \neq 0$. The spaces $Z_\pm(x', \xi')$ and $N_\pm(x', \xi')$ defined in (11.32)–(11.34) have dimension $m_\pm(x', \xi')$ (cf. Example 10.39).*

The principal boundary symbol operators for $K^\pm$ and $C^\pm$ in Theorem 11.8 are determined as

$$k^{\pm,0}(x', \xi', D_n) = K^\pm(x', \xi'), \quad c^{\pm,0}(x', \xi') = \varrho^\pm k^{\pm,0}(x', \xi', D_n)$$

(through formulas given in (11.35)–(11.37)); here

$$k^{\pm,0}(x', \xi', D_n) : N_\pm(x', \xi') \xrightarrow{\sim} Z_\pm(x', \xi'),$$

and $c^{\pm,0}(x', \xi')$ project $\mathbb{C}^{Nd}$ onto its complementing subspaces $N_\pm(x', \xi')$, respectively.

11.2 Application to boundary value problems

The Calderón projectors are very useful in a treatment of boundary value problems for A. We now return to the notation ϱ for ϱ^+. Let there be given a problem:

$$Au = f \text{ on } X, \quad S\varrho u = \varphi \text{ on } X', \quad (11.38)$$

where S is a system of ψdo's S_{jk} of order $j - k$ ($j, k = 0, \ldots, d-1$) going from E_1' to bundles F_j of dimension ≥ 0 over X'; $M = \sum_{0 \leq j < d} \dim F_j$. (The zero-dimensional bundles could be omitted; they are just included for notational convenience.)

In the following, we consider $\{A, S\varrho\}$ as a mapping from $H^s(E_1)$ to $H^{s-d}(E_2) \times \mathcal{H}^s(F)$ (recall (11.1)), for some $s > d - \frac{1}{2}$, and discuss right/left inverses that are continuous in the opposite direction; here S is considered as a mapping from $\mathcal{H}^s(E_1'^d)$ to $\mathcal{H}^s(F)$ and the $C^{\pm}$ act in $\mathcal{H}^s(E_1'^d)$. We assume as above that A is invertible on $\widetilde{X}$, with inverse Q.

Theorem 11.10. $1°$ *If* SC^+ *has a right inverse* S_1, *then* $\left(\begin{smallmatrix} A \\ S\varrho \end{smallmatrix}\right)$ *has the right inverse*

$$\left(R_S \ \ K_S\right) = \left(Q_+ - K^+ S_1 S\varrho Q_+ \ \ K^+ S_1\right). \tag{11.39}$$

Conversely, if $\left(\begin{smallmatrix} A \\ S\varrho \end{smallmatrix}\right)$ *has a right inverse* $\left(R_S \ \ K_S\right)$, *then* SC^+ *has the right inverse*

$$S_1 = \varrho K_S. \tag{11.40}$$

$2°$ *If* $\left(\begin{smallmatrix} S \\ C^- \end{smallmatrix}\right)$ *has a left inverse* $\left(S_1 \ \ S_2\right)$, *then* $\left(\begin{smallmatrix} A \\ S\varrho \end{smallmatrix}\right)$ *has the left inverse* (11.39).

Conversely, if $\left(\begin{smallmatrix} A \\ S\varrho \end{smallmatrix}\right)$ *has a left inverse* $\left(R_S \ \ K_S\right)$, *then* $\left(\begin{smallmatrix} S \\ C^- \end{smallmatrix}\right)$ *has the left inverse*

$$\left(S_1 \ \ S_2\right) = \left(\varrho K_S \ \ I - \varrho K_S S\right). \tag{11.41}$$

Proof. We first observe some auxiliary formulas:

$$AQ_+ = I, \quad Q_+ A = I - K^+ \varrho, \quad K^+ C^- = 0. \tag{11.42}$$

The first formula holds since $AQ = I$ on $\widetilde{X}$ and A is local. Next, we note that Green's formula (11.9) can be written in distributional form (compare with (10.82)):

$$e^+ r^+ A\tilde{u} = A e^+ r^+ \tilde{u} + \widetilde{\varrho}^*(\mathfrak{A}\varrho u) \text{ for } \tilde{u} \in H^{s+d}(\widetilde{E}_1), \ u = r^+\tilde{u}, \tag{11.43}$$

for $s > -\frac{1}{2}$. The second formula follows from this by composition with $r^+ Q$, using (11.28) and the fact that $QA = I$ on $\widetilde{X}$; it holds on $H^{s+d}(E_1)$, $s > -\frac{1}{2}$. The third formula holds since $K^+ C^+ = K^+ \varrho K^+ = K^+$, cf. Proposition 11.5.

For statement $1°$, let S_1 be a right inverse of SC^+. Then, by the above rules,

$$A(Q_+ - K^+ S_1 S\varrho Q_+) = I,$$
$$S\varrho(Q_+ - K^+ S_1 S\varrho Q_+) = S\varrho Q_+ - SC^+ S_1 S\varrho Q_+ = 0,$$
$$AK^+ S_1 = 0,$$
$$S\varrho K^+ S_1 = SC^+ S_1 = I.$$

Conversely, when $\left(R_S \ \ K_S\right)$ is a right inverse of $\left(\begin{smallmatrix} A \\ S\varrho \end{smallmatrix}\right)$, then $AK_S = 0$, $S\varrho K_S = I$, so K_S maps into Z_+^s, whereby $C^- \varrho K_S = 0$ and consequently $SC^+ \varrho K_S = S\varrho K_S - SC^- \varrho K_S = I$. Thus ϱK_S is a right inverse of SC^+. This proves $1°$.

For $2°$, we check the composition of (11.39) to the left with $\left(\begin{smallmatrix} A \\ S\varrho \end{smallmatrix}\right)$ as follows, using (11.42) and the fact that $C^- C^+ = 0$:

$$\left(Q_+ - K^+ S_1 S \varrho Q_+ \ K^+ S_1\right) \left(\begin{smallmatrix} A \\ S \varrho \end{smallmatrix}\right) = (I - K^+ S_1 S \varrho) Q_+ A + K^+ S_1 S \varrho$$
$$= (I - K^+ S_1 S \varrho)(I - K^+ \varrho) + K^+ S_1 S \varrho = I - K^+ (I - S_1 S C^+) \varrho$$
$$= I - K^+ (I - (I - S_2 C^-) C^+) \varrho = I - K^+ C^- \varrho = I. \tag{11.44}$$

Conversely, define $(S_1 \ S_2)$ by (11.41) and check its left composition with $\left(\begin{smallmatrix} S \\ C^- \end{smallmatrix}\right)$:

$$\left(\varrho K_S \ I - \varrho K_S S\right) \left(\begin{smallmatrix} S \\ C^- \end{smallmatrix}\right) = \varrho K_S S + C^- - \varrho K_S S C^- = \varrho K_S S C^+ + I - C^+. \tag{11.45}$$

When $w = K^+ C^+ \varphi$ for some $\varphi \in C^\infty(E_1'^d)$, then $Aw = 0$, $\varrho w = C^+ C^+ \varphi = C^+ \varphi$ and $S \varrho w = S C^+ \varphi$, so since $(R_S \ K_S)$ is a left inverse of $\left(\begin{smallmatrix} A \\ S \varrho \end{smallmatrix}\right)$,

$$w = K_S S \varrho w = K_S S C^+ \varphi.$$

It follows that $\varrho K_S S C^+ \varphi = \varrho w = C^+ \varphi$ for $\varphi \in C^\infty(E_1'^d)$. Then the expression in (11.45) equals I. This ends the proof of 2°. $\qquad \square$

The statements have generalizations where the word "inverse" is replaced by "parametrix", also when Q is merely a parametrix of A (here one can keep track of the smoothing terms as in [G77]).

The result holds in particular on the principal boundary symbol level, when we discuss solutions in $\mathscr{S}_+$. We can extend the terminology from Chapter 7 of surjectively elliptic, resp. injectively elliptic, operators and symbols, to boundary value problems, calling the systems with surjectiveness, resp. injectiveness, of the model operator (for all x', all $|\xi'| = 1$) *surjectively elliptic*, resp. *injectively elliptic*.

Theorem 11.11. *Consider the model operator (principal boundary symbol operator)*

$$\begin{pmatrix} a^0(x',0,\xi',D_{x_n}) \\ s^0(x',\xi') \varrho \end{pmatrix} : \mathscr{S}_+^N \to \begin{matrix} \mathscr{S}_+^N \\ \times \\ \mathbb{C}^M \end{matrix},$$

for $\{A, S\varrho\}$ in Theorem 11.10. For all x', all $|\xi'| \geq 1$, the statements of Theorem 11.10 hold for the principal boundary symbol operators:
$\begin{pmatrix} a^0(x',0,\xi',D_{x_n}) \\ s^0(x',\xi') \varrho \end{pmatrix}$ *has a right inverse if and only if* $\begin{pmatrix} s^0(x',\xi') \\ c^{-,0}(x',\xi') \end{pmatrix}$ *does so,*
and $\begin{pmatrix} a^0(x',0,\xi',D_{x_n}) \\ s^0(x',\xi') \varrho \end{pmatrix}$ *has a left inverse if and only if $s^0(x',\xi') c^{+,0}(x',\xi')$ does*
so (with the corresponding versions of (11.39)–(11.41)).

In particular,

$$\begin{matrix} \left(\begin{smallmatrix} A \\ S \varrho \end{smallmatrix}\right) \text{ is injectively elliptic} \iff \left(\begin{smallmatrix} S \\ C^- \end{smallmatrix}\right) \text{ is injectively elliptic;} \\ \left(\begin{smallmatrix} A \\ S \varrho \end{smallmatrix}\right) \text{ is surjectively elliptic} \iff SC^+ \text{ is surjectively elliptic.} \end{matrix} \tag{11.46}$$

Here $\begin{pmatrix} s^0(x',\xi') \\ c^{-,0}(x',\xi') \end{pmatrix}$ is an $(M+Nd) \times Nd$-matrix, whereas $s^0(x',\xi') c^{+,0}(x',\xi')$ is an $M \times Nd$-matrix. In this way, the question of existence of a parametrix is reduced to a matrix question.

In view of Proposition 11.9, the injectively resp. surjectively elliptic problems can also be characterized by *injectiveness resp. surjectiveness of* $s^0(x', \xi')$ *from* $N_+(x', \xi')$ *to* $\mathbb{C}^M$ for all x', $|\xi'| = 1$. In particular, this requires $M \geq m_+(x', \xi')$ resp. $M \leq m_+(x', \xi')$. Thus for two-sided elliptic problems, M must equal $m_+(x', \xi')$ (which must be constant in (x', ξ') then). In the properly elliptic case (in all cases when $n \geq 3$), one has $m_+(x', \xi') = m_-(x', \xi') = \frac{1}{2}Nd$, so two-sided ellipticity is possible only when $M = \frac{1}{2}Nd$.

We observe that injective ellipticity holds if and only if

$$v \in \mathbb{C}^{Nd}, \ s^0(x', \xi')v = 0, \ c^{-,0}(x', \xi')v = 0 \implies v = 0; \quad (11.47)$$

i.e., the nullspaces of s^0 and $c^{-,0}$ are *linearly independent*.

Example 11.12. The systems $\begin{pmatrix} A \\ \varrho \end{pmatrix}$ and $\begin{pmatrix} A \\ C^+\varrho \end{pmatrix}$ are injectively elliptic; they both have the left inverse $(Q_+ \ K^+)$. In fact, by (11.42),

$$Q_+A + K^+\varrho = I; \quad Q_+A + K^+C^+\varrho = I.$$

This left inverse is also found from (11.39), when we use that $\begin{pmatrix} I \\ LiC^- \end{pmatrix}$ and $\begin{pmatrix} C^+ \\ C^- \end{pmatrix}$ both have the left inverse $(C^+ \ C^-)$. See also Exercises 11.22 and 11.23. The case $S = C^+$ for first-order systems is closely related to the Atiyah-Patodi-Singer problem [APS75]. Links to the abundant literature on this are found e.g. in [BW93], [G99] and [G03].

(11.42) also shows that Q_+ is a right inverse of A *without boundary condition*; i.e., in the case $F = 0$. This is also confirmed by the formulas in the theorem.

The boundary conditions in this example are too strong, resp. too weak, to have unique solvability for all data.

The cases in Example 11.12 are somewhat atypical. The main aim has been to consider two-sided elliptic cases, and there are numerous studies of such cases in the literature. As a fundamental example, we consider the Dirichlet problem for strongly elliptic operators; some other cases are treated in the exercises.

11.3 The solution operator for the Dirichlet problem

In this section we consider a strongly elliptic differential operator A of order $d = 2m$ on $\widetilde{X}$. It satisfies a Gårding inequality (cf. Theorem 7.23) on $\widetilde{X}$ as well as on X (on $C_0^\infty(X^\circ)$), and it defines the Dirichlet realization A_γ of A on X with domain $D(A_\gamma) = D(A_{\max}) \cap H_0^m(X)$ (cf. Theorem 7.24). We shall show that in fact $D(A_\gamma) = H^{2m}(X) \cap H_0^m(X)$, and we shall describe the solution operator for the nonhomogeneous Dirichlet problem under a hypothesis of invertibility.

We assume that A is invertible on $\widetilde{X}$, and that

$$A_\gamma : D(A_\gamma) \to L_2(X) \text{ is bijective.} \tag{11.48}$$

This can always be obtained by adding a large enough constant to A such that

$$\operatorname{Re}(Au, u) \geq c_0 \|u\|_m^2 \text{ for } u \in C^\infty(X), \tag{11.49}$$

with $c_0 > 0$; then A on $\widetilde{X}$ and its Dirichlet realizations on X_+ and X_- have positive lower bound.

Remark 11.13. Using the compactness of these inverses as operators in $L_2(\widetilde{X})$ resp. $L_2(X_\pm)$ (which holds since the injections of H^m-spaces into L_2-spaces are compact, cf. Section 8.2), and the fact that the spectra are contained in proper subsectors of $\mathbb{C}$, one can show that the resolvent sets are complements of a countable set of eigenvalues with finite multiplicities and no accumulation points — in short, the spectrum is discrete. Then in fact it suffices to add a small constant to a strongly elliptic operator, to get invertible Dirichlet realizations on $X_\pm$. Without the assumption of invertibility, the results in the following hold modulo negligible operators.

The inverse of A_γ will as in Chapter 9 be denoted R_γ.

The operator R_γ solves the semihomogeneous Dirichlet problem

$$Au = f \text{ on } X, \quad \gamma u = 0 \text{ at } X', \tag{11.50}$$

where $\gamma = \{\gamma_0, \ldots, \gamma_{m-1}\}$, f given in $L_2(X)$. The other semihomogeneous Dirichlet problem

$$Az = 0 \text{ on } X, \quad \gamma z = \varphi \text{ at } X', \tag{11.51}$$

can be solved as follows, for $\varphi \in \prod_{0 \leq j < m} H^{2m-j-\frac{1}{2}}(X')$: Let $v = \mathcal{K}_{(m)}\varphi$ according to Lemma 11.2; it lies in $H^{2m}(X)$ and has $\gamma v = \varphi$. Setting $u = z - v$, we can then reduce problem (11.51) to the problem

$$Au = -Av \text{ on } X, \quad \gamma u = 0 \text{ at } X', \tag{11.52}$$

which has the unique solution $u = -R_\gamma A \mathcal{K}_{(m)}\varphi$ in $D(A_\gamma)$. Then $z = u + v = (\mathcal{K}_{(m)} - R_\gamma A \mathcal{K}_{(m)})\varphi$ solves (11.51). We conclude that the fully nonhomogeneous Dirichlet problem

$$Au = f \text{ on } X, \quad \gamma u = \varphi \text{ at } X', \tag{11.53}$$

has the solution operator

$$
\begin{pmatrix} R_\gamma & K_\gamma \end{pmatrix} : \begin{array}{c} L_2(X) \\ \times \\ \prod_{0\le j<m} H^{2m-j-\frac{1}{2}}(X') \end{array} \to D(A_{\max}) \cap H^m(X); \tag{11.54}
$$

$$
R_\gamma = A_\gamma^{-1}, \quad K_\gamma = \mathcal{K}_{(m)} - R_\gamma A \mathcal{K}_{(m)}.
$$

These considerations also work for the boundary symbol operator, showing that

$$
\begin{pmatrix} a^0(x',\xi',D_n) \\ \gamma \end{pmatrix} : \begin{array}{c} \mathscr{S}_+ \\ \times \\ \mathbb{C}^m \end{array} \to \mathscr{S}_+
$$

has an inverse $\begin{pmatrix} r_\gamma^0(x',\xi',D_n) & k_\gamma^0(x',\xi',D_n) \end{pmatrix}$, and it belongs to the calculus, by Theorem 10.37. Using this, we have according to Theorem 10.38 a *parametrix*, continuous for $s \ge 0$,

$$
\begin{pmatrix} R_\gamma' & K_\gamma' \end{pmatrix} : \begin{array}{c} H^s(X) \\ \times \\ \prod_{0\le j<m} H^{s+2m-j-\frac{1}{2}}(X') \end{array} \to H^{s+2m}(X). \tag{11.55}
$$

It satisfies

$$
\begin{pmatrix} A \\ \gamma \end{pmatrix} \begin{pmatrix} R_\gamma' & K_\gamma' \end{pmatrix} = \begin{pmatrix} I & 0 \\ 0 & I \end{pmatrix} + \mathcal{R}_1,
$$

$$
\begin{pmatrix} R_\gamma' & K_\gamma' \end{pmatrix} \begin{pmatrix} A \\ \gamma \end{pmatrix} = I + \mathcal{R}_2, \tag{11.56}
$$

where $\mathcal{R}_1$ and $\mathcal{R}_2$ are negligible Green operators; $\mathcal{R}_1$ of class 0 and $\mathcal{R}_2$ of class m (since γ is of class m).

The parametrix can be used to show *regularity of solutions*, as follows: Let $u = R_\gamma f$ for some $f \in L_2(X)$. Since $u \in H^m(X)$, $\mathcal{R}_2$ applies to it, and we can write, in view of (11.56),

$$
u = \left[\begin{pmatrix} R_\gamma' & K_\gamma' \end{pmatrix} \begin{pmatrix} A \\ \gamma \end{pmatrix} - \mathcal{R}_2 \right] u = \begin{pmatrix} R_\gamma' & K_\gamma' \end{pmatrix} \begin{pmatrix} f \\ 0 \end{pmatrix} - \mathcal{R}_2 u = R_\gamma' f - \mathcal{R}_2 u. \tag{11.57}
$$

This lies in $H^{2m}(X)$ since $R_\gamma' f$ does so and $\mathcal{R}_2$ maps into $C^\infty(X)$. It follows that

$$
D(A_\gamma) = H^{2m}(X) \cap H_0^m(X). \tag{11.58}
$$

Going back to the definition of K_γ, we see moreover that it maps $\prod_{0\le j<m} H^{2m-j-\frac{1}{2}}(X')$ into $H^{2m}(X)$, so the exact solution operator has a continuity property as in (11.55) for $s = 0$.

It is not hard to pursue this technique still further to show that the full mapping property as in (11.55) holds for $\begin{pmatrix} R_\gamma & K_\gamma \end{pmatrix}$ (this is left to the reader in Exercise 11.6).

In this way we obtain:

Theorem 11.14. *When A is strongly elliptic and satisfies* (11.48), *the Dirichlet problem*

$$Au = f \text{ on } X, \quad \gamma u = \varphi \text{ at } X', \tag{11.59}$$

has a solution operator

$$\begin{pmatrix} A \\ \gamma \end{pmatrix}^{-1} = \begin{pmatrix} R_\gamma & K_\gamma \end{pmatrix}, \tag{11.60}$$

with the mapping property

$$\begin{pmatrix} R_\gamma & K_\gamma \end{pmatrix} : \begin{matrix} H^s(X) \\ \times \\ \prod_{0 \le j < m} H^{s+2m-j-\frac{1}{2}}(X') \end{matrix} \to H^{s+2m}(X), \ s \ge 0. \tag{11.61}$$

Hereby we have finally obtained the general statement on the regularity of solutions of the Dirichlet problem, mentioned several times earlier in this book. The above considerations work also for strongly elliptic *systems A*.

If we do not assume (11.48), Theorem 11.14 assures that $D(A_\gamma + k) = H^{2m}(X) \cap H_0^m(X)$ for a suitable constant k, and it follows that $D(A_\gamma) = H^{2m}(X) \cap H_0^m(X)$.

One can moreover show that $\begin{pmatrix} R_\gamma & K_\gamma \end{pmatrix}$ itself belongs to the ψdbo calculus. On one hand, this is a consequence of a general principle, that we cannot find space to include the proof of here: When an elliptic element has an inverse operator, then the inverse belongs to the calculus (this is part of a property called *spectral invariance*). But it can also be shown for the Dirichlet problem in a more elementary way, passing via the Dirichlet-to-Neumann operator and using that we already have the ψdo properties of the Calderón projector; this will be done below.

The Dirichlet-to-Neumann operator (called the Steklov-Poincaré operator in some texts) enters both in modern and older literature on PDEs, e.g., in the applications of the abstract theory of Chapter 13 that were developed in [G68]–[G74]; see also [BGW08] and its references.

Definition 11.15. Let A be given as above, strongly elliptic and satisfying (11.48), with $\begin{pmatrix} R_\gamma & K_\gamma \end{pmatrix}$ solving the Dirichlet problem, and denote

$$\nu = \{\gamma_m, \dots, \gamma_{2m-1}\}, \tag{11.62}$$

the Neumann trace operator. Then the Dirichlet-to-Neumann operator is defined by

$$P_{\gamma,\nu} = \nu K_\gamma. \tag{11.63}$$

In other words, $P_{\gamma,\nu}$ maps the Dirichlet data into the Neumann data for solutions of $Au = 0$ on X. In view of (11.61), it maps

$$P_{\gamma,\nu} : \prod_{0 \le j < m} H^{s+2m-j-\frac{1}{2}}(X') \to \prod_{0 \le j < m} H^{s+m-j-\frac{1}{2}}(X'), \tag{11.64}$$

for $s \geq 0$. We shall show that it is a pseudodifferential operator over X', and is elliptic.

Note that the Dirichlet-to-Neumann operator is $m \times m$-matrix formed, whereas the Calderón projector for A is $2m \times 2m$-matrix formed. They are in fact closely related.

Let us write C^+ in blocks according to the splitting of $\{0, 1, \ldots, 2m - 1\}$ into $\{0, 1, \ldots, m - 1\}$ and $\{m, \ldots, 2m - 1\}$:

$$C^+ = \begin{pmatrix} C_{00}^+ & C_{01}^+ \\ C_{10}^+ & C_{11}^+ \end{pmatrix}, \tag{11.65}$$

then

$$C_{10}^+ \psi_0 + C_{11}^+ \psi_1 = P_{\gamma,\nu}(C_{00}^+ \psi_0 + C_{01}^+ \psi_1) \tag{11.66}$$

holds for all

$$\{\psi_0, \psi_1\} \in \prod_{0 \leq j < m} H^{2m-j-\frac{1}{2}}(X') \times \prod_{0 \leq j < m} H^{m-j-\frac{1}{2}}(X').$$

Lemma 11.16. *Each of the blocks C_{ij}^+ in* (11.65) *is elliptic.*

The proof is developed in Exercises 11.7–11.10 (originally shown in [G71]). Now, note that in particular,

$$C_{10}^+ \psi_0 = P_{\gamma,\nu} C_{00}^+ \psi_0, \tag{11.67}$$

for $\psi_0 \in \prod_{0 \leq j < m} H^{2m-j-\frac{1}{2}}(X')$. Let S be a parametrix of C_{00}^+, so that $C_{00}^+ S = I - \mathcal{R}$ with a negligible ψdo $\mathcal{R}$. Then

$$C_{10}^+ S = P_{\gamma,\nu} C_{00}^+ S = P_{\gamma,\nu}(I - \mathcal{R}),$$

from which it follows that

$$P_{\gamma,\nu} = C_{10}^+ S + P_{\gamma,\nu} \mathcal{R}. \tag{11.68}$$

Here $\mathcal{R}$ maps distributions to C^∞-functions, and $P_{\gamma,\nu}$ has the mapping properties (11.64), so it follows that $P_{\gamma,\nu}\mathcal{R}$ maps $\mathscr{D}'(X')$ continuously into $C^\infty(X')$, hence is a negligible ψdo. Thus $P_{\gamma,\nu}$ is a ψdo. Finally, since both C_{10}^+ and S are elliptic, it follows from (11.68) that $P_{\gamma,\nu}$ is elliptic. We have then obtained:

Theorem 11.17. *$P_{\gamma,\nu}$ is an elliptic ψdo.*

Behind Lemma 11.16 lies a consideration of both the Dirichlet problem on X_+ and that on X_-. It always works for the model problems (which is what is used in Exercises 11.7–11.10), but let us now look at the full operators. Assuming that (11.48) holds also for the Dirichlet realization on X_-, we can introduce the notation $K_\gamma^\pm$ for the operators solving

$$Au = 0 \text{ in } X_+, \ \gamma^+ u = \varphi, \quad \text{resp.} \quad Au = 0 \text{ in } X_-, \ \gamma^- u = \varphi$$

(so $K_\gamma^+ = K_\gamma$), and denote

$$\nu^\pm K_\gamma^\pm = P_{\gamma,\nu}^\pm; \tag{11.69}$$

here

$$\nu^\pm = \{\gamma_m^\pm, \dots, \gamma_{2m-1}^\pm\}, \tag{11.70}$$

and $P_{\gamma,\nu}^+ = P_{\gamma,\nu}$.

Then the above argumentation for $P_{\gamma,\nu}^+$ works also for $P_{\gamma,\nu}^-$, showing that it is a ψdo. The calculations in Exercises 11.7–11.10 now imply:

Theorem 11.18. *When* (11.48) *holds also for the Dirichlet realization on* X_- *(so that $P_{\gamma,\nu}^-$ is well-defined), then $P_{\gamma,\nu}^-$ and $P_{\gamma,\nu}^+ - P_{\gamma,\nu}^-$ are elliptic ψdo's.*

In this case, one can moreover show that $P_{\gamma,\nu}^+ - P_{\gamma,\nu}^-$ is invertible. Then C^+ can be described by an explicit formula from $P_{\gamma,\nu}^+$ and $P_{\gamma,\nu}^-$, and vice versa (details are worked out in Exercise 11.25):

$$\begin{pmatrix} C_{00}^+ & C_{01}^+ \\ C_{10}^+ & C_{11}^+ \end{pmatrix} = \begin{pmatrix} -(P_{\gamma,\nu}^+ - P_{\gamma,\nu}^-)^{-1}P_{\gamma,\nu}^- & (P_{\gamma,\nu}^+ - P_{\gamma,\nu}^-)^{-1} \\ -P_{\gamma,\nu}^+(P_{\gamma,\nu}^+ - P_{\gamma,\nu}^-)^{-1}P_{\gamma,\nu}^- & P_{\gamma,\nu}^+(P_{\gamma,\nu}^+ - P_{\gamma,\nu}^-)^{-1} \end{pmatrix}, \tag{11.71}$$

$$P_{\gamma,\nu}^+ = C_{11}^+(C_{01}^+)^{-1}. \tag{11.72}$$

Returning to K_γ, we note that when $\varphi \in \prod_{0 \le j < m} H^{2m-j-\frac{1}{2}}(X')$, then, with K^+ as in Theorem 11.8,

$$K^+ \begin{pmatrix} \varphi \\ P_{\gamma,\nu}\varphi \end{pmatrix}$$

solves the Dirichlet problem (11.51), so

$$K_\gamma = K^+ \begin{pmatrix} I \\ P_{\gamma,\nu} \end{pmatrix}. \tag{11.73}$$

This shows that K_γ is a *Poisson operator*. In particular, the mapping property mentioned in Theorem 11.14 extends to all s:

$$K_\gamma : \prod_{0 \le j < m} H^{s-j-\frac{1}{2}}(X') \to H^s(X) \text{ for all } s \in \mathbb{R}. \tag{11.74}$$

As for R_γ, it can be expressed in terms of $Q = A^{-1}$ and K_γ by

$$R_\gamma = Q_+ - K_\gamma \gamma Q_+, \tag{11.75}$$

exactly as in the calculations leading to Theorem 9.18, so we conclude that R_γ likewise belongs to the ψdbo calculus.

Theorem 11.19. *When A is strongly elliptic, invertible on $\widetilde{X}$, and satisfies (11.48), then the solution operator $\left(R_\gamma \; K_\gamma\right)$ of the Dirichlet problem satisfies:*
K_γ *is a Poisson operator (mapping as in (11.61)), and R_γ is the sum of a truncated ψdo Q_+ and a singular Green operator $-K_\gamma \gamma Q_+$, of order $-2m$.*

One can moreover show that the continuity of R_γ from $H^s(X)$ to $H^{s+2m}(X)$ for $s \geq 0$ extends down to $s > -m - \frac{1}{2}$, see e.g. the analysis in [G90].

Exercises for Chapter 11

11.1. Let A be as in Section 11.1, with $E_1 = E_2 = E$. Define $A_{\max}$ as the operator in $L_2(E) = H^0(E)$ acting like A and with domain

$$D(A_{\max}) = \{u \in L_2(E) \mid Au \in L_2(E)\}.$$

Show that ϱ can be defined on $D(A_{\max})$, mapping it continuously into $\mathcal{H}^0(E'^d)$ (here $D(A_{\max})$ is provided with the graph norm).
(*Hint.* Begin by showing that when $u \in D(A_{\max})$, Q_+Au is in $H^d(E)$, and $z = u - Q_+Au$ is in Z_+^0.)

11.2. Let A and $S\varrho$ be as in the beginning of Section 11.2, with $E_1 = E_2 = E$. Define the realization A_S of A as the operator in $L_2(E)$ with domain (using Exercise 11.1)
$$D(A_S) = \{u \in D(A_{\max}) \mid S\varrho u = 0\}.$$

Show that A_S is closed, densely defined.

11.3. Continuation of Exercise 11.2. Assume moreover that

$$\begin{pmatrix} A \\ S\varrho \end{pmatrix} \text{ has the inverse } \left(R_S \; K_S\right),$$

with all the properties in Theorem 11.10; in particular it belongs to the ψdbo calculus.
(a) Show that $D(A_S) \subset H^d(E)$.
(b) Show that when $u \in D(A_S)$ with $Au \in H^s(E)$ for some $s \geq 0$, then $u \in H^{s+d}(E)$.
(*Comment.* It is possible to show similar regularity results when $\{A, S\varrho\}$ is just assumed to be elliptic, e.g., by constructing a parametrix that acts as an inverse on suitable subspaces with finite codimension, using Fredholm theory. A systematic treatment can be found in [G90], including general Green operators and L_p-based spaces.)

11.4. Let A be a second-order scalar differential operator on $\mathbb{R}^n$ with principal symbol independent of x_n:

$$a^0(x', \xi', \xi_n) = s_2(x')\xi_n^2 + s_1(x', \xi')\xi_n + s_0(x', \xi').$$

(a) Assume that A is properly elliptic, such that for all (x', ξ') with $\xi' \neq 0$, the two roots $\tau^{\pm}(x', \xi')$ of the second-order polynomial $a^0(x', \xi', \tau)$ in τ lie in $\mathbb{C}_{\pm}$, respectively. Denote $\sigma^+ = -i\tau^+$, $\sigma^- = i\tau^-$; show that σ^+ and σ^- have positive real part and are homogeneous in ξ' of degree 1.

(b) Show that the principal symbol of a parametrix Q of A is (for $|\xi'| \geq 1$)

$$q^0(x', \xi', \xi_n) = \frac{1}{s_2(\xi_n - \tau^+)(\xi_n - \tau^-)}$$

$$= \frac{1}{s_2(\sigma^+ + \sigma^-)}\left(\frac{1}{\sigma^+ + i\xi_n} + \frac{1}{\sigma^- - i\xi_n}\right).$$

(c) Find (at each (x', ξ'), $|\xi'| \geq 1$) the Poisson operator $r^+ q^0(x', \xi', D_n)\widetilde{\varrho}^*$ and the ψdo symbol $\varrho^+ q^0(x', \xi', D_n)\widetilde{\varrho}^*$, expressed in terms of $\sigma^{\pm}$.
(*Hint.* One can apply the rules in Theorem 10.25 and Lemma 10.18, see also Exercise 10.5.)

(d) Find the principal symbol of the Calderón projector C^+ for A.

11.5. Continuation of Exercise 11.4. Consider a boundary condition

$$b_1(x')D_1 u + \cdots + b_n(x')D_n u = 0 \text{ for } x_n = 0,$$

where the b_j are complex C^∞-functions, with $b_n(x') \neq 0$ for all x'. Give examples of choices of coefficients b_j where the condition defines an elliptic problem for A, and where it does not.

11.6. In the setting of Section 11.3, show how (11.57) can be used to conclude that $(R_\gamma \ K_\gamma)$ has the mapping property as in (11.61).

11.7. Let A be as in Section 11.3, and define ν by (11.62). Show that the Neumann problem

$$Au = f \text{ in } X, \quad \nu u = \varphi \text{ on } X',$$

is elliptic.
(*Hint.* Consider the boundary symbol operator at an (x', ξ') with $\xi' \neq 0$. For dimensional reasons, it suffices to show that

$$\left(\begin{matrix} a^0(x', \xi', D_n) \\ \nu \end{matrix}\right) z = \left(\begin{smallmatrix} 0 \\ 0 \end{smallmatrix}\right) \implies z = 0,$$

when $z \in \mathscr{S}_+$. Show that $u = D_n^m z$ solves the Dirichlet problem with zero data, hence is 0. Use Theorem 4.19 to conclude that $z = 0$.)

11.8. For the boundary symbol operator $a^0(x', \xi', D_n)$ for A considered in Section 11.3 (at an (x', ξ') with $\xi' \neq 0$), let k_γ^+ and k_γ^- be the solution operators for the Dirichlet problems on $\mathbb{R}_+$ and $\mathbb{R}_-$:

$$k_\gamma^\pm : \mathbb{C}^m \to \mathscr{S}_\pm \quad \text{solves} \quad \left(\underset{\gamma}{a^0(x',\xi',D_n)} \right) z = \begin{pmatrix} 0 \\ \varphi \end{pmatrix} \text{ on } \mathbb{R}_\pm.$$

With $\nu^\pm$ defined in (11.70), let

$$p^\pm = \nu^\pm k_\gamma^\pm,$$

respectively. Show that $p^\pm$ are invertible matrices.
(*Hint.* One can use the result of Exercise 11.7.)

11.9. Continuation of Exercise 11.8. Show that

$$p^+ - p^-$$

is an invertible matrix. (It is important for this calculation to recall that $\gamma_j^+ u = \gamma_0^+ D_n^j u$, $\gamma_j^- u = \gamma_0^- D_n^j u$, where D_n^j has the same meaning in the two expressions, namely, $(-i\partial_{x_n})^j$ — with the same direction of x_n.)
(*Hint.* Let $\varphi \in \mathbb{C}^m$ be such that $(p^+ - p^-)\varphi = 0$, and let $u^\pm \in Z_\pm$ with $\gamma^\pm u^\pm = \varphi$. Since $p^+\varphi = p^-\varphi$, $\nu^+ u^+ = \nu^- u^-$. Then $u = e^+ u^+ + e^- u^-$ is a solution in $H^{2m}(\mathbb{R})$ of $a^0(D_n)u = 0$.)

11.10. Continuation of Exercises 11.8 ff. Show that the Calderón projector for $a^0(x', \xi', D_n)$ (equal to the principal symbol of the Calderón projector for A) satisfies

$$c^{+,0} = \begin{pmatrix} -(p^+ - p^-)^{-1}p^- & (p^+ - p^-)^{-1} \\ -p^+(p^+ - p^-)^{-1}p^- & p^+(p^+ - p^-)^{-1} \end{pmatrix}.$$

In particular, the four blocks are invertible.
(More information can be found in [G71, Appendix].)

11.11. Let $A = I - \Delta$ on $\mathbb{R}_+^n$. With the notation of Exercise 11.8, show that $p^+ = -\langle\xi'\rangle$ and $p^- = \langle\xi'\rangle$, and that

$$c^{+,0} = \begin{pmatrix} \frac{1}{2} & -\frac{1}{2}\langle\xi'\rangle^{-1} \\ -\frac{1}{2}\langle\xi'\rangle & \frac{1}{2} \end{pmatrix}.$$

Moreover, C^+ is the ψdo on $\mathbb{R}^{n-1}$ with this symbol.

11.12. Show that the calculations in Exercises 11.7–11.10 are valid also for strongly elliptic *systems* (acting on vector-valued functions), cf. (7.57).

11.13. Let A be as in Section 11.3.
(a) Show that for any $s \in \mathbb{R}$, N_+^s (cf. (11.14)) identifies with the graph of $P_{\gamma,\nu}$ going from $\prod_{0 \leq j < m} H^{s-j-\frac{1}{2}}(X')$ to $\prod_{0 \leq j < m} H^{s-m-j-\frac{1}{2}}(X')$. In other words,

$$\begin{pmatrix} I \\ P_{\gamma,\nu} \end{pmatrix} \text{ maps } \prod_{0\leq j<m} H^{s-j-\frac{1}{2}}(X') \text{ onto } N_s^+$$

with the inverse $\mathrm{pr}_1 = (I\ 0)$, for any $s \in \mathbb{R}$.

(b) Show that K_γ defines a homeomorphism

$$K_\gamma : \prod_{0\leq j<m} H^{s-j-\frac{1}{2}}(X') \xrightarrow{\sim} Z_+^s$$

with inverse

$$\gamma : Z_+^s \to \prod_{0\leq j<m} H^{s-j-\frac{1}{2}}(X'),$$

for all $s \in \mathbb{R}$.

(*Hint.* Use (11.73), (11.29) and the mapping properties of the operators in the ψdbo calculus.)

11.14. Continuation of Exercise 11.13.

(a) Show that $Z_+^\infty = \bigcap_{s\in\mathbb{R}} Z_+^s$ is dense in Z_+^t for all $t \in \mathbb{R}$, and that $N_+^\infty = \bigcap_{s\in\mathbb{R}} N_+^s$ is dense in N_+^t for all $t \in \mathbb{R}$.
(*Hint.* One can combine the results of Exercise 11.13 with the fact that $C^\infty(X')$ is dense in $H^s(X')$ for any s.)

(b) Show that $D(A_{\max})$ is the direct sum of $D(A_\gamma)$ and Z_+^0.

(c) Show that $H^{2m}(X)$ is the direct sum of $D(A_\gamma)$ and Z_+^{2m}.

(d) Show that $H^{2m}(X)$ is dense in $D(A_{\max})$ (with respect to the graph norm).
(*Hint.* One can use that (a) implies the denseness of Z_+^{2m} in Z_+^0.)

11.15. Let A be as in Section 11.3, now with $m = 1$ but acting on N-vectors $u = (u_1,\ldots,u_N)$. In order to use ideas from Chapter 9, where X has a different meaning, we now denote $\widetilde{X} = \Xi$, $X_\pm = \Omega_\pm$, $X' = \Gamma = \partial\Omega_\pm$, so $X = \overline{\Omega}_+$, $U = \Gamma\times\,]-1,1[$. We also denote $\Omega_+ = \Omega$. Let $A_0 = A_{\min}$, $A_1 = A_{\max}$; they are operators in $L_2(\Omega)^N$.

Show that the results of Exercise 11.13 imply that

$$K_\gamma : H^{-\frac{1}{2}}(\Gamma)^N \xrightarrow{\sim} Z(A_1)$$

with inverse

$$\gamma : Z(A_1) \xrightarrow{\sim} H^{-\frac{1}{2}}(\Gamma)^N;$$

the latter operator can more specifically be called γ_Z.

11.16. Continuation of Exercise 11.15. Write A on $U = \Gamma \times [-1,1[$ as

$$A = S_2 D_n^2 + S_1 D_n + S_0,$$

as in Proposition 11.3, where S_2, S_1 and S_0 are x_n-dependent $N\times N$-matrices of functions, first-order differential operators, resp. second-order differential operators on Γ.

(a) Verify that there is a Green's formula for $u, v \in H^{2m}(\Omega)^N$:

$$(Au, v)_\Omega - (u, A'v)_\Omega = (\mathfrak{A}\varrho u, \varrho v)_\Gamma = (iS_2\gamma_1 u, \gamma_0 v)_\Gamma - (\gamma_0 u, iS_2^*\gamma_1 v - \mathfrak{A}_{00}^*\gamma_0 v)_\Gamma,$$

where

$$\mathfrak{A} = \begin{pmatrix} \mathfrak{A}_{00} & iS_2 \\ iS_2 & 0 \end{pmatrix},$$

with $\mathfrak{A}_{00}$ equal to iS_1 plus a matrix of functions (S_2 and S_1 taken at $x_n = 0$); A' denotes the adjoint of A as a differential operator on Ξ. Show that Green's formula extends to $u \in H^{2m}(\Omega)^N$, $v \in D(A_1')$, with Sobolev space dualities in the right-hand side. (*Hint.* Use that ϱ is well-defined on the nullspace Z_+^0 for A'.)

(b) Define the modified Neumann trace operators

$$\beta u = iS_2\gamma_1 u, \quad \beta'v = iS_2^*\gamma_1 v - \mathfrak{A}_{00}^*\gamma_0 v,$$

which allow writing Green's formula as

$$(Au, v)_\Omega - (u, A'v)_\Omega = (\beta u, \gamma_0 v)_\Gamma - (\gamma_0 u, \beta'v)_\Gamma;$$

and define the modified Dirichlet-to-Neumann operators

$$P_{\gamma,\beta} = iS_2\gamma_1 P_{\gamma_0,\gamma_1}, \quad P'_{\gamma,\beta'} = iS_2^*\gamma_1 P'_{\gamma_0,\gamma_1} - \mathfrak{A}_{00}^*,$$

where P_{γ_0,γ_1} and P'_{γ_0,γ_1} stand for $P_{\gamma,\nu}$ for A resp. A' (with $m = 1$). Define moreover the trace operators

$$\mu u = \beta u - P_{\gamma,\beta}\gamma_0 u, \quad \mu'v = \beta'v - P'_{\gamma,\beta'}\gamma_0 v.$$

Show that $\mu = \beta A_\gamma^{-1} Au$, continuous from $D(A_1)$ to $H^{\frac{1}{2}}(\Gamma)^N$, with a similar result for μ', and that

$$(Au, v)_\Omega - (u, A'v)_\Omega = \langle \mu u, \overline{\gamma_0 v} \rangle_{H^{\frac{1}{2}}, H^{-\frac{1}{2}}} - \langle \gamma_0 u, \overline{\mu'v} \rangle_{H^{-\frac{1}{2}}, H^{\frac{1}{2}}}$$

holds for all $u \in D(A_1)$, $v \in D(A_1')$.
(*Hint.* Proceed as in Proposition 9.27.)

(c) Carry the construction in Section 9.4, up to and including Theorem 9.29, over to the present situation, showing a 1–1 correspondence between closed realizations $\widetilde{A}$ of A and closed densely defined operators $L : X \to Y^*$, where X and Y are closed subspaces of $H^{-\frac{1}{2}}(\Gamma)^N$.

11.17. Continuation of Exercise 11.15 ff. Show that in the 1–1 correspondence established in Exercise 11.16, the Dirichlet realization A_γ corresponds to the case $X = Y = \{0\}$, $L = 0$.

11.18. Continuation of Exercise 11.15 ff. Show that in the 1–1 correspondence established in Exercise 11.16, the realization A_M with domain

$D(A_M) = D(A_0) \dotplus Z(A_{1'})$ corresponds to the case $X = Y = H^{-\frac{1}{2}}(\Gamma)^N$, $L = 0$.

(*Comment.* This is a realization with no regularity. When A is formally self-adjoint, it is the von Neumann realization or Kreĭn's soft realization, cf. Example 13.10 and the remarks after (13.67).)

11.19. Continuation of Exercise 11.15 ff. Let A_ν be the realization defined by the Neumann condition $\gamma_1 u = 0$,

$$D(A_\nu) = \{u \in D(A_{\max}) \mid \gamma_1 u = 0\},$$

cf. Exercises 11.1 and 11.2.

(a) Show that $D(A_\nu) \subset H^2(\Omega)^N$.
(*Hint.* Show first that when $u \in D(A_\nu)$, $z = u - Q_+ Au$ has $\nu z \in H^{\frac{1}{2}}(\Gamma)^N$. Then use the ellipticity of P_{γ_0,γ_1} to conclude that $\varrho z \in H^{\frac{3}{2}}(\Gamma)^N \times H^{\frac{1}{2}}(\Gamma)^N$.)
(b) Show that in the 1–1 correspondence established in Exercise 11.16, A_ν corresponds to the case $X = Y = H^{-\frac{1}{2}}(\Gamma)^N$, $L = -P_{\gamma,\beta}$ with domain $D(L) = H^{\frac{3}{2}}(\Gamma)^N$.

11.20. Continuation of Exercise 11.15 ff. Consider the realization $\widetilde{A}$ defined by a Neumann-type boundary condition

$$\gamma_1 u + B \gamma_0 u = 0,$$

where B is a first-order ψdo on N-vectors of functions on Γ such that $B + P_{\gamma_0,\gamma_1}$ is elliptic of order 1; here

$$D(\widetilde{A}) = \{u \in D(A_{\max}) \mid \gamma_1 u + B \gamma_0 u = 0\}.$$

Show that in the 1–1 correspondence established in Exercise 11.16, $\widetilde{A}$ corresponds to the case $X = Y = H^{-\frac{1}{2}}(\Gamma)^N$, $L = -iS_2(B + P_{\gamma_0,\gamma_1})$ with domain $D(L) = H^{\frac{3}{2}}(\Gamma)^N$; in particular, $D(\widetilde{A}) \subset H^2(\Omega)^N$.

11.21. Continuation of Exercise 11.15 ff. For simplicity of the calculations in the following, assume that $S_2 = I$. Let J be an integer in $]1, N[$, and consider the realization $\widetilde{A}$ defined by a boundary condition

$$\gamma_0(u_1, \ldots, u_J) = 0, \quad \gamma_1(u_{J+1}, \ldots, u_N) + B\gamma_0(u_{J+1}, \ldots, u_N) = 0,$$

applied to column vectors $u = (u_1, \ldots, u_N)$; here B is an $(N-J) \times (N-J)$-matrix of ψdo's on Γ of order 1. (This is sometimes called a mixed condition.) Assume that the first-order ψdo

$$\mathcal{L} = -iB - i \begin{pmatrix} 0_{N-J,J} & I_{N-J} \end{pmatrix} P_{\gamma_0,\gamma_1} \begin{pmatrix} 0_{J,N-J} \\ I_{N-J} \end{pmatrix}$$

is elliptic. Here I_J stands for the $J \times J$-identity-matrix, and $0_{J,J'}$ stands for the $J \times J'$-zero-matrix.

Show that in the 1–1 correspondence established in Exercise 11.15, $\widetilde{A}$ corresponds to the case $X = Y = H^{-\frac{1}{2}}(\Gamma)^{N-J}$, with L acting like $\mathcal{L}$ with domain $D(L) = \dot{H}^{\frac{3}{2}}(\Gamma)^{N-J}$. In particular, $D(\widetilde{A}) \subset H^2(\Omega)^N$.

(*Comment.* Similar boundary conditions, where the splitting into the first J and last $N - J$ components is replaced by more general projections, are considered e.g. in Avramidi and Esposito [AE99], [G03] (and earlier). For higher order studies, see e.g. Fujiwara and Shimakura [FS70], [G70]–[G74]. Also positivity issues are treated in these papers, see in particular [G74]; then β and β' in Exercise 11.15 are replaced by more symmetric choices of modified Neumann operators, letting each of them carry half of $\mathfrak{A}_{00}$. See also [BGW08].)

11.22. Consider the boundary value problem

$$Au = f, \quad C^+\varrho u = 0$$

(cf. Example 11.12), with f given in $H^0(E_2)$. Show that it has the unique solution

$$u = Q_+ f - K^+ C^+ \varrho Q_+ f,$$

lying in $H^d(E_1)$. (Note, however, that the nonhomogeneous problem

$$Au = f, \quad C^+\varrho u = \varphi,$$

does *not* have a solution for every $f \in H^0(E_2)$, $\varphi \in \mathcal{H}^d(E_1'^d)$. Why?)

11.23. Let A be a first-order differential operator on $\widetilde{X}$ such that $A = I_N \partial_{x_n} + P$ near X', where P is an $N \times N$-matrix-formed first-order selfadjoint elliptic differential operator on X'. Let $n \geq 3$.

(a) Show that the principal symbol $p^0(x', \xi')$ of P (considered at an (x', ξ') with $|\xi'| \geq 1$) has $N/2$ positive eigenvalues $\lambda_1^+, \ldots, \lambda_{N/2}^+$ and $N/2$ negative eigenvalues $\lambda_1^-, \ldots, \lambda_{N/2}^-$ (repeated according to multiplicities); in particular, N is even.

(*Hint.* Note that $p^0(x', -\xi') = -p^0(x', \xi')$.)

(b) Show that the roots of $\det(i\tau I_N + p^0(x', \xi'))$ are: $\tau_j^+ = i\lambda_j^+$ lying in $\mathbb{C}_+$, $j = 1, \ldots, N/2$, and $\tau_j^- = i\lambda_j^-$ lying in $\mathbb{C}_-$, $j = 1, \ldots, N/2$.

(c) Assume from now on that all eigenvalues $\lambda_j^\pm$ are simple. Show that the functions in $Z_+(x', \xi')$ are linear combinations of functions $e^{-\lambda_j^+(x', \xi')x_n} v_j(x', \xi')$, where v_j is an eigenvector for the eigenvalue λ_j^+.

(d) Show that $N^+(x', \xi')$ equals the space spanned by the eigenvectors for eigenvalues λ_j^+, $j = 1, \ldots, N/2$, also called the positive eigenspace for $p^0(x', \xi')$.

(e) Show that $c^{+,0}(x', \xi')$ is the *orthogonal* projection onto the positive eigen-space for $p^0(x', \xi')$.

11.24. Consider the biharmonic operator $A = \Delta^2$ on a smooth bounded set $\Omega \subset \mathbb{R}^n$.

(a) Find the principal symbol of the Dirichlet-to-Neumann operator ($|\xi'| \geq 1$). (*Hint.* Show first that in the model problem on $\mathbb{R}_+$, the null solutions are linear combinations of $e^{-\sigma x_n}$ and $x_n e^{-\sigma x_n}$, $\sigma = |\xi'|$.)

(b) Find the principal symbol of the Calderón projector ($|\xi'| \geq 1$). (*Hint.* One can use the results of Exercises 11.9 and 11.10.)

11.25. Assumptions of Theorem 11.18.

(a) Show that $P^+_{\gamma,\nu} - P^-_{\gamma,\nu}$ is injective. (*Hint.* One can use a generalization of the method of Exercise 11.9.)

(b) Show that $P^+_{\gamma,\nu} - P^-_{\gamma,\nu}$ is surjective. (*Hint.* Apply C^+ and C^-, written in blocks as in (11.65), to functions $\binom{0}{\psi}$. Conclude that

$$P^+_{\gamma,\nu} C^+_{01} \psi = C^+_{11} \psi, \quad P^-_{\gamma,\nu} C^-_{01} \psi = C^-_{11} \psi,$$

and use the fact that $C^+ + C^- = I$ to see that $(P^+_{\gamma,\nu} - P^-_{\gamma,\nu}) C^+_{01} \psi = \psi$.

(c) Show that $P^+_{\gamma,\nu} - P^-_{\gamma,\nu}$ has the inverse C^+_{01}.

(d) Show (11.71) and (11.72).

Part V
Topics on Hilbert space operators

Chapter 12
Unbounded linear operators

12.1 Unbounded operators in Banach spaces

In the elementary theory of Hilbert and Banach spaces, the linear operators that are considered acting on such spaces — or from one such space to another — are taken to be *bounded*, i.e., when T goes from X to Y, it is assumed to satisfy

$$\|Tx\|_Y \leq C\|x\|_X, \text{ for all } x \in X; \qquad (12.1)$$

this is the same as being continuous. We denote the space of these operators $\mathbf{B}(X,Y)$; $\mathbf{B}(X,X)$ is also denoted $\mathbf{B}(X)$. Recall that $\mathbf{B}(X,Y)$ is a Banach space when provided with the operator norm

$$\|T\| = \sup\{\|Tx\|_Y \mid x \in X, \|x\|_X = 1\}.$$

(We generally consider complex vector spaces; most of the theory holds word for word also for real spaces.)

But when one deals with differential operators, one discovers the need to consider also unbounded linear operators. Here $T : X \to Y$ need not be defined on all of X but may be so on a linear subset, $D(T)$, which is called the domain of T. Of special interest are the operators with *dense* domain in X (i.e., with $\overline{D(T)} = X$). When T is bounded and densely defined, it extends by continuity to an operator in $\mathbf{B}(X,Y)$, but when it is not bounded, there is no such extension. For such operators, another property of interest is the property of being *closed*:

Definition 12.1. A linear operator $T : X \to Y$ is said to be closed when the graph $G(T)$,

$$G(T) = \{\{x, Tx\} \mid x \in D(T)\}, \qquad (12.2)$$

is a closed subspace of $X \times Y$.

Since X and Y are metric spaces, we can reformulate the criterion for closedness as follows:

Lemma 12.2. $T : X \to Y$ *is closed if and only if the following holds: When* $(x_n)_{n \in \mathbb{N}}$ *is a sequence in* $D(T)$ *with* $x_n \to x$ *in* X *and* $T x_n \to y$ *in* Y, *then* $x \in D(T)$ *with* $y = Tx$.

The closed graph theorem (recalled in Appendix B, Theorem B.16) implies that if $T : X \to Y$ is closed and has $D(T) = X$, then T is bounded. Thus for closed, densely defined operators, $D(T) \neq X$ is equivalent with unboundedness.

Note that a subspace G of $X \times Y$ is the graph of a linear operator $T :$ $X \to Y$ if and only if the set $\mathrm{pr}_1 G$,

$$\mathrm{pr}_1 G = \{ x \in X \mid \exists y \in Y \text{ so that } \{x, y\} \in G \},$$

has the property that for any $x \in \mathrm{pr}_1 G$ there is *at most one* y so that $\{x, y\} \in G$; then $y = Tx$ and $D(T) = \mathrm{pr}_1 G$. In view of the linearity we can also formulate the criterion for G being a graph as follows:

Lemma 12.3. *A subspace* G *of* $X \times Y$ *is a graph if and only if* $\{0, y\} \in G$ *implies* $y = 0$.

All operators in the following are assumed to be linear, this will not in general be repeated.

When S and T are operators from X to Y, and $D(S) \subset D(T)$ with $Sx = Tx$ for $x \in D(S)$, we say that T is an extension of S and S is a restriction of T, and we write $S \subset T$ (or $T \supset S$). One often wants to know whether a given operator T has a closed extension. If T is bounded, this always holds, since we can simply take the operator $\overline{T}$ with graph $\overline{G(T)}$; here $\overline{G(T)}$ is a graph since $x_n \to 0$ implies $Tx_n \to 0$. But when T is unbounded, one cannot be certain that it has a closed extension (cf. Exercise 12.1). But *if* T has a closed extension T_1, then $G(T_1)$ is a closed subspace of $X \times Y$ containing $G(T)$, hence also containing $\overline{G(T)}$. In that case $\overline{G(T)}$ is a graph (cf. Lemma 12.3). It is in fact the graph of the smallest closed extension of T (the one with the smallest domain); we call it *the closure of* T and denote it $\overline{T}$. (Observe that when T is unbounded, then $D(\overline{T})$ is a proper subset of $\overline{D(T)}$.)

When S and T are operators from X to Y, the sum $S + T$ is defined by

$$\begin{aligned} D(S + T) &= D(S) \cap D(T) , \\ (S + T)x &= Sx + Tx \text{ for } x \in D(S + T) ; \end{aligned} \tag{12.3}$$

and when R is an operator from Y to Z, the product (or composition) RT is defined by

$$\begin{aligned} D(RT) &= \{ x \in D(T) \mid Tx \in D(R) \} , \\ (RT)x &= R(Tx) \text{ for } x \in D(RT) . \end{aligned} \tag{12.4}$$

As shown in Exercise 12.4, $R(S + T)$ need not be the same as $RS + RT$. Concerning closures of products of operators, see Exercise 12.6. When S and T are invertible, one has $(ST)^{-1} = T^{-1} S^{-1}$.

Besides the norm topology we can provide $D(T)$ with the so-called *graph topology*. For Banach spaces it is usually defined by the norm

$$\|x\|'_{D(T)} = \|x\|_X + \|Tx\|_Y , \tag{12.5}$$

called the *graph norm*, and for Hilbert spaces by the equivalent norm (also called the graph norm)

$$\|x\|_{D(T)} = (\|x\|_X^2 + \|Tx\|_Y^2)^{\frac{1}{2}} , \tag{12.6}$$

which has the associated scalar product

$$(x,y)_{D(T)} = (x,y)_X + (Tx,Ty)_Y .$$

(These conventions are consistent with (A.10)–(A.12).) The graph norm on $D(T)$ is clearly stronger than the X-norm on $D(T)$; the norms are equivalent if and only if T is a bounded operator. Observe that the operator T is closed if and only if $D(T)$ is complete with respect to the graph norm (Exercise 12.3).

Recall that when X is a Banach space, the dual space $X^* = \mathbf{B}(X, \mathbb{C})$ consists of the bounded linear functionals x^* on X; it is a Banach space with the norm

$$\|x^*\|_{X*} = \sup\{ |x^*(x)| \mid x \in X, \|x\| = 1 \}.$$

When $T : X \to Y$ is *densely defined*, we can define *the adjoint operator* $T^* : Y^* \to X^*$ as follows: The domain $D(T^*)$ consists of the $y^* \in Y^*$ for which the linear functional

$$x \mapsto y^*(Tx) , \quad x \in D(T) , \tag{12.7}$$

is *continuous* (from X to $\mathbb{C}$). This means that there is a constant c (depending on y^*) such that

$$|y^*(Tx)| \le c\|x\|_X, \text{ for all } x \in D(T).$$

Since $D(T)$ is dense in X, the mapping extends by continuity to X, so there is a uniquely determined $x^* \in X^*$ so that

$$y^*(Tx) = x^*(x) \quad \text{for } x \in D(T) . \tag{12.8}$$

Since x^* is determined from y^*, we can define the operator T^* from Y^* to X^* by

$$T^*y^* = x^* , \text{ for } y^* \in D(T^*). \tag{12.9}$$

Lemma 12.4. *Let T be densely defined. Then there is an adjoint operator $T^* : Y^* \to X^*$, uniquely defined by (12.7)–(12.9). Moreover, T^* is closed.*

Proof. The definition of T^* is accounted for above; it remains to show the closedness.

Let $y_n^* \in D(T^*)$ for $n \in \mathbb{N}$, with $y_n^* \to y^*$ and $T^*y_n^* \to z^*$ for $n \to \infty$; then we must show that $y^* \in D(T^*)$ with $T^*y^* = z^*$ (cf. Lemma 12.2). Now we have for all $x \in D(T)$:

$$y^*(Tx) = \lim_{n\to\infty} y_n^*(Tx) = \lim_{n\to\infty} (T^*y_n^*)(x) = z^*(x) \,.$$

This shows that $y^* \in D(T^*)$ with $T^*y^* = z^*$. $\square$

Here is some more notation: We denote the range of T by $R(T)$, and we denote the kernel of T (i.e., the nullspace) by $Z(T)$,

$$Z(T) = \{x \in D(T) \mid Tx = 0\} \,.$$

When $X = Y$, it is of interest to consider the operators $T - \lambda I$ where $\lambda \in \mathbb{C}$ and I is the identity operator (here $D(T - \lambda I) = D(T)$). The *resolvent set* $\varrho(T)$ is defined as the set of $\lambda \in \mathbb{C}$ for which $T - \lambda I$ is a bijection of $D(T)$ onto X with bounded inverse $(T - \lambda I)^{-1}$; the *spectrum* $\sigma(T)$ is defined as the complement $\mathbb{C} \setminus \varrho(T)$. $T - \lambda I$ is also written $T - \lambda$.

12.2 Unbounded operators in Hilbert spaces

We now consider the case where X and Y are complex Hilbert spaces. Here the norm on the dual space X^* of X is a Hilbert space norm, and the Riesz representation theorem assures that for any element $x^* \in X^*$ there is a unique element $v \in X$ such that

$$x^*(x) = (x, v) \text{ for all } x \in X;$$

and here $\|x^*\|_{X^*} = \|v\|_X$. In fact, the mapping $x^* \mapsto v$ is a bijective isometry, and one usually *identifies* X^* with X by this mapping.

With this identification, the adjoint operator T^* of a densely defined operator $T : X \to Y$ is defined as the operator from Y to X for which

$$(Tx, y)_Y = (x, T^*y)_X \text{ for all } x \in D(T) \,, \tag{12.10}$$

with $D(T^*)$ equal to the set of all $y \in Y$ for which *there exists* a $z \in X$ such that z can play the role of T^*y in (12.10).

Observe in particular that $y \in Z(T^*)$ if and only if $y \perp R(T)$, so we always have

$$Y = \overline{R(T)} \oplus Z(T^*) \,. \tag{12.11}$$

It is not hard to show that when $S : X \to Y$, $T : X \to Y$ and $R : Y \to Z$ are densely defined, with $D(S + T)$ and $D(RT)$ dense in X, then

$$S^* + T^* \subset (S + T)^* \text{ and } T^*R^* \subset (RT)^* \,; \tag{12.12}$$

these inclusions can be sharp (cf. Exercise 12.7). Note in particular that for $\alpha \in \mathbb{C} \setminus \{0\}$,

$$(T + \alpha I)^* = T^* + \overline{\alpha} I \text{ , and } (\alpha T)^* = \overline{\alpha} T^* \text{ .} \qquad (12.13)$$

(But $(0T)^* = 0 \in \mathbf{B}(Y, X)$ is different from $0T^*$ when $D(T^*) \neq Y$.)

The following theorem gives an efficient tool to prove some important facts on closedness and denseness, in connection with taking adjoints.

Theorem 12.5. *Let $T : X \to Y$ be a densely defined operator between two Hilbert spaces X and Y. Then*

$$X \oplus Y = \overline{G(T)} \oplus U G(T^*) \text{ ,} \qquad (12.14)$$

where U is the operator from $Y \oplus X$ to $X \oplus Y$ given by $U\{v, w\} = \{-w, v\}$. If in addition T is closed, then T^* is densely defined and $T^{**} = T$.

Proof. Let $\{v, w\} \in X \oplus Y$. The following statements are equivalent:

$$\{v, w\} \in U G(T^*) \iff \{w, -v\} \in G(T^*)$$
$$\iff (Tx, w)_Y = -(x, v)_X \quad \forall x \in D(T)$$
$$\iff (\{x, Tx\}, \{v, w\})_{X \oplus Y} = 0 \quad \forall x \in D(T)$$
$$\iff \{v, w\} \perp G(T) \text{ .}$$

Since $\overline{G(T)} = G(T)^{\perp\perp}$ (by a standard rule for subspaces), this shows the identity (12.14). U is clearly an isometry of $Y \oplus X$ onto $X \oplus Y$, and preserves orthogonality, so we have moreover:

$$Y \oplus X = U^{-1}(X \oplus Y) = U^{-1}\overline{G(T)} \oplus G(T^*) \text{ .} \qquad (12.15)$$

Now assume that T is closed, i.e., $G(T) = \overline{G(T)}$. We can then show that $D(T^*)$ is dense in Y: If $y \in Y \ominus \overline{D(T^*)}$, then $\{y, 0\} \perp G(T^*)$, hence $\{y, 0\} \in U^{-1}\overline{G(T)}$ by (12.15) and then also $\{0, y\} \in G(T)$. By Lemma 12.3 we must have $y = 0$, which shows that $Y \ominus \overline{D(T^*)} = \{0\}$.

In this case, $T^* : Y \to X$ has an adjoint operator T^{**}, and we have from what was already shown:

$$Y \oplus X = \overline{G(T^*)} \oplus U^{-1} G(T^{**}) = G(T^*) \oplus U^{-1} G(T^{**}) \text{ ,}$$

since T^* is closed according to Lemma 12.4. This implies

$$X \oplus Y = U(Y \oplus X) = U G(T^*) \oplus G(T^{**}) \text{ ,}$$

which gives, by comparison with (12.14), that $G(T^{**}) = \overline{G(T)} = G(T)$ (since T was closed). Thus $T^{**} = T$. $\qquad \square$

Note that when S is densely defined, then the following holds:

$$S \subset T \text{ implies } S^* \supset T^* \text{ .} \qquad (12.16)$$

Corollary 12.6. *Let* $T : X \to Y$ *be densely defined. Then* T *has a closed extension if and only if* T^* *is densely defined, and in the affirmative case,*

$$T^* = (\overline{T})^* \quad \text{and} \quad T^{**} = \overline{T} .$$

Proof. If T has a closure, then in particular $\overline{G(T)} = G(\overline{T})$. Then $T^* = (\overline{T})^*$ by (12.14), and $(\overline{T})^*$ is densely defined according to Theorem 12.5, with $T^{**} = (\overline{T})^{**} = \overline{T}$. Conversely, it is clear that if T^* is densely defined, then T^{**} is a closed extension of T. □

We can also show an important theorem on the relation between adjoints and inverses:

Theorem 12.7. *Assume that* $T : X \to Y$ *has the properties:*

(1) T *is densely defined,*
(2) T *is closed,*
(3) T *is injective,*
(4) T *has range dense in* Y.

Then T^* *and* T^{-1} *also have the properties* (1)–(4), *and*

$$(T^*)^{-1} = (T^{-1})^* . \tag{12.17}$$

Proof. T^{-1} is clearly injective, densely defined and closed (cf. Lemma 12.2) with dense range, and the same holds for T^* by Theorem 12.5, Corollary 12.6 and (12.11) (applied to T and T^*). It then follows moreover that $(T^*)^{-1}$ and $(T^{-1})^*$ have the same properties. Using the linearity of the operators we find, with notation as in Theorem 12.5, that

$$X \oplus Y = G(-T) \oplus UG(-T^*)$$

implies

$$Y \oplus X = U^{-1}(X \oplus Y) = U^{-1}G(-T) \oplus G(-T^*)$$
$$= G(T^{-1}) \oplus G(-T^*) = G(T^{-1}) \oplus U^{-1}G((T^*)^{-1}) .$$

An application of Theorem 12.5 to $T^{-1} : Y \to X$ then shows that $(T^{-1})^* = (T^*)^{-1}$. □

We end this section with a remark that is useful when discussing various Hilbert space norms:

It follows from the open mapping principle (recalled e.g. in Theorem B.15) that if a linear space X is a Hilbert space with respect to two norms $\|x\|$ and $\|x\|'$, and there is a constant $c > 0$ such that $\|x\| \leq c\|x\|'$ for all $x \in X$, then the two norms are equivalent:

$$\|x\| \leq c\|x\|' \leq C\|x\| \text{ for all } x \in X,$$

for some $C > 0$. In particular, if the domain $D(T)$ of a closed operator $T : X \to Y$ is a Hilbert space with respect to a norm $\|x\|'$ such that $\|x\|_X + \|Tx\|_Y \le c\|x\|'$ for all $x \in D(T)$, then $\|x\|'$ is equivalent with the graph norm on $D(T)$. (There is a similar result for Banach spaces.)

12.3 Symmetric, selfadjoint and semibounded operators

When X and Y equal the same Hilbert space H, and T is a linear operator in H, we say that T is *symmetric* if

$$(Tx, y) = (x, Ty) \text{ for } x \text{ and } y \in D(T) . \tag{12.18}$$

We say that T is *selfadjoint*, when T is densely defined (so that the adjoint T^* exists) and $T^* = T$. (It is a matter of taste whether the assumption $\overline{D(T)} = H$ should also be included in the definition of symmetric operators — we do not do it here, but the operators we consider will usually have this property.)

Lemma 12.8. *Let T be an operator in the complex Hilbert space H.*
1° *T is symmetric if and only if (Tx, x) is real for all x.*
2° *When T is densely defined, T is symmetric if and only if $T \subset T^*$. In the affirmative case, T has a closure $\overline{T}$, which is likewise symmetric, and*

$$T \subset \overline{T} \subset \overline{T}^* = T^* . \tag{12.19}$$

3° *When T is densely defined, T is selfadjoint if and only if T is closed and both T and T^* are symmetric.*

Proof. When T is symmetric,

$$(Tx, x) = (x, Tx) = \overline{(Tx, x)} \quad \text{for } x \in D(T) ,$$

whereby $(Tx, x) \in \mathbb{R}$. Conversely, when $(Tx, x) \in \mathbb{R}$ for all $x \in D(T)$, then we first conclude that $(Tx, x) = (x, Tx)$ for $x \in D(T)$; next, we obtain for x and $y \in D(T)$:

$$4(Tx, y) = \sum_{\nu=0}^{3} i^{\nu}(T(x + i^{\nu}y), x + i^{\nu}y)$$
$$= \sum_{\nu=0}^{3} i^{\nu}(x + i^{\nu}y, T(x + i^{\nu}y)) = 4(x, Ty).$$

This shows 1°.

The first assertion in 2° is seen from the definition; the second assertion follows by use of Corollary 12.6.

For 3° we have on one hand that when T is selfadjoint, $T = T^*$, so T is closed and both T and T^* are symmetric. For the other direction, observe that we have according to Corollary 12.6 for a densely defined operator T with densely defined adjoint T^* that T is closed if and only if $T = T^{**}$. Thus,when T is closed and T and T^* are symmetric, then $T \subset T^*$ and $T^* \subset T^{**} = T$; hence $T = T^*$.   □

An operator T for which $\overline{T}$ exists and is selfadjoint, is said to be *essentially selfadjoint* (sometimes in the physics literature such operators are simply called selfadjoint).

A symmetric operator T is called *maximal symmetric* if $S \supset T$ with S symmetric implies $S = T$. Selfadjoint operators are maximal symmetric, but the converse does not hold, cf. Exercise 12.12.

It is useful to know that when S is symmetric and $\lambda = \alpha + i\beta$ with α and $\beta \in \mathbb{R}$, then

$$
\begin{aligned}
\|(S - \lambda I)x\|^2 &= (Sx - \alpha x - i\beta x, Sx - \alpha x - i\beta x) \\
&= \|(S - \alpha I)x\|^2 + \beta^2 \|x\|^2 \quad \text{for } x \in D(S) .
\end{aligned}
\tag{12.20}
$$

For an arbitrary linear operator T in H we define the *numerical range* $\nu(T)$ by

$$
\nu(T) = \{(Tx, x) \mid x \in D(T) , \ \|x\| = 1\} \subset \mathbb{C} ,
\tag{12.21}
$$

and the *lower bound* $m(T)$ by

$$
m(T) = \inf\{\operatorname{Re}(Tx, x) \mid x \in D(T) , \ \|x\| = 1\} \geq -\infty .
\tag{12.22}
$$

T is said to be *lower (semi)bounded* when $m(T) > -\infty$, and *upper (semi)-bounded* when $m(-T) > -\infty$; $-m(-T)$ is called the upper bound for T. When a notation is needed, we write $-m(-T) = u(T)$.

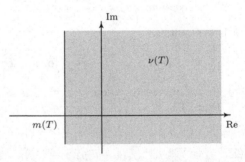

From a geometric point of view, $m(T) > -\infty$ means that the numerical range $\nu(T)$ is contained in the half-space $\{\lambda \in \mathbb{C} \mid \operatorname{Re}\lambda \geq m(T)\}$.

Note in particular that the symmetric operators S are precisely those whose numerical range is contained in the real axis (Lemma 12.8 1°). They are also characterized by the property $m(iS) = m(-iS) = 0$.

Bounded operators are clearly upper and lower semibounded. As for a converse, see Theorem 12.12 and Exercise 12.9. For symmetric operators T, we express $m(T) \geq \alpha$ and $m(-T) \geq -\beta$ briefly by saying that $T \geq \alpha$, resp. $T \leq \beta$. T is called positive (resp. nonnegative) when $m(T) > 0$ (resp. $m(T) \geq 0$).

Theorem 12.9. 1° *If $m(T) \geq \alpha > 0$, then T is injective, and T^{-1} (with $D(T^{-1}) = R(T)$) is a bounded operator in H with norm $\|T^{-1}\| \leq \alpha^{-1}$.*
2° *If furthermore T is closed, then $R(T)$ is closed.*
3° *If T is closed and densely defined, and both $m(T)$ and $m(T^*)$ are $\geq \beta$, then the half-space $\{\lambda \mid \mathrm{Re}\,\lambda < \beta\}$ is contained in the resolvent set for T and for T^*.*

Proof. The basic observation is that $m(T) \geq \alpha$ implies the inequality

$$\|Tx\| \, \|x\| \geq |(Tx, x)| \geq \mathrm{Re}(Tx, x) \geq \alpha \|x\|^2 \quad \text{for} \quad x \in D(T) \,, \qquad (12.23)$$

from which we obtain (by division by $\|x\|$ if $x \neq 0$) that

$$\|Tx\| \geq \alpha \|x\| \quad \text{for} \quad x \in D(T) \,.$$

If $\alpha > 0$, T is then injective. Inserting $Tx = y \in R(T) = D(T^{-1})$, we see that

$$\|T^{-1}y\| = \|x\| \leq \alpha^{-1} \|y\| \,,$$

which shows 1°.

When T is closed, T^{-1} is then a closed bounded operator, so $D(T^{-1})$ ($= R(T)$) is closed; this shows 2°.

For 3° we observe that when $\mathrm{Re}\,\lambda = \beta - \alpha$ for some $\alpha > 0$, then $m(T - \lambda I) \geq \alpha$ and $m(T^* - \bar{\lambda}I) \geq \alpha$, and $T^* - \bar{\lambda}I$ and $T - \lambda I$ are surjective, by 2° and (12.11), which shows 3°.   □

Operators L such that $m(-L) \geq 0$ (i.e., $u(L) \leq 0$) are called dissipative in works of R. S. Phillips, see e.g. [P59], where it is shown that L is maximal dissipative (maximal with respect to the property of being dissipative) precisely when $-L$ satisfies Theorem 12.9 3° with $\beta = 0$; also the problem of finding maximal dissipative extensions of a given dissipative operator is treated there. The maximal dissipative operators are of interest in applications to partial differential equations, because they are the generators of contraction semigroups e^{tL}, used to solve Cauchy problems; an extensive book on semigroups is Hille and Phillips [HP57]. There is more on semigroups in Chapter 14.

In the book of Kato [K66], the T such that $m(T) \geq 0$ are called accretive (i.e., $-T$ is dissipative), and the maximal such operators are called m-accretive.

We can in particular apply Theorem 12.9 to $i(S - \lambda I)$ and $-i(S - \lambda I)$ where S is a symmetric operator. This gives that $S - \lambda I$ is injective with

$$\|(S - \lambda I)^{-1}\| \leq |\mathrm{Im}\,\lambda|^{-1} \,, \quad \text{when} \quad \mathrm{Im}\,\lambda \neq 0 \qquad (12.24)$$

(which could also have been derived directly from (12.20)).

Theorem 12.10. *Let S be densely defined and symmetric. Then S is selfadjoint if and only if*

$$R(S + iI) = R(S - iI) = H \; ; \tag{12.25}$$

and in the affirmative case, $\mathbb{C} \setminus \mathbb{R} \subset \varrho(S)$.

Proof. Let S be selfadjoint. Then iS and $-iS$ satisfy the hypotheses of Theorem 12.9 3° with $\beta = 0$, and hence the half-spaces

$$\mathbb{C}_\pm = \{\lambda \in \mathbb{C} \mid \operatorname{Im} \lambda \gtrless 0\} \tag{12.26}$$

are contained in the resolvent set; in particular, we have (12.25).

Conversely, if (12.25) holds, we see from (12.11) that the operators $S^* \pm iI$ are injective. Here $S^* + iI$ is an injective extension of $S + iI$, which is a bijection of $D(S)$ onto H; this can only happen if $S = S^*$. $\qquad\square$

See also Exercise 12.32.

What a symmetric, densely defined operator "lacks" in being selfadjoint can be seen from what $S+iI$ and $S-iI$ lack in being surjective. The *deficiency indices* of S are defined by

$$\operatorname{def}_+(S) = \dim R(S + iI)^\perp \quad \text{and} \quad \operatorname{def}_-(S) = \dim R(S - iI)^\perp \,. \tag{12.27}$$

It is very interesting to study the possible selfadjoint extensions of a symmetric, densely defined operator S. It can be shown that S has a selfadjoint extension if and only if the two deficiency indices in (12.27) are equal (with suitable interpretations of infinite dimensions, cf. Exercise 12.26); and in the case of equal indices the family of selfadjoint extensions may be parametrized by the linear isometries of $R(S + iI)^\perp$ onto $R(S - iI)^\perp$; cf. Exercise 12.19. One can also use the Cayley transformation $S \mapsto U = (S + iI)(S - iI)^{-1}$ (see e.g. Rudin [R74]), carrying the study over to a study of isometries U. We shall later (in Chapter 13) consider cases where S in addition is injective or positive, by different methods.

The next theorem gives a type of examples of selfadjoint unbounded operators, using the technique of Theorem 12.5 in a clever way.

Theorem 12.11. *Let H and H_1 be Hilbert spaces, and let $T : H \to H_1$ be densely defined and closed. Then $T^*T : H \to H$ is selfadjoint and ≥ 0. In particular, $T^*T + I \geq 1$ and is bijective from $D(T^*T)$ to H, and the inverse has norm ≤ 1 and lower bound ≥ 0. Moreover, $D(T^*T)$ is dense in $D(T)$ with respect to the graph norm on $D(T)$.*

Proof. The operator T^*T is clearly symmetric and ≥ 0, since

$$(T^*Tx, x)_H = (Tx, Tx)_{H_1} \geq 0 \quad \text{for} \quad x \in D(T^*T) \,, \tag{12.28}$$

cf. Lemma 12.8 1°. Since T is densely defined and closed,

$$H \oplus H_1 = G(T) \oplus UG(T^*)$$

by Theorem 12.5. Every $\{x, 0\} \in H \times H_1$ then has a unique decomposition

$$\{x, 0\} = \{y, Ty\} + \{-T^*z, z\} ,$$

where y and z are determined from x. Since this decomposition is linear, it defines two bounded linear operators $S : H \to H$ and $P : H \to H_1$ such that $y = Sx$ and $z = Px$. Note that $R(S) \subset D(T)$ and $R(P) \subset D(T^*)$. We will show that S equals $(T^*T + 1)^{-1}$ and is selfadjoint, bounded and ≥ 0; this will imply the assertions on T^*T.

In view of the orthogonality,

$$\|x\|_H^2 = \|\{y, Ty\}\|_{H \oplus H_1}^2 + \|\{-T^*z, z\}\|_{H \oplus H_1}^2$$
$$= \|Sx\|_H^2 + \|TSx\|_{H_1}^2 + \|T^*Px\|_H^2 + \|Px\|_{H_1}^2 ,$$

which implies that S and P have norm ≤ 1. Since

$$x = Sx - T^*Px , \quad 0 = TSx + Px ,$$

we see that $TSx = -Px \in D(T^*)$ and

$$x = (1 + T^*T)Sx ; \tag{12.29}$$

hence S maps the space H into $D(T^*T)$, and $(1 + T^*T)S = I$ on H. The bounded operator S^* is now seen to satisfy

$$(S^*x, x) = (S^*(1 + T^*T)Sx, x) = (Sx, Sx) + (TSx, TSx) \geq 0 \text{ for } x \in H ,$$

which implies that S^* is symmetric ≥ 0, and $S = S^{**} = S^*$ is likewise symmetric ≥ 0.

Since S is injective (cf. (12.29)), selfadjoint, closed and densely defined and has dense range (since $Z(S^*) = \{0\}$, cf. (12.11)), Theorem 12.7 implies that S^{-1} has the same properties. According to (12.29), $1 + T^*T$ is a symmetric extension of S^{-1}. Since $1 + T^*T$ is injective, $1 + T^*T$ must equal S^{-1}. Hence $I + T^*T$ and then also T^*T is selfadjoint.

The denseness of $D(T^*T)$ in $D(T)$ with respect to the graph norm (cf. (12.6)) is seen as follows: Let $x \in D(T)$ be orthogonal to $D(T^*T)$ with respect to the graph norm, i.e.,

$$(x, y)_H + (Tx, Ty)_{H_1} = 0 \text{ for all } y \in D(T^*T) .$$

Since $(Tx, Ty)_{H_1} = (x, T^*Ty)_H$, we see that

$$(x, y + T^*Ty)_H = 0 \text{ for all } y \in D(T^*T) ,$$

from which it follows that $x = 0$, since $I + T^*T$ is surjective. □

We have the following connection between semiboundedness and boundedness.

Theorem 12.12. 1° *When S is symmetric ≥ 0, one has the following version of the Cauchy-Schwarz inequality:*

$$|(Sx,y)|^2 \leq (Sx,x)(Sy,y) \text{ for } x, y \in D(S) .\qquad(12.30)$$

2° *If S is a densely defined, symmetric operator with $0 \leq S \leq \alpha$, then S is bounded with $\|S\| \leq \alpha$.*

Proof. 1°. When $t \in \mathbb{R}$, we have for x and $y \in D(S)$,

$$0 \leq (S(x + ty), x + ty) = (Sx,x) + t(Sx,y) + t(Sy,x) + t^2(Sy,y)$$
$$= (Sx,x) + 2\operatorname{Re}(Sx,y)t + (Sy,y)t^2 .$$

Since this polynomial in t is ≥ 0 for all t, the discriminant must be ≤ 0, i.e.,

$$|2\operatorname{Re}(Sx,y)|^2 \leq 4(Sx,x)(Sy,y) .$$

When x is replaced by $e^{i\theta}x$ where $-\theta$ is the argument of (Sx,y), we get (12.30).

2°. It follows from 1° that

$$|(Sx,y)|^2 \leq (Sx,x)(Sy,y) \leq \alpha^2\|x\|^2\|y\|^2 \text{ for } x, y \in D(S) .$$

Using that $D(S)$ is dense in H, we derive:

$$|(Sx,z)| \leq \alpha\|x\| \|z\| \text{ for } x \in D(S) \text{ and } z \in H .\qquad(12.31)$$

Recall that the Riesz representation theorem defines an *isometry* between H^* and H such that the norm of an element $y \in H$ satisfies

$$\|y\| = \sup\left\{\frac{|(y,z)|}{\|z\|} \mid z \in H \setminus \{0\}\right\} .\qquad(12.32)$$

In particular, we can conclude from the information on S that $\|Sx\| \leq \alpha\|x\|$ for $x \in D(S)$, which shows that S is bounded with norm $\leq \alpha$. $\square$

A similar result is obtained for slightly more general operators in Exercise 12.9.

An important case of a complex Hilbert space is the space $L_2(\Omega)$, where Ω is an open subset of $\mathbb{R}^n$ and the functions are complex valued. The following gives a simple special case of not necessarily bounded operators in $L_2(\Omega)$, where one can find the adjoint in explicit form and there is an easy criterion for selfadjointness.

Theorem 12.13. *Let Ω be an open subset of $\mathbb{R}^n$, and let $p : \Omega \to \mathbb{C}$ be a measurable function. The multiplication operator M_p in $L_2(\Omega)$ defined by*

$$D(M_p) = \{u \in L_2(\Omega) \mid pu \in L_2(\Omega)\},$$
$$M_p u = pu \ \text{ for } u \in D(M_p),$$

is densely defined and closed, and the adjoint operator M_p^* is precisely the multiplication operator $M_{\bar{p}}$.

Here M_p is selfadjoint if p is real.

If $|p(x)| \leq C$ (for a constant $C \in [0, \infty[$), then M_p is everywhere defined and bounded, with norm $\leq C$.

Proof. Clearly, the operator is linear. Observe that a measurable function f on Ω lies in $D(M_p)$ if and only if $(1 + |p|)f \in L_2(\Omega)$. Hence

$$D(M_p) = \Big\{ \frac{\varphi}{1 + |p|} \ \Big| \ \varphi \in L_2(\Omega) \Big\}.$$

It follows that $D(M_p)$ is dense in $L_2(\Omega)$, for if $\overline{D(M_p)}$ were $\neq L_2(\Omega)$, then there would exist an $f \in L_2(\Omega) \setminus \{0\}$ such that $f \perp \overline{D(M_p)}$, and then one would have

$$0 = \Big(\frac{\varphi}{1 + |p|}, f \Big) = \Big(\varphi, \frac{f}{1 + |p|} \Big), \ \text{ for all } \varphi \in L_2(\Omega),$$

and this would imply $\dfrac{f}{1 + |p|} = 0$ in contradiction to $f \neq 0$.

By definition of the adjoint, we have that $f_1 \in D(M_p{}^*)$ if and only if there exists a $g \in L_2(\Omega)$ such that

$$(M_p f, f_1) = (f, g) \text{ for all } f \in D(M_p),$$

i.e.,

$$\Big(\frac{p\varphi}{1 + |p|}, f_1 \Big) = \Big(\frac{\varphi}{1 + |p|}, g \Big) \text{ for all } \varphi \in L_2(\Omega).$$

We can rewrite this as

$$\Big(\varphi, \frac{\bar{p} f_1}{1 + |p|} \Big) = \Big(\varphi, \frac{g}{1 + |p|} \Big), \ \text{ for all } \varphi \in L_2(\Omega); \qquad (12.33)$$

here we see, since $\dfrac{\bar{p} f_1}{1 + |p|}$ and $\dfrac{g}{1 + |p|}$ belong to $L_2(\Omega)$, that (12.33) holds if and only if they are the same element, and then

$$\bar{p} f_1 = g.$$

This shows that $M_p{}^* = M_{\bar{p}}$. It is a closed operator; then so is M_p.

It follows immediately that if $p = \bar{p}$, then M_p is selfadjoint.

Finally, it is clear that when $|p(\xi)| \leq C$ for $\xi \in X$, then $D(M_p) = L_2(\Omega)$ and

$$\|M_p f\|_{L_2} \leq C \|f\|_{L_2};$$

and then M_p is bounded with $\|M_p\| \leq C$. $\qquad\qquad\qquad\qquad$ □

Note that $D(M_p) = D(M_{\overline{p}}) = D(M_{|p|})$; here we have a case where the operator and its adjoint have the same domain (which is not true in general).

We can also observe that when p and q are bounded functions, then $M_{pq} = M_p M_q$. (For unbounded functions, the domains of M_{pq} and $M_p M_q$ may not be the same; consider e.g. $p(x) = x$ and $q(x) = 1/x$ on $\Omega = \mathbb{R}_+$.)

It is not hard to see that $M_p M_{\overline{p}} = M_{\overline{p}} M_p$, so that M_p is *normal* (i.e., commutes with its adjoint).

Since $M_p u$ and $M_{p_1} u$ define the same element of L_2 if p and p_1 differ on a null-set only, the definition of M_p is easily extended to almost everywhere defined functions p. We can also observe that if $p \in L_\infty(\Omega)$, then $M_p \in \mathbf{B}(L_2(\Omega))$, and

$$\|M_p\| = \operatorname{ess\,sup}_{x \in \Omega} |p(x)|. \tag{12.34}$$

Here the inequality $\|M_p\| \leq \operatorname{ess\,sup} |p(x)|$ is easily seen by choosing a representative p_1 for p with $\sup |p_1| = \operatorname{ess\,sup} |p|$. On the other hand, one has for $\varepsilon > 0$ that the set $\omega_\varepsilon = \{x \in \Omega \mid |p(x)| \geq \operatorname{ess\,sup} |p| - \varepsilon\}$ has positive measure, so $\|M_p u\| \geq (\sup |p| - \varepsilon)\|u\|$ for $u = 1_K$ (cf. (A.27)), where K denotes a measurable subset of ω_ε with finite, positive measure; here $u \in L_2(\Omega)$ with $\|u\| > 0$.

It is seen in a similar way that $p = \overline{p}$ a.e. is necessary for the selfadjointness of M_p.

12.4 Operators associated with sesquilinear forms

A complex function $a : \{x, y\} \mapsto a(x, y) \in \mathbb{C}$, defined for x, y in a vector space V, is said to be a *sesquilinear form*, when it is linear in x and conjugate linear — also called antilinear or semilinear — in y:

$$a(\alpha x_1 + \beta x_2, y) = \alpha a(x_1, y) + \beta a x_2, y),$$
$$a(x, \gamma y_1 + \delta y_2) = \overline{\gamma} a(x, y_1) + \overline{\delta} a(x, y_2).$$

(The word "sesqui" is Latin for $1\frac{1}{2}$.) For example, the scalar product on a complex Hilbert space is a sesquilinear form.

Let H be a complex Hilbert space, and let $s(x, y)$ be a sesquilinear form defined for x and y in a subspace $D(s)$ of H; $D(s)$ is called the domain of s, and we may say that s is a sesquilinear form on H (even if $D(s) \neq H$). The *adjoint* sesquilinear form s^* is defined to have $D(s^*) = D(s)$ and

$$s^*(x, y) = \overline{s(y, x)} \quad \text{for} \quad x, y \in D(s), \tag{12.35}$$

and we call s *symmetric* when $s = s^*$. A criterion for symmetry is that $s(x, x)$ is real for all x; this is shown just like in Lemma 12.8 1°.

s is said to be *bounded* (on H) when there is a constant C so that

$$|s(x,y)| \leq C\|x\|_H\|y\|_H \ , \quad \text{for } x \text{ and } y \in D(s) \ .$$

Any (linear, as always) operator T in H gives rise to a sesquilinear form t_0 by the definition

$$t_0(x,y) = (Tx,y)_H \ , \quad \text{with } D(t_0) = D(T) \ .$$

If T is bounded, so is t_0. The converse holds when $D(T)$ is dense in H, for then the boundedness of t_0 implies that when $x \in D(T)$,

$$|(Tx,y)_H| \leq C\|x\|_H\|y\|_H$$

for y in a dense subset of H and hence for all y in H; then (cf. (12.32))

$$\|Tx\|_H = \sup\{\frac{|(Tx,y)_H|}{\|y\|_H} | y \in H \setminus \{0\}\} \leq C\|x\|_H \ .$$

In this case, T and t_0 are extended in a trivial way to a bounded operator, resp. a bounded sesquilinear form, defined on all of H.

The unbounded case is more challenging.

Definition 12.14. Let $t(x,y)$ be a sesquilinear form on H (not assumed to be bounded), with $D(t)$ dense in H. The **associated operator** T in H is defined as follows:

$D(T)$ consists of the elements $x \in D(t)$ for which there exists $y \in H$ such that

$$t(x,v) = (y,v)_H \ \text{ for all } \ v \in D(t) \ .$$

When the latter equations hold, y is uniquely determined from x since $D(t)$ is dense in H, and we set

$$Tx = y \ .$$

When t is unbounded, $D(T)$ will usually be a proper subset of $D(t)$ and T an unbounded operator in H.

The construction leads to a useful class of operators in a special case we shall now describe.

Let V be a linear subspace of H, which is dense in H and which is a Hilbert space with a norm that is stronger than the norm in H:

$$\|v\|_V \geq c\|v\|_H \ \text{ for } \ v \in V \ , \tag{12.36}$$

with $c > 0$. We then say that $V \subset H$ *algebraically, topologically and densely*. Let $a(u,v)$ be a sesquilinear form with $D(a) = V$ and a bounded on V, i.e.,

$$|a(u,v)| \leq C\|u\|_V\|v\|_V, \ \text{for all } u,v \in V. \tag{12.37}$$

The form a induces the following two different operators: a bounded operator in V and a (usually) unbounded operator in H, obtained by applying Definition 12.14 to a as a form on V resp. H. Let us denote them $\mathcal{A}$ resp. A.

The form a is called V-*elliptic* if there is a constant $c_0 > 0$ so that

$$\operatorname{Re} a(v, v) \geq c_0 \|v\|_V^2 \quad \text{for} \quad v \in V . \tag{12.38}$$

It will be called V-*coercive* if an inequality as in (12.38) can be obtained by adding a multiple of $(u, v)_H$ to a, i.e., if there exists $c_0 > 0$ and $k \in \mathbb{R}$ so that

$$\operatorname{Re} a(v, v) + k\|v\|_H^2 \geq c_0 \|v\|_V^2 \quad \text{for} \quad v \in V . \tag{12.39}$$

For $\mu \in \mathbb{C}$ we denote

$$a_\mu(u, v) = a(u, v) + \mu(u, v)_H, \text{ with } D(a_\mu) = D(a). \tag{12.40}$$

In view of (12.36), a_μ is bounded on V. When (12.39) holds, then a_μ is V-elliptic whenever $\operatorname{Re} \mu \geq k$.

Note that when a is V-elliptic or V-coercive, then the same holds for the adjoint form a^* (recall (12.35)), with the same constants.

The results in the following are known as "the Lax-Milgram lemma", named after Lax and Milgram's paper [LM54]. One finds there the bounded version:

Lemma 12.15. *Let a be a bounded everywhere defined sesquilinear form on V, and let $\mathcal{A}$ be the associated operator in V. Then $\mathcal{A} \in \mathbf{B}(V)$ with norm $\leq C$ (cf. (12.37)), and its adjoint $\mathcal{A}^*$ in V is the operator in V associated with a^*.*

Moreover, if a is V-elliptic, then $\mathcal{A}$ and $\mathcal{A}^*$ are homeomorphisms of V onto V, the inverses having norms $\leq c_0^{-1}$ (cf. (12.38)).

Proof. The boundedness of $\mathcal{A}$ was shown above. That the adjoint is the operator in V associated with a^* follows from (12.10) and (12.35). Now (12.38) implies that $m(\mathcal{A}) \geq c_0$ as well as $m(\mathcal{A}^*) \geq c_0$, where $c_0 > 0$. Then $\mathcal{A}$ and $\mathcal{A}^*$ are bijective from V to V with bounded inverses by Theorem 12.9 3°, which also gives the bound on the inverses. $\qquad\square$

By the Riesz representation theorem, there is an identification of H with the dual space of continuous linear functionals on H, such that $x \in H$ is identified with the functional $y \mapsto (y, x)_H$. To avoid writing x to the right, we can instead identify H with the *antidual* space, where x corresponds to the *antilinear* (conjugate linear) functional $y \mapsto (x, y)_H$. (This follows from the usual Riesz theorem by conjugation.)

From now on, *we denote by H^* the antidual space of H.* Stated in details: H^* is the space of continuous antilinear functionals on H, and we identify $x \in H$ with the functional $y \mapsto (x, y)_H$. The prefix anti- is generally left out.

In the situation of real Hilbert spaces, the antidual and dual spaces are of course the same.

The space V likewise has an antidual space V^* consisting of the continuous antilinear functionals on V. By the Riesz representation theorem, every element $w \in V$ corresponds to an element $Jw \in V^*$ such that

$$(Jw)(v) = (w, v)_V, \text{ all } v \in V, \tag{12.41}$$

and the map J is an isometry of V onto V^*. But rather than using this isometry to identify V and V^*, we want to focus on the embedding $V \subset H$.

Since $V \subset H$ densely and (12.36) holds, we can define a map from H to V^* sending $f \in H$ over into the antilinear functional $\ell_f \in V^*$ for which

$$\ell_f(v) = (f, v)_H \text{ for all } v \in V. \tag{12.42}$$

(12.36) assures that ℓ_f is continuous on V. The mapping from f to ℓ_f is injective by the denseness of V in H (if ℓ_{f_1} acts like ℓ_{f_2}, $f_1 - f_2$ is H-orthogonal to V, hence is zero). Thus the map $f \mapsto \ell_f$ can be regarded as an embedding of H into V^*, and we henceforth denote ℓ_f by f, writing $\ell_f(v)$ as $(f, v)_H$.

Lemma 12.16. *There are continuous injections*

$$V \hookrightarrow H \hookrightarrow V^*; \tag{12.43}$$

here, when $f \in H$,

$$\|f\|_{V^*} \leq c^{-1}\|f\|_H. \tag{12.44}$$

Proof. The injections are accounted for above, and (12.44) follows from the calculation

$$\|f\|_{V^*} = \sup\{\frac{|\ell_f(v)|}{\|v\|_V} \mid v \in V \setminus \{0\}\} = \sup\{\frac{|(f, v)_H|}{\|v\|_V} \mid v \in V \setminus \{0\}\}$$

$$\leq \sup\{\frac{\|f\|_H\|v\|_H}{\|v\|_V} \mid v \in V \setminus \{0\}\} \leq c^{-1}\|f\|_H,$$

using the definition of the norm of a functional, the denseness of V in H, (12.42) and (12.36). $\qquad\square$

Note that in the identification of ℓ_f with f when $f \in H$, we have obtained that the duality between V^* and V *extends the scalar product in* H, cf. (12.42).

When $\mathcal{A}$ is defined as in Lemma 12.15, we let $\widetilde{\mathcal{A}} = J\mathcal{A}$ (recall (12.41)); it is the operator in $\mathbf{B}(V, V^*)$ that satisfies

$$(\widetilde{\mathcal{A}}u)(v) = (\mathcal{A}u, v)_V = a(u, v) \text{ for all } u, v \in V.$$

We also define $\widetilde{\mathcal{A}'} = J\mathcal{A}^*$. Lemma 12.15 implies immediately:

Corollary 12.17. *When a is bounded on V and V-elliptic (satisfying (12.37), (12.38)), then $\widetilde{A} = JA$ is a homeomorphism of V onto V^*, with $\|\widetilde{A}\|_{\mathbf{B}(V,V^*)} \leq C$ and $\|\widetilde{A}^{-1}\|_{\mathbf{B}(V^*,V)} \leq c_0^{-1}$. $\widetilde{A}' = JA^*$ is similar.*

Now we take A into the picture.

Theorem 12.18. *Consider a triple (H, V, a) where H and V are complex Hilbert spaces with $V \subset H$ algebraically, topologically and densely (satisfying (12.36)), and where a is a bounded sesquilinear form on V with $D(a) = V$ (satisfying (12.37)). Let A be the operator associated with a in H:*

$$D(A) = \{\, u \in V \mid \exists f \in H \text{ so that } a(u,v) = (f,v)_H \text{ for all } v \in V \,\}, \qquad (12.45)$$
$$Au = f.$$

When a is V-elliptic (satisfying (12.38)), then A is a closed operator with $D(A)$ dense in H and in V, and with lower bound $m(A) \geq c_0 c^2 > 0$. It is a bijection of $D(A)$ onto H, and

$$\{\lambda \mid \operatorname{Re}\lambda < c_0 c^2\} \subset \varrho(A) . \qquad (12.46)$$

Moreover, the operator associated with a^* in H equals A^*; it likewise has the properties listed for A. In particular, if a is symmetric, A is selfadjoint > 0.

Proof. By Corollary 12.17, a gives rise to a bijection $\widetilde{A}$ from V to V^*, such that

$$a(u,v) = (\widetilde{A}u)(v) \text{ for all } u, v \in V.$$

By the definition of A, the elements of $D(A)$ are those $u \in V$ for which the functional $a(u,v)$ has the form $\ell_f(v)$ for some $f \in H$. In other words, $u \in D(A)$ precisely when $\widetilde{A}u$ identifies with an element $f \in H$ such that $(\widetilde{A}u)(v) = (f,v)_H$, and then $f = Au$. Consider the maps

$$
\begin{array}{ccccc}
& \widetilde{A} \to & & & \\
V & \subset & H & \subset & V^* \\
\| & & \| & & \\
D(A) & \subset & V & \subset & H \\
& A \to & & &
\end{array}
$$

Here $D(A)$ consists of exactly those $u \in V$ for which $\widetilde{A}u$ belongs to the subspace H of V^*, and then $Au = \widetilde{A}u$. Since $\widetilde{A}$ is invertible, we can simply write

$$D(A) = \widetilde{A}^{-1}(H), \text{ with } Au = \widetilde{A}u \text{ for } u \in D(A).$$

Similarly, a^* gives rise to the bijection $\widetilde{A}'$ from V to V^* such that

$$a^*(u,v) = (\widetilde{A}'u)(v) \text{ for all } u, v \in V.$$

The operator A' associated with a^* by Definition 12.14 then satisfies:

$$D(A') = (\widetilde{\mathcal{A}}')^{-1}(H), \text{ with } A'u = \widetilde{\mathcal{A}}'u \text{ for } u \in D(A').$$

Thus A and A' are *bijective from their domains onto* H, with inverses $T = A^{-1}$ and $T' = (A')^{-1}$ defined on all of H as the restrictions of $\widetilde{\mathcal{A}}^{-1}$ resp. $(\widetilde{\mathcal{A}}')^{-1}$ to H:

$$T = A^{-1} = \widetilde{\mathcal{A}}^{-1}|_H; \quad T' = (A')^{-1} = (\widetilde{\mathcal{A}}')^{-1}|_H.$$

Here T and T' are bounded from H to V and a fortiori from H to H, since

$$\|Tf\|_H \le c^{-1}\|Tf\|_V = c^{-1}\|\widetilde{\mathcal{A}}^{-1}f\|_V \le c^{-1}c_0^{-1}\|f\|_{V^*} \le c^{-2}c_0^{-1}\|f\|_H,$$

for $f \in H$, cf. (12.44); there is a similar calculation for T'.

Now we have for all $f, g \in H$, setting $u = Tf$, $v = T'g$ so that $f = Au$, $g = A'v$:

$$(f, T'g)_H = (f, v)_H = a(u, v) = \overline{a^*(v, u)} = \overline{(A'v, u)_H} = (u, A'v)_H = (Tf, g)_H;$$

this shows that the bounded operators T and T' in H are adjoints of one another. Their ranges are dense in H since their nullspaces are 0 (the operators are injective), cf. (12.11).

Since T and $T' = T^*$ are closed densely defined injective operators in H with dense ranges, we can apply Theorem 12.7 to conclude that their inverses are also each other's adjoints. So $A = T^{-1}$ and $A' = (T^*)^{-1}$ are each other's adjoints, as unbounded operators in H, and they are closed and densely defined there.

From (12.38) and (12.36) follows moreover:

$$\mathrm{Re}(Au, u)_H = \mathrm{Re}\, a(u, u) \ge c_0\|u\|_V^2 \ge c_0 c^2 \|u\|_H^2$$

for all $u \in D(A)$, so $m(A) \ge c_0 c^2$. We likewise find that $m(A^*) \ge c_0 c^2$, and (12.46) then follows from Theorem 12.9 3°.

To see that the set $D(A)$ is dense in V, let $v_0 \in V$ be such that $(u, v_0)_V = 0$ for all $u \in D(A)$. Then we can let $w_0 = (\mathcal{A}^*)^{-1}v_0$ (using Lemma 12.15) and calculate:

$$0 = (u, v_0)_V = (u, \mathcal{A}^*w_0)_V = a(u, w_0) = (Au, w_0)_H, \text{ for all } u \in D(A),$$

which implies $w_0 = 0$ and hence $v_0 = 0$. Similarly, $D(A^*)$ is dense in V. $\square$

Corollary 12.19. *Hypotheses as in Theorem* 12.18, *except that* V-*ellipticity is replaced by* V-*coercivity* (12.39). *Then* A *is a closed operator with* $D(A)$ *dense in* H *and in* V, *and with* $m(A) \ge c_0 c^2 - k$. *Moreover,*

$$\{\lambda \mid \mathrm{Re}\,\lambda < c_0 c^2 - k\} \subset \varrho(A) . \tag{12.47}$$

The operator associated with a^* *in* H *equals* A^* *and has the same properties.*

Proof. Note that for $\mu \in \mathbb{C}$, $A + \mu I$ (with $D(A + \mu I) = D(A)$) is the operator in H associated with the sesquilinear form a_μ defined in (12.40). When (12.39) holds, we replace a by a_k. Theorem 12.18 applies to this form and shows that the associated operator $A + kI$ and its adjoint $A^* + kI$ have the properties described there. This gives for A itself the properties we have listed in the corollary. □

We shall call the operator A defined from the triple (H, V, a) by Theorem 12.18 or Corollary 12.19 the *variational operator* associated with (H, V, a) (it can also be called the Lax-Milgram operator). The above construction has close links with the calculus of variations, see the remarks at the end of this section.

The construction and most of the terminology here is based on works of J.-L. Lions, as presented e.g. in his book [L63] and subsequent papers and books. The operators are also studied in the book of T. Kato [K66], where they are called m-sectorial.

Example 12.20. Variational operators are studied in many places in this book; abstract versions enter in Chapter 13 and concrete versions enter in Chapter 4 (and 14), see in particular Section 4.4 where the Dirichlet and Neumann realizations of the Laplace operator are studied. The distribution theory, or at least the definition of Sobolev spaces, is needed to give a satisfactory interpretation of the operators that arise from the construction. In fact, H is then usually a space $L_2(\Omega)$ over a subset $\Omega \subset \mathbb{R}^n$, and V is typically a Sobolev space such as $H^1(\Omega)$ or $H_0^1(\Omega)$. (There is then also an interpretation of V^* as a Sobolev space with exponent -1 — for the case $\Omega = \mathbb{R}^n$, we explain such spaces in Section 6.3.)

Let us at present just point to the one-dimensional example taken up in Exercise 12.25. Here V is the closure of $C^1(\overline{I})$ in the norm $\|u\|_1 = (\|u\|_{L_2}^2 + \|u'\|_{L_2}^2)^{\frac{1}{2}}$ (identified with $H^1(I)$ in Section 4.3), and $\frac{d}{dt}$ can be defined by extension as a continuous operator from V to H. Let $q(t)$ be real ≥ 1, then $a(u, u) \geq \|u\|_1^2$, and Theorem 12.18 applies, defining a selfadjoint operator A in H. When $u \in C^2(\overline{I})$, $v \in C^1(\overline{I})$, we have by integration by parts that

$$a(u, v) = (-u'' + qu, v) + u'(\beta)\overline{v}(\beta) - u'(\alpha)\overline{v}(\alpha).$$

Then if $u \in C^2(\overline{I})$ with $u'(\beta) = u'(\alpha) = 0$, it satisfies the requirement $a(u, v) = (f, v)$ for $v \in V$, *with* $f = -u'' + qu$. Hence it is in $D(A)$ with $Au = -u'' + qu$.

The typical information here is that A acts as a *differential operator*, namely, $-\frac{d^2}{dt^2} + q$ (of order 2, while $a(u, v)$ is of order 1), and the domain $D(A)$ involves a *boundary condition*, namely, $u'(\beta) = u'(\alpha) = 0$.

The example, and many more, can be completely worked out with the tools from distribution theory established in other parts of this book. One finds that $D(A) = \{u \in H^2(I) \mid u'(\alpha) = u'(\beta) = 0\}$.

We note that $D(A)$ and $D(A^*)$ need not be the same set even though $D(a) = D(a^*)$ (cf. e.g. Exercise 12.37).

We can use the boundedness of a on (12.37) to show that when a satisfies (12.39), then

$$|\operatorname{Im} a(u,u)| \leq |a(u,u)| \leq C\|u\|_V^2 \leq Cc_0^{-1}(\operatorname{Re} a(u,u) + k\|u\|_H^2) \,,$$

and hence

$$|\operatorname{Im}(Au,u)_H| \leq Cc_0^{-1}(\operatorname{Re}(Au,u)_H + k\|u\|_H^2) \,, \qquad (12.48)$$

when $u \in D(A)$. This shows that the *numerical range* $\nu(A)$ for A — and correspondingly the numerical range $\nu(A^*)$ for A^* — satisfy

$$\nu(A) \text{ and } \nu(A^*) \subset M' = \{\lambda \in \mathbb{C} \mid |\operatorname{Im}\lambda| \leq Cc_0^{-1}(\operatorname{Re}\lambda + k)\} \,. \qquad (12.49)$$

But this means that certain *rotations* of A and A^*, namely, $e^{\pm i\theta}A$ and $e^{\pm i\theta}A^*$ for a suitable θ (see the figure), are semibounded below, which implies by Theorem 12.9 3° that the spectra $\sigma(A)$ and $\sigma(A^*)$ are likewise contained in M'.

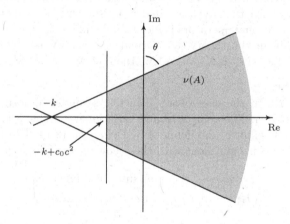

Thus we have:

Corollary 12.21. *When A and A^* are defined from the triple (H,V,a) as in Corollary 12.19, then the spectra $\sigma(A)$ and $\sigma(A^*)$ and the numerical ranges $\nu(A)$ and $\nu(A^*)$ are contained in the angular set with opening $< \pi$:*

$$M = \{\lambda \in \mathbb{C} \mid \operatorname{Re}\lambda \geq -k + c_0c^2 , \, |\operatorname{Im}\lambda| \leq Cc_0^{-1}(\operatorname{Re}\lambda + k)\} \,, \qquad (12.50)$$

where the constants are taken from (12.36), (12.39), (12.37).

Finally we observe that when A is a variational operator in a Hilbert space H, then V and a are *uniquely determined* by A. Indeed, if A is the variational operator associated with two triples (H,V_1,a_1) and (H,V_2,a_2), then we have for u and $v \in D(A)$:

$$(Au, v)_H = a_1(u, v) = a_2(u, v).$$

When k is sufficiently large, $[\mathrm{Re}((A + k)v, v)_H]^{\frac{1}{2}}$ is a norm on $D(A)$ which is equivalent with the V_1-norm and with the V_2-norm. Since $D(A)$ is dense in V_1 and in V_2 (with respect to their norms), we get by completion an identification between V_1 and V_2. Since a_1 and a_2 coincide on the dense subset $D(A)$, they must be equal. This shows:

Corollary 12.22. *When A is a variational operator in H, then A stems from one and only one triple (H, V, a); here V is determined as the completion of $D(A)$ under the norm $[\mathrm{Re}((A + k)v, v)_H]^{\frac{1}{2}}$ for a suitably large k, and a is defined on V by closure of $(Au, v)_H$.*

In this result it is assumed that A stems from a triple (H, V, a). One can show that such a triple exists, when A is closed densely defined and there exists a sector M' as in (12.49) such that A is *maximal* with regards to the property $\nu(A) \subset M'$. We return to this alternative description in Section 12.6.

The variational operators are a useful generalization of selfadjoint lower bounded operators, which enter for example in the study of partial differential equations; they have the advantage in comparison with normal operators that the class of variational operators is more stable under the perturbations that occur naturally in the theory. (A normal operator N is a closed, densely defined operator with $NN^* = N^*N$; then $D(N^*) = D(N)$, cf. Exercise 12.20.)

Remark 12.23. In the case of a symmetric sesquilinear form $a(u, v)$, the connection between Theorem 12.18 and variational calculus is as follows: Assume that a is V-elliptic, and define A by Definition 12.14. Then the problem of solving $Au = f$ is equivalent with the following variational problem:

$$\text{For a given } f \in H, \text{ minimize the functional} \qquad (12.51)$$
$$J(u) = a(u, u) - 2\,\mathrm{Re}(f, u)_H \text{ for } u \in V.$$

For, the equation $Au = f$ is (in this abstract setting) the Euler-Lagrange equation associated with J. More precisely, we have: Let $u, v \in V$ and let $\varepsilon \in \mathbb{R}$. Then

$$I(\varepsilon) \equiv J(u + \varepsilon v) = a(u, u) + 2\varepsilon\,\mathrm{Re}\,a(u, v) + \varepsilon^2 a(v, v)$$
$$-2\,\mathrm{Re}(f, u)_H - 2\varepsilon\,\mathrm{Re}(f, v)_H, \qquad (12.52)$$

so in particular $\frac{d}{d\varepsilon}I(0)$ (the so-called *first variation*) satisfies

$$\tfrac{d}{d\varepsilon}I(0) = 2\,\mathrm{Re}(a(u, v) - (f, v)_H).$$

If u solves (12.51), then $\frac{d}{d\varepsilon}I(0) = 0$ for all $v \in V$, so since we can insert αv instead of v for any complex α, it follows that

$$a(u,v) - (f,v)_H = 0 \text{ for all } v \in V. \tag{12.53}$$

By definition of A, this means precisely that

$$u \in D(A) \text{ with } Au = f. \tag{12.54}$$

Conversely, if u satisfies (12.54), (12.53) holds, so $\frac{d}{d\varepsilon}I(0) = 0$ for all $v \in V$. Then

$$J(u + \varepsilon v) = a(u,u) + \varepsilon^2 a(v,v) - 2\operatorname{Re}(f,u)_H \geq a(u,u) - 2\operatorname{Re}(f,u)_H,$$

for all $v \in V$ and $\varepsilon \in \mathbb{R}$, so u solves (12.51) (uniquely).

12.5 The Friedrichs extension

Let S be a symmetric, densely defined operator. When $S \geq c > 0$, it is easy to find a selfadjoint extension of S.

Let us first consider $\overline{S}$, the closure of S, which is also symmetric, cf. (12.19). It is easily seen that $m(S) = m(\overline{S})$, which is then $\geq c > 0$. According to Theorem 12.9, $\overline{S}$ has a bounded, symmetric inverse (the closure of S^{-1}), so if $R(S)$ and hence $R(\overline{S})$ is dense in H, $R(\overline{S})$ must equal H, so that $\overline{S}^{-1}$ is selfadjoint and $\overline{S}$ is a selfadjoint extension of S by Theorem 12.7. (This is the case where S is essentially selfadjoint.)

If, on the other hand, $Z(S^*) = R(S)^{\perp}$ is $\neq \{0\}$, we can introduce the operator P with

$$
\begin{aligned}
D(P) &= D(\overline{S}) \dotplus Z(S^*)\,, \\
P(v + z) &= \overline{S}v \quad \text{for } v \in D(\overline{S})\,,\ z \in Z(S^*)\,;
\end{aligned}
\tag{12.55}
$$

it is a selfadjoint extension (Exercise 12.23). This extension has $m(P) = 0$ in contrast to $m(S) > 0$. It is J. von Neumann's solution [N29] of the problem of finding a selfadjoint semibounded extension of S. A more refined (and useful) extension was found by K. Friedrichs [F34]; it has the same lower bound as S, and we shall explain its construction in the following. M. G. Krein later showed [K47] that the full set of selfadjoint extensions $\widetilde{T} \geq 0$, via the associated sesquilinear forms, can be characterized at the operators "lying between" T and P in a certain way. Here T ("the hard extension") is closest to S, whereas P ("the soft extension") is farthest from S. In practical applications, T is of great interest whereas P is somewhat exotic. More about the role of P in Section 13.2, see in particular Corollary 13.22, (13.67) and the surrounding text. Concrete interpretations to realizations of elliptic operators are given in Chapter 9 in a constant-coefficient case, see Example 9.31, and for variable-coefficient operators in Exercise 11.18.

Theorem 12.24 (FRIEDRICHS). *Let S be densely defined, symmetric and lower bounded in H. There exists a selfadjoint extension T with the same lower bound $m(T) = m(S)$, and with $D(T)$ contained in the completion of $D(S)$ in the norm $((Sv, v) + (1 - m(S))\|v\|^2)^{\frac{1}{2}}$.*

Proof. Assume first that $m(S) = c > 0$. The sesquilinear form

$$s_0(u, v) = (Su, v)$$

is then a scalar product on $D(S)$ (cf. Theorem 12.12), and we denote the completion of $D(S)$ with respect to this scalar product by V. Hereby $s_0(u, v)$ is extended to a sesquilinear form $s(u, v)$ with $D(s) = V$ (s is the scalar product itself on V).

We would like to use the Lax-Milgram construction (Theorem 12.18), but this requires that we show that there is an injection of V into H. The inequality

$$\|v\|_V^2 = s_0(v, v) \geq c\|v\|_H^2 \text{ for } v \in D(S),$$

implies that the injection $J_0 : D(S) \hookrightarrow H$ extends to a continuous map J from V to H, but does not assure that the extended map J is *injective*. This point (which is sometimes overlooked in other texts) can be treated as follows: From $(Su, v)_H = (u, v)_V$ for $u, v \in D(S)$ it follows by passage to the limit that

$$(Su, Jv)_H = (u, v)_V \quad \text{for } u \in D(S), \, v \in V.$$

If $Jv = 0$, v is orthogonal to the dense subspace $D(S)$ of V, and then $v = 0$. Hence J is injective, and we may identify V with the subspace $J(V)$ of H.

Since clearly

$$s(v, v) = \|v\|_V^2 \geq c\|v\|_H^2 \text{ for } v \in V,$$

the conditions for the Lax-Milgram construction are fulfilled by the triple (H, V, s) with $c_0 = 1$, and we obtain a selfadjoint operator T in H associated with s, satisfying $m(T) = c = m(S)$. It is clear that $T \supset S$.

When $m(S)$ is arbitrary, we define a selfadjoint extension T' of $S' = S + (1 - m(S))I$ by the procedure above; then $T = T' - (1 - m(S))I$ is the desired extension of S.   □

The proof of the Lax-Milgram construction can of course be simplified in the case where the sesquilinear form equals the scalar product on V, but the basic principles are the same.

The constructed operator T is called the *Friedrichs extension* of S. It is uniquely determined by the properties in Theorem 12.24, for we even have, in view of Corollary 12.22:

Corollary 12.25. *Let S be a densely defined, symmetric, lower bounded operator in H. For $k \geq -m(S)$, the completion V of $D(S)$ with respect to the scalar product*

$$(u, v)_V = (Su + (1 + k)u, v)_H \tag{12.56}$$

can be identified with a subspace of H independent of k, and the Friedrichs extension T of S is characterized by being the only *lower bounded selfadjoint extension of S which has its domain D(T) contained in V.*

When $\overline{S}$ is not selfadjoint, there exist many more selfadjoint extensions of S; more on this in Section 13.2. When $T \geq 0$, one can moreover show that V equals $D(T^{\frac{1}{2}})$, where $T^{\frac{1}{2}}$ is defined by spectral theory.

12.6 More on variational operators

In this section, we take up a converse question in connection with Corollary 12.19 ff.: When is an operator with numerical range in a sector (12.49) variational? This discussion is not needed for the immediate applications of the variational construction. (Theorem 12.26 is used in the proofs of Corollaries 13.16 and 9.36.) We shall show:

Theorem 12.26. *When A is a closed densely defined operator in H such that A and A^* both have their numerical ranges in a sector:*

$$\nu(A) \text{ and } \nu(A^*) \subset M' = \{\lambda \in \mathbb{C} \mid |\operatorname{Im}\lambda| \leq c_1(\operatorname{Re}\lambda + k)\}, \qquad (12.57)$$

for some $c_1 \geq 0$, $k \in \mathbb{R}$, then A is variational with spectrum in M' (and so is A^*). The associated sesquilinear form $a(u,v)$ is the closure of $a_0(u,v) = (Au,v)$, defined on the completion of $D(A)$ with respect to the norm $(\operatorname{Re} a_0(u,u) + (1+k)\|u\|_H^2)^{\frac{1}{2}}$.

Before giving the proof we make some preparations. When s is a given sesquilinear form, we define the sesquilinear forms s_{Re} and s_{Im}, with domain $D(s)$, by

$$\begin{aligned} s_{\operatorname{Re}}(u,v) &= \tfrac{1}{2}(s(u,v) + s^*(u,v)), \\ s_{\operatorname{Im}}(u,v) &= \tfrac{1}{2i}(s(u,v) - s^*(u,v)) \, ; \end{aligned} \qquad (12.58)$$

note that they are both symmetric (and can take complex values), and that

$$s(u,v) = s_{\operatorname{Re}}(u,v) + i s_{\operatorname{Im}}(u,v), \; u,v \in D(s). \qquad (12.59)$$

We define the numerical range and lower bound:

$$\begin{aligned} \nu(s) &= \{s(u,u) \in \mathbb{C} \mid u \in D(s), \|u\|_H = 1\}, \\ m(s) &= \inf\{\operatorname{Re} s(u,u) \mid u \in D(s), \|u\|_H = 1\}; \end{aligned} \qquad (12.60)$$

then s is said to be lower bounded, positive, resp. nonnegative, when $m(s)$ is $> -\infty$, > 0, resp. ≥ 0.

Note that when A is variational, defined from a as in Corollary 12.19, then

$$\overline{\nu(a)} = \overline{\nu(A)}, \tag{12.61}$$

since $D(A)$ is dense in $D(a)$.

Lemma 12.27. *When b and b' are symmetric sesquilinear forms with the same domain $D(b)$, satisfying*

$$|b(u,u)| \le b'(u,u), \ \ all \ u \in D(b), \tag{12.62}$$

then

$$|b(u,v)| \le b'(u,u)^{\frac{1}{2}} b'(v,v)^{\frac{1}{2}}, \ \ all \ u,v \in D(b). \tag{12.63}$$

Proof. Note that b' is a nonnegative sesquilinear form. Let $u,v \in D(b)$. If $b'(u,u)$ or $b'(v,v)$ is 0, so is $b(u,u)$ resp. $b(v,v)$ according to (12.62), so (12.63) is valid then. We now assume that $b'(u,u)$ and $b'(v,v) \ne 0$. By multiplication of u by $e^{i\theta}$ for a suitable θ, we can obtain that $b(u,v)$ is real, equal to $b(v,u)$. Then

$$b(u,v) = \tfrac{1}{4}(b(u+v,u+v) - b(u-v,u-v)), \tag{12.64}$$

as is easily checked. It follows by use of (12.62) that

$$|b(u,v)| \le \tfrac{1}{4}(b'(u+v,u+v) + b'(u-v,u-v)) = \tfrac{1}{2}(b'(u,u) + b'(v,v)).$$

Then for any $\alpha > 0$,

$$|b(\alpha u, \alpha^{-1} v)| \le \tfrac{1}{2}(\alpha^2 b'(u,u) + \alpha^{-2} b'(v,v)).$$

Taking $\alpha^2 = b'(v,v)^{\frac{1}{2}} b'(u,u)^{-\frac{1}{2}}$, we obtain (12.63). $\qquad\square$

Proof (of Theorem 12.26). By adding a constant to A, we reduce to the case where (12.57) holds with $k = -1$, so that

$$\operatorname{Re} a_0(u,u) \ge \|u\|_H^2. \tag{12.65}$$

Let us define $a_0(u,v) = (Au,v)$, with $D(a_0) = D(A)$. Then $a_{0,\text{Re}}$ is a positive symmetric sesquilinear form defining a scalar product and norm on $D(a_0)$; we denote the completion of $D(a_0)$ in this norm by V.

The crucial part of the proof is to show that V identifies with a subspace of H and that a_0 extends to a bounded positive sesquilinear form a on V. (This resembles a step in the proof of Theorem 12.24, but demands more effort.) In view of (12.57) with $k = -1$,

$$\begin{aligned}|a_0(u,u)| &\le |\operatorname{Re} a_0(u,u)| + |\operatorname{Im} a_0(u,u)| \\ &\le (1 + c_1) \operatorname{Re} a_0(u,u) = (1 + c_1) a_{0,\text{Re}}(u,u).\end{aligned} \tag{12.66}$$

Moreover, by an application of Lemma 12.27 with $b = a_{0,\text{Im}}$, $b' = (1+c_1)a_{0,\text{Re}}$,

$$|a_0(u,v)| \le |a_{0,\mathrm{Re}}(u,v)| + |a_{0,\mathrm{Im}}(u,v)|$$
$$\le a_{0,\mathrm{Re}}(u,u)^{\frac{1}{2}} a_{0,\mathrm{Re}}(v,v)^{\frac{1}{2}} + (1+c_1)a_{0,\mathrm{Re}}(u,u)^{\frac{1}{2}} a_{0,\mathrm{Re}}(v,v)^{\frac{1}{2}}$$
$$= (2+c_1)a_{0,\mathrm{Re}}(u,u)^{\frac{1}{2}} a_{0,\mathrm{Re}}(v,v)^{\frac{1}{2}}. \tag{12.67}$$

In view of (12.65), the map $J : D(a_0) \hookrightarrow H$ extends to a continuous map $J : V \to H$ with $\|Jv\|_H \le \|v\|_V$. To show that J is injective, let $v \in V$ with $Jv = 0$; we must show that $v = 0$. There exists a sequence $v_k \in D(a_0)$ such that v_k converges to v in V and v_k converges to 0 in H; then we are through if we can show that $\|v_k\|_V \to 0$, i.e., $\mathrm{Re}\, a_0(v_k, v_k) \to 0$, for $k \to \infty$. We know that $\mathrm{Re}\, a_0(v_k - v_l, v_k - v_l) \to 0$ for $k,l \to \infty$, and that $\mathrm{Re}\, a_0(v_l, v_l) \le C$ for some $C > 0$, all l. Now

$$| \mathrm{Re}\, a_0(v_k, v_k)| \le |a_0(v_k, v_k)| \le |a_0(v_k, v_k - v_l)| + |a_0(v_k, v_l)|$$
$$\le |a_0(v_k, v_k - v_l)| + |(Av_k, v_l)_H|,$$

where

$$|a_0(v_k, v_k - v_l)| \le (2+c_1)\|v_k\|_V \|v_k - v_l\|_V$$

in view of (12.67). For any $\varepsilon > 0$ we can find N such that $\|v_k - v_l\|_V \le \varepsilon$ for $k, l \ge N$. Let $l \to \infty$, then since $\|v_l\|_H \to 0$, we get the inequality

$$| \mathrm{Re}\, a_0(v_k, v_k)| \le (2+c_1)C\varepsilon, \text{ for } k \ge N.$$

Since the constant $(2+c_1)C$ is independent of k, this shows the desired fact, that $\|v_k\|_V \to 0$. Thus J is injective from V to H, so that V can be identified with $JV \subset H$. Here $V \subset H$ algebraically, topologically and densely.

Next, since $a_{0,\mathrm{Re}}$ and $a_{0,\mathrm{Im}}$ are bounded in the V-norm, they extend uniquely to bounded sesquilinear forms on V, so $a_0(u,v)$ does so too, and we can denote the extension by $a(u,v)$. The information in (12.57) with $k = -1$ implies

$$| \mathrm{Im}\, a(v,v)| \le c_1(\mathrm{Re}\, a(v,v) - \|v\|_H^2). \tag{12.68}$$

So a is V-elliptic, and hence defines a variational operator A_1; clearly

$$A \subset A_1, \tag{12.69}$$

and $\nu(A_1) \subset M'$.

We can do the whole procedure for A^* too. It is easily checked that the space V is the same as for A, and that the extended sesquilinear form is a^*. Then the variational operator defined from (H, V, a^*) is equal to A_1^* and extends A^*; in view of (12.69) this implies $A = A_1$. So A is the variational operator defined from (H, V, a). $\square$

Remark 12.28. An equivalent way to formulate the conditions on A for being variational, is to say that it is closed, densely defined and *maximal* with respect to the property $\nu(A) \subset M'$. Then the construction of V and a go through as in the above proof, leading to a variational operator A_1 such that (12.69) holds. Here A_1 is an extension of A with $\nu(A_1) \subset M'$, so in view of the maximality, A_1 must equal A.

The result is due to Schechter and Kato, see e.g. [K66, Th. VI.1.27].

Exercises for Chapter 12

12.1. Let H be a separable Hilbert space, with the orthonormal basis $(e_j)_{j\in\mathbb{N}}$. Let V denote the subspace of (finite) linear combinations of the basis-vectors. Define the operator T in H with $D(T) = V$ by

$$T(\sum_{j=1}^{n} c_j e_j) = \sum_{j=1}^{n} c_j e_1 .$$

Show that T has no closure. Find T^*, and find the closure of $G(T)$.

12.2. With H and T as in the preceding exercise, let T_1 be the restriction of T with $D(T_1) = V_1$, where V_1 is the subspace of linear combinations of the basis-vectors e_j with $j \geq 2$. Show that T_1 is a symmetric operator which has no closure.
Find an isometry U of a subspace of H into H such that $U - I$ is injective, but $\overline{U} - I$ is not injective.

12.3. Show that an operator $T : X \to Y$ is closed if and only if $D(T)$ is complete with respect to the graph norm.

12.4. Show that when R, S and T are operators in a Banach space X, then $RS + RT \subset R(S+T)$. Investigate the example

$$S = \frac{d}{dx} , \quad T = -\frac{d}{dx} , \quad R = \frac{d}{dx} ,$$

for the Banach space $C^0([0,1])$.

12.5. Let H be a Hilbert space, and let B and T be operators in H, with $B \in \mathbf{B}(H)$. Show the following assertions:

(a) If T is closed, then TB is closed.

(b) If T is densely defined, T^*B^* is closed but not necessarily densely defined; and BT is densely defined but not necessarily closable. Moreover, $T^*B^* = (BT)^*$.

(c) If T and T^* are densely defined and BT is closed, then T is closed and T^*B^* is densely defined, and $BT = (T^*B^*)^*$.

(d) If T is densely defined and closed, and TB is densely defined, then $(TB)^* = \overline{B^*T^*}$.

12.6. Find, for example for $H = L_2(]0,1[)$, selfadjoint operators $B \in \mathbf{B}(H)$ and T in H such that TB is not densely defined and BT is not closable. (*Hint.* Using results from Chapter 4, one can take as T the realization $A_\#$ defined in Theorem 4.16 and take a suitable B with $\dim R(B) = 1$.)

12.7. Investigate (12.12) by use of the examples in Exercises 12.4 and 12.6.

12.8. Show that if the operator T in H is densely defined, and $(Tx, x) = 0$ for all $x \in D(T)$, then $T = 0$. Does this hold without the hypothesis $\overline{D(T)} = H$?

12.9. Let T be a densely defined operator in H with $D(T) \subset D(T^*)$. Show that if $\nu(T)$ is bounded, then T is bounded.
(*Hint.* One can consider the symmetric operators $\operatorname{Re} T = \frac{1}{2}(T + T^*)$ and $\operatorname{Im} T = \frac{1}{2i}(T - T^*)$.)

12.10. With H equal to the complex Hilbert space $L_2(]0,1[)$, consider the operator T defined by

$$D(T) = \{u \in H \mid (u, 1) = 0\},$$

$$T : u(t) \mapsto f(t) = \int_0^t u(s)\, ds.$$

(a) Show that T is bounded.
(b) Show that T is skew-symmetric (i.e., iT is symmetric).
(*Hint.* The illustration in Section 4.3 may be helpful.)
(c) Is iT selfadjoint?

12.11. Consider the intervals

$$I_1 =]0,1[, \quad I_2 =]0,\infty[= \mathbb{R}_+, \quad I_3 = \mathbb{R},$$

and the Hilbert spaces $H_j = L_2(I_j)$, $j = 1, 2, 3$. Let T_j be the multiplication operator M_{t^3} (the mapping $u(t) \mapsto t^3 u(t)$) defined for functions in H_j, $j = 1, 2, 3$, respectively.
Find out in each case whether T_j is

(1) bounded,
(2) lower bounded,
(3) selfadjoint.

12.12. Let $I =]0,\infty[$ and consider the Hilbert space $H = L_2(I)$ (where $C_0^\infty(I)$ is a dense subset). Let T be the operator acting like $D = \frac{1}{i}\frac{d}{dx}$ with $D(T) = C_0^\infty(I)$.
(a) Show that T is symmetric and has a closure $\overline{T}$ (as an operator in H).
(b) Show that the equation $u'(t) + u(t) = f(t)$ has a solution in H for every $f \in C_0^\infty(I)$. (The solution method is known from elementary calculus.)

(c) Show that $R(\overline{T} - i)$ contains $C_0^\infty(I)$, and conclude that $\overline{T} - i$ is surjective.

(d) Show that the function e^{-t} is in $Z(T^* - i)$.

(e) Show that $\overline{T}$ is maximal symmetric, but does not have a selfadjoint extension.

12.13. Let $J = [\alpha, \beta]$. Show that the operator $\frac{d}{dx}$ in the Banach space $C^0(J)$, with domain $C^1(J)$, is closed.

12.14. Show the assertions in Corollary 12.21 on the spectra $\sigma(A)$ and $\sigma(A^*)$ in details.

12.15. Let H be a separable Hilbert space, with the orthonormal basis $(e_j)_{j \in \mathbb{Z}}$. For each $j > 0$, let $f_j = e_{-j} + je_j$. Let V and W be the closures of the spaces of linear combinations of, respectively, the vectors $(e_j)_{j \geq 0}$ and the vectors $(f_j)_{j > 0}$. Show that $V + W$ is dense in H but not closed. (*Hint.* Consider for example the vector $x = \sum_{j>0} \frac{1}{j} e_{-j}$.)

12.16. Let H be a Hilbert space over $\mathbb{C}$ or $\mathbb{R}$, and let V be a dense subspace. Show the following assertions:

(a) For each $x \in H$, $\{x\}^\perp \cap V$ is dense in $\{x\}^\perp$. (*Hint.* Choose for example sequences x_n and y_n from V which converge to x resp. $y \in \{x\}^\perp$ and consider $(x_n, x)y_n - (y_n, x)x_n$ for $\|x\| = 1$.)

(b) For each finite dimensional subspace K of H, $K^\perp \cap V$ is dense in $K^\perp$.

(c) To x and y in H with $x \perp y$ there exist sequences $(x_n)_{n \in \mathbb{N}}$ and $(y_n)_{n \in \mathbb{N}}$ in V such that $x_n \to x$, $y_n \to y$, and

$$(x_n, y_m) = 0 \quad \text{for all } n \text{ and } m.$$

(*Hint.* Choose successively x_{n+1} such that $\|x - x_{n+1}\| \leq 2^{-n-1}$ and $x_{n+1} \perp \{y_1, y_2, \ldots, y_n, y\}$, with y_{n+1} chosen similarly.)

12.17. Let X and Y be Banach spaces and let $(T_n)_{n \in \mathbb{N}}$ be a sequence of operators in $\mathbf{B}(X, Y)$. Assume that there is a constant $c > 0$ so that $\|T_n\| \leq c$ for all n, and that for x in a dense subspace V of X, $T_n x$ is convergent in Y. Show that there is a uniquely determined operator $T \in \mathbf{B}(X, Y)$ such that $T_n x \to Tx$ for all $x \in X$. (One can use an $\varepsilon/3$-argument.)

12.18. Let A be an operator in a Hilbert space H such that $D(A^2) = D(A)$ and $A^2 x = -x$ for $x \in D(A)$. Show that $D(A)$ is a direct sum of eigenspaces for A corresponding to the eigenvalues $+i$ and $-i$, that $V_+ \perp V_-$ and that

$$G(A) = V_+ + V_-,$$

where
$$V_\pm = \{\{x, y\} \in G(A) \mid y = \pm ix\}.$$

Show that the spaces $V_\pm$ are closed when A is closed.

12.19. Let T be a densely defined, closed symmetric operator in a Hilbert space H.

(a) Show that
$$G(T^*) = G(T) \oplus W_+ \oplus W_- \, ,$$
where $W_\pm = \{ \{x,y\} \in G(T^*) \mid y = \pm ix \}$, and that
$$D(T^*) = D(T) \dotplus Z(T^* - iI) \dotplus Z(T^* + iI) \, .$$

(One can use Exercise 12.18.)

(b) Let S be a closed, symmetric operator which extends T, i.e., $T \subset S$. Show that if u and $v \in D(T^*)$ with $T^*u = iu$, $T^*v = -iv$ and $u + v \in D(S)$, then $\|u\| = \|v\|$. Show that there exists an isometry U of a closed subspace K of $Z(T^* - iI)$ into $Z(T^* + iI)$ such that
$$D(S) = \{ x+u+Uu \mid x \in D(T), \, u \in K \} \quad \text{with} \quad S(x+u+Uu) = Tx+iu-iUu \, .$$

(*Hint.* For K one can take $\{u \in Z(T^*-i) \mid \exists v \in Z(T^*+i) \text{ s. t. } u+v \in D(S)\}$, and let U map u to v.)

Conversely, every operator defined in this way is a closed, symmetric extension of T.

(c) Show that S is a selfadjoint extension of T if and only if $K = Z(T^* - iI)$ and U is an isometry of $Z(T^* - iI)$ onto $Z(T^* + iI)$.

(d) Show that there exists a selfadjoint extension S of T if and only if $Z(T^* - iI)$ and $Z(T^* + iI)$ have the same Hilbert dimension (cf. Exercise 12.26).

(*Comment.* If Exercise 12.26 has not been covered, assume H separable; then any closed infinite dimensional subspace has a countable orthonormal basis. Applications of this result are found e.g. in Naimark's book [N68], and works of Everitt and Markus, cf. [EM99].)

12.20. An operator N in a Hilbert space H is called *normal* if N is closed and densely defined, and $NN^* = N^*N$. Show that N is normal if and only if N is a densely defined operator and N and N^* are metrically alike, i.e., $D(N) = D(N^*)$ and $\|Nx\| = \|N^*x\|$ for all $x \in D(N)$.

12.21. Let H be a separable Hilbert space. Show that there exists a densely defined, closed unbounded operator A in H satisfying $A^2 = I|_{D(A)}$. (Cf. Exercises 12.15 and 12.18.) Show that such an A cannot be selfadjoint, nor symmetric.

12.22. Let X and Y be vector spaces, and let $A : X \to Y$ and $B : Y \to X$ be operators with $R(A) \subset D(B)$ and $R(B) \subset D(A)$. Let $\lambda \in \mathbb{C} \setminus \{0\}$ and let $k \in \mathbb{N}$. Show that λ is an eigenvalue of AB with multiplicity k if and only if λ is an eigenvalue for BA with multiplicity k.

12.23. Let S be a densely defined, closed symmetric operator in a Hilbert space H, with $m(S) > 0$. Show that $D(S) \cap Z(S^*) = \{0\}$. Show that the operator P defined by

$$D(P) = D(S) \dotplus Z(S^*),$$
$$P(v + z) = Sv, \quad \text{when } v \in D(S) \text{ and } z \in Z(S^*),$$

is a selfadjoint extension of S with $m(P) \geq 0$ (the von Neumann extension [N29]). Show that P is different from the Friedrichs extension T of S if and only if $Z(S^*) \neq \{0\}$.

(*Hint.* One can show first that P is symmetric. Next, one can show that P^* has the same range as S, and use S^{-1} to get functions in $D(P^*)$ decomposed into a term in $D(S)$ and a term in $Z(S^*)$.)

12.24. Consider S, T and P as in Exercise 12.23, and let $t(u, v)$ and $p(u, v)$ denote the sesquilinear forms associated with T and P. Here $D(t) = V$ as described in Corollary 12.25.

Show that $D(p)$ equals the direct sum

$$D(p) = V \dotplus Z(S^*),$$

and that

$$p(v + z, v' + z') = t(v, v'), \quad \text{when } v, v' \in V, \ z, z' \in Z(S^*).$$

12.25. (This exercise uses definitions from Chapter 4, see in particular Exercise 4.14.) Let $H = L_2(I)$, where $I =]\alpha, \beta[$, and let $V = H^1(I)$. Let $a(u, v)$ be the sesquilinear form on V,

$$a(u, v) = \int_\alpha^\beta \left(u'(t)\overline{v'(t)} + q(t)u(t)\overline{v}(t) \right) dt,$$

where q is a function in $C^0(\overline{I})$. Show that a is bounded on V and V-coercive. Show that the operator A associated with (H, V, a) by the Lax-Milgram lemma is the operator

$$Au = -\frac{d^2}{dt^2}u + qu,$$

with domain $D(A)$ consisting of the functions $u \in H^2(I)$ with

$$u'(\alpha) = u'(\beta) = 0.$$

Show that if one replaces V by $V_0 = H_0^1(I)$ and a by $a_0 = a|_{V_0}$, one obtains an operator A_0 acting like A but with domain consisting of the functions $H^2(I)$ with

$$u(\alpha) = u(\beta) = 0.$$

12.26. Let H be a pre-Hilbert space. The set of orthonormal systems in H is inductively ordered and therefore has maximal elements. Let $(e_i)_{i \in I}$ and $(f_j)_{j \in J}$ be two maximal orthonormal systems. Show that I and J have the same cardinality. (*Hint.* If I is finite, H is a finite dimensional vector space with the basis $(e_i)_{i \in I}$; since the vectors f_j, $j \in J$, are linearly independent, card $J \le$ card I. If I is infinite, let $J_i = \{ j \in J \mid (f_j, e_i) \ne 0 \}$ for each $i \in I$; since all J_i are denumerable and $J = \bigcup_{i \in I} J_i$, one finds that card $J \le$ card $(I \times \mathbb{N}) =$ card I.)

When H is a Hilbert space, the subspace H_0 of linear combinations of a maximal orthonormal system $(e_i)_{i \in I}$ is dense in H, since $\overline{H_0} = \{ e_i \mid i \in I \}^{\perp\perp} = \{ 0 \}^{\perp}$. Here card I is called the Hilbert dimension of H.

12.27. Let H be a Hilbert space and let E and $F \in \mathbf{B}(H)$ be two orthogonal projections (i.e., $E = E^* = E^2$ and $F = F^* = F^2$). Show that if EH has a larger Hilbert dimension (cf. Exercise 12.26) than FH, then EH contains a unit vector which is orthogonal to FH. Show that if $\|E - F\| < 1$, then EH and FH have the same Hilbert dimension.

12.28. Let T be an operator in a complex Hilbert space H. Let $\Omega_e = \Omega_e(T) = \{ \lambda \in \mathbb{C} \mid \exists c_\lambda > 0 \ \forall x \in D(T) : \|(T - \lambda)x\| \ge c_\lambda \|x\| \}$.

(a) If $S \subset T$, then $\Omega_e(T) \subset \Omega_e(S)$. If T is closable, then $\Omega_e(\overline{T}) = \Omega_e(T)$.

(b) If T is closed, then $\lambda \in \Omega_e$ if and only if $T - \lambda$ is injective and $R(T - \lambda)$ is closed.

(c) If $\lambda \notin \Omega_e(S)$, and T is an extension of S, then $T - \lambda$ does not have a bounded inverse.

(d) For $\lambda \in \Omega_e$ and $|\mu - \lambda| < c_\lambda$ one has that $\mu \in \Omega_e$, and

$$R(T - \lambda) \cap R(T - \mu)^{\perp} = R(T - \lambda)^{\perp} \cap R(T - \mu) = \{ 0 \}.$$

(*Hint.* For $(T - \lambda)z \perp (T - \mu)z$ one can show that $c_\lambda \|z\| \, \|(T - \lambda)z\| \le \|(T - \lambda)z\|^2 = |\mu - \lambda| \, |(z, (T - \lambda)z)| \le |\mu - \lambda| \, \|z\| \, \|(T - \lambda)z\|$.)

(e) Assume that T is closed. The Hilbert dimension of $R(T - \lambda)^{\perp}$ is constant on each connected component of Ω_e. (Use for example Exercise 12.27.)

(f) If T is symmetric, then $\mathbb{C} \setminus \mathbb{R} \subset \Omega_e$.

(g) If T is lower bounded with lower bound c, then $]-\infty, c[\subset \Omega_e$.

(h) Show that if T is densely defined and symmetric, and $\mathbb{R} \cap \Omega_e \ne \emptyset$, then T can be extended to a selfadjoint operator. (The deficiency indices for $\overline{T}$ are equal, by (a), (e) and (f).)

12.29. Let H be a Hilbert space, F a closed subspace. Let S and T be closed injective operators in H with $R(S)$ and $R(T)$ closed.

(a) ST is closed.

(b) Show that $R(S|_{F \cap D(S)})$ is closed. Define $Q \in \mathbf{B}(H)$ by $Qx = S^{-1}x$ for $x \in R(S)$, $Qx = 0$ for $x \perp R(S)$. Show that $R(S|_{F \cap D(S)})$ and $R(S)^{\perp}$ and $\overline{Q^*(F^{\perp})}$ are pairwise orthogonal subspaces of H, with sum H.

(c) Assume that S is densely defined. Show that the Hilbert dimension of $R(S|_{F\cap D(S)})^\perp$ is the sum of the Hilbert dimensions of $R(S)^\perp$ and $F^\perp$.

(d) Assume that S and T are densely defined, and that $R(T)^\perp$ has finite dimension. Show that ST is a closed densely defined operator (use for example Exercise 12.16). Show that $R(ST)$ is closed and that the Hilbert dimension of $R(ST)^\perp$ is the sum of the Hilbert dimensions of $R(S)^\perp$ and $R(T)^\perp$.

(e) Let A be a closed, densely defined, symmetric operator on H with deficiency indices m and n. If m or n is finite, then A^2 is a closed, densely defined, symmetric, lower bounded operator with deficiency indices $m+n$ and $m+n$. (*Hint.* Use for example $A^2 + I = (A+i)(A-i) = (A-i)(A+i)$, and Exercise 12.28.)

12.30.   Let there be given two Hilbert spaces H_1 and H_2. Let φ_i denote the natural injection of H_i in $H_1 \oplus H_2$, $i = 1, 2$. Find φ_i^*. An operator $A \in \mathbf{B}(H_1 \oplus H_2)$ is described naturally by a matrix $(A_{ij})_{i,j=1,2}$, $A_{ij} = \varphi_i^* A \varphi_j$. Find conversely A expressed by the matrix elements, and find the matrix for A^*.

12.31.   Let H be a Hilbert space. Let T be a densely defined closed operator in H, and let $P(T)$ denote the orthogonal projection ($\in \mathbf{B}(H \oplus H)$) onto the graph $G(T)$.

(a) Show that $P(T^*) = 1 + VP(T)V$, where $V \in \mathbf{B}(H \oplus H)$ has the matrix (cf. Exercise 12.30) $\left(\begin{smallmatrix} 0 & -1 \\ 1 & 0 \end{smallmatrix}\right)$.

(b) Show that $P(T)$ has the matrix

$$\begin{pmatrix} rr(1+T^*T)^{-1} & T^*(1+TT^*)^{-1} \\ T(1+T^*T)^{-1} & TT^*(1+TT^*)^{-1} \end{pmatrix} .$$

12.32.   Show that when S is a symmetric operator in H with $R(S+i) = R(S-i) = H$, then S is densely defined. (Hence the requirement on dense domain in Theorem 12.10 is superfluous.)

12.33.   Let S be a densely defined, closed symmetric operator in H. Show that S is maximal symmetric if and only if either $S+i$ or $S-i$ is surjective. (One can use Exercise 12.19.)

12.34.   Let K denote the set of complex selfadjoint 2×2 matrices $A \geq 0$ with trace $\operatorname{Tr} A = 1$. (The trace of a matrix $\left(\begin{smallmatrix} \alpha & \beta \\ \gamma & \delta \end{smallmatrix}\right)$ equals $\alpha + \delta$.) Define $\partial K = \{ A \in K \mid A^2 = A \}$. Set $P_u v = (v, u)u$ for u and $v \in \mathbb{C}^2$.

(a) Show that $\partial K = \{ P_u \mid u \in \mathbb{C}^2,\ \|u\| = 1 \}$.

(b) Show that

$$\varphi(a, b, c) = \begin{pmatrix} \frac{1}{2} & 0 \\ 0 & \frac{1}{2} \end{pmatrix} + \begin{pmatrix} a & b-ic \\ b+ic & -a \end{pmatrix}$$

defines an affine homeomorphism of $\mathbb{R}^3$ onto the set of selfadjoint 2×2 matrices with trace 1; show that

$$\varphi(\{\,(a,b,c) \in \mathbb{R}^3 \mid a^2 + b^2 + c^2 \leq \tfrac{1}{4}\,\}) = K;$$
$$\varphi(\{\,(a,b,c) \in \mathbb{R}^3 \mid a^2 + b^2 + c^2 = \tfrac{1}{4}\,\}) = \partial K.$$

(c) Let T be a 2×2 matrix; show that $A \mapsto \operatorname{Tr}(TA)$ is a continuous affine map ψ of K onto a compact convex set $N \subset \mathbb{C}$; and show that $\psi(\partial K) = N$.

(d) Show that $\operatorname{Tr}(TP_u) = (Tu, u)$ for $u \in \mathbb{C}^2$; show that $N = \nu(T)$.

Now let H be an arbitrary complex Hilbert space and S an operator in H.

(e) Show that $\nu(S)$ is convex (the Toeplitz-Hausdorff theorem). (For $x, y \in D(S)$, consider $P_K S|_K$, where P_K is the projection onto the subspace K spanned by x and y.)

(f) Assume that S is closed and densely defined with $D(S^*) = D(S)$. Show that $\nu(S^*) = \overline{\nu(S)}$ ($= \{\,\overline{\lambda} \mid \lambda \in \nu(S)\,\}$); show that the spectrum $\sigma(S)$ is contained in any closed half-plane that contains $\nu(S)$; show that $\sigma(S)$ is contained in the closure of $\nu(S)$.

(g) Show that for $H = l^2(\mathbb{N})$ and a suitably chosen $S \in \mathbf{B}(H)$, $\nu(S)$ is not closed (use for example $S(x_1, x_2, \dots) = (\lambda_1 x_1, \lambda_2 x_2, \dots)$, where $\lambda_n > 0$ for $n \in \mathbb{N}$, and $\lambda_n \to 0$).

12.35. Let Ω be an open subset of $\mathbb{R}^n$ and let M_p be the multiplication operator defined from a continuous function p on Ω by Theorem 12.13.

(a) Show that $m(M_p) \geq \alpha$ if and only if $\operatorname{Re} p(x) \geq \alpha$ for all x.

(b) Show that $\nu(M_p)$ and $\sigma(M_p)$ are contained in the intersection of all closed half-planes which contain the range of p (and hence $\nu(M_p)$ and $\sigma(M_p)$ are contained in the closed convex hull of the range of p).

12.36. Let M_p be as in Exercise 12.35, and assume that the range of p is contained in a sector

$$\{\, z \in \mathbb{C} \mid |\operatorname{Im} z| \leq c \operatorname{Re} z \,\}$$

where $c > 0$. Show that M_p is the variational operator defined from the triple $(L_2(\Omega), V, s)$, where

$$s(u, v) = \int p u \overline{v} \, dx \quad \text{on} \quad V = D(M_{|p|^{\frac{1}{2}}}) \,,$$

which is a Hilbert space with norm

$$(\| |p|^{\frac{1}{2}} u \|_{L_2}^2 + \| u \|_{L_2}^2)^{\frac{1}{2}} \,.$$

12.37. (This exercise uses definitions from Chapter 4.) Let $H = L_2(\mathbb{R}_+)$, and let $V = H^1(\mathbb{R}_+)$. Let $a(u, v)$ be the sesquilinear form on V,

$$a(u, v) = \int_0^\infty (u'(t)\overline{v}'(t) + u'(t)\overline{v}(t) + u(t)\overline{v}(t)) \, dt.$$

Show that a is continuous on V and V-elliptic. Show that the operator A associated with (H, V, a) by the Lax-Milgram lemma is of the form

$$Au = -u'' + u' + u,$$

with domain $D(A)$ consisting of the functions $u \in H^2(\mathbb{R}_+)$ with

$$u'(0) = 0.$$

Show that A^* is of the form

$$A^*v = -v'' - v' + v,$$

with domain $D(A^*)$ consisting of the functions $v \in H^2(\mathbb{R}_+)$ with

$$v'(0) + v(0) = 0.$$

So this is a case where $D(A^*) \neq D(A)$.

12.38. (Communicated by Peter Lax, May 2006.) Let H be a Hilbert space provided with the norm $\|x\|$, and let $|x|$ be another norm on H such that

$$\|x\| \leq C|x| \text{ for all } x \in H,$$

with $C > 0$. Let $T : H \to H$ be a symmetric operator with $D(T) = H$ such that

$$|Tx| \leq k|x| \text{ for all } x \in H,$$

for some $k > 0$. Show that $T \in \mathbf{B}(H)$.
(*Hint.* Use calculations such as

$$\|Tx\|^2 = (Tx, Tx) = (x, T^2x) \leq \|x\| \|T^2x\|,$$

to show successively that

$$\|Tx\|^4 \leq \|x^2\| \|T^2x\|^2 \leq \|x\|^3 \|T^4x\|,$$

$$\vdots$$

$$\|Tx\|^m \leq \|x\|^{m-1} \|T^mx\| \quad \text{for } m = 2^j, j = 1, 2, \ldots.$$

Conclude that with $m = 2^j$,

$$\|Tx\| \leq \|x\|^{\frac{m-1}{m}} C^{\frac{1}{m}} k|x|^{\frac{1}{m}} \to \|x\|k$$

for $j \to \infty$.)

Chapter 13
Families of extensions

13.1 A universal parametrization of extensions

We have seen in Chapter 12 that a closed, densely defined symmetric operator S in general has several selfadjoint extensions: When S is lower bounded with a nontrivial nullspace, there is the Friedrichs extension established in Section 12.5, but there is also the von Neumann or Kreĭn extension defined in Exercise 12.23 when $m(S) > 0$. When S has equal deficiency indices (finite or infinite), cf. (12.27), there is a family of selfadjoint extensions parametrized by the family of linear isometries (unitary operators) of $R(S + iI)^{\perp}$ onto $R(S - iI)^{\perp}$, cf. Exercise 12.19.

The latter family has been used in the study of boundary conditions for ordinary differential operators (where the deficiency indices are finite) in the classical book of Naimark [N68], and more recently e.g. by Everitt and Markus in a number of papers and surveys, see for example [EM99].

For partial differential operators of elliptic type, one is often in a position of having a symmetric and injective operator to depart from; here there is another family of extensions that has proved to be particularly adequate, as initiated by Kreĭn [K47], Vishik [V52] and Birman [B56], and more fully developed by Grubb [G68], [G70].

We shall give an introduction to this theory below. Since the theory works also in nonsymmetric cases, we begin with a more general situation than the search for selfadjoint extensions of symmetric operators.

There is given a complex Hilbert space H with scalar product $(u, v)_H$ and norm $\|u\|_H$, usually denoted (u, v) and $\|u\|$, and two densely defined closed operators A_0 and A_0' satisfying

$$A_0 \subset (A_0')^*, \quad A_0' \subset (A_0)^*. \tag{13.1}$$

We denote

$$(A_0')^* = A_1, \quad (A_0)^* = A_1'. \tag{13.2}$$

Moreover, there is given a closed densely defined operator A_β having a bounded, everywhere defined inverse A_β^{-1} and satisfying

$$A_0 \subset A_\beta \subset A_1; \text{ then also } A_0' \subset A_\beta^* \subset A_1', \qquad (13.3)$$

and $(A_\beta^*)^{-1} = (A_\beta^{-1})^*$ (cf. Theorem 12.7). We shall denote by $\mathcal{M}$, resp. $\mathcal{M}'$, the family of linear operators $\widetilde{A}$ (resp. $\widetilde{A}'$) satisfying

$$A_0 \subset \widetilde{A} \subset A_1, \text{ resp. } A_0' \subset \widetilde{A}' \subset A_1'; \qquad (13.4)$$

note that a closed operator $\widetilde{A}$ is in $\mathcal{M}$ if and only if $\widetilde{A}^* \in \mathcal{M}'$.

Observe in particular that when A_0 is symmetric and A_β is selfadjoint, we can obtain this situation by taking

$$A_0' = A_0, \quad A_\beta^* = A_\beta, \quad A_1' = A_1 = A_0^*; \quad \mathcal{M} = \mathcal{M}'. \qquad (13.5)$$

we call this situation *the symmetric case*.

In the applications to partial differential operators A on an open set $\Omega \subset \mathbb{R}^n$, A_0 will be the minimal operator in $L_2(\Omega)$ defined from A, and A_1 will be the maximal operator in $L_2(\Omega)$ defined from A, cf. Chapter 4, and the $\widetilde{A}$ are the realizations of A. In elliptic cases one can often establish the existence of an operator A_β satisfying (13.3). The family of realizations $\widetilde{A}$ are viewed as the general extensions of A_0.

For simplicity in the formulas, we shall use the convention that for all the operators $\widetilde{A}$ in $\mathcal{M}$ (acting like A_1), we can write $\widetilde{A}u$ as Au without extra markings. Similarly, when $\widetilde{A}' \in \mathcal{M}'$, we can write $A'v$ instead of $\widetilde{A}'v$.

The domains $D(\widetilde{A})$ of closed operators are closed with respect to the graph norm $\|u\|_A = (\|u\|^2 + \|Au\|^2)^{\frac{1}{2}}$.

By use of A_β we can establish the following *basic decompositions* of $D(A_1)$ and $D(A_1')$.

Lemma 13.1. *There are decompositions into direct sums:*

$$D(A_1) = D(A_\beta) \dotplus Z(A_1), \quad D(A_1') = D(A_\beta^*) \dotplus Z(A_1'); \qquad (13.6)$$

here we write $u \in D(A_1)$ resp. $v \in D(A_1')$ in decomposed form as

$$u = u_\beta + u_\zeta, \quad v = v_{\beta'} + v_{\zeta'}, \qquad (13.7)$$

with the notation for the hereby defined decomposition maps

$$\begin{aligned} \mathrm{pr}_\beta : D(A_1) \to D(A_\beta), \quad \mathrm{pr}_\zeta : D(A_1) \to Z(A_1), \\ \mathrm{pr}_{\beta'} : D(A_1') \to D(A_\beta^*), \quad \mathrm{pr}_{\zeta'} : D(A_1') \to Z(A_1'). \end{aligned} \qquad (13.8)$$

The maps are projections and satisfy

$$\mathrm{pr}_\beta : u \mapsto A_\beta^{-1} A u, \quad \mathrm{pr}_{\beta'} : v \mapsto (A_\beta^*)^{-1} A' v,$$
$$\mathrm{pr}_\zeta = I - \mathrm{pr}_\beta, \quad \mathrm{pr}_{\zeta'} = I - \mathrm{pr}_{\beta'}, \tag{13.9}$$

and they are continuous with respect to the graph norms.

Proof. Clearly, $D(A_\beta) + Z(A_1) \subset D(A_1)$. Define pr_β by (13.9); it sends $D(A_1)$ into $D(A_\beta)$ and is continuous with respect to the graph norm:

$$\| \mathrm{pr}_\beta u \|^2 + \| A \, \mathrm{pr}_\beta u \|^2 = \| A_\beta^{-1} A u \|^2 + \| A u \|^2 \le (\| A_\beta^{-1} \|^2 + 1) \| A u \|^2.$$

Then pr_ζ defined in (13.9) is also continuous in the graph norm; and this operator maps $D(A_1)$ into $Z(A_1)$ since

$$A(u - A_\beta^{-1} A u) = A u - A u = 0.$$

It is immediate to check that $\mathrm{pr}_\beta \mathrm{pr}_\beta = \mathrm{pr}_\beta$. Then also $\mathrm{pr}_\zeta \mathrm{pr}_\zeta = (I - \mathrm{pr}_\beta)(I - \mathrm{pr}_\beta) = \mathrm{pr}_\zeta$, and $\mathrm{pr}_\beta \mathrm{pr}_\zeta = \mathrm{pr}_\beta - \mathrm{pr}_\beta \mathrm{pr}_\beta = 0$, so the operators are projections onto complementing subspaces.

There is a similar proof for $D(A_1')$. □

Note that the injections of $D(A_\beta)$ and $Z(A_1)$ into $D(A_1)$ are likewise continuous in the graph norm (they are closed subspaces), so the mapping

$$u \mapsto \{\mathrm{pr}_\beta u, \mathrm{pr}_\zeta u\} \text{ from } D(A_1) \text{ to } D(A_\beta) \times Z(A_1) \tag{13.10}$$

is a homeomorphism with respect to graph norms. The notation is such that

$$\mathrm{pr}_\beta u = u_\beta, \quad \mathrm{pr}_\zeta u = u_\zeta,$$
$$\mathrm{pr}_{\beta'} v = v_{\beta'}, \quad \mathrm{pr}_{\zeta'} v = v_{\zeta'}. \tag{13.11}$$

These projections, called the *basic projections*, are generally *not* orthogonal in H.

In the following, the notation U_V is used for the *orthogonal* projection of an element or subspace U of H into a closed subspace V of H. The orthogonal projection operator from H to V is denoted pr_V.

For simplicity in the notation, we shall henceforth denote

$$Z(A_1) = Z, \quad Z(A_1') = Z'.$$

Lemma 13.2. *For $u \in D(A_1)$, $v \in D(A_1')$, one has*

$$(A u, v) - (u, A' v) = (A u, v_{\zeta'}) - (u_\zeta, A' v)$$
$$= ((A u)_{Z'}, v_{\zeta'}) - (u_\zeta, (A' v)_Z). \tag{13.12}$$

Proof. Since $A u = A u_\beta$, $A' v = A' v_{\beta'}$, we can write

$$(A u, v) - (u, A' v) = (A u_\beta, v_{\beta'} + v_{\zeta'}) - (u_\beta + u_\zeta, A' v_{\beta'})$$
$$= (A u_\beta, v) - (u, A' v_{\beta'}) + (A u, v_{\zeta'}) - (u_\zeta, A' v).$$

Here

$$(Au_\beta, v_{\beta'}) - (u_\beta, A'v_{\beta'}) = 0,$$

since u_β and $v_{\beta'}$ belong to the domains of the adjoint operators A_β and A_β^*. In the remaining terms, $(Au, v_{\zeta'}) = ((Au)_{Z'}, v_{\zeta'})$ since $v_{\zeta'} \in Z'$, and similarly $(u_\zeta, Av) = (u_\zeta, (Av)_Z)$.   □

We say that S_1, S_2 is a pair of adjoint operators, when $S_1 = S_2^*$, $S_2 = S_1^*$; in particular, they are closed.

The heart of our construction lies in the following result:

Proposition 13.3. *Let* $\widetilde{A} \in \mathcal{M}$, $\widetilde{A}^* \in \mathcal{M}'$ *be a pair of adjoint operators. Define*

$$D(T) = \mathrm{pr}_\zeta D(\widetilde{A}), \quad V = \overline{D(T)},$$
$$D(T_1) = \mathrm{pr}_{\zeta'} D(\widetilde{A}^*), \quad W = \overline{D(T_1)}, \tag{13.13}$$

closures in H. *Then the equations*

$$Tu_\zeta = (Au)_W, \ \textit{for } u \in D(\widetilde{A}), \tag{13.14}$$

define an operator $T : V \to W$ *with domain* $D(T)$; *and*

$$T_1 v_{\zeta'} = (A'v)_V, \ \textit{for } v \in D(\widetilde{A}^*), \tag{13.15}$$

define an operator $T_1 : W \to V$ *with domain* $D(T_1)$. *Moreover, the operators* T *and* T_1 *are adjoints of one another.*

Proof. Note that V and W are closed subspaces of Z resp. Z'. When $u \in D(\widetilde{A})$ and $v \in D(\widetilde{A}^*)$, then in view of Lemma 13.2,

$$0 = (Au, v) - (u, A'v) = (Au, v_{\zeta'}) - (u_\zeta, A'v),$$

so that, since $u_\zeta \in V$ and $v_{\zeta'} \in W$,

$$((Au)_W, v_{\zeta'}) = (u_\zeta, (A'v)_V), \ \text{when } u \in D(\widetilde{A}), \ v \in D(\widetilde{A}^*). \tag{13.16}$$

We shall show that the set $G \subset V \times W$ defined by

$$G = \{\, \{u_\zeta, (Au)_W\} \mid u \in D(\widetilde{A})\} \tag{13.17}$$

is a graph, of an operator T from V to W with domain $D(T)$ defined in (13.13). For this, we have to show that for the pairs in (13.17), $u_\zeta = 0$ implies $(Au)_W = 0$ (recall Lemma 12.3). So let $u_\zeta = 0$ for such a pair. Then by (13.16),

$$((Au)_W, v_{\zeta'}) = 0 \text{ for all } v \in D(\widetilde{A}^*). \tag{13.18}$$

But here $v_{\zeta'}$ runs through a dense subset of W, so it follows that $(Au)_W = 0$. Thus G is the graph of an operator T from V to W; its domain is clearly $D(T)$ defined in (13.13).

The proof that (13.15) defines an operator $T_1 : W \to V$ with domain $D(T_1)$ defined in (13.13) is similar.

Note that T and T_1 are densely defined, so that they have well-defined adjoints $T^* : W \to V$ and $T_1^* : V \to W$. By (13.16),

$$(Tu_\zeta, v_{\zeta'})_W = (u_\zeta, T_1 v_{\zeta'})_V \text{ for all } u_\zeta \in D(T), \ v_{\zeta'} \in D(T_1), \qquad (13.19)$$

so $T \subset T_1^*$. We shall show that this inclusion is an equality.

Let $z \in D(T_1^*)$ (which is contained in $V \subset Z$). Define

$$x = z + A_\beta^{-1} T_1^* z; \qquad (13.20)$$

clearly $x \in D(A_1)$ with

$$x_\beta = A_\beta^{-1} T_1^* z, \ x_\zeta = z, \ Ax = T_1^* z. \qquad (13.21)$$

We have for all $v \in D(\widetilde{A}^*)$:

$$
\begin{aligned}
(Ax, v) - (x, A'v) &= (Ax, v_{\zeta'}) - (z, A'v) \text{ by Lemma 13.2} \\
&= (T_1^* z, v_{\zeta'}) - (z, (A'v)_V) \text{ since } z \in V \\
&= (T_1^* z, v_{\zeta'}) - (z, T_1 v_{\zeta'}) = 0, \text{ by definition of } T_1.
\end{aligned}
$$

Since v is arbitrary in $D(\widetilde{A}^*)$, it follows that $x \in D(\widetilde{A}^{**}) = D(\widetilde{A})$, and then $z = \mathrm{pr}_\zeta \, x \in D(T)$, as was to be shown.

There is a similar proof that $T_1 = T^*$. $\qquad\qquad\qquad\qquad\qquad\qquad$ $\square$

Proposition 13.3 shows how every pair of adjoints $\widetilde{A} \in \mathcal{M}$, $\widetilde{A}^* \in \mathcal{M}'$, gives rise to a pair of adjoints $T : V \to W$, $T^* : W \to V$, with V and W being closed subspaces of Z, resp. Z'. The next proposition shows that all choices of such pairs of adjoints are reached in a one-to-one way, and gives formulas for the domains $D(\widetilde{A})$, $D(\widetilde{A}^*)$ corresponding to T, T^*.

Proposition 13.4. *Let V and W be closed subspaces of Z, resp. Z', and let $T : V \to W$, $T^* : W \to V$ be a pair of adjoint operators (generally unbounded). Define the operators $\widetilde{A} \subset A_1$, $\widetilde{A}' \subset A_1'$ by*

$$D(\widetilde{A}) = \{u \in D(A_1) \mid u_\zeta \in D(T), \ (Au)_W = Tu_\zeta\}, \qquad (13.22)$$

$$D(\widetilde{A}') = \{v \in D(A_1') \mid v_{\zeta'} \in D(T^*), \ (A'v)_V = T^* v_{\zeta'}\}; \qquad (13.23)$$

they are in $\mathcal{M}$ resp. $\mathcal{M}'$. They are adjoints of one another, and the operators derived from $\widetilde{A}$, $\widetilde{A}^*$ by Proposition 13.3 are exactly $T : V \to W$, $T^* : W \to V$.

If $T : V \to W$, $T^* : W \to V$, are operators derived from a pair of adjoints $\widetilde{A} \in \mathcal{M}$, $\widetilde{A}^* \in \mathcal{M}'$ by Proposition 13.3, then the formulas (13.22), (13.23) give back the domains $D(\widetilde{A})$, $D(\widetilde{A}^*)$.

Proof. Let $\widetilde{A}$ and $\widetilde{A}'$ be given by (13.22), (13.23); they clearly extend A_0 resp. A'_0. It follows by use of Lemma 13.2 that for $u \in D(\widetilde{A})$, $v \in D(\widetilde{A}')$,

$$(Au,v)-(u,A'v) = ((Au)_W, v_{\zeta'})-(u_\zeta,(A'v)_V) = (Tu_\zeta, v_{\zeta'})-(u_\zeta, T^*v_{\zeta'}) = 0,$$

in view of the definitions of $\widetilde{A}$ and $\widetilde{A}'$. So $\widetilde{A}$ and $\widetilde{A}'$ are contained in each other's adjoints. We have to show equality, by symmetry it suffices to show that $\widetilde{A}^* \subset \widetilde{A}'$.

Let $v \in D(\widetilde{A}^*)$. Since A_0 is closed with a bounded inverse, its range is closed, so

$$H = R(A_0) \oplus Z', \tag{13.24}$$

orthogonal direct sum. In particular, $R(A_0) \perp W$. Therefore any element $u = z + A_\beta^{-1}Tz + x$, where $z \in D(T)$ and $x \in D(A_0)$, is in $D(\widetilde{A})$, since $u_\zeta = z$ and $(Au)_W = (Tz + Ax)_W = Tz$. We have for all such u, using Lemma 13.2 again:

$$0 = (Au,v) - (u,A'v) = (Tz + Ax, v_{\zeta'}) - (z,A'v) = (Tz, v_{\zeta'}) - (z,(A'v)_V).$$

Since z is arbitrary in $D(T)$, this shows that $v_{\zeta'} \in D(T^*)$ with $T^*v_{\zeta'} = (A'v)_V$, and hence, by definition, $v \in D(\widetilde{A}')$. So $D(\widetilde{A}^*) \subset D(\widetilde{A}')$, and since $\widetilde{A}' \subset \widetilde{A}^*$, it follows that $\widetilde{A}' = \widetilde{A}^*$. It is now also obvious that T and T^* are determined from $\widetilde{A}$ and $\widetilde{A}^*$ as in Proposition 13.3. This shows the first statement in the proposition.

Finally to see that the pair that gives rise to $T : V \to W$ and $T^* : W \to V$ is unique, let $\widetilde{A}$, $\widetilde{A}^*$ be the pair defined by (13.22), (13.23), and let $\widetilde{B}$, $\widetilde{B}^*$ be another pair that gives rise to $T : V \to W$ and $T^* : W \to V$ as in Proposition 13.3. Then according to (13.13)–(13.15) and (13.22)–(13.23),

$$D(\widetilde{A}) \supset D(\widetilde{B}), \quad D(\widetilde{A}^*) \supset D(\widetilde{B}^*).$$

It follows that $\widetilde{A} \supset \widetilde{B}$ and $\widetilde{A}^* \supset \widetilde{B}^*$, but this can only happen when $\widetilde{A} = \widetilde{B}$. □

The two propositions together imply:

Theorem 13.5. *There is a one-to-one correspondence between all pairs of adjoint operators $\widetilde{A} \in \mathcal{M}$, $\widetilde{A}^* \in \mathcal{M}'$, and all pairs of adjoint operators $T : V \to W$, $T^* : W \to V$, where V and W run through the closed subspaces of Z resp. Z'; the correspondence being given by*

$$D(\widetilde{A}) = \{u \in D(A_1) \mid u_\zeta \in D(T), (Au)_W = Tu_\zeta\}, \tag{13.25}$$

$$D(\widetilde{A}^*) = \{v \in D(A'_1) \mid v_{\zeta'} \in D(T^*), (A'v)_V = T^*v_{\zeta'}\}. \tag{13.26}$$

In this correspondence,

$$D(T) = \mathrm{pr}_\zeta \, D(\widetilde{A}), \quad V = \overline{D(T)}, \tag{13.27}$$

$$D(T^*) = \mathrm{pr}_{\zeta'} \, D(\widetilde{A}^*), \quad W = \overline{D(T^*)}. \tag{13.28}$$

The formulation is completely symmetric in $\widetilde{A}$ and $\widetilde{A}^*$, and in T and T^*. The pair $\widetilde{A}$, $\widetilde{A}^*$ is of course completely determined by the closed operator $\widetilde{A} \in \mathcal{M}$, so we need only mention $\widetilde{A}$; similarly we need only mention T : $V \to W$. Before we formulate this in a corollary, we shall show another useful description of $\widetilde{A}$ in terms of T.

Theorem 13.6. *When $\widetilde{A}$ corresponds to $T : V \to W$ as above, the mapping*

$$\Psi : \{z, f, v\} \mapsto u = z + A_\beta^{-1}(Tz + f) + v \tag{13.29}$$

defines a bijection

$$\Psi : D(T) \times (Z' \ominus W) \times D(A_0) \xrightarrow{\sim} D(\widetilde{A}), \tag{13.30}$$

homeomorphic with respect to graph norms on $D(T)$, $D(A_0)$ and $D(\widetilde{A})$ and the H-norm on $Z' \ominus W$ (in a sense a graph norm too).

Proof. Let $u = z + A_\beta^{-1}(Tz + f) + v$, with $\{z, f, v\} \in D(T) \times (Z' \ominus W) \times D(A_0)$. Clearly, u belongs to $D(A_1)$, with $u_\zeta = z$, $u_\beta = A_\beta^{-1}(Tz + f) + v$ according to the decomposition in Lemma 13.1. Moreover, $Au = Tz + f + Av$, with $Tz \in W$, $f \in Z' \ominus W$ and $Av \in R(A_0)$, so since

$$H = W \oplus (Z' \ominus W) \oplus R(A_0) \tag{13.31}$$

(recall (13.24)), $(Au)_W = Tz$. Thus $u \in D(\widetilde{A})$ according to (13.25).

Conversely, when $u \in D(\widetilde{A})$, we can decompose Au according to (13.31), setting $f = (Au)_{Z' \ominus W}$, $v = A_\beta^{-1}[(Au)_{R(A_0)}]$. Then $v \in D(A_0)$. Here $u = u_\zeta + u_\beta$, where $u_\beta = A_\beta^{-1} Au$, and $u_\zeta \in D(T)$, $(Au)_W = Tu_\zeta$ by assumption. Then

$$u = u_\zeta + A_\beta^{-1} Au = u_\zeta + A_\beta^{-1}(Tu_\zeta + f + Av) = u_\zeta + A_\beta^{-1}(Tu_\zeta + f) + v,$$

where $u_\zeta \in D(T)$, $f \in Z' \ominus W$ and $v \in D(A_0)$. This shows that the mapping Ψ is surjective. The injectiveness is easily shown by use of (13.31).

The continuity (with respect to graph norms) is easily verified, and then the inverse is likewise continuous (in view of the open mapping theorem or by a direct verification).   □

As a consequence of Theorems 13.5 and 13.6 we can now formulate:

Theorem 13.7. *There is a one-to-one correspondence between all closed operators $\widetilde{A} \in \mathcal{M}$ and all operators $T : V \to W$, where V and W are closed* subspaces of Z resp. Z', and T is closed with domain $D(T)$ dense in V; the correspondence is defined by (13.25), (13.27).

The domain $D(\widetilde{A})$ is also described by (13.29), (13.30).

There are many results on how properties of $\widetilde{A}$ are reflected in properties of T in this correspondence. We begin with the following, that follow straightforwardly from the formulas:

Theorem 13.8. *Let $\widetilde{A} \in \mathcal{M}$ correspond to $T : V \to W$ as in Theorem 13.7. Then:*

$1°$ $Z(\widetilde{A}) = Z(T)$. *In particular,* $\dim Z(\widetilde{A}) = \dim Z(T)$, *and $\widetilde{A}$ is injective if and only if T is so.*

$2°$ $R(\widetilde{A})$ *has the following decomposition:*

$$R(\widetilde{A}) = R(T) + [(Z' \ominus W) \oplus R(A_0)], \qquad (13.32)$$

where also the first sum is orthogonal. In particular, $R(\widetilde{A})$ is closed if and only if $R(T)$ is closed, and in the affirmative case, $H \ominus R(\widetilde{A}) = W \ominus R(T)$. The ranges of $\widetilde{A}$ and T (in H resp. W) have the same codimension, and $\widetilde{A}$ is surjective if and only if T is so.

$3°$ $\widetilde{A}$ *is Fredholm, if and only if T is so, and then they have the same index.*

Proof. $1°$. Let $u \in Z(\widetilde{A})$. Then $u_\zeta = u$, and $Tu_\zeta = (Au)_W = 0$, so $u \in Z(T)$. Conversely, if $u \in Z(T)$, then $u \in D(\widetilde{A})$ (take f and v equal to 0 in (13.29)), so since $u \in D(\widetilde{A}) \cap Z$, $\widetilde{A}u = 0$.

$2°$. By the proof of Theorem 13.6 and (13.31), the general element of $R(\widetilde{A})$ is orthogonally decomposed as $g = Tz + f + Av$, where $z \in D(T)$, $f \in Z' \ominus W$ and $Av \in R(A_0)$; this gives the decomposition (13.32). Here $Z' \ominus W$ and $R(A_0)$ are closed, so closedness of $R(\widetilde{A})$ holds if and only if it holds for $R(T)$. Moreover, if they are closed, the orthogonal complement of $R(\widetilde{A})$ in H equals the orthogonal complement of $R(T)$ in W.

The remaining statements are straightforward consequences. (Fredholm operators are explained in Section 8.3.) $\square$

Note that

$$[(Z' \ominus W) \oplus R(A_0)] = H \ominus W. \qquad (13.33)$$

In the case where $\widetilde{A}$ and T are injective, we have a simple formula linking their inverses.

Theorem 13.9. *Let $\widetilde{A} \in \mathcal{M}$ correspond to $T : V \to W$ as in Theorem 13.7. Assume that $\widetilde{A}$ is injective, then so is T. Define $T^{(-1)}$ as the linear extension of*

$$T^{(-1)}f = \begin{cases} T^{-1}f & when\ f \in R(T), \\ 0 & when\ f \in H \ominus W. \end{cases} \qquad (13.34)$$

Then

$$\widetilde{A}^{-1} = A_\beta^{-1} + T^{(-1)}, \quad \text{defined on } R(\widetilde{A}). \tag{13.35}$$

Proof. Let $f \in R(\widetilde{A})$. Let $u = \widetilde{A}^{-1}f$, $v = A_\beta^{-1}f$. Then $u - v = z$, where $z \in Z$. By the definition of T, z belongs to $D(T)$ and $Tz = (Au)_W = f_W$. Therefore $f_W \in R(T)$, and

$$T^{(-1)}f = T^{-1}f_W = z.$$

Inserting this in $u = v + z$, we find

$$\widetilde{A}^{-1}f = A_\beta^{-1}f + T^{(-1)}f,$$

which proves the theorem. □

When $U \subset X$, we use the notation $i_{U \to X}$ for the injection of U into X. Then (13.35) may be written

$$\widetilde{A}^{-1} = A_\beta^{-1} + i_{V \to H} T^{-1} \operatorname{pr}_W \quad \text{on } R(\widetilde{A}). \tag{13.36}$$

Example 13.10. A simple example of the choice of T is to take $V = Z$, $W = Z'$, $T = 0$ on $D(T) = Z$. This corresponds to an operator in $\mathcal{M}$ that we shall denote A_M (M here indicates that it is maximal in a certain sense). Since $Z' \ominus W = \{0\}$, we see from Theorem 13.6 that $D(A_M) = D(A_0) + Z$. The adjoint A_M^* corresponds to T^* equal to the zero operator from Z' to Z, so it is completely analogous to A_M, with $D(A_M^*) = D(A_0') + Z'$.

In the symmetric situation where (13.5) holds, we see that A_M is selfadjoint. Moreover, it is the only selfadjoint extension of A_0 containing all of Z in its domain. Namely, any extension of A_0 having Z in its domain must extend A_M, hence cannot be selfadjoint unless it equals A_M. In the case where $m(A_0) > 0$, A_M is the extension of J. von Neumann, see Exercise 12.23.

Example 13.11. As an elementary illustration of Theorem 13.7, consider a Sturm-Liouville operator (also studied in Exercises 12.25 and 4.11–4.15):

Let A be the differential operator $Au = -u'' + qu$ on $I =]\alpha, \beta[$, where $q(t) \geq 0$, continuous on $[\alpha, \beta]$. Let A_0 be the closure of $A|_{C_0^\infty}$ (as an operator in $H = L_2(I)$); it is symmetric. Let $A_1 = A_0^*$. In other words, A_0 is the minimal operator $A_{\min}$, and A_1 is the maximal operator $A_{\max}$ (the L_2 weak definition of A). It is known from Chapter 4 that $D(A_1) = H^2(I)$ and

$$D(A_0) = H_0^2(I) = \{u \in H^2(I) \mid u(\alpha) = u'(\alpha) = u(\beta) = u'(\beta) = 0\}.$$

Together with A we can consider the sesquilinear form

$$a(u, v) = (u', v') + (qu, v), \quad u, v \in H^1(I)$$

(scalar products in $L_2(I)$). Let $V_1 = H^1(I)$, $V_0 = H_0^1(I)$. Clearly, a is V_1-coercive and ≥ 0; moreover, its restriction a_0 to $H_0^1(I)$ is V_0-elliptic in view of the Poincaré inequality (Theorem 4.29). The variational operator A_γ defined from the triple (H, V_0, a_0), acts like A_1 with domain $D(A_\gamma) = H^2(I) \cap H^1_0(I)$; it is invertible with $A_\gamma^{-1} \in \mathbf{B}(H)$, and is the Friedrichs extension of A_0. We take $A_\beta = A_\gamma$.

The nullspace Z of A_1 is spanned by two real null solutions $z_1(t)$ and $z_2(t)$ with $z_1(\alpha) = 0$, $z_1'(\alpha) \neq 0$, resp. $z_2(\beta) = 0$, $z_2'(\beta) \neq 0$ (they are linearly independent since $D(A_\gamma) \cap Z = \{0\}$). Defining

$$\gamma_0 u = \begin{pmatrix} u(\alpha) \\ u(\beta) \end{pmatrix}, \quad \gamma_1 u = \begin{pmatrix} u'(\alpha) \\ -u'(\beta) \end{pmatrix},$$

we have the Lagrange (or Green's) formula

$$(Au, v) - (u, Av) = \gamma_1 u \cdot \overline{\gamma_0 v} - \gamma_0 u \cdot \overline{\gamma_1 v}, \text{ for } u, v \in H^2(I). \qquad (13.37)$$

A simple calculation shows that the mapping $\gamma_0 : Z \to \mathbb{C}^2$ has the inverse $K : \mathbb{C}^2 \to Z$ defined by

$$K : \begin{pmatrix} x_1 \\ x_2 \end{pmatrix} \mapsto (z_1(t)\ z_2(t)) \begin{pmatrix} 0 & 1/z_1(\beta) \\ 1/z_2(\alpha) & 0 \end{pmatrix} \begin{pmatrix} x_1 \\ x_2 \end{pmatrix}$$

(called a Poisson operator in higher dimensional theories). Let us also define

$$P = \gamma_1 K = \begin{pmatrix} z_2'(\alpha)/z_2(\alpha) & z_1'(\alpha)/z_1(\beta) \\ -z_2'(\beta)/z_2(\alpha) & -z_1'(\beta)/z_1(\beta) \end{pmatrix};$$

note that the off-diagonal entries are nonzero. An application of (13.36) to $z, w \in Z$,

$$0 = (Az, w) - (z, Aw) = \gamma_1 z \cdot \overline{\gamma_0 w} - \gamma_0 z \cdot \overline{\gamma_1 w} = P\gamma_0 z \cdot \overline{\gamma_0 w} - \gamma_0 z \cdot \overline{P\gamma_0 w},$$

shows that $P^* = P$. (P is often called the Dirichlet-to-Neumann operator.)

Consider an operator $\widetilde{A}$ for which $V = W = Z$; we shall investigate the meaning of T. In the defining equation

$$(Au, z) = (Tu_\zeta, z), \text{ for } u \in D(\widetilde{A}), \ z \in Z, \qquad (13.38)$$

we rewrite the left-hand side as

$$(Au, z) = (Au, z) - (u, Az) = \gamma_1 u \cdot \overline{\gamma_0 z} - \gamma_0 u \cdot \overline{\gamma_1 z} = (\gamma_1 u - P\gamma_0 u) \cdot \overline{\gamma_0 z},$$

and the right-hand side (using that $\gamma_0 u_\zeta = \gamma_0 u$) as

$$(Tu_\zeta, z) = (TK\gamma_0 u_\zeta, K\gamma_0 z) = (TK\gamma_0 u, K\gamma_0 z) = K^* TK\gamma_0 u \cdot \overline{\gamma_0 z}.$$

Set $K^* TK = L$ (this carries the operator $T : Z \to Z$ over into a 2×2-matrix L), then (13.38) is turned into

$$(\gamma_1 u - P\gamma_0 u) \cdot \overline{\gamma_0 z} = L\gamma_0 u \cdot \overline{\gamma_0 z}, \text{ for } u \in D(\widetilde{A}),\ z \in Z.$$

Since $\gamma_0 z$ runs through $\mathbb{C}^2$, we conclude that for $u \in D(\widetilde{A})$, $\gamma_1 u - P\gamma_0 u = L\gamma_0 u$, or,

$$\gamma_1 u = (L + P)\gamma_0 u. \tag{13.39}$$

Conversely, when $u \in H^2(I)$ satisfies this, we see that (13.38) holds, so $u \in D(\widetilde{A})$. Thus $\widetilde{A}$ is the realization of A determined by the boundary condition (13.39). Note that (13.39) is a general type of Neumann condition.

Some examples: If $T = 0$, corresponding to $L = 0$, (13.39) is a nonlocal boundary condition (nonlocal in the sense that it mixes information from the two endpoints α and β). The local Neumann-type boundary conditions $u'(\alpha) = b_1 u(\alpha)$, $-u'(\beta) = b_2 u(\beta)$, are obtained by taking $L = \begin{pmatrix} b_1 & 0 \\ 0 & b_2 \end{pmatrix} - P$. Properties of $\widetilde{A}$ are reflected in properties of L, as shown in the analysis of consequences of Theorem 13.7 (also as followed up in the next sections).

As a case where V and W are nontrivial subspaces of Z, take for example $V = W = \operatorname{span} z_1$. Then T is a multiplication operator, and $\widetilde{A}$ represents a boundary condition of Neumann type at α, Dirichlet type at β $(u(\beta) = 0)$.

For PDE in dimension $n > 1$, the nullspaces Z and Z' will in general have infinite dimension, and there are many more choices of $T : V \to W$. Elliptic cases have some of the same flavor as above, but need extensive knowledge of boundary maps and Sobolev spaces, cf. [G68]–[G74], [BGW08].

13.2 The symmetric case

In this section, we restrict the attention to symmetric A_0. It can be shown in general that when A_0 is a symmetric, closed, densely defined operator in H with a bounded inverse, then there exists a selfadjoint extension A_β (J. W. Calkin [C40], see also the book of F. Riesz and B. Sz.-Nagy [RN53], p. 336). Then we can obtain the situation where (13.5) holds, *the symmetric case*. If A_0 is lower bounded, we can take for A_β the Friedrichs extension of A_0 (cf. Theorem 12.24); for this particular choice, A_β will be called A_γ.

When (13.5) holds, there is just one nullspace Z and one set of decomposition projections pr_β, pr_ζ to deal with. Then we get as an immediate corollary of Theorems 13.5 and 13.7:

Corollary 13.12. *Assume that* (13.5) *holds.*

Let $\widetilde{A}$ correspond to $T : V \to W$ as in Theorem 13.7. *Then $\widetilde{A}$ is selfadjoint if and only if: $V = W$ and T is selfadjoint.*

In more detail: If V is a closed subspace of Z and $T : V \to V$ is selfadjoint, then the operator $\widetilde{A} \subset A_1$ with domain

$$D(\widetilde{A}) = \{u \in D(A_1) \mid u_\zeta \in D(T),\ (Au)_V = Tu_\zeta\} \tag{13.40}$$

is a selfadjoint extension of A_0. Conversely, if $\widetilde{A} \in \mathcal{M}$ is selfadjoint, then the operator $T : V \to V$, where

$$D(T) = \operatorname{pr}_\zeta D(\widetilde{A}), \quad V = \overline{D(T)}, \quad Tu_\zeta = (Au)_V \text{ for } u \in D(\widetilde{A}), \quad (13.41)$$

is selfadjoint.

It is also easy to see that $\widetilde{A}$ is symmetric if and only if $V \subset W$ and $T : V \to W$ is symmetric as an operator in W, and that $\widetilde{A}$ is maximal symmetric if and only if T is so.

In the following, we consider throughout the situation where $\widetilde{A}$ corresponds to $T : V \to W$ as in Theorem 13.7, and shall not repeat this every time. The following observation will be important:

Lemma 13.13. *When* (13.5) *holds and $V \subset W$, then*

$$(Au, v) = (Au_\beta, v_\beta) + (Tu_\zeta, v_\zeta), \qquad (13.42)$$

for all $u, v \in D(\widetilde{A})$.

Proof. This follows from the calculation

$$(Au, v) = (Au, v_\beta) + (Au, v_\zeta) = (Au_\beta, v_\beta) + (Au, v_\zeta)$$
$$= (Au_\beta, v_\beta) + ((Au)_W, v_\zeta) = (Au_\beta, v_\beta) + (Tu_\zeta, v_\zeta),$$

where we used that $v_\zeta \in V \subset W$ for the third equality, and that $(Au)_W = Tu_\zeta$ for the last equality. □

Now we shall consider lower bounded extensions, in the case where A_β is lower bounded. We can assume that its lower bound is positive. Part of the results can be obtained for general A_β (cf. [G68], [G70]), but the most complete results are obtained when A_β equals the Friedrichs extension A_γ. Recall from Section 12.5 that A_γ is the variational operator defined from (H, V_0, a_γ), where V_0 (called V in Section 12.5) is the completion of $D(A_0)$ in the norm $\|u\|_{V_0} = (A_0u, u)^{\frac{1}{2}}$, and $a_\gamma(u, v)$ is the extension of (A_0u, v) to V_0. In particular, $D(A_0)$ is dense in V_0 (and A_γ is the only selfadjoint positive extension of A_0 for which $D(A_0)$ is dense in the sesquilinear form domain). To simplify the presentation, we go directly to this case, assuming

$$m(A_0) > 0, \quad A_\beta = A_\gamma, \qquad (13.43)$$

and recalling from Theorem 12.24 that $m(A_\gamma) = m(A_0)$. The results in the following are from [G70].

Lemma 13.14. *Assume that* (13.5) *and* (13.43) *hold.*
If for some $\lambda \in \mathbb{C}$, $c > 0$,

$$|(Au, u) - \lambda\|u\|^2| \geq c\|u\|^2 \text{ for all } u \in D(\widetilde{A}), \qquad (13.44)$$

then

$$V \subset W, \qquad (13.45)$$

$$|(Tz, z) - \lambda \|z\|^2| \geq c\|z\|^2 \text{ for all } z \in D(T). \qquad (13.46)$$

Proof. By Theorem 13.6, the general element of $D(\widetilde{A})$ is $u = z + A_\gamma^{-1}(Tz+f) + v$, with $\{z, f, v\} \in D(T) \times (Z \ominus W) \times D(A_0)$; here $u_\zeta = z$, $u_\gamma = A_\gamma^{-1}(Tz+f) + v$. For such u,

$$\begin{aligned}
(Au, u) &= (Au, u_\gamma + u_\zeta) = (Au_\gamma, u_\gamma) + (Tz + f + Av, u_\zeta) \\
&= (Au_\gamma, u_\gamma) + (Tz, z) + (f, z),
\end{aligned} \qquad (13.47)$$

where we used in the last step that $Av \in R(A_0) \perp Z$. Now assume that (13.44) holds, i.e.,

$$\left| \frac{(Au, u)}{\|u\|^2} - \lambda \right| \geq c, \text{ all } u \in D(\widetilde{A}) \setminus \{0\},$$

then in view of (13.47),

$$\left| \frac{(Au_\gamma, u_\gamma) + (Tz, z) + (f, z)}{\|u_\gamma + z\|^2} - \lambda \right| \geq c, \text{ all } \{z, f, v\} \neq 0. \qquad (13.48)$$

We shall now use that $D(A_0)$ is dense in $V_0 = D(a_\gamma)$, where $V_0 \subset H$ algebraically, topologically and densely (cf. (12.36) ff.). For any given $z \in D(T)$ and $f \in Z \ominus W$, let v_k be a sequence in $D(A_0)$ converging to $A_\gamma^{-1}(Tz+f)$ in V_0-norm. Then $u_k = A_\gamma^{-1}(Tz + f) - v_k \in D(A_\gamma)$ and satisfies

$$\|u_k\|_{V_0}^2 = (Au_k, u_k) \to 0, \text{ and a fortiori } \|u_k\|_H \to 0, \text{ for } k \to \infty.$$

Inserting the terms defined from the triple $\{z, f, -v_k\}$ in (13.48) and letting $k \to \infty$, we conclude that

$$\left| \frac{(Tz, z) + (f, z)}{\|z\|^2} - \lambda \right| \geq c, \text{ when } z \neq 0. \qquad (13.49)$$

This can be further improved: Note that if (f, z) is nonzero for some choice of $z \in D(T)$, $f \in Z \ominus W$, a replacement of f by $\frac{a}{(f,z)} f$ in the inequality (13.49) gives

$$\left| \frac{(Tz, z) + a}{\|z\|^2} - \lambda \right| \geq c, \text{ any } a \in \mathbb{C},$$

which cannot hold since some choice of a will give 0 in the left-hand side. Thus $(f, z) = 0$ for all $z \in D(T)$, $f \in Z \ominus W$, so the closure V of $D(T)$ in Z must satisfy $V \subset W$. This shows (13.45), and then (13.49) simplifies to give (13.46). $\qquad \square$

This has an immediate consequence for the numerical ranges (cf. (12.21)):

Theorem 13.15. *Assume that* (13.5) *and* (13.43) *hold, and let* $\widetilde{A}$ *correspond to* $T : V \to W$ *as in Theorem* 13.7.
   If $\nu(\widetilde{A})$ *is not all of* $\mathbb{C}$, *then* $V \subset W$, *and*

$$\overline{\nu(T)} \subset \overline{\nu(\widetilde{A})}. \tag{13.50}$$

Proof. Let $\nu(\widetilde{A}) \neq \mathbb{C}$; then since $\nu(\widetilde{A})$ is convex (Exercise 12.34(e)), it is contained in a half-plane, and so is $\overline{\nu(\widetilde{A})}$. By definition, a number $\lambda \in \mathbb{C}$ has a positive distance to $\nu(\widetilde{A})$ if and only if (13.44) holds for some $c > 0$, i.e., $\lambda \in \mathbb{C} \setminus \overline{\nu(\widetilde{A})}$. Such λ's exist, so $V \subset W$, by Lemma 13.14. The lemma shows moreover that $\lambda \in \mathbb{C} \setminus \nu(\widetilde{A})$ implies $\lambda \in \mathbb{C} \setminus \overline{\nu(T)}$, so (13.50) holds.   □

One can also get information on spectra:

Corollary 13.16. *Assumptions as in Theorem* 13.15.
   1° *If* $\widetilde{A}$ *and* $\widetilde{A}^*$ *both have the lower bound* $a \in \mathbb{R}$, *then* $V = W$, *and* T *and* T^* *have the lower bound* a, *their spectra being contained in* $\{\lambda \in \mathbb{C} \mid \operatorname{Re}\lambda \geq a\}$.
   2° *Assume moreover that* $\widetilde{A}$ *is variational, with numerical range (and hence spectrum) contained in an angular set*

$$M = \{\lambda \in \mathbb{C} \mid |\operatorname{Im}\lambda| \leq c_1(\operatorname{Re}\lambda + k), \operatorname{Re}\lambda \geq a\} \tag{13.51}$$

(recall that then $\widetilde{A}^*$ *has the same properties). Then* $V = W$, *and* T *and* T^* *are variational, with numerical ranges and spectra contained in* M.

Proof. 1°. The preceding theorem applied to $\widetilde{A}$ gives that $V \subset W$ and $m(T) \geq a$, and when it is applied to $\widetilde{A}^*$ it gives that $W \subset V$ and $m(T^*) \geq a$. The statement on the spectra follows from Theorem 12.9.
   2°. Since $\widetilde{A}$ is variational (cf. Corollary 12.19 ff.), so is $\widetilde{A}^*$. When $\nu(\widetilde{A}) \subset M$, the associated sesquilinear form $\tilde{a}$ has $\nu(\tilde{a}) \subset M$ (cf. (12.60), (12.61)), so $\widetilde{A}^*$ likewise has its numerical range (and both $\widetilde{A}$ and $\widetilde{A}^*$ have their spectra) in M; cf. Theorem 12.18. Theorem 13.15 implies immediately that T and T^* have their numerical ranges in M. Now Theorem 12.26 implies that T and T^* are variational, with spectra in M.   □

In the converse direction we find:

Theorem 13.17. *Assume that* (13.5) *and* (13.43) *hold, and let* $\widetilde{A}$ *correspond to* $T : V \to W$ *as in Theorem* 13.7.
   Assume that $V \subset W$.
   1° *If* $m(T) > -m(A_\gamma)$, *then*

$$m(\widetilde{A}) \geq \frac{m(A_\gamma)m(T)}{m(A_\gamma) + m(T)}. \tag{13.52}$$

In particular, $m(T) \geq 0$ (> 0) implies $m(\widetilde{A}) \geq 0$ (resp. > 0).

2° *If, for some $\theta \in] - \pi/2, \pi/2[$, $m(e^{i\theta}T) > -\cos\theta \, m(A_\gamma)$, then*

$$m(e^{i\theta}\widetilde{A}) \geq \frac{\cos\theta \, m(A_\gamma) m(e^{i\theta}T)}{\cos\theta \, m(A_\gamma) + m(e^{i\theta}T)}. \tag{13.53}$$

Proof. 1°. For $u \in D(\widetilde{A}) \setminus \{0\}$, we have since $V \subset W$ (cf. Lemma 13.13),

$$\frac{\mathrm{Re}(Au, u)}{\|u\|^2} = \frac{\mathrm{Re}((Au_\gamma, u_\gamma) + (Tu_\varsigma, u_\varsigma))}{\|u_\gamma + u_\varsigma\|^2} \geq \frac{m(A_\gamma)\|u_\gamma\|^2 + m(T)\|u_\varsigma\|^2}{\|u_\gamma + u_\varsigma\|^2}. \tag{13.54}$$

When $m(T) \geq 0$, the numerator is ≥ 0 for all u, so we can continue the estimate by

$$\frac{m(A_\gamma)\|u_\gamma\|^2 + m(T)\|u_\varsigma\|^2}{\|u_\gamma + u_\varsigma\|^2} \geq \frac{m(A_\gamma)\|u_\gamma\|^2 + m(T)\|u_\varsigma\|^2}{(\|u_\gamma\| + \|u_\varsigma\|)^2}. \tag{13.55}$$

For the function $f(x, y) = (\alpha x^2 + \beta y^2)(x+y)^{-2}$ with $\alpha > 0$, $\beta \geq 0$, considered for $x \geq 0$, $y \geq 0$, $(x, y) \neq (0, 0)$, one easily shows the inequality

$$f(x, y) \geq \alpha\beta(\alpha + \beta)^{-1},$$

by finding the minimum of $f(t, 1)$ for $t \geq 0$. This implies (13.52) when $m(T) \geq 0$.

When $m(T) < 0$, we cannot get (13.55) in general, since the numerator can be negative. But when $0 > m(T) > -m(A_\gamma)$, we can instead proceed as follows:

If $m(A_\gamma)\|u_\gamma\|^2 + m(T)\|u_\varsigma\|^2 \geq 0$, we have that $\mathrm{Re}(Au, u) \geq 0$, confirming (13.52) for such u. It remains to consider the u for which $m(A_\gamma)\|u_\gamma\|^2 + m(T)\|u_\varsigma\|^2 < 0$. This inequality implies that $u_\varsigma \neq 0$, and that $t = \|u_\gamma\|/\|u_\varsigma\|$ satisfies

$$t = \|u_\gamma\|/\|u_\varsigma\| < (-m(T))^{\frac{1}{2}} m(A_\gamma)^{-\frac{1}{2}} < 1.$$

Then since $\|u_\gamma + u_\varsigma\|^2 \geq (\|u_\gamma\| - \|u_\varsigma\|)^2 > 0$, we find that

$$\frac{m(A_\gamma)\|u_\gamma\|^2 + m(T)\|u_\varsigma\|^2}{\|u_\gamma + u_\varsigma\|^2} \geq \frac{m(A_\gamma)\|u_\gamma\|^2 + m(T)\|u_\varsigma\|^2}{(\|u_\gamma\| - \|u_\varsigma\|)^2}$$

$$= \frac{m(A_\gamma)t^2 + m(T)}{(t - 1)^2},$$

since the numerator is negative. The function $g(t) = (\alpha t^2 + \beta)(t - 1)^2$, $\alpha > 0$, $0 > \beta > -\alpha$, defined for $0 \leq t < (-\beta)^{\frac{1}{2}}\alpha^{-\frac{1}{2}}$, obtains its minimum at $t = -\beta/\alpha$ with value $\alpha\beta(\alpha + \beta)^{-1}$. This confirms (13.52) for such u. The last statement in 1° is an immediate consequence.

2°. The same proof is here applied to the operators multiplied by $e^{i\theta}$, and leads to the result since $m(e^{i\theta}A_\gamma) = \cos\theta \, m(A_\gamma) > 0$. $\qquad\square$

The condition $m(e^{i\theta}T) > -\cos\theta\, m(A_\gamma)$ means geometrically that $\nu(T)$ is contained in a half-plane in $\mathbb{R}^2$ lying to the right of the point $(-m(A_\gamma),0)$ on the real axis; more precisely it is the half-plane lying to the right of the line through $(m(e^{i\theta}T)/\cos\theta,0)$ with angle $\theta + \pi/2$ to the real axis.

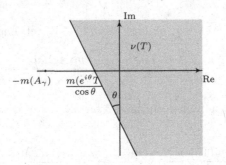

Two inequalities $m(e^{\pm i\theta}T) \geq b$ valid together, with $b > -\cos\theta\, m(A_\gamma)$, $\theta \in\,]0,\pi/2[$, mean that $\nu(T)$ is contained in

$$M' = \{\lambda \in \mathbb{C} \mid |\operatorname{Im}\lambda| \leq \cot\theta\,(\operatorname{Re}\lambda - b/\cos\theta)\} \qquad (13.56)$$

with $b/\cos\theta > -m(A_\gamma)$ (a sector to the right of $(-m(A_\gamma),0)$).

Remark 13.18. This theorem only applies when the lower bound of T resp. $e^{i\theta}T$ is "not too negative". It can be shown by a greater effort that if A_γ^{-1} is a compact operator in H (as it often is in the applications), then one gets lower boundedness of $e^{i\theta}\widetilde{A}$ for any lower bounded $e^{i\theta}T$, see [G74].

Theorem 13.17 also holds when A_γ is replaced by a general A_β with positive lower bound, cf. [G68], [G70]. However, the extended version in [G74], where the condition $m(T) > -m(A_\gamma)$ is removed when A_γ^{-1} is compact, uses the particular properties of A_γ.

We shall now study variational extensions in terms of the associated sesquilinear forms.

Let $\widetilde{A}$ be variational, associated with the triple $(H, D(\tilde{a}), \tilde{a})$. (We here denote by $D(\tilde{a})$ the Hilbert space where $\tilde{a}$ is defined and bounded, earlier called V in Sections 12.4, 12.6.) As observed in Corollary 13.16, $V = W$, and $\widetilde{A}^*$, T and T^* are likewise variational, with numerical ranges and spectra in the set M (13.51) defined for $\widetilde{A}$.

Theorem 13.19. *Assume that (13.5) and (13.43) hold, let $\widetilde{A}$ be a variational operator associated with a triple $(H, D(\tilde{a}), \tilde{a})$, and let $T : V \to V$ be the corresponding variational operator as in Corollary 13.16, associated with the triple $(V, D(t), t)$.*
Then

$$D(\tilde{a}) = D(a_\gamma)\dotplus D(t), \quad with\ D(t) = D(\tilde{a}) \cap Z, \qquad (13.57)$$

the projections pr_γ *and* pr_ζ *extending continuously to this direct sum, and*

$$\tilde{a}(u,v) = a_\gamma(u_\gamma, v_\gamma) + t(u_\zeta, v_\zeta), \quad \text{for } u,v \in D(\tilde{a}). \tag{13.58}$$

Proof. We have from Lemma 13.13 that (13.58) holds for $u,v \in D(\tilde{A})$, and we want to extend this to $D(\tilde{a})$. To see that $D(t) \subset D(\tilde{a})$, let $z \in D(t)$ and let $z_k \in D(T)$, converging to z in $D(t)$ for $k \to \infty$; then it also converges in H. Since $A_\gamma^{-1} T z_k \in D(A_\gamma)$, and $D(A_0)$ is dense in $D(a_\gamma)$, we can choose $v_k \in D(A_0)$ such that $w_k = v_k + A_\gamma^{-1} T z_k \to 0$ in $D(a_\gamma)$, hence in H. Now $u_k = w_k + z_k \in D(\tilde{A})$ (cf. Theorem 13.6), and $u_k \to z$ in H. Moreover, since $\mathrm{pr}_\gamma u_k = w_k$, $\mathrm{pr}_\zeta u_k = z_k$,

$$\tilde{a}(u_k - u_l, u_k - u_l) = a_\gamma(w_k - w_l, w_k - w_l) + t(z_k - z_l, z_k - z_l) \to 0, \quad \text{for } k,l \to \infty,$$

which implies that u_k has a limit in $D(\tilde{a})$; this must equal z. Hence $z \in D(\tilde{a})$. Since $A_0 \subset \tilde{A}$, $D(a_\gamma) \subset D(\tilde{a})$ and we conclude that

$$D(a_\gamma) + D(t) \subset D(\tilde{a}). \tag{13.59}$$

We next show that

$$D(a_\gamma) \cap D(t) = \{0\}. \tag{13.60}$$

Here $D(t) \subset V \subset Z$, so it suffices to show that

$$D(a_\gamma) \cap Z = \{0\}. \tag{13.61}$$

But if $z \in D(a_\gamma) \cap Z$, let $u_k \in D(A_0)$ converge to z in $D(a_\gamma)$; then

$$a_\gamma(z,z) = \lim_{k \to \infty} a_\gamma(u_k, z) = \lim_{k \to \infty} (A_0 u_k, z)_H = 0,$$

since $R(A_0) \perp Z$. Then $z = 0$ since $m(a_\gamma) > 0$.

If $m(\tilde{A}) > 0$, we are almost through, for the real part of the identity (13.58) that we know is valid for $u = v \in D(\tilde{A})$,

$$\mathrm{Re}\,\tilde{a}(u,u) = a_\gamma(u_\gamma, u_\gamma) + \mathrm{Re}\,t(u_\zeta, u_\zeta) \tag{13.62}$$

shows, since all three terms are squares of norms (on $D(\tilde{a})$, $D(a_\gamma)$ resp. $D(t)$), that the maps $\mathrm{pr}_\gamma : D(\tilde{a}) \to D(a_\gamma)$ and $\mathrm{pr}_\zeta : D(\tilde{a}) \to D(t)$ extend from the dense subset $D(\tilde{A})$ to continuous maps from $D(\tilde{a})$ to $D(a_\gamma)$ resp. $D(t)$; they remain projections, and must have linearly independent ranges in view of (13.59) and (13.60). The second statement in (13.57) is obvious from (13.60) ff. The identity (13.58) follows by extension by continuity.

If $m(\tilde{A}) \leq 0$, we have to do a little more work: Choosing a large constant μ such that $m(\tilde{A} + \mu) > 0$, we can apply the preceding argument to the setup with $\tilde{A}$, A_γ resp. $T : V \to V$ replaced by $\tilde{A} + \mu$, $A_\gamma + \mu$ resp. $T_\mu : V_\mu \to V_\mu$ ($V_\mu \subset Z(A_1 + \mu)$); here $D(\tilde{A} + \mu) = D(\tilde{A})$, $D(A_\gamma + \mu) = D(A_\gamma)$,

$D(\tilde{a} + \mu) = D(\tilde{a})$, $D(a_\gamma + \mu) = D(a_\gamma)$. Let t_μ be the sesquilinear form with domain $D(t_\mu) \subset V_\mu$ associated with T_μ.

Let $u \in D(\tilde{a})$ and decompose it first using $D(\tilde{a} + \mu) = D(a_\gamma + \mu) \dotplus D(t_\mu)$ (by the first part of the proof), and next using $D(A_1) = D(A_\gamma) \dotplus Z$ for z_μ:

$$u = v + z_\mu, \quad v \in D(a_\gamma), \; z_\mu \in D(t_\mu) \subset Z(A_1 + \mu) \subset D(A_1),$$
$$= v + z_1 + z_2, \quad z_1 \in D(A_\gamma), \; z_2 \in Z. \tag{13.63}$$

Then we have inequalities (with positive constants)

$$\|u\|_{D(\tilde{a})} \geq c(\|v\|_{D(a_\gamma)} + \|z_\mu\|_{D(t_\mu)})$$
$$\geq c\|v\|_{D(a_\gamma)} + c'(\|z_1\|_{D(A_\gamma)} + \|z_2\|_H)$$
$$\geq c''\|v + z_1\|_{D(a_\gamma)} + c'\|z_2\|_H,$$

since the norm in $D(\widetilde{A})$ is stronger than that in $D(a_\gamma)$, and the norm in $D(t_\mu)$ is stronger than the H-norm. Let $u_k \in D(\widetilde{A})$, $u_k \to u$ in $D(\tilde{a})$; then in the decomposition (13.63), $u_{k,\gamma} = v_k + z_{k,1} \to u_\gamma$ in $D(a_\gamma)$ and $u_{k,\zeta} = z_{k,2} \to u_\zeta$ in H. Now we combine this with (13.62) applied to $u_k - u_l$; it shows that

$$\operatorname{Re} t(u_{k,\zeta} - u_{l,\zeta}, u_{k,\zeta} - u_{l,\zeta}) = \operatorname{Re} \tilde{a}(u_k - u_l, u_k - u_l) - a_\gamma(u_{k,\gamma} - u_{l,\gamma}, u_{k,\gamma} - u_{l,\gamma})$$

goes to 0 for $k, l \to \infty$, so $z_{k,2}$ is a Cauchy sequence in $D(t)$.

This shows that $D(\tilde{a}) \subset D(a_\gamma) \dotplus D(t)$ by the decomposition $u = (v + z_1) + z_2$ in (13.63), with continuous maps. So we can conclude that (13.57) holds, and the proof is completed as above.   □

The theorem gives a description of all variational operators $\widetilde{A} \in \mathcal{M}$. In view of Corollary 13.16, the variational operators $T : V \to V$ include all choices $(V, D(t), t)$, where V is closed $\subset Z$, $D(t) \subset V$ algebraically, topologically and densely, and t is $D(t)$-coercive with lower bound $m(t) > -m(a_\gamma)$. Namely, such t have

$$\nu(t) \subset \{\lambda \in \mathbb{C} \mid |\operatorname{Im} \lambda| \leq c_1(\operatorname{Re} \lambda + k), \; \operatorname{Re} \lambda \geq m(t)\}$$
$$\subset \{\lambda \in \mathbb{C} \mid |\operatorname{Im} \lambda| \leq c_1'(\operatorname{Re} \lambda - m(t) + \varepsilon)\} \tag{13.64}$$

for any ε, with a possibly larger c_1' (the angular set M is contained in a right sector with rays emanating from $m(t) - \varepsilon$). Then Corollary 13.16 2° applies to T.

As mentioned in Remark 13.18, the restriction on how negative $m(t)$ is allowed to be, is removed in [G74] when A_γ^{-1} is compact.

Note that the theorem shows a very simple connection between the forms $\tilde{a}$ and t: The domain of $\tilde{a}$ is simply the direct sum of $D(a_\gamma)$ and $D(t)$, and the value of $\tilde{a}$ on $D(a_\gamma)$ is fixed. One can choose any V, closed subspace of Z, and take as $D(t)$ any Hilbert space V_0 that is algebraically, topologically and densely injected in V, and among the possible choices of t there are at least the bounded and V_0-elliptic sesquilinear forms on V_0. So we have:

Corollary 13.20. *Assume* (13.5) *and* (13.43). *The Hilbert spaces U such that there exists a variational $\tilde{A} \in \mathcal{M}$ with $D(\tilde{a}) = U$ are the spaces satisfying*

$$D(a_\gamma) \subset U \subset D(a_\gamma) \dot{+} Z \tag{13.65}$$

with continuous injections.

The largest possible space U here is $D(a_\gamma) \dot{+} Z$. Let t be the zero sesquilinear form on Z; then T is the zero operator on Z, so $\tilde{A}$ is a selfadjoint extension of A_0 that has all of Z included in its domain and nullspace, since $Z(\tilde{A}) = Z(T)$, cf. Theorem 13.8 1°. This $\tilde{A}$ equals the von Neumann extension A_M considered in Example 13.10. The associated sesquilinear form a_M satisfies

$$D(a_M) = D(a_\gamma) \dot{+} Z, \quad a_M(u,v) = a_\gamma(u_\gamma, v_\gamma), \tag{13.66}$$

in view of (13.58). We can extend Corollary 13.20 as follows:

Corollary 13.21. *Assume* (13.5) *and* (13.43). *A variational operator $\tilde{A}$ belongs to $\mathcal{M}$ if and only if the associated sesquilinear form $\tilde{a}$ satisfies* (i)–(iii):
  (i) $D(a_\gamma) \subset D(\tilde{a}) \subset D(a_M)$, *with continuous injections;*
  (ii) $\tilde{a} \supset a_\gamma$;
  (iii) $\tilde{a}(w,z) = \tilde{a}(z,w) = 0$, *for $w \in D(a_\gamma)$, $z \in D(\tilde{a}) \cap Z$.*

Proof. The necessity of (i) and (ii) is immediate from (13.57) and (13.58), and we get (iii) by applying (13.58) with $u = w$, $v = z$, or $u = z$, $v = w$.

For the converse direction, let $\tilde{A}$ be the variational operator associated with a form $\tilde{a}$ satisfying (i)–(iii). Then $\tilde{A}$ extends A_0, since for all $w \in D(A_0)$, $v \in D(\tilde{a})$,

$$(A_0 w, v) = (A_0 w, v_\gamma + v_\zeta) = (A_0 w, v_\gamma) = a_\gamma(w, v_\gamma) = \tilde{a}(w, v),$$

where we used $R(A_0) \perp Z$ in the second step and (iii) in the last step. Moreover, the adjoint $\tilde{A}^*$ has the analogous properties with $\tilde{a}$ replaced by $\tilde{a}^*$, so it also extends A_0. Then $\tilde{A} \in \mathcal{M}$.   □

Restricting the attention to selfadjoint nonnegative operators, we get in particular:

Corollary 13.22. *Assume* (13.5) *and* (13.43). *A selfadjoint nonnegative operator $\tilde{A}$ belongs to $\mathcal{M}$ if and only if the associated sesquilinear form $\tilde{a}$ satisfies* (i)–(iii):
  (i) $D(a_\gamma) \subset D(\tilde{a}) \subset D(a_M)$, *with continuous injections;*
  (ii) $\tilde{a} \supset a_\gamma$;
  (iii) $\tilde{a}(u,u) \geq a_M(u,u)$, *for $u \in D(\tilde{a})$.*

Proof. Here we note that if $\tilde{A} \in \mathcal{M}$, then the corresponding form t is ≥ 0, so (iii) follows from (13.58).

In the converse direction, we reduce to an application of Corollary 13.21 by observing that (ii) and (iii) imply, for $u \in D(\tilde{a})$, $u = w + z$ according to the decomposition (13.66),

$$a_\gamma(w, w) = a_M(u, u) \leq \tilde{a}(w + z, w + z)$$
$$= a_\gamma(w, w) + \tilde{a}(w, z) + \tilde{a}(z, w) + \tilde{a}(z, z),$$

hence

$$\tilde{a}(w, z) + \tilde{a}(z, w) + \tilde{a}(z, z) \geq 0.$$

Here $u = w + z$ can be replaced by $u = \alpha w + z$, any $\alpha \in \mathbb{C}$, in view of (i). Taking $\alpha = s\,\tilde{a}(w, z)$, $s \in \mathbb{R}$, gives

$$2s\,|\tilde{a}(w, z)| + \tilde{a}(z, z) \geq 0 \text{ for all } s \in \mathbb{R},$$

which clearly cannot hold if $\tilde{a}(w, z) \neq 0$. Similarly, $\tilde{a}(z, w)$ must equal 0. Then (iii) of Corollary 13.21 holds, and we can apply that corollary to conclude that $\tilde{A} \in \mathcal{M}$.   □

Some historical remarks: Corollary 13.22 was shown by M. G. Kreĭn in [K47]. Kreĭn introduced a different viewpoint, deducing the result from another equivalent result: The selfadjoint nonnegative extensions of A_0 are precisely the selfadjoint nonnegative operators $\tilde{A}$ satisfying

$$(A_\gamma + 1)^{-1} \leq (\tilde{A} + 1)^{-1} \leq (A_M + 1)^{-1}; \qquad (13.67)$$

i.e., $((A_\gamma + 1)^{-1}f, f) \leq ((\tilde{A} + 1)^{-1}f, f) \leq ((A_M + 1)^{-1}f, f)$ for all $f \in H$. The two operators A_γ and A_M are extreme in this scale of operators; Kreĭn called A_γ the "hard extension" and A_M the "soft extension". In the sense of (13.67), or in the sense expressed in Corollary 13.22, the operators $\tilde{A}$ are the selfadjoint operators "lying between" the hard extension A_γ and the soft extension A_M.

Birman [B56] showed a version of the inequality in Theorem 13.17 for selfadjoint operators. The parametrization we have presented in Section 13.1 is related to that of Vishik [V52], except that he sets the $\tilde{A}$ in relation to operators on the nullspaces going the opposite way of our T's, and in this context focuses on *normally solvable* extensions (those with closed range).

One of the primary interests of the extension theory is its application to the study of boundary value problems for elliptic differential operators; this has been developed in a rather complete way in papers by Grubb [G68]–[G74]. The basic idea relies on the fact that the nullspace Z is in such cases isomorphic to a suitable Sobolev space over the boundary of the set where the differential operator acts; then the operator $T : V \to W$ can be carried over to an operator $L : X \to Y^*$ where X and Y are spaces defined over the boundary, and $\tilde{A}$ is seen to represent a boundary condition.

Such studies use, among other things, the distribution definition of boundary values established in Lions and Magenes [LM68], and pseudodifferential operators. To illustrate the main idea, we show in Chapter 9 how it is applied in a relatively simple, constant-coefficient situation, that can be understood on the basis of Chapters 1–6. More general variable-coefficient applications are included in the exercises to Chapter 11, which builds on Chapters 7, 8 and 10.

Extension theories have been developed also in other directions, often with a focus on applications to ordinary differential equations and other cases with finite dimensional nullspaces. A central work in this development is the book of Gorbachuk and Gorbachuk, [GG91], with references to many other works. Extensions are here characterized in terms of so-called boundary triplets. An important trend in the extension theories has been to replace *operators* by *relations*, as e.g. in Kochubei [K75], Derkach and Malamud [DM91], Malamud and Mogilevskii [MM99], and many other works. Many contributions deal with extensions of symmetric operators, fewer with extensions of adjoint couples as in Section 13.1. One of the important issues in such studies is spectral theory, in particular the description of the spectrum and the resolvent in terms of the associated "boundary operators".

We shall show below how the results of Section 13.1 can be extended to cover also resolvent questions. The direct consequences are given in Section 13.3, and the connection with considerations of boundary triplets is explained in Section 13.4. Here we also describe a recent development where boundary triplets are introduced in subspace situations; in the study of closed extensions as in Section 13.1, this replaces the need to discuss relations.

13.3 Consequences for resolvents

We now return to the general setup of Section 13.1, and consider the situation where a spectral parameter $\lambda \in \mathbb{C}$ is subtracted from the operators in $\mathcal{M}$. When $\lambda \in \varrho(A_\beta)$, we have a similar situation as in Section 13.1:

$$A_0 - \lambda \subset A_\beta - \lambda \subset A_1 - \lambda, \quad A_0' - \bar\lambda \subset A_\beta^* - \bar\lambda \subset A_1' - \bar\lambda, \qquad (13.68)$$

and we use the notation $\mathcal{M}_\lambda$, $\mathcal{M}_{\bar\lambda}'$ for the families of operators $\widetilde{A} - \lambda$ resp. $\widetilde{A}' - \bar\lambda$, and define

$$Z(A_1 - \lambda) = Z_\lambda, \quad Z(A_1' - \bar\lambda) = Z_{\bar\lambda}';$$
$$\mathrm{pr}_\beta^\lambda = (A_\beta - \lambda)^{-1}(A - \lambda), \quad \mathrm{pr}_{\beta'}^{\bar\lambda} = (A_\beta^* - \bar\lambda)^{-1}(A' - \bar\lambda), \qquad (13.69)$$
$$\mathrm{pr}_\zeta^\lambda = I - \mathrm{pr}_\beta^\lambda, \quad \mathrm{pr}_{\zeta'}^{\bar\lambda} = I - \mathrm{pr}_{\beta'}^{\bar\lambda},$$

the *basic λ-dependent projections*. We denote $\mathrm{pr}_\beta^\lambda u = u_\beta^\lambda$, $\mathrm{pr}_\zeta^\lambda u = u_\zeta^\lambda$, etc.

The following is an immediate corollary of the results in Section 13.1:

Theorem 13.23. *Let* $\lambda \in \varrho(A_\beta)$. *There is a 1-1 correspondence between the closed operators* $\widetilde{A} - \lambda$ *in* $\mathcal{M}_\lambda$ *and the closed, densely defined operators* $T^\lambda : V_\lambda \to W_{\bar\lambda}$, *where* V_λ *and* $W_{\bar\lambda}$ *are closed subspaces of* Z_λ *resp.* $Z'_{\bar\lambda}$; *here*

$$D(T^\lambda) = \operatorname{pr}_\zeta^\lambda D(\widetilde{A}), \quad V_\lambda = \overline{D(T^\lambda)}, \quad W_{\bar\lambda} = \overline{\operatorname{pr}_{\zeta'}^{\bar\lambda} D(\widetilde{A}^*)},$$

$$T^\lambda u_\zeta^\lambda = ((A - \lambda)u)_{W_{\bar\lambda}} \text{ for } u \in D(\widetilde{A}), \tag{13.70}$$

$$D(\widetilde{A}) = \{ u \in D(A_1) \mid u_\zeta^\lambda \in D(T^\lambda), ((A - \lambda)u)_{W_{\bar\lambda}} = T^\lambda u_\zeta^\lambda \}.$$

In this correspondence,

$$\begin{aligned} Z(\widetilde{A} - \lambda) &= Z(T^\lambda), \\ R(\widetilde{A} - \lambda) &= R(T^\lambda) + (H \ominus W_{\bar\lambda}), \end{aligned} \tag{13.71}$$

orthogonal sum. In particular, if $\widetilde{A} - \lambda$ *is injective,*

$$(\widetilde{A} - \lambda)^{-1} = (A_\beta - \lambda)^{-1} + i_{V_\lambda \to H}(T^\lambda)^{-1} \operatorname{pr}_{W_{\bar\lambda}} \tag{13.72}$$

on $R(\widetilde{A} - \lambda)$.

Proof. The first part is covered by Theorem 13.7, and the formulas in (13.71) and (13.72) follow from Theorems 3.8 and 3.9; see also (13.36). □

In the literature on extensions, formulas describing the difference between two resolvents in terms of associated operators in other spaces are often called Kreĭn resolvent formulas. We see here how Section 13.1 gives rise to a simple and universally valid Kreĭn resolvent formula (13.72).

The next results have been developed from calculations in [G74, Section 2].

Define, for $\lambda \in \varrho(A_\beta)$, the bounded operators on H:

$$\begin{aligned} E^\lambda &= A_1(A_\beta - \lambda)^{-1} = I + \lambda(A_\beta - \lambda)^{-1}, \\ F^\lambda &= (A_1 - \lambda)A_\beta^{-1} = I - \lambda A_\beta^{-1}, \\ E'^{\bar\lambda} &= A_1'(A_\beta^* - \bar\lambda)^{-1} = I + \bar\lambda(A_\beta^* - \bar\lambda)^{-1} = (E^\lambda)^*, \\ F'^{\bar\lambda} &= (A_1' - \bar\lambda)(A_\beta^*)^{-1} = I - \bar\lambda(A_\beta^*)^{-1} = (F^\lambda)^*. \end{aligned} \tag{13.73}$$

Lemma 13.24. E^λ *and* F^λ *are inverses of one another, and so are* $E'^{\bar\lambda}$ *and* $F'^{\bar\lambda}$. *In particular, the operators restrict to homeomorphisms*

$$\begin{aligned} E_Z^\lambda : Z \xrightarrow{\sim} Z_\lambda, \quad F_Z^\lambda : Z_\lambda \xrightarrow{\sim} Z, \\ E_{Z'}'^{\bar\lambda} : Z' \xrightarrow{\sim} Z'_{\bar\lambda}, \quad F_{Z'}'^{\bar\lambda} : Z'_{\bar\lambda} \xrightarrow{\sim} Z'. \end{aligned} \tag{13.74}$$

Moreover, for $u \in D(A_1)$, $v \in D(A_1')$,

$$\operatorname{pr}_\zeta^\lambda u = E^\lambda \operatorname{pr}_\zeta u, \quad \operatorname{pr}_\beta^\lambda u = \operatorname{pr}_\beta u - \lambda(A_\beta - \lambda)^{-1}\operatorname{pr}_\zeta u,$$
$$\operatorname{pr}_{\zeta'}^{\bar\lambda} v = E'^\lambda \operatorname{pr}_{\zeta'} v, \quad \operatorname{pr}_{\beta'}^{\bar\lambda} v = \operatorname{pr}_{\beta'} v - \bar\lambda(A_\beta^* - \bar\lambda)^{-1}\operatorname{pr}_{\zeta'} v. \tag{13.75}$$

Proof. To show that E^λ and F^λ are inverses of one another, note that for

$$v \in A_\beta^{-1}H = D(A_\beta) = D(A_\beta - \lambda) = (A_\beta - \lambda)^{-1}H,$$

one has that $v = A_\beta^{-1}A_1 v = (A_\beta - \lambda)^{-1}(A_1 - \lambda)v$. Hence

$$E^\lambda F^\lambda f = A_1(A_\beta - \lambda)^{-1}(A_1 - \lambda)A_\beta^{-1}f = A_1 A_\beta^{-1}f = f;$$
$$F^\lambda E^\lambda f = (A_1 - \lambda)A_\beta^{-1}A_1(A_\beta - \lambda)^{-1}f = (A_1 - \lambda)(A_\beta - \lambda)^{-1}f = f.$$

There is the analogous proof for the primed operators.

Now E^λ maps Z into Z_λ, and F^λ maps the other way, since

$$A_1 u = 0 \implies (A_1 - \lambda)E^\lambda u = (A_1 - \lambda)(u + \lambda(A_\beta - \lambda)^{-1}u)$$
$$= -\lambda u + \lambda u = 0,$$
$$(A_1 - \lambda)v = 0 \implies A_1 F^\lambda v = A_1(v - \lambda A_\beta^{-1}v) = \lambda v - \lambda v = 0,$$

so since E^λ and F^λ are bijective in H, E^λ maps Z homeomorphically onto $Z(A_1 - \lambda)$, with inverse F^λ. This justifies the first line in (13.74), and the second line is similarly justified.

The relation of the λ-dependent decompositions in (13.69) to the original basic decompositions is found by observing that for $u \in D(A_1)$,

$$u = u_\beta + u_\zeta = [u_\beta + (I - E^\lambda)u_\zeta] + E^\lambda u_\zeta,$$

where $E^\lambda u_\zeta \in Z(A_1 - \lambda)$ and

$$(I - E^\lambda)u_\zeta = -\lambda(A_\beta - \lambda)^{-1}u_\zeta \in D(A_\beta - \lambda);$$

hence, by the uniqueness in the decomposition $D(A_1 - \lambda) = D(A_\beta - \lambda) \dotplus Z_\lambda$,

$$u_\zeta^\lambda = E^\lambda u_\zeta, \quad u_\beta^\lambda = u_\beta - \lambda(A_\beta - \lambda)^{-1}u_\zeta. \tag{13.76}$$

Similarly for $v \in D(A_1')$,

$$v_{\zeta'}^{\bar\lambda} = E'^\lambda v_{\zeta'}, \quad v_{\beta'}^{\bar\lambda} = v_{\beta'} - \bar\lambda(A_\beta^* - \bar\lambda)^{-1}v_{\zeta'}. \tag{13.77}$$

$\square$

Now we can find the relation between T and T^λ (similarly to [G74, Prop. 2.6]):

Theorem 13.25. *Let* $T : V \to W$ *correspond to* $\widetilde{A}$ *by Theorem 13.7, let* $\lambda \in \varrho(A_\beta)$ *and let* $\widetilde{A} - \lambda$ *correspond to* $T^\lambda : V_\lambda \to W_{\bar\lambda}$ *by Theorem 13.23.*

For $\lambda \in \varrho(A_\beta)$, define the operator G^λ from Z to Z' by

$$G^\lambda z = -\lambda \operatorname{pr}_{Z'} E^\lambda z, \quad z \in Z. \tag{13.78}$$

Then

$$D(T^\lambda) = E^\lambda D(T), \quad V_\lambda = E^\lambda V, \quad W_{\bar\lambda} = E'^{\bar\lambda} W,$$
$$(T^\lambda E^\lambda v, E'^{\bar\lambda} w) = (Tv, w) + (G^\lambda v, w), \text{ for } v \in D(T), \ w \in W. \tag{13.79}$$

Proof. The first line in (13.79) follows from (13.70) in view of (13.75). The second line is calculated as follows: For $u \in D(\widetilde{A})$, $w \in W$,

$$(Tu_\zeta, w) = (Au, w) = (Au, F'^{\bar\lambda} E'^{\bar\lambda} w) = (F^\lambda Au, E'^{\bar\lambda} w)$$
$$= ((A - \lambda)(A_\beta)^{-1} Au, E'^{\bar\lambda} w) = ((A - \lambda)u_\beta, E'^{\bar\lambda} w)$$
$$= ((A - \lambda)u, E'^{\bar\lambda} w) - ((A - \lambda)u_\zeta, E'^{\bar\lambda} w)$$
$$= (T^\lambda u_\zeta^\lambda, E'^{\bar\lambda} w) + (\lambda u_\zeta, E'^{\bar\lambda} w) = (T^\lambda E^\lambda u_\zeta, E'^{\bar\lambda} w) + (\lambda E^\lambda u_\zeta, w).$$

This shows the equation in (13.79) when we set $u_\zeta = v$. $\qquad\square$

Denote by E_V^λ the restriction of E^λ to a mapping from V to V_λ, with inverse F_V^λ, and let similarly $E_W'^{\bar\lambda}$ be the restriction of $E'^{\bar\lambda}$ to a mapping from W to $W_{\bar\lambda}$, with inverse $F_W'^{\bar\lambda}$. Then the second line of (13.79) can be written

$$(E_W'^{\bar\lambda})^* T^\lambda E_V^\lambda = T + G_{V,W}^\lambda \text{ on } D(T) \subset V, \tag{13.80}$$

where

$$G_{V,W}^\lambda = \operatorname{pr}_W G^\lambda \operatorname{i}_{V \to Z}. \tag{13.81}$$

(The reader is warned that the adjoint $(E_W'^{\bar\lambda})^*$ is a mapping from $W_{\bar\lambda}$ to W, which is not derived by restriction from the formulas in (13.73).) Equivalently,

$$T^\lambda = (F_W'^{\bar\lambda})^* (T + G_{V,W}^\lambda) F_V^\lambda \text{ on } D(T^\lambda) \subset V_\lambda. \tag{13.82}$$

Then the Kreĭn resolvent formula (13.72) can be made more explicit as follows:

Corollary 13.26. *When $\lambda \in \varrho(\widetilde{A}) \cap \varrho(A_\beta)$, T^λ is bijective, and*

$$(T^\lambda)^{-1} = E_V^\lambda (T + G_{V,W}^\lambda)^{-1} (E_W'^{\bar\lambda})^*. \tag{13.83}$$

Hence

$$(\widetilde{A} - \lambda)^{-1} = (A_\beta - \lambda)^{-1} + \operatorname{i}_{V_\lambda \to H} E_V^\lambda (T + G_{V,W}^\lambda)^{-1} (E_W'^{\bar\lambda})^* \operatorname{pr}_{W_{\bar\lambda}}. \tag{13.84}$$

Proof. (13.83) follows from (13.82) by inversion, and insertion in (13.72) shows (13.84). $\qquad\square$

Note that $G_{V,W}^\lambda$ depends in a simple way on V and W and is independent of T.

13.4 Boundary triplets and M-functions

We shall now consider the relation of our theory to constructions of extensions in terms of boundary triplets. In the nonsymmetric case, the setup is the following, according to Brown, Marletta, Naboko and Wood [BMNW08] (with reference to Vainerman [V80], Lyantze and Storozh [LS83], Malamud and Mogilevskii [MM02]):

A_0, A_1, A_0' and A_1' are given as in the beginning of Section 13.1, and there is given a pair of Hilbert spaces $\mathcal{H}$, $\mathcal{K}$ and two pairs of "boundary operators"

$$\begin{pmatrix} \Gamma_1 \\ \Gamma_0 \end{pmatrix} : D(A_1) \to \begin{matrix} \mathcal{H} \\ \times, \\ \mathcal{K} \end{matrix} \qquad \begin{pmatrix} \Gamma_1' \\ \Gamma_0' \end{pmatrix} : D(A_1') \to \begin{matrix} \mathcal{K} \\ \times, \\ \mathcal{H} \end{matrix} \qquad (13.85)$$

bounded with respect to the graph norm and surjective, and satisfying

$$(Au, v) - (u, A'v) = (\Gamma_1 u, \Gamma_0' v)_{\mathcal{H}} - (\Gamma_0 u, \Gamma_1' v)_{\mathcal{K}}, \text{ all } u \in D(A_1), v \in D(A_1'). \quad (13.86)$$

Moreover it is assumed that

$$D(A_0) = D(A_1) \cap Z(\Gamma_1) \cap Z(\Gamma_0), \quad D(A_0') = D(A_1') \cap Z(\Gamma_1') \cap Z(\Gamma_0'). \quad (13.87)$$

In our setting, these hypotheses are satisfied by the choice

$$\begin{aligned} \mathcal{H} &= Z', \quad \mathcal{K} = Z, \\ \Gamma_1 &= \mathrm{pr}_{Z'} A_1, \quad \Gamma_0 = \mathrm{pr}_\zeta, \\ \Gamma_1' &= \mathrm{pr}_Z A_1', \quad \Gamma_0' = \mathrm{pr}_{\zeta'}. \end{aligned} \qquad (13.88)$$

The surjectiveness follows since $u \in D(A_1)$ has the form $u = v + z$ with independent v and z, where $z = \mathrm{pr}_\zeta u = \mathrm{pr}_\zeta z$ runs through Z, and v runs through $D(A_\beta)$ so that $Au = Av$ runs through H, hence $(Av)_{Z'}$ runs through Z'. For (13.86), cf. Lemma 13.2. For (13.87), note that $(Au)_{Z'} = 0$ implies $Au \in R(A_0)$, and $\mathrm{pr}_\zeta u = 0$ implies $u \in D(A_\beta)$, so $u = A_\beta^{-1} Au \in D(A_0)$. (The choice (13.88) is relevant for applications to elliptic PDE, cf. [BGW08].)

Following [BMNW08], the boundary triplet is used to define operators $A_B \in \mathcal{M}$ and $A_{B'}' \in \mathcal{M}'$ for any pair of operators $B \in \mathbf{B}(\mathcal{K}, \mathcal{H})$, $B' \in \mathbf{B}(\mathcal{H}, \mathcal{K})$ by

$$D(A_B) = \{u \in D(A_1) \mid \Gamma_1 = B\Gamma_0\}, \quad D(A_{B'}') = \{v \in D(A_1') \mid \Gamma_1' = B'\Gamma_0'\}. \quad (13.89)$$

We see that for the present choice of boundary triplet (13.88),

$$D(A_B) = \{u \in D(A_1) \mid (Au)_{Z'} = Bu_\zeta\}.$$

Thus the operator $T : V \to W$ that A_B corresponds to by Theorem 13.7 is

$$T = B, \text{ with } V = Z, \quad W = Z'.$$

This covers some particular cases of the operators appearing in Theorem 13.7. For one thing, V and W are the full spaces Z and Z'. Second, T is bounded. When Z and Z' are finite dimensional, all operators from Z to Z' will be bounded, but our theory allows $\dim Z = \dim Z' = \infty$, and then we must also allow unbounded T's, to cover general extensions. Therefore we in the following replace B by a closed, densely defined and possibly unbounded operator T from Z to Z', and define A_T, now called $\widetilde{A}$, by

$$D(\widetilde{A}) = \{u \in D(A_1) \mid u_\zeta \in D(T), (Au)_{Z'} = Tu_\zeta\}; \qquad (13.90)$$

this is consistent with the notation in Theorem 13.7.

In the discussion of resolvents via boundary triplets, a central object is the so-called M-function, originally introduced by Weyl and Titchmarsh in connection with Sturm-Liouville problems. We define (as an extension of the definition in [BMNW08] to unbounded B's):

Definition 13.27. For $\lambda \in \varrho(\widetilde{A})$, the M-function $M_{\widetilde{A}}(\lambda)$ is defined as a mapping from $R(\Gamma_1 - T\Gamma_0)$ to $\mathcal{K}$, by

$$M_{\widetilde{A}}(\lambda)(\Gamma_1 - T\Gamma_0)x = \Gamma_0 x \text{ for all } x \in Z_\lambda \text{ with } \Gamma_0 x \in D(T). \qquad (13.91)$$

In terms of (13.88), $M_{\widetilde{A}}(\lambda)$ is defined as a mapping from $R(\mathrm{pr}_{Z'} A_1 - T \mathrm{pr}_\zeta)$ to Z by

$$M_{\widetilde{A}}(\lambda)(\mathrm{pr}_{Z'} A_1 - T \mathrm{pr}_\zeta)x = \mathrm{pr}_\zeta x \text{ for all } x \in Z_\lambda \text{ with } \mathrm{pr}_\zeta x \in D(T). \quad (13.92)$$

There is the analogous definition of $M'_{\widetilde{A}'}(\lambda)$ in the adjoint setting.

We shall show that $M_{\widetilde{A}}(\lambda)$ is well-defined. First we observe:

Lemma 13.28. $R(\Gamma_1 - T\Gamma_0) = R(\mathrm{pr}_{Z'} A_1 - T \mathrm{pr}_\zeta)$ *is equal to* Z'. *In fact, any* $f \in Z'$ *can be written as*

$$f = (\mathrm{pr}_{Z'} A_1 - T \mathrm{pr}_\zeta)v = \mathrm{pr}_{Z'} A_1 v, \text{ for } v = A_\beta^{-1} f. \qquad (13.93)$$

Proof. Let v run through $D(A_\beta)$. Then $\Gamma_0 v = \mathrm{pr}_\zeta v = 0 \in D(T)$, and Av runs through H, so $\Gamma_1 v - T\Gamma_0 v = (Av)_{Z'}$ runs through Z'. In other words, for any $f \in Z'$ we can take $v = A_\beta^{-1} f$. $\qquad \square$

Proposition 13.29. *For any* $\lambda \in \varrho(\widetilde{A})$, $M_{\widetilde{A}}(\lambda)$ *is well-defined as a bounded map from* Z' *to* Z *by* (13.91) *or* (13.92); *and ranges in* $D(T)$. *In fact,*

$$M_{\widetilde{A}}(\lambda) = \mathrm{pr}_\zeta (I - (\widetilde{A} - \lambda)^{-1}(A_1 - \lambda))A_\beta^{-1}\, \mathrm{i}_{Z' \to H}\,. \qquad (13.94)$$

This is a holomorphic family of operators in $\mathbf{B}(Z',Z)$ *for* $\lambda \in \varrho(\widetilde{A})$.

Proof. The mapping $\Phi = \Gamma_1 - T\Gamma_0 = \mathrm{pr}_{Z'}\, A_1 - T\,\mathrm{pr}_\zeta$ is defined for those $u \in D(A_1)$ for which $\mathrm{pr}_\zeta\, u \in D(T)$. Let

$$Z_{\lambda,T} = \{ z^\lambda \in Z_\lambda \mid \mathrm{pr}_\zeta\, z^\lambda \in D(T) \},$$

then (13.92) means that $M_{\widetilde{A}}(\lambda)$ should satisfy

$$M_{\widetilde{A}}(\lambda)\Phi z^\lambda = \mathrm{pr}_\zeta\, z^\lambda \text{ for } z^\lambda \in Z_{\lambda,T}.$$

We first show that Φ maps $Z_{\lambda,T}$ bijectively onto Z', with inverse

$$\Psi = (I - (\widetilde{A} - \lambda)^{-1}(A_1 - \lambda))A_\beta^{-1}\, \mathrm{i}_{Z' \to H}\,. \qquad (13.95)$$

For the surjectiveness, let $f \in Z'$. Let $v = A_\beta^{-1}f$, then $f = \Phi v$ according to Lemma 13.28. Next, let $w = (\widetilde{A} - \lambda)^{-1}(A_1 - \lambda)v$, then since $w \in D(\widetilde{A})$, $\mathrm{pr}_\zeta\, w \in D(T)$ and

$$\Phi w = \mathrm{pr}_{Z'}\, A_1 w - T\,\mathrm{pr}_\zeta\, w = 0,$$

by (13.90). Let $z^\lambda = v - w$, it lies in Z_λ and has $\mathrm{pr}_\zeta\, z^\lambda = -\mathrm{pr}_\zeta\, w \in D(T)$, so $z^\lambda \in Z_{\lambda,T}$. It follows that

$$\Phi z^\lambda = \Phi v = f.$$

Thus Φ is indeed surjective from $Z_{\lambda,T}$ to Z'. It is also injective, for if $z^\lambda \in Z_{\lambda,T}$ is such that $\Phi z^\lambda = 0$, then, by (13.90), $z^\lambda \in D(\widetilde{A}) \cap Z_\lambda = \{0\}$ (recall that $\lambda \in \varrho(\widetilde{A})$). Following the steps in the construction of z^λ from f, we see that the inverse of $\Phi : Z_{\lambda,T} \to Z'$ is indeed given by (13.95).

We can now rewrite the defining equation (13.92) as

$$M_{\widetilde{A}}(\lambda)f = \mathrm{pr}_\zeta\, \Psi f \text{ for all } f \in Z'.$$

This shows that $M_{\widetilde{A}}(\lambda) = \mathrm{pr}_\zeta\, \Psi$ is the desired operator, and clearly (13.94) holds. Since $\widetilde{A}$ is closed, $M_{\widetilde{A}}(\lambda)$ is closed as a mapping from Z' to Z, hence continuous. $\qquad \square$

For a further analysis of $M_{\widetilde{A}}(\lambda)$, assume $\lambda \in \varrho(A_\beta) \cap \varrho(\widetilde{A})$. Then the maps E^λ, F^λ etc. in (13.73) are defined. Let $z^\lambda \in Z_\lambda$, and consider the defining equation

$$M_{\widetilde{A}}(\lambda)((Az^\lambda)_{Z'} - T\,\mathrm{pr}_\zeta\, z^\lambda) = \mathrm{pr}_\zeta\, z^\lambda, \qquad (13.96)$$

where $\mathrm{pr}_\zeta\, z^\lambda$ is required to lie in $D(T)$. By (13.74) and (13.75), there is a unique $z \in Z$ such that $z^\lambda = E_Z^\lambda z$; in fact

$$z^\lambda = E_Z^\lambda z = \mathrm{pr}_\zeta^\lambda z; \quad z = F_Z^\lambda z^\lambda = \mathrm{pr}_\zeta z^\lambda,$$

so the requirement is that $z^\lambda \in E_Z^\lambda D(T)$.

Writing (13.96) in terms of z, and using that $Az^\lambda = \lambda z^\lambda$, we find

$$
\begin{aligned}
z = \mathrm{pr}_\zeta z^\lambda &= M_{\widetilde{A}}(\lambda)((Az^\lambda)_{Z'} - Tz) = M_{\widetilde{A}}(\lambda)((\lambda z^\lambda)_{Z'} - Tz) \\
&= M_{\widetilde{A}}(\lambda)((\lambda E^\lambda z)_{Z'} - Tz) = M_{\widetilde{A}}(\lambda)(-G^\lambda - T)z,
\end{aligned}
\tag{13.97}
$$

cf. (13.78). Here we observe that the operator to the right of $M_{\widetilde{A}}(\lambda)$ equals T^λ from Section 13.3 up to homeomorphisms:

$$-G^\lambda - T = -(E_{Z'}^{\prime\lambda})^* T^\lambda E_Z^\lambda, \tag{13.98}$$

by (13.80); here T^λ is invertible from $E_Z^\lambda D(T)$ onto Z'_λ. We conclude that $M_{\widetilde{A}}(\lambda)$ is the inverse of the operator in (13.98). We have shown:

Theorem 13.30. *Let $\widetilde{A}$ be defined by (13.90), where $T : Z \to Z'$. When the boundary triplet is chosen as in (13.88) and $\lambda \in \varrho(\widetilde{A}) \cap \varrho(A_\beta)$, $-M_{\widetilde{A}}(\lambda)$ equals the inverse of $T + G^\lambda$, also equal to the inverse of T^λ modulo homeomorphisms:*

$$-M_{\widetilde{A}}(\lambda)^{-1} = T + G^\lambda = (E_{Z'}^{\prime\lambda})^* T^\lambda E_Z^\lambda. \tag{13.99}$$

In particular, $M_{\widetilde{A}}(\lambda)$ has range $D(T)$.

With this insight we have access to the straightforward resolvent formula (13.84), which implies in this case:

Corollary 13.31. *For $\lambda \in \varrho(\widetilde{A}) \cap \varrho(A_\beta)$,*

$$(\widetilde{A} - \lambda)^{-1} = (A_\beta - \lambda)^{-1} - \mathrm{i}_{Z_\lambda \to H}\, E_Z^\lambda M_{\widetilde{A}}(\lambda)(E_{Z'}^{\prime\lambda})^*\, \mathrm{pr}_{Z'_\lambda}. \tag{13.100}$$

We also have the direct link between nullspaces and ranges (13.71), when merely $\lambda \in \varrho(A_\beta)$.

Corollary 13.32. *For any $\lambda \in \varrho(A_\beta)$,*

$$
\begin{aligned}
Z(\widetilde{A} - \lambda) &= E_Z^\lambda Z(T + G^\lambda), \\
R(\widetilde{A} - \lambda) &= (F_{Z'}^{\prime\lambda})^* R(T + G^\lambda) + R(A_0 - \lambda).
\end{aligned}
\tag{13.101}
$$

There is a result in [BMNW08] on the relation between poles of $M_{\widetilde{A}}(\lambda)$ and eigenvalues of $\widetilde{A}$; for the points in $\varrho(A_\beta)$, Corollary 13.32 is more informative.

The analysis moreover implies that $M_{\widetilde{A}}(\lambda)$ and $M'_{\widetilde{A}_*}(\bar{\lambda})$ are adjoints, at least when $\lambda \in \varrho(A_\beta)$.

Note that T^λ is well-defined for all $\lambda \in \varrho(A_\beta)$, whereas $M_{\widetilde{A}}(\lambda)$ is well-defined for all $\lambda \in \varrho(\widetilde{A})$; the latter fact is useful for other purposes. In this way, the two operator families supply each other.

So far, we have only discussed M-functions for elements of $\mathcal{M}$ with $V = Z$, $W = Z'$ in Theorem 13.7. But, inspired by the result in Theorem 13.30, we can in fact establish useful M-functions for all closed $\widetilde{A}$. They will be homeomorphic to the inverses of the operators T^λ that exist for $\lambda \in \varrho(\widetilde{A}) \cap \varrho(A_\beta)$, and extend to exist for all $\lambda \in \varrho(\widetilde{A})$.

Theorem 13.33. *Let $\widetilde{A}$ be an arbitrary closed densely defined operator between A_0 and A_1, and let $T : V \to W$ be the corresponding operator according to Theorem 13.7. For any $\lambda \in \varrho(\widetilde{A})$ there is a bounded operator $M_{\widetilde{A}}(\lambda) : W \to V$, depending holomorphically on $\lambda \in \varrho(\widetilde{A})$, such that when $\lambda \in \varrho(A_\beta)$, $-M_{\widetilde{A}}(\lambda)$ is the inverse of $T + G^\lambda_{V,W}$, and is homeomorphic to T^λ (as defined in Section 13.3). It satisfies*

$$M_{\widetilde{A}}(\lambda)\big((Az^\lambda)_W - T\operatorname{pr}_\zeta z^\lambda\big) = \operatorname{pr}_\zeta z^\lambda, \qquad (13.102)$$

for all $z^\lambda \in Z_\lambda$ such that $\operatorname{pr}_\zeta z^\lambda \in D(T)$. Its definition extends to all $\lambda \in \varrho(\widetilde{A})$ by the formula

$$M_{\widetilde{A}}(\lambda) = \operatorname{pr}_\zeta\big(I - (\widetilde{A} - \lambda)^{-1}(A_1 - \lambda)\big)A_\beta^{-1}\,i_{W \to H}. \qquad (13.103)$$

In particular, the Kreĭn resolvent formula

$$(\widetilde{A} - \lambda)^{-1} = (A_\beta - \lambda)^{-1} - i_{V_\lambda \to H}\,E^\lambda_V M_{\widetilde{A}}(\lambda)(E'^{\bar\lambda}_W)^*\,\operatorname{pr}_{W_{\bar\lambda}} \qquad (13.104)$$

holds when $\lambda \in \varrho(\widetilde{A}) \cap \varrho(A_\beta)$. For all $\lambda \in \varrho(A_\beta)$,

$$
\begin{aligned}
Z(\widetilde{A} - \lambda) &= E^\lambda_V Z(T + G^\lambda_{V,W}), \\
R(\widetilde{A} - \lambda) &= (F'^\lambda_W)^* R(T + G^\lambda_{V,W}) + H \ominus W_{\bar\lambda},
\end{aligned}
\qquad (13.105)
$$

where $W_{\bar\lambda} = E'^{\bar\lambda}W$.

Proof. Following the lines of proofs of Lemma 13.28 and Proposition 13.29, we define $M_{\widetilde{A}}(\lambda)$ satisfying (13.102) as follows: Let $f \in W$. Let $v = A_\beta^{-1}f$; then $\operatorname{pr}_\zeta v = 0 \in D(T)$, and

$$(Av)_W - T\operatorname{pr}_\zeta v = Av = f. \qquad (13.106)$$

Next, let $w = (\widetilde{A} - \lambda)^{-1}(A - \lambda)v$, then $z^\lambda = v - w$ lies in Z_λ and satisfies $\operatorname{pr}_\zeta z^\lambda \in D(T)$ (since $\operatorname{pr}_\zeta v = 0$ and $\operatorname{pr}_\zeta w \in D(T)$). This z^λ satisfies

$$(Az^\lambda)_W - T\operatorname{pr}_\zeta z^\lambda = f, \qquad (13.107)$$

in view of (13.106) and the fact that $w \in D(\widetilde{A})$.

Next, observe that for any vector $z^\lambda \in Z_\lambda$ with $\operatorname{pr}_\zeta z^\lambda \in D(T)$ such that (13.107) holds, $f = 0$ implies $z^\lambda = 0$, since such a z^λ lies in the two linearly independent spaces $D(\widetilde{A} - \lambda)$ and Z_λ. So there is indeed a mapping Ψ from

f to $\mathrm{pr}_\zeta z^\lambda$ solving (13.107), for any $f \in W$. Then $M_{\widetilde{A}}(\lambda)$ is the composition $\mathrm{pr}_\zeta \Psi$; it is described by (13.103). The holomorphicity in $\lambda \in \varrho(\widetilde{A})$ is seen from this formula.

The mapping connects with T^λ (cf. Theorem 13.23) as follows:

When $\lambda \in \varrho(\widetilde{A}) \cap \varrho(A_\beta)$, then $z = \mathrm{pr}_\zeta z^\lambda = F_V^\lambda z^\lambda$, and $z^\lambda = E_V^\lambda z$, so the vectors z^λ with $\mathrm{pr}_\zeta z^\lambda \in D(T)$ constitute the space $E_V^\lambda D(T)$. Calculating as in (13.97) we then find that

$$
\begin{aligned}
z = \mathrm{pr}_\zeta z^\lambda &= M_{\widetilde{A}}(\lambda)((Az^\lambda)_W - Tz) = M_{\widetilde{A}}(\lambda)((\lambda z^\lambda)_W - Tz) \\
&= M_{\widetilde{A}}(\lambda)((\lambda E^\lambda z)_W - Tz) = M_{\widetilde{A}}(\lambda)(-G_{V,W}^\lambda - T)z,
\end{aligned}
$$

so $M_{\widetilde{A}}(\lambda)$ is the inverse of $-(T + G_{V,W}^\lambda) : D(T) \to W$. The remaining statements follow from Theorem 13.23 and Corollary 13.26.   □

Note that when $M_{\widetilde{A}}(\lambda)$ is considered in a neighborhood of a spectral point of $\widetilde{A}$ in $\varrho(A_\beta)$, then we have not only information on the possibility of a pole of $M_{\widetilde{A}}(\lambda)$, but we have an inverse T^λ, from which $Z(\widetilde{A} - \lambda)$ and $R(\widetilde{A} - \lambda)$ can be read off.

An application of these concepts to elliptic PDE is given in [BGW08].

Exercises for Chapter 13

13.1.   Work out the statement and proof of Theorem 13.8 3° in detail.

13.2.   Show the assertions in Example 13.11 concerning the case $V = W = \mathrm{span}\, z_1$.

13.3.   Work out the details of Example 13.11 in the case where $\alpha = 0$, $\beta = 1$, $q = 0$.

Consider the realization $\widetilde{A}$ defined by the Neumann-type condition $u'(0) = b_1 u(0)$, $-u'(1) = b_2 u(1)$. Show that $\widetilde{A}$ is selfadjoint positive if and only if

$$
b_1 > -1, \quad b_2 > -1 + (b_1 + 1)^{-1}.
$$

13.4.   Work out the details of Example 13.11 in the case where $\alpha = 0$, $\beta = 1$, $q = 1$.

For the realization $\widetilde{A}$ defined by the Neumann condition $u'(0) = b_1 u(0)$, $-u'(1) = b_2 u(1)$, find the choices of b_1, b_2 for which $\widetilde{A}$ is selfadjoint positive.

13.5.   Under the assumption of (13.5), assume that $\widetilde{A}$ corresponds to $T : V \to W$ as in Theorem 13.7.

(a) Show that $\widetilde{A}$ is symmetric if and only if $V \subset W$ and $T : V \to W$ is symmetric as an operator in W.

(b) Show that $\widetilde{A}$ is maximal symmetric if and only if $V \subset W$ and $T : V \to W$ is maximal symmetric as an operator in W.

In the next exercises, it is checked how the theory in this chapter looks for a very simple example.

13.6. (Notation from Chapter 4 is used here.) Consider the differential operator $Au = -u'' + \alpha^2 u$ on $\mathbb{R}_+$, where α is a positive constant. With $A_{\max}$, $A_{\min}$ and A_γ defined as in Exercise 4.14, denote $A_{\max} = A_1$ and $A_{\min} = A_0$.

(a) Show that the hypotheses of Sections 13.1 and 13.2 are satisfied for these operators in $H = L_2(\mathbb{R}_+)$, as a symmetric case, with A_β equal to the Dirichlet realization A_γ.

(b) Show that Z is one-dimensional, spanned by the vector $e^{-\alpha t}$.

(c) Show that $A_\gamma^{-1} e^{-\alpha t} = \frac{1}{2\alpha} t e^{-\alpha t}$.

(d) Show that the choice $V = W = 0$, $T = 0$ (necessarily), corresponds to the operator $\widetilde{A} = A_\gamma$.

(e) From now on we let $V = W = Z$, and let T be the multiplication by a complex number τ. For the corresponding operator $\widetilde{A}$, describe the elements of $D(\widetilde{A})$ by use of the formula (13.29).

(f) Show that $\widetilde{A}$ is a realization of A determined by a boundary condition $u'(0) = bu(0)$, where

$$\tau = 2\alpha(b + \alpha).$$

Which choice of τ gives the Neumann condition $u'(0) = 0$?

13.7. (Continuation of Exercise 13.6.)

(a) Show that the numerical range of T is the point

$$\nu(T) = \{2\alpha(b + \alpha)\}.$$

(b) Show that $m(\widetilde{A}) \geq 0$ if and only if $\text{Re}(b + \alpha) \geq 0$.

(c) Let $\text{Im}\, b > 0$, and describe the convex hull M of the set

$$[\alpha^2, \infty[\, \cup \{2\alpha(b + \alpha)\}.$$

Using the result of Exercise 12.34(e), show that the closure of the numerical range of $\widetilde{A}$ contains M.

(d) Show that if $b = -\alpha$, $\widetilde{A}$ is not bijective.

13.8. (Continuation of Exercise 13.6.) The next point is to identify the operators from Sections 13.3 and 13.4 for the present example. Let $\lambda \in \mathbb{C} \backslash \overline{\mathbb{R}}_+$, and let

$$\sigma = (\alpha^2 - \lambda)^{\frac{1}{2}}, \quad \sigma_1 = (\alpha^2 - \bar{\lambda})^{\frac{1}{2}},$$

13 Families of extensions

where the square root is taken to be holomorphic on $\mathbb{C} \setminus \overline{\mathbb{R}_-}$ and positive on $\mathbb{R}_+$. Consider now $\widetilde{A} - \lambda$, where $\widetilde{A}$ is the realization of A defined in Exercise 13.6 by the boundary condition $u'(0) = bu(0)$.

(a) Show that $\sigma_1 = \bar{\sigma}$, and that $\operatorname{Re} \sigma > 0$.

(b) Show that $Z_\lambda = \operatorname{span}(e^{-\sigma t})$, and that $Z'_{\bar{\lambda}} = \operatorname{span}(e^{-\sigma_1 t})$.

(c) Show that when $\operatorname{Im} \lambda \neq 0$,

$$(A_\gamma - \lambda)^{-1} e^{-\sigma_1 t} = (\lambda - \bar{\lambda})^{-1}(e^{-\sigma t} - e^{-\sigma_1 t}).$$

(d) For $\operatorname{Im} \lambda > 0$, find the τ^λ such that the mapping $T^\lambda : Z_\lambda \to Z'_{\bar{\lambda}}$, defined by sending $e^{-\sigma t}$ over into $\tau^\lambda e^{-\sigma_1 t}$, is the operator corresponding to $\widetilde{A} - \lambda$ by Theorem 13.23.

13.9. (Continuation of Exercise 13.6.)

(a) Show that the mapping $E_Z^\lambda : Z \to Z_\lambda$ sends $e^{-\alpha t}$ over into $e^{-\sigma t}$. Similarly, $E_{Z'}^{\prime \bar{\lambda}} : Z \to Z'_{\bar{\lambda}}$ sends $e^{-\alpha t}$ over into $e^{-\sigma_1 t}$.

(b) Show that the adjoint $(E_{Z'}^{\prime \bar{\lambda}})^* : Z'_{\bar{\lambda}} \to Z$ sends $e^{-\sigma_1 t}$ over into $\alpha(\operatorname{Re} \sigma)^{-1} e^{-\alpha t}$.

(c) Show that $G^\lambda : Z \to Z$ maps $e^{-\alpha t}$ into $g^\lambda e^{-\alpha t}$, where

$$g^\lambda = -2\alpha\lambda(\alpha + \sigma)^{-1} = 2\alpha(\sigma - \alpha).$$

(*Hint.* For the last equality, use that $(\sigma - \alpha)(\sigma + \alpha) = \sigma^2 - \alpha^2 = -\lambda$.)

(d) Conclude that $T + G^\lambda$ is the multiplication by $2\alpha(b + \sigma)$.

(e) Find $M_{\widetilde{A}}(\lambda)$.

(*Comment.* The multiplication by $-\sigma$ can be regarded as the Dirichlet-to-Neumann operator for $A - \lambda$, since it maps $u(0)$ to $u'(0)$ for solutions of $(A - \lambda)u = 0$; in particular, $-\alpha$ is the Dirichlet-to-Neumann operator for A. Point (d) shows that the subtraction of $-\alpha$ is replaced by the subtraction of $-\sigma$ in the passage from T to $T + G^\lambda$.)

13.10. (Continuation of Exercise 13.6.)

Show that when $b = ir$ with $r > 0$, then

$$\mathbb{C} \setminus [\alpha^2, \infty[\subset \varrho(\widetilde{A}),$$

and yet the closure of the numerical range of $\widetilde{A}$ contains the set

$$\{x + iy \mid x, y \in \mathbb{R}, \ x \geq 2\alpha^2, 0 \leq y \leq 2\alpha r\}.$$

(*Comment.* This shows an interesting case of an operator that has its spectrum contained in $[\alpha^2, \infty[$ but is far from being selfadjoint.)

Chapter 14
Semigroups of operators

14.1 Evolution equations

The investigations in Chapter 13 are designed particularly for the concretization of elliptic operators, in terms of boundary conditions. But they have a wider applicability. In fact, operators with suitable semiboundedness properties are useful also in the concretization of parabolic or hyperbolic problems, where there is a first- or second-order time derivative in addition to the elliptic operator. We present in the following a basic method for the discussion of such time-dependent problems. More refined methods, building on microlocal techniques (using not only pseudodifferential operators but also Fourier integral operators and wave front sets), have been introduced since the time this method was worked out, but we still think that it can have an interest as a first introduction to time-dependent equations.

The Laplace operator $\Delta = \partial_{x_1}^2 + \cdots + \partial_{x_n}^2$ is used in physics, e.g., in the equation for a potential field u in an open subset of $\mathbb{R}^3$

$$\Delta u(x) = 0 \text{ for } x \in \Omega . \tag{14.1}$$

It enters together with a time parameter t, e.g., in the *heat equation*

$$\partial_t u(x,t) - \Delta_x u(x,t) = 0 \text{ for } x \in \Omega \text{ and } t \geq 0 , \tag{14.2}$$

the *Schrödinger equation*

$$\frac{1}{i}\partial_t u(x,t) - \Delta_x u(x,t) = 0 \text{ for } x \in \Omega \text{ and } t \in \mathbb{R} , \tag{14.3}$$

and the *wave equation*

$$\partial_t^2 u(x,t) - \Delta_x u(x,t) = 0 \text{ for } x \in \Omega \text{ and } t \in \mathbb{R} . \tag{14.4}$$

The last three equations have the common property that they can formally be considered as equations of the form

$$\partial_t u(t) = Bu(t) ,\qquad(14.5)$$

where $t \mapsto u(t)$ is a function from the time axis into a space of functions of x, where B operates. For (14.2), B acts like Δ, for (14.3) like $i\Delta$. For (14.4) we obtain the form (14.5) by introducing the vector

$$v(t) = \begin{pmatrix} u(t) \\ \partial_t u(t) \end{pmatrix} ,$$

which must satisfy

$$\partial_t v(t) = Bv(t) ,\quad \text{with}\ B = \begin{pmatrix} 0 & I \\ \Delta & 0 \end{pmatrix} .\qquad(14.6)$$

The equation (14.5) is called an *evolution equation*. We get a very simple version in the case where u takes its values in $\mathbb{C}$, and B is just multiplication by the constant $\lambda \in \mathbb{C}$

$$\partial_t u(t) = \lambda u(t) .$$

It is well-known that the solutions of this equation are

$$u(t) = e^{t\lambda} c ,\quad c \in \mathbb{C}.\qquad(14.7)$$

We shall show in the following how the abstract functional analysis allows us to define similar solutions $\exp(tB)u_0$, when λ is replaced by an operator B in a Banach space X, under suitable hypotheses.

In preparation for this, we need to consider Banach space valued functions $v(t)$. Let $v : I \to X$, where I is an interval of $\mathbb{R}$ and X is Banach space. The function $v(t)$ is said to be *continuous at* $t_0 \in I$, when $v(t) \to v(t_0)$ in X for $t \to t_0$ in I. In details, this means: For any $\varepsilon > 0$ there is a $\delta > 0$ such that $|t - t_0| < \delta$ implies $\|v(t) - v(t_0)\| < \varepsilon$. When continuity holds for all $t_0 \in I$, v is said to be continuous on I.

The function $v : I \to X$ is said to be *differentiable* at t, if $\lim_{h \to 0} \frac{1}{h}(v(t + h) - v(t))$ exists in X (for t and $t + h \in I$); the limit is denoted $v'(t)$ (or $\partial_t v$ or $\frac{dv}{dt}$). More precisely, one says here that $v(t)$ is *norm differentiable* or *strongly differentiable*, to distinguish this property from the property of being *weakly differentiable*. The latter means that for any continuous linear functional $x^* \in X^*$, the function

$$f_{x^*}(t) = x^*(v(t))$$

is differentiable, in such a way that there exists a $v'(t) \in X$ so that $\partial_t f_{x^*}(t) = x^*(v'(t))$ for all x^*. A norm differentiable function is clearly weakly differentiable, with the same derivative $v'(t)$, but there exist weakly differentiable functions that are not norm differentiable.

The integral of a continuous vector function $v : I \to X$ is defined by the help of middlesums precisely as for real or complex functions (here $\mathbb{R}$ or $\mathbb{C}$ is replaced by X, the modulus is replaced by the norm). One has the usual rules (where the proofs are straightforward generalizations of the proofs for real functions):

$$\int_a^b (\alpha v(t) + \beta w(t))\, dt = \alpha \int_a^b v(t)\, dt + \beta \int_a^b w(t)\, dt\ ,$$

$$\int_a^b v(t)\, dt + \int_b^c v(t)\, dt = \int_a^c v(t)\, dt\ ,$$

for arbitrary points a, b and c in I, and

$$\| \int_a^b v(t)\, dt \|_X \leq \int_a^b \|v(t)\|_X\, dt \quad \text{when}\ \ a \leq b\ . \tag{14.8}$$

Moreover, $\int_a^t v(s)\, ds$ is differentiable, with

$$\frac{d}{dt} \int_a^t v(s)\, ds = v(t). \tag{14.9}$$

Like for integrals of complex functions there is not a genuine mean value theorem, but one does have that

$$\frac{1}{h} \int_c^{c+h} v(t)\, dt \to v(c) \quad \text{in}\ \ X \quad \text{for}\ \ h \to 0\ . \tag{14.10}$$

When the vector function $v : I \to X$ is differentiable with a continuous derivative $v'(t)$, one has the identity

$$\int_a^b v'(t)\, dt = v(b) - v(a) \tag{14.11}$$

(since the corresponding identity holds for all functions $f_{x^*}(t) = x^*(v(t))$).

We now return to the possible generalizations of (14.7) to functions valued in a Banach space X. The easiest case is where B is a *bounded* operator on X. Here we can simply put

$$\exp(tB) = \sum_{n \in \mathbb{N}_0} \frac{1}{n!}(tB)^n \tag{14.12}$$

for all $t \in \mathbb{R}$, since the series converges (absolutely) in the operator norm to a bounded operator; this is seen e.g. by noting that

$$\| \sum_{N \leq n < N'} \frac{1}{n!}(tB)^n \| \leq \sum_{N \leq n < N'} \frac{1}{n!}|t|^n \|B\|^n \, ,$$

where the right-hand side is a partial sum in the convergent series

$$\exp(|t|\|B\|) = \sum_{n \in \mathbb{N}_0} \frac{1}{n!}(|t|\|B\|)^n.$$

It follows in particular that $\| \exp(tB) \| \leq \exp(|t|\|B\|)$.

The operator family satisfies, for $s, t \in \mathbb{R}$,

$$\exp(sB)\exp(tB) = \sum_{n=0}^{\infty} \frac{(sB)^n}{n!} \sum_{m=0}^{\infty} \frac{(tB)^m}{m!} = \sum_{l=0}^{\infty} \sum_{n+m=l} \frac{s^n t^m}{n!m!}B^l \qquad (14.13)$$

$$= \sum_{l=0}^{\infty} \frac{(s+t)^l}{l!}B^l = \exp((s+t)B) \, ,$$

where we have used the corresponding identity for the exponential series; the reorganization is allowed since the norms of the terms form a convergent series. Moreover, we have for $C \geq s \geq t \geq 0$ that

$$\| \exp(sB) - \exp(tB) \| = \| \sum_{n=0}^{\infty} \frac{s^n - t^n}{n!}B^n \| \leq \sum_{n=0}^{\infty} \frac{s^n - t^n}{n!}\|B\|^n$$

$$= \exp(s\|B\|) - \exp(t\|B\|) \to 0 \ \text{ for } \ s - t \to 0 \, ,$$

$$\| \exp(-sB) - \exp(-tB) \| = \| \exp(-tB)(\exp(tB) - \exp(sB))\exp(-sB) \|$$

$$\leq \exp(|t|\|B\|)\exp(|s|\|B\|)\| \exp(tB) - \exp(sB) \|$$

$$\to 0 \ \text{ for } \ (-s) - (-t) \to 0 \, ,$$

which shows that the operator family is continuous in t with respect to the operator norm.

That the operator function $\exp(tB)$ is differentiable with derivative $B \exp(tB)$ can for example be seen as follows. Integration of the continuous function $\exp(tB)$ and composition with B gives

$$B \int_0^t \exp(sB) \, ds = B \int_0^t \sum_{n \in \mathbb{N}_0} \frac{1}{n!}(sB)^n \, ds = \sum_{n \in \mathbb{N}_0} \frac{1}{n!}B^{n+1} \int_0^t s^n \, ds$$

$$= \sum_{n \in \mathbb{N}_0} \frac{1}{(n+1)!}t^{n+1}B^{n+1} = \exp(tB) - I,$$

from which it follows by differentiation of both sides that

$$B \exp(tB) = \frac{d}{dt}\exp(tB). \qquad (14.14)$$

(We could interchange integration and summation in view of the majorized convergence.)

Exponential functions can also be set up for suitable unbounded operators B in X, more on this below. If X is in particular a Hilbert space, one can define the exponential function via spectral theory if B is a selfadjoint or normal operator. This is used e.g. in the following cases (that we mention without proof to begin with):

$1°$ When B is *selfadjoint* and *upper semibounded*, $\exp(tB)$ is well-defined for $t \geq 0$.

$2°$ When B is *skew-selfadjoint*, i.e., $B^* = -B$, $\exp(tB)$ is well-defined for $t \in \mathbb{R}$.

The case $1°$ is relevant for the heat equation on $\mathbb{R}^n$, since $-\Delta$ can be given a sense as a selfadjoint unbounded operator $-B \geq 0$ in $L_2(\mathbb{R}^n)$. The case $2°$ then pertains to the Schrödinger equation, since $i\Delta$ hereby becomes skew-selfadjoint in $L_2(\mathbb{R}^n)$.

For the wave equation in the form (14.6), the skew-selfadjointness can be obtained by use of some other particular Hilbert spaces. If one wants to work with the wave equation in $L_2(\mathbb{R}^n)$, one can instead interpret the solutions of the abstract equation

$$\partial_t^2 u = -Au \tag{14.15}$$

as combinations of the solutions $\cos(tA^{\frac{1}{2}})u_0$ and $A^{-\frac{1}{2}} \sin(tA^{\frac{1}{2}})u_1$ (where $A = -\Delta$ is ≥ 0); also this can be achieved by use of spectral theory.

The spectral theory that is needed here is an extension of the standard theory to unbounded operators.

We now turn to a more general definition of the exponential function, not requiring selfadjointness or normality of the operator B, namely, the theory of semigroups of operators. It also covers the cases $1°$ and $2°$.

14.2 Contraction semigroups in Banach spaces

The following account builds on Appendix 1 in the book of Lax and Phillips [LP67].

A *semigroup* of operators in a Banach space X is a family of operators $G(t) \in \mathbf{B}(X)$, parametrized by $t \in \overline{\mathbb{R}}_+$ and satisfying

(a) $G(0) = I$, and $G(s+t) = G(s)G(t)$ for all s and $t \geq 0$.

A *group* of operators is a family of operators $G(t) \in \mathbf{B}(X)$ parametrized by $t \in \mathbb{R}$ and such that (a) holds for all s and $t \in \mathbb{R}$. Here all the operators are invertible, since the second condition implies $G(t)^{-1} = G(-t)$. Note that both $G(t)$ and $G(-t)$ are semigroups for $t \geq 0$.

Note that we have both for semigroups and groups that $G(s)G(t) = G(t)G(s)$.

The semigroups and groups we consider will furthermore be required to satisfy the following condition:

$$\text{(b)} \quad G(t)x \to x \text{ for } t \to 0+, \text{ for all } x \in X .$$

There are other special conditions imposed, that lead to various classes of semigroups, see the comprehensive treatise of Hille and Phillips [HP57]. For the present purposes it will suffice to consider semigroups (and groups) of *contractions*; they are the ones that in addition satisfy

$$\text{(c)} \quad \|G(t)\| \le 1 \text{ for all } t .$$

Lemma 14.1. *When $G(t)$ satisfies (a)–(c), the map $t \mapsto G(t)x$ is for any $x \in X$ a continuous function from $\overline{\mathbb{R}}_+$ to X (resp. from $\mathbb{R}$ to X in case of a group).*

Proof. By (a) and (c) we have for $t_1 \le t_2$ that

$$\|G(t_2)x - G(t_1)x\| = \|G(t_1)(G(t_2 - t_1)x - x)\| \le \|G(t_2 - t_1)x - x\| .$$

If we let t_1 and t_2 converge to $t_0 \in [t_1, t_2]$, the expression goes to 0 according to (b).   □

The property in Lemma 14.1 is called strong continuity. We can hereafter call the (semi)groups which satisfy (a), (b) and (c) the *strongly continuous contraction (semi)groups*. (For strongly continuous semigroups in general one requires (a) and strong continuity, then (b) follows.)

The *infinitesimal generator B* is now introduced as the operator defined by

$$Bx = \lim_{h \to 0} \frac{1}{h}(G(h) - I)x , \tag{14.16}$$

with $D(B)$ consisting of those x for which the limit exists.

The following theorem shows that the vector function $u(t) = G(t)x$ satisfies the differential equation (14.5) when $x \in D(B)$.

Theorem 14.2. *For $x \in D(B)$, the function $G(t)x : \overline{\mathbb{R}}_+ \to X$ is differentiable, and takes its values in $D(B)$:*

$$\lim_{h \to 0} \frac{1}{h}(G(t+h)x - G(t)x) = G(t)Bx = BG(t)x \text{ for all } t \ge 0 . \tag{14.17}$$

Proof. When $h > 0$,

$$\frac{1}{h}(G(t+h)x - G(t)x) = G(t)\frac{G(h) - I}{h}x = \frac{G(h) - I}{h}G(t)x .$$

When $x \in D(B)$, the expression in the middle converges to $G(t)Bx$ for $h \to 0+$. This shows that $G(t)D(B) \subset D(B)$, and that (14.17) holds, when $h \to 0$ is replaced by $h \to 0+$. If $t > 0$, we must also investigate the passage to the limit through negative h; here we use that

$$\frac{1}{h}(G(t+h)x - G(t)x) - G(t)Bx = G(t+h)\frac{G(-h) - I}{-h}x - G(t)Bx$$

$$= G(t+h)\left(\frac{G(-h) - I}{-h}x - Bx\right) + (G(t+h) - G(t))Bx .$$

For a given ε we first choose $\delta > 0$ such that $\|\frac{1}{-h}(G(-h) - I)x - Bx\| < \varepsilon$ for $-\delta \leq h < 0$. Then the first term is $< \varepsilon$ because of (c); next we can choose $0 < \delta' \leq \min\{\delta, t\}$ so that the second term is $< \varepsilon$ for $-\delta' \leq h < 0$, in view of Lemma 14.1. $\qquad \square$

Corollary 14.3. *For $x \in D(B)$,*

$$G(t)x - x = \int_0^t G(s)Bx\, ds . \qquad (14.18)$$

This follows from the general property (14.11). We can also show:

Lemma 14.4. *For all $x \in X$, $t > 0$, $\int_0^t G(s)x\, ds$ belongs to $D(B)$ and*

$$G(t)x - x = B \int_0^t G(s)x\, ds . \qquad (14.19)$$

Proof. It follows from the continuity and the semigroup property (a) that for $h > 0$:

$$\frac{G(h) - I}{h}\int_0^t G(s)x\, dx = \frac{1}{h}\int_0^t (G(s+h) - G(s))x\, ds$$

$$= \frac{1}{h}\int_h^{t+h} G(s)x\, ds - \frac{1}{h}\int_0^t G(s)x\, ds$$

$$= \frac{1}{h}\int_t^{t+h} G(s)x\, ds - \frac{1}{h}\int_0^h G(s)x\, ds,$$

which converges to $G(t)x - x$ for $h \to 0$, by (14.10). $\qquad \square$

Lemma 14.5. *B is closed and densely defined.*

Proof. According to Lemma 14.4, $\frac{1}{h}\int_0^h G(s)x\, ds \in D(B)$ for all $x \in X$, $h > 0$, so since this converges to x for $h \to 0$, $D(B)$ is dense in X. Now if $x_n \in D(B)$ with $x_n \to x$ and $Bx_n \to y$, then $G(s)Bx_n \to G(s)y$ *uniformly in s* (by (c)), such that we have for any $h > 0$ that

$$\frac{1}{h}(G(h)x - x) = \lim_{n \to \infty} \frac{1}{h}(G(h)x_n - x_n)$$

$$= \lim_{n \to \infty} \frac{1}{h} \int_0^h G(s)Bx_n ds = \frac{1}{h} \int_0^h G(s)y\, ds \;,$$

by Corollary 14.3. However, $\frac{1}{h} \int_0^h G(s)y\, ds \to y$ for $h \to 0$, from which we conclude that $x \in D(B)$ with $Bx = y$. $\square$

Lemma 14.6. *A contraction semigroup is uniquely determined from its infinitesimal generator.*

Proof. Assume that $G_1(t)$ and $G_2(t)$ have the same infinitesimal generator B. For $x \in D(B)$, $G_2(t)x \in D(B)$, and we have by a generalization of the Leibniz formula:

$$\frac{d}{ds}G_1(t-s)G_2(s)x = -G_1(t-s)BG_2(s)x + G_1(t-s)G_2(s)Bx = 0 \;.$$

Integration over intervals $[0,t]$ gives

$$G_1(0)G_2(t)x - G_1(t)G_2(0)x = 0 \;,$$

hence $G_1(t)x = G_2(t)x$ for $x \in D(B)$. Since $D(B)$ is dense in X, and these operators are bounded, we conclude that $G_1(t) = G_2(t)$. $\square$

We can now show an important property of B, namely, that the half-plane $\{\lambda \in \mathbb{C} \mid \operatorname{Re}\lambda > 0\}$ lies in the resolvent set. Moreover, the resolvent can be obtained directly from the semigroup, and it satisfies a convenient norm estimate.

Theorem 14.7. *Let $G(t)$ be a strongly continuous contraction semigroup with infinitesimal generator B. Any λ with $\operatorname{Re}\lambda > 0$ belongs to the resolvent set $\rho(B)$, and then*

$$(B - \lambda I)^{-1}x = -\int_0^\infty e^{-\lambda t}G(t)x\, dt \quad for \; x \in X \;, \tag{14.20}$$

with

$$\|(B - \lambda I)^{-1}\| \le (\operatorname{Re}\lambda)^{-1} \;. \tag{14.21}$$

Proof. Let $\operatorname{Re}\lambda > 0$. Note first that $e^{-\lambda t}G(t)$ is a strongly continuous contraction semigroup with infinitesimal generator $B - \lambda I$. An application of Corollary 14.3 and Lemma 14.4 to this semigroup gives that

$$e^{-\lambda s}G(s)x - x = (B - \lambda I)\int_0^s e^{-\lambda t}G(t)x\, dt \quad for \; x \in X \;, \tag{14.22}$$

$$e^{-\lambda s}G(s)x - x = \int_0^s e^{-\lambda t}G(t)(B - \lambda I)x\, dt \quad for \; x \in D(B) \;. \tag{14.23}$$

For any $y \in X$ we have that $\|e^{-\lambda t}G(t)y\| \leq e^{-\mathrm{Re}\,\lambda t}\|y\|$; therefore the limit

$$Ty = \lim_{s \to \infty} \int_0^s e^{-\lambda t}G(t)y\,dt = \int_0^\infty e^{-\lambda t}G(t)y\,dt$$

exists, and T is clearly a linear operator on X with norm

$$\|T\| \leq \int_0^\infty e^{-\mathrm{Re}\,\lambda t}\,dt = (\mathrm{Re}\,\lambda)^{-1}.$$

In particular, $e^{-\lambda s}G(s)x \to 0$ for $s \to \infty$. Then (14.22) and (14.23) imply after a passage to the limit:

$$-x = (B - \lambda I)Tx \quad \text{for } x \in X, \tag{14.24}$$
$$-x = T(B - \lambda I)x \quad \text{for } x \in D(B), \tag{14.25}$$

which shows that λ is in the resolvent set, with resolvent equal to $-T$. $\quad\square$

We have found some properties of the infinitesimal generator B for a contraction semigroup $G(t)$, and now turn to the question (of interest for applications) of how the operators look that *can* be infinitesimal generators of a contraction semigroup. The question was answered by Hille and Yoshida (around 1945, in noncommunicating ends of the world) by different proofs of the following theorem.

Theorem 14.8 (HILLE-YOSHIDA). *When B is a densely defined, closed operator in X with $\mathbb{R}_+$ contained in $\rho(B)$, and*

$$\|(B - \lambda I)^{-1}\| \leq \lambda^{-1} \ \text{for } \lambda \in \mathbb{R}_+, \tag{14.26}$$

then B is the infinitesimal generator of a strongly continuous contraction semigroup.

Proof. The operators

$$B_\lambda = -\lambda^2(B - \lambda I)^{-1} - \lambda I,$$

defined for each $\lambda > 0$, are bounded with norm $\leq 2\lambda$, and we can form the operator families

$$G_\lambda(t) = \exp(tB_\lambda) \ \text{for } t \in \mathbb{R},$$

by (14.12) ff.; they are continuous in t with respect to the operator norm. Now observe that for $x \in D(B)$,

$$\lambda(B - \lambda I)^{-1}x + x = (B - \lambda I)^{-1}(\lambda x + (B - \lambda I)x) = (B - \lambda I)^{-1}Bx,$$

so that (14.26) implies that when $x \in D(B)$,

$$-\lambda(B - \lambda I)^{-1}x \to x \ \text{ for } \lambda \to \infty \,. \tag{14.27}$$

Since $\| - \lambda(B - \lambda I)^{-1}\| \leq 1$ for all $\lambda > 0$ and $D(B)$ is dense in X, (14.27) extends to all $x \in X$. We then get furthermore:

$$B_\lambda x = -\lambda(B - \lambda)^{-1}Bx \to Bx \ \text{ for } x \in D(B) , \ \lambda \to \infty \,, \tag{14.28}$$

and we want to show that $G_\lambda(t)$ converges strongly, for each t, toward a semigroup $G(t)$ with B as generator, for $\lambda \to \infty$.

To do this, note first that an application of the product formula, as in (14.13), gives that $G_\lambda(t) = \exp(-\lambda^2(B - \lambda)^{-1}t)\exp(-\lambda t)$, where

$$\| \exp(-\lambda^2(B - \lambda)^{-1}t)\| \leq \sum_{n=0}^{\infty} \frac{\| - \lambda^2(B - \lambda)^{-1}t\|^n}{n!} \leq \exp(\lambda t),$$

by (14.26), so that

$$\|G_\lambda(t)\| \leq \exp(-\lambda t)\exp\lambda t = 1 \tag{14.29}$$

for all $t \geq 0$ and $\lambda > 0$. Since all the bounded operators B_λ, B_μ, $G_\lambda(t)$, $G_\mu(s)$ for λ and $\mu > 0$, s and $t \geq 0$, commute, we find that

$$\begin{aligned}
G_\lambda(t) - G_\mu(t) &= \int_0^t \frac{d}{ds}[G_\lambda(s)G_\mu(t - s)]\, ds \\
&= \int_0^t G_\lambda(s)G_\mu(t - s)(B_\lambda - B_\mu)\, ds
\end{aligned}$$

which implies, using (14.29), that

$$\|G_\lambda(t)x - G_\mu(t)x\| \leq t\|B_\lambda x - B_\mu x\| \ \text{ for } x \in X \,. \tag{14.30}$$

When $x \in D(B)$, we know that $B_\mu x \to Bx$ for $\mu \to \infty$ according to (14.28); then we see in particular that the sequence $\{G_n(t)x\}_{n \in \mathbb{N}}$ is a Cauchy sequence in X. Denoting the limit by $G(t)x$, we obtain that

$$\|G_\lambda(t)x - G(t)x\| \leq t\|B_\lambda x - Bx\| \ \text{ for } x \in D(B) \,. \tag{14.31}$$

This shows that when $x \in D(B)$, then $G_\lambda(t)x \to G(t)x$ for $\lambda \to \infty$, *uniformly for t in bounded intervals* $[0, a] \subset \mathbb{R}_+$. Since we also have that $\|G_\lambda(t)x\| \leq \|x\|$ for all t and λ (cf. (14.29)), it follows that $\|G(t)x\| \leq \|x\|$, so that the operator $x \mapsto G(t)x$ defined for $x \in D(B)$ extends by closure to an operator $G(t) \in \mathbf{B}(X)$ with norm $\|G(t)\| \leq 1$. For the extended operator we now also find that

$$G_\lambda(t)x \to G(t)x \ \text{ for each } x \in X \,,$$

uniformly for t in bounded intervals $[0, a]$, for if we let $x_k \in D(B)$, $x_k \to x$, we have when $t \in [0, a]$,

$$\|G_\lambda(t)x - G(t)x\|$$
$$\leq \|G_\lambda(t)(x - x_k)\| + \|G_\lambda(t)x_k - G(t)x_k\| + \|G(t)(x - x_k)\|$$
$$\leq 2\|x - x_k\| + a\|B_\lambda x_k - B x_k\| ,$$

where the last expression is seen to go to 0 for $\lambda \to \infty$, independently of t, by choosing first x_k close to x and then adapting λ.

The semigroup property (a) carries over to $G(t)$ from the $G_\lambda(t)$'s. That $G(t)$ satisfies (b) is seen from

$$\|G(t)x - x\| \leq \|G(t)x - G_\lambda(t)x\| + \|G_\lambda(t)x - x\| ,$$

where one first chooses λ so large that $\|G(t)x - G_\lambda(t)x\| < \varepsilon$ for $t \in [0,1]$ and next lets $t \to 0$.

The semigroup $G(t)$ now has an infinitesimal generator C, and it remains to show that $C = B$. For $x \in D(B)$ we have

$$\|G_\lambda(s)B_\lambda x - G(s)Bx\| \leq \|G_\lambda(s)\|\|B_\lambda x - Bx\| + \|(G_\lambda(s) - G(s))Bx\|$$
$$\leq \|B_\lambda x - Bx\| + \|(G_\lambda(s) - G(s))Bx\|$$
$$\to 0 \text{ for } \lambda \to \infty,$$

uniformly for s in a bounded interval, by (14.28) and the proved convergence of $G_\lambda(s)y$ at each $y \in X$. This gives by use of Corollary 14.3 and (14.30), that for $x \in D(B)$,

$$\frac{1}{h}(G(h)x - x) = \lim_{\lambda \to \infty} \frac{1}{h}(G_\lambda(h)x - x) = \lim_{\lambda \to \infty} \frac{1}{h} \int_0^h G_\lambda(s)B_\lambda x \, ds$$
$$= \frac{1}{h} \int_0^h G(s)Bx \, dx .$$

If we here let $h \to 0$, the last expression will converge to $G(0)Bx = Bx$, from which we conclude that $x \in D(C)$ with $Cx = Bx$. Since $B - 1$ and $C - 1$ are bijections of $D(B)$ and $D(C)$, respectively, onto X, B must equal C. $\square$

The operator family $G(t)$ defined from B in this way is also called $\exp(tB)$.

The theory can be extended to semigroups which do not necessarily consist of contractions. For one thing, there is the obvious generalization where we to the operator $B + \mu I$ associate the semigroup

$$\exp(t(B + \mu I)) = \exp(t\mu)\exp(tB) ; \qquad (14.32)$$

this only gives contractions when $\text{Re}\,\mu \leq 0$. Here $\|\exp(t(B + \mu I))\| \leq |\exp(t\mu)| = \exp(t\,\text{Re}\,\mu)$. More general strongly continuous semigroups can only be expected to satisfy inequalities of the type $\|G(t)\| \leq c_1 \exp tc_2$ with $c_1 \geq 1$, and need a more general theory — see e.g. the books of E. Hille and R. Phillips [HP57] and of N. Dunford and J. Schwartz [DS58].

14.3 Contraction semigroups in Hilbert spaces

We now consider the case where X is a Hilbert space H. Here we shall use the notation $u(T)$ for the upper bound of an upper semibounded operator, cf. (12.21) ff.

Lemma 14.9. *When $G(t)$ is a semigroup in H satisfying* (a), (b) *and* (c), *then its infinitesimal generator B is upper semibounded, with upper bound ≤ 0.*

Proof. For $x \in X$ one has that

$$\operatorname{Re} \tfrac{1}{h}(G(h)x - x, x) = \tfrac{1}{h}\big(\operatorname{Re}(G(h)x, x) - \|x\|^2\big)$$
$$\leq \tfrac{1}{h}\big(\|G(h)x\|\, \|x\| - \|x\|^2\big) \leq 0$$

according to (c), from which we conclude for $x \in D(B)$ by a passage to the limit:

$$\operatorname{Re}(Bx, x) \leq 0 \quad \text{for } x \in D(B) . \tag{14.33}$$

Thus $u(B) \leq 0$, which shows the lemma.   □

In the Hilbert space case we can also consider the family of adjoint operators $G^*(t)$.

Theorem 14.10. *When $G(t)$ is a semigroup in H satisfying* (a), (b) *and* (c), *then $G^*(t)$ is likewise a semigroup in H satisfying* (a), (b) *and* (c); *and when the generator for $G(t)$ is B, then the generator for $G^*(t)$ is precisely B^*.*

Proof. It is seen immediately that $G^*(t)$ satisfies (a) and (c). For (b) we observe that one for x and $y \in H$ has:

$$(G^*(t)x, y) = (x, G(t)y) \to (x, y) \quad \text{for } t \to 0 .$$

This implies that

$$0 \leq \|G^*(t)x - x\|^2 = (G^*(t)x, G^*(t)x) + \|x\|^2 - (G^*(t)x, x) - (x, G^*(t)x)$$
$$\leq \|x\|^2 - (G^*(t)x, x) + \|x\|^2 - (x, G^*(t)x)$$
$$\to 0 \quad \text{for } t \to 0 ,$$

from which we conclude that $G^*(t)x - x \to 0$ for $t \to 0$.

Now let C be the infinitesimal generator of $G^*(t)$, and let $x \in D(B)$, $y \in D(C)$. Then

$$(Bx, y) = \lim_{h \to 0}\big(\tfrac{1}{h}(G(h)x - x), y\big) = \lim_{h \to 0}\big(x, \tfrac{1}{h}(G^*(h)y - y)\big) = (x, Cy) ;$$

thus $C \subset B^*$. By Theorem 14.7, $R(C - I) = H$; and $B^* - I$ is injective since $Z(B^* - I) = R(B - I)^\perp = \{0\}$. Then $C \subset B^*$ cannot hold unless $C = B^*$.   □

Corollary 14.11. *An operator B in a Hilbert space H is the infinitesimal generator of a strongly continuous contraction semigroup if and only if B is densely defined and closed, has $u(B) \leq 0$, and has $\mathbb{R}_+$ contained in its resolvent set.*

Proof. The necessity of the conditions follows from what we have just shown, together with Lemma 14.5 and Theorem 14.7. The sufficiency is seen from the fact that (14.33) implies that $-(B - \lambda I) = -B + \lambda I$ for $\lambda > 0$ has a bounded inverse with norm $\leq \lambda^{-1}$, by Theorem 12.9 1°; then the Hille-Yoshida Theorem (Theorem 14.8) can be applied. □

Operators satisfying (14.33) are in part of the literature called *dissipative operators*. There is a variant of the above theorems:

Corollary 14.12. *Let B be a closed, densely defined operator in a Hilbert space H. Then the following properties are equivalent:*

(i) *B is the infinitesimal generator of a strongly continuous contraction semigroup.*

(ii) *B is dissipative (i.e., $u(B) \leq 0$) and $\mathbb{R}_+ \subset \rho(B)$.*

(iii) *B and B^* are dissipative.*

Proof. The equivalence of (i) and (ii) is shown above. Condition (i) implies (iii) by Theorem 14.10, and (iii) implies (ii) by Theorem 12.9 3°. □

Remark 14.13. It is shown in [P59] that the conditions (i)–(iii) in Corollary 14.12 are equivalent with

(iv) *B is maximal dissipative.*

Let us here observe that (ii) easily implies (iv), for if B satisfies (ii) and B' is a dissipative extension, then $B' - I$ is injective (by Theorem 12.9 1° applied to $-(B' - I)$); hence, since $B - I$ is already bijective from its domain to H, B' must equal B. For the direction from (iv) to (ii), [P59] carries the problem over to $J = (I + B)(I - B)^{-1}$; this is a contraction with $D(J) = R(I - B)$ and $D(B) = R(I + J)$. Here B is maximal dissipative if and only if J is maximal with respect to being a contraction. Since we take B closed ([P59] considers more general operators), $D(J) = R(I - B)$ is closed (by Theorem 12.9 2°); then there exists a proper contraction extension of J unless $D(J) = H$. We see that maximality of B implies surjectiveness of $I - B$, and similarly of $I - \frac{1}{\lambda} B$ for all $\lambda > 0$, assuring (ii).

So, the infinitesimal generators of contraction semigroups are the maximal dissipative operators, as described by (ii), (iii) or (iv). Note that these operators have

$$\nu(B),\ \nu(B^*),\ \sigma(B),\ \sigma(B^*) \subset \{\lambda \in \mathbb{C} \mid \operatorname{Re} \lambda \leq 0\}. \tag{14.34}$$

We now consider some special cases:

1° $B = -A$, where A is selfadjoint ≥ 0. Here A and A^* are ≥ 0, so that B and B^* are dissipative.

2° $B = -A$, where A is a variational operator with $m(A) \geq 0$ (Section 12.4). Here $m(A^*)$ is likewise ≥ 0, so B and B^* are dissipative. This case is more general than 1°, but less general than the full set of operators satisfying (14.34), for, as we recall from (12.50) (applied to $-B$), we here have $\nu(B)$, $\nu(B^*)$, $\sigma(B)$ and $\sigma(B^*)$ contained in an angular set

$$\widetilde{M} = \{\lambda \in \mathbb{C} \mid \operatorname{Re}\lambda \leq 0, \, |\operatorname{Im}\lambda| \leq c(-\operatorname{Re}\lambda + k)\},$$

for some $c \geq 0$, $k \in \mathbb{R}$. The semigroups generated by such operators belong to the so-called *holomorphic semigroups* (where $G(t)x$ extends holomorphically to t in a sector around $\mathbb{R}_+$); they have particularly convenient properties, for example that $G(t)x \in D(B)$ for $t \neq 0$, any $x \in X$. (Besides [HP57], they enter e.g. in Kato [K66], Friedman [F69] and many other works.)

3° $B = iA$ where A is selfadjoint, i.e., $B = -B^*$, B is *skew-selfadjoint*. Here $\nu(B)$ and $\nu(B^*)$ are contained in the imaginary axis, so B and B^* are dissipative.

In the last case we can introduce

$$U(t) = \exp(tB) \qquad \text{for } t \geq 0 , \qquad (14.35)$$
$$U(t) = \exp(-tB^*) = U(t)^* \quad \text{for } t \leq 0 , \qquad (14.36)$$

writing $U(t) = \exp(tB)$ also for $t \leq 0$. We shall now show that $U(t)$ is a *strongly continuous group of unitary operators*.

That $U(t)x$ is continuous from $t \in \mathbb{R}$ into H follows from the continuity for $t \geq 0$ and $t \leq 0$. Next, we can show that $U(t)$ is an isometry for each $t \in \mathbb{R}_+$ or $t \in \mathbb{R}_-$, since we have for $x \in D(B)$:

$$\frac{d}{dt}\|U(t)x\|^2 = \lim_{h \to 0} \frac{1}{h}\big[(U(t+h)x, U(t+h)x) - (U(t)x, U(t)x)\big]$$
$$= \lim_{h \to 0}\Big(\frac{U(t+h)x - U(t)x}{h}, U(t)x\Big) + \lim_{h \to 0}\Big(U(t)x, \frac{U(t+h)x - U(t)x}{h}\Big)$$
$$+ \lim_{h \to 0}\Big(\frac{U(t+h)x - U(t)x}{h}, U(t+h)x - U(t)x\Big)$$
$$= (BU(t)x, U(t)x) + (U(t)x, BU(t)x) = 0 ;$$

where it was used after the second equality sign that there for any $\varepsilon > 0$ exists h_0 such that $\|U(t+h)x - U(t)x\| < \varepsilon$ for $|h| \leq h_0$; then we could let $h \to 0$. Hence $\|U(t)x\|^2$ is constant in t and thus equal to $\|U(0)x\|^2 = \|x\|^2$ for all t; the identity extends by continuity to $x \in H$. In a similar way it is seen that for $x \in D(B)$, $t \in \mathbb{R}_+$ or $t \in \mathbb{R}_-$,

$$\frac{d}{dt}U(t)U(-t)x = U(t)BU(-t)x + U(t)(-B)U(-t)x = 0 ,$$

so that $U(t)U(-t)x$ is constant for $t \geq 0$ and for $t \leq 0$, and hence equal to $U(0)U(0)x = x$. The identity

$$U(t)U(-t)x = x \quad \text{for } t \in \mathbb{R}$$

extends by continuity to all $x \in H$, and shows that

$$U(t)^{-1} = U(-t) \quad \text{for } t \in \mathbb{R} \; ;$$

hence U is unitary. It also implies the group property, since one has e.g. for $s \geq 0, 0 \geq t \geq -s$:

$$U(s)U(t) = U(s+t)U(-t)U(t) = U(s+t) \; .$$

Let now conversely $U(t)$ be a strongly continuous group of contractions. If $\{U(t)\}_{t \geq 0}$ and $\{U(-t)\}_{t \geq 0}$ have the infinitesimal generators B resp. C, then for $x \in D(B)$,

$$\lim_{h \to 0+} \frac{U(-h) - I}{h} x = \lim_{h \to 0+} U(-h) \frac{I - U(h)}{h} x = Bx \; ,$$

i.e., $-B \subset C$. Similarly, $-C \subset B$, so $B = -C$. For B it then holds that $m(-B) = m(B) = 0$, and both $\{\lambda \mid \operatorname{Re} \lambda > 0\}$ and $\{\lambda \mid \operatorname{Re} \lambda < 0\}$ are contained in the resolvent set. This shows that B is skew-selfadjoint, by Corollary 14.12 (or by Theorem 12.10 applied to iB). It is now seen from the preceding analysis that the operators $U(t)$ are unitary.

We have hereby obtained the theorem of M. H. Stone:

Theorem 14.14 (STONE). *An operator B in H is the infinitesimal generator of a strongly continuous group of unitary operators if and only if B is skew-selfadjoint.*

14.4 Applications

When B is the infinitesimal generator of a strongly continuous contraction semigroup (or group) $G(t)$ in a Banach space or Hilbert space X, the vector function

$$u(t) = G(t)u_0 \qquad (14.37)$$

is, according to Theorem 14.2, a solution of the abstract Cauchy problem (initial value problem)

$$\begin{cases} u'(t) = Bu(t) \; , & t > 0 \; (t \in \mathbb{R}) \; , \\ u(0) = u_0 \; , \end{cases} \qquad (14.38)$$

for any given initial value $u_0 \in D(B)$. (One can furthermore show that $G(t)u_0$ is the only continuously differentiable solution of (14.38).) If $G(t)$ is a holomorphic semigroup, (14.37) solves (14.38) even when $u_0 \in X$.

The semigroup theory can in this way be used to get solutions of the problem (14.38) for various types of operators B.

We have a wealth of examples:

First, we have in Chapter 12 defined the selfadjoint, the semibounded and the variational operators entering in 1°, 2° and 3° above.

Next, we have in Chapter 13 described operators of these types that are given to act in a particular way, namely, belonging to the set of closed operators lying between A_0 and A_1 with given properties.

Concrete interpretations to particular differential operators are given in Chapter 4, most prominently for the Laplace operator Δ on bounded sets, but also including Δ on $\mathbb{R}^n$, some variable-coefficient cases, and second-order ordinary differential equations on intervals (considered in some exercises). Here $X = L_2(\mathbb{R}^n)$ or $L_2(\Omega)$ for an open set $\Omega \subset \mathbb{R}^n$.

For example, if $B = -A_\gamma =$ the Dirichlet realization of the Laplace operator (called $-T$ in Theorem 4.27), the problem (14.38) becomes:

$$\partial_t u(x,t) = \Delta_x u(x,t) \text{ for } x \in \Omega, \ t > 0,$$
$$\gamma_0 u(x,t) = 0 \text{ for } t > 0,$$
$$u(x,0) = u_0(x) \text{ for } x \in \Omega,$$

and the function $\exp(-tA_\gamma)u_0$ solves this problem, the heat equation with homogeneous Dirichlet boundary condition.

An interpretation of the general study (from Chapter 13) of closed extensions of a minimal elliptic realization is given in Chapter 9, with details for an accessible example and introductory remarks on the general case. The analysis of lower bounded operators in Chapter 13 is particularly suited for application to evolution problems using semigroup theory. Variable-coefficient cases can be studied on the basis of Chapter 11, where the application of abstract results of Chapter 13 is followed up in Exercises 11.16–11.21 (a full account is given in [BGW08]). Semiboundedness for general elliptic boundary value problems is studied systematically in [G70]–[G74].

We can moreover get solutions of the Schrödinger equation with an initial condition, as mentioned in the beginning of the present chapter, when $i\Delta$ is concretized as a skew-selfadjoint operator. Also the wave equation can be studied using (14.6); here the matrix is taken to act e.g. in $H_0^1(\Omega) \times L_2(\Omega)$ (as defined in Chapter 4).

The solutions defined in this way are of course somewhat abstract and need further investigation, and one can show much more precisely which spaces the solutions belong to, and discuss their uniqueness and other properties. Further questions can be asked in a framework of functional analysis (as in scattering theory). *Parabolic* problems generalizing the heat equation have been widely studied. Classical references are Ladyzhenskaya, Solonnikov and Uraltseva [LSU68] and Friedman [F69]; recently there have been studies by refined methods using pseudodifferential techniques, as in [G95] and its references.

For the constructive analysis of solutions of *hyperbolic* equations, generalizing the wave equation, the most modern tools come from microlocal analysis (based on refined Fourier analysis). There are many interesting works on this; here we shall just point to the four volumes of L. Hörmander (cf. [H83], [H85]), which form a cornerstone in this development.

Exercises for Chapter 14

14.1. Show that if S is a densely defined and maximal symmetric operator in a Hilbert space H, then either iS or $-iS$ is the infinitesimal generator of a strongly continuous contraction semigroup.

14.2. Let B be a closed, densely defined operator in a Hilbert space H, and assume that there is a constant $\alpha \geq 0$ so that $m(B)$, $m(-B)$, $m(B^*)$ and $m(-B^*)$ are $\geq -\alpha$.

Let $G(t)$ be the family of operators defined by

$$G(t) = e^{\alpha t} \exp(t(B - \alpha I)) \qquad \text{for } t \geq 0 , \tag{14.39}$$

$$G(t) = e^{-\alpha t} \exp(-t(-B - \alpha I)) \text{ for } t \leq 0 . \tag{14.40}$$

Show that $G(t)$ is a strongly continuous group, satisfying $\|G(t)\| \leq \exp(\alpha|t|)$ for all t.

14.3. Let $G(t)$ be a strongly continuous contraction semigroup on a Banach space X, and let $f(t)$ be a continuous function of $t \in \overline{\mathbb{R}}_+$, valued in X. Show that $G(t)f(t)$ is a continuous function of $t \in \overline{\mathbb{R}}_+$.
(*Hint.* To $G(t)f(t) - G(t_0)f(t_0)$ one can add and subtract $G(t)f(t_0)$ and use that $\|G(t)\| \leq 1$ for all t.)

14.4. Consider the nonhomogeneous initial value problem for functions $u(t)$ taking values in a Banach space X:

$$\begin{aligned} u'(t) - Bu(t) &= f(t), \text{ for } t > 0, \\ u(0) &= 0. \end{aligned} \tag{14.41}$$

It is assumed that B is the infinitesimal generator of a strongly continuous contraction semigroup $G(t)$.

Let $T > 0$. Show that if $f \in C^1([0, T], X)$, then the function

$$u(t) = \int_0^t G(t - s)f(s) \, ds \tag{14.42}$$

is a solution of (14.41), with $u \in C^1([0,T], X) \cap C^0([0,T], D(B))$.
(*Hints.* The formula (14.42) can also be written

$$u(t) = \int_0^t G(s') f(t - s') \, ds',$$

from which it can be deduced that $u \in C^1([0,T], X)$. To verify (14.41), let
$h > 0$ and write

$$\frac{u(t+h) - u(t)}{h} = \frac{1}{h} \int_t^{t+h} G(t + h - s) f(s) \, ds$$

$$+ \frac{G(h) - I}{h} \int_0^t G(t - s) f(s) \, ds = I_1 + I_2.$$

Show that $I_1 \to f(t)$ for $h \to 0$. Use the differentiability of u to show that
$u(t) \in D(B)$ and $I_2 \to Bu(t)$.)

Give an example of an application to a PDE problem, where B is a realization
of an elliptic operator.

(*Comment.* The use of the semigroup in formula (14.42) is sometimes called
the Duhamel principle.)

Appendix A
Some notation and prerequisites

We denote by $\mathbb{Z}$ the integers, by $\mathbb{N}$ the positive integers and by $\mathbb{N}_0$ the nonnegative integers. $\mathbb{R}$ denotes the real numbers, $\mathbb{R}_+$ and $\mathbb{R}_-$ the positive resp. negative real numbers. $\mathbb{R}^n$ is the n-dimensional real Euclidean space, with points $x = (x_1, \ldots, x_n)$ and distance $\mathrm{dist}\,(x,y) = |x-y|$, where $|x| = (x_1^2 + \cdots + x_n^2)^{\frac{1}{2}}$. $\mathbb{R}_+^n$ and $\mathbb{R}_-^n$ denote the subsets, respectively,

$$\mathbb{R}_\pm^n = \{x \in \mathbb{R}^n \mid x_n \gtrless 0\}, \tag{A.1}$$

whose boundary $\{x \in \mathbb{R}^n \mid x_n = 0\}$ is identified with $\mathbb{R}^{n-1}$. The points in $\mathbb{R}^{n-1}$ are then often denoted x',

$$x' = (x_1, \ldots, x_{n-1}), \tag{A.2}$$

so that $x = (x', x_n)$.

We denote

$$\{t \in \mathbb{R} \mid a \le t \le b\} = [a,b], \quad \{t \in \mathbb{R} \mid a < t \le b\} =]a,b],$$
$$\{t \in \mathbb{R} \mid a \le t < b\} = [a,b[, \quad \{t \in \mathbb{R} \mid a < t < b\} =]a,b[$$

(to avoid conflict between the use of (x,y) for an open interval, for a point in $\mathbb{R}^2$ and for a scalar product).

The space of complex numbers is denoted $\mathbb{C}$; $\mathbb{C}_\pm$ denote the complex numbers with positive resp. negative imaginary part:

$$\mathbb{C}_\pm = \{z \in \mathbb{C} \mid \mathrm{Im}\, z \gtrless 0\}. \tag{A.3}$$

$\mathbb{C}^n$ denotes the n-dimensional complex Euclidean space. The functions we consider are usually functions on (subsets of) $\mathbb{R}^n$ taking values in $\mathbb{C}$. (Vector valued functions, valued in $\mathbb{C}^N$, can also occur, or we can consider real functions.)

Set inclusions are denoted by $\subset$, whether or not the sets are equal.

Differentiation of functions on $\mathbb{R}$ is indicated by $\dfrac{d}{dx}$, ∂_x or ∂. Moreover, we write $\dfrac{1}{i}\dfrac{d}{dx} = D_x$ or D (here i is the imaginary unit $i = \sqrt{-1}$); the factor $\dfrac{1}{i}$ is included for convenience in the use of the Fourier transformation. Partial differentiation of functions on $\mathbb{R}^n$ is indicated by

$$\frac{\partial}{\partial x_j} = \partial_{x_j} \text{ or } \partial_j\,; \quad \frac{1}{i}\frac{\partial}{\partial x_j} = D_{x_j} \text{ or } D_j \,. \tag{A.4}$$

In more complicated expressions we use multiindex notation: When $\alpha \in \mathbb{N}_0^n$, $\alpha = (\alpha_1, \ldots, \alpha_n)$, then

$$\partial^\alpha = \partial_{x_1}^{\alpha_1} \ldots \partial_{x_n}^{\alpha_n}, \quad D^\alpha = D_{x_1}^{\alpha_1} \ldots D_{x_n}^{\alpha_n} = (-i)^{|\alpha|}\partial_{x_1}^{\alpha_1} \ldots \partial_{x_n}^{\alpha_n}\,; \tag{A.5}$$

here the length of α is $|\alpha| = \alpha_1 + \cdots + \alpha_n$. The notation is used for instance for functions having continuous partial derivatives up to order $|\alpha|$, such that differentiations in different directions (up to that order) are interchangeable. Using the conventions

$$\alpha \leq \beta \text{ means } \alpha_1 \leq \beta_1, \ldots, \alpha_n \leq \beta_n \,,$$
$$\alpha! = \alpha_1! \ldots \alpha_n! \,, \tag{A.6}$$
$$\alpha \pm \beta = (\alpha_1 \pm \beta_1, \ldots, \alpha_n \pm \beta_n) \,,$$

we have for u and v with continuous derivatives up to order N the *Leibniz formula*

$$\partial^\alpha(uv) = \sum_{\beta \leq \alpha} \frac{\alpha!}{\beta!(\alpha - \beta)!}\partial^\beta u\, \partial^{\alpha-\beta}v \,, \text{ for } |\alpha| \leq N \,,$$
$$D^\alpha(uv) = \sum_{\beta \leq \alpha} \frac{\alpha!}{\beta!(\alpha - \beta)!}D^\beta u\, D^{\alpha-\beta}v \,, \text{ for } |\alpha| \leq N \,, \tag{A.7}$$

and the *Taylor formula*

$$u(x+y) = \sum_{|\alpha|<N} \frac{y^\alpha}{\alpha!}\partial^\alpha u(x) + \sum_{|\alpha|=N} \frac{N}{\alpha!}y^\alpha \int_0^1 (1-\theta)^{N-1}\partial^\alpha u(x+\theta y)d\theta \tag{A.8}$$

(this is an exact version from which the other well-known formulations can be deduced).

When $x \in \mathbb{R}^n$ or $\mathbb{C}^n$, we write

$$x^\alpha = x_1^{\alpha_1} \ldots x_n^{\alpha_n} \,, \text{ and }$$
$$x \cdot y = x_1 y_1 + \cdots + x_n y_n \,, \quad |x| = (x \cdot \bar{x})^{\frac{1}{2}} \,.$$

The norm $|x|$ (the Euclidean norm) makes $\mathbb{R}^n$ and $\mathbb{C}^n$ Hilbert spaces over $\mathbb{R}$ resp. $\mathbb{C}$, with scalar product $x \cdot \bar{y}$. The overline indicates complex conjugation. We also define

$$\langle x \rangle = (1 + |x|^2)^{\frac{1}{2}}, \text{ which satisfies, for } m \in \mathbb{N}:$$
$$\sum_{|\alpha| \leq m} x^{2\alpha} \leq (1 + |x|^2)^m = \sum_{|\alpha| \leq m} C_{m,\alpha} x^{2\alpha}; \tag{A.9}$$

here $C_{m,\alpha} = \frac{m!}{\alpha!(m-|\alpha|)!}$, it is integer ≥ 1.

When X and Y are topological spaces, $X \times Y$ denotes the product space, consisting of pairs $\{x, y\}$ where $x \in X$ and $y \in Y$, provided with the product topology (having as a subbasis the sets $U \times V$ where U resp. V run through a subbasis of the topology of X resp. Y). When X and Y are vector spaces, $X \times Y$ is a vector space in the obvious way. If X and Y are normed spaces, one can provide $X \times Y$ by the norm

$$\|\{x, y\}\|_{X \times Y} = \|x\|_X + \|y\|_Y, \tag{A.10}$$

making $X \times Y$ a normed space. When X and Y are Hilbert spaces, it is more convenient to use the equivalent norm

$$\|\{x, y\}\|_{X \oplus Y} = (\|x\|_X^2 + \|y\|_Y^2)^{\frac{1}{2}}, \tag{A.11}$$

associated with the scalar product

$$(\{x, y\}, \{x', y'\})_{X \oplus Y} = (x, x')_X + (y, y')_Y, \tag{A.12}$$

with which $X \times Y$ is a Hilbert space, denoted $X \oplus Y$. We use this notation also for the direct sum of two orthogonal closed subspaces X and Y of a Hilbert space H. For L_p-spaces it can be convenient to use $(\|x\|^p + \|y\|^p)^{\frac{1}{p}}$ as the norm on the product space.

We generally define

$$X \pm Y = \{x \pm y \mid x \in X \text{ and } y \in Y\},$$
$$\Omega X = \{\alpha x \mid \alpha \in \Omega \text{ and } x \in X\}, \tag{A.13}$$

when X and Y are subsets of a vector space V with scalar field $\mathbb{L}$ ($\mathbb{L} = \mathbb{R}$ or $\mathbb{C}$), and $\Omega \subset \mathbb{L}$. In particular, we write

$$\{x\} + Y = x + Y,$$
$$\{\alpha\} Y = \alpha Y, \tag{A.14}$$

when $x \in X$ and $\alpha \in \mathbb{L}$. When X and Y are subspaces of a vector space V, $X + Y$ is denoted $X \dotplus Y$ if X and Y are linearly independent. (There is also the notation $X \oplus Y$ for orthogonal closed subspaces of a Hilbert space.)

When X is a closed subspace of a Hilbert space H, the orthogonal complement is denoted $H \ominus X$; it can also be denoted $X^{\perp}$.

Integration by parts in one variable is generalized to functions of several variables by the Gauss and Green formulas, which we briefly recall:

Let $\Omega \subset \mathbb{R}^n$ be an open set with C^1 boundary $\partial\Omega$ and let $\nu(x)$ denote the interior unit normal vector field at $\partial\Omega$.

To explain this further: Ω is said to have a C^1 boundary, when every boundary point has a neighborhood V such that — after a relabeling of the coordinates if necessary —

$$\Omega \cap V = \{\,(x_1,\ldots,x_n) \in V \mid x_n > f(x_1,\ldots,x_{n-1})\,\}, \qquad (A.15)$$

where $f : \mathbb{R}^{n-1} \to \mathbb{R}$ is a C^1-function (continuous with continuous first-order derivatives). Here

$$\partial\Omega \cap V = \{\,x \in V \mid x_n = f(x_1,\ldots,x_{n-1})\,\}, \qquad (A.16)$$

and the interior unit normal vector at the point $x \in \partial\Omega \cap V$ equals (with the notation (A.4))

$$\nu(x', f(x')) = \frac{(-\partial_1 f(x'),\ldots,-\partial_{n-1}f(x'),1)}{\sqrt{(\partial_1 f(x'))^2 + \cdots + (\partial_{n-1}f(x'))^2 + 1}}. \qquad (A.17)$$

For a C^1-function u defined on a neighborhood of $\overline{\Omega}$ one has the *Gauss formula* (when u has compact support or the integrability is assured in some other way):

$$\int_\Omega \partial_k u \, dx = -\int_{\partial\Omega} \nu_k(x)u(x)\,d\sigma, \quad k = 1,\ldots,n, \qquad (A.18)$$

where $d\sigma$ is the surface measure on $\partial\Omega$. In the situation of (A.16),

$$d\sigma = \frac{1}{|\nu_n(x)|}\,dx' = \sqrt{(\partial_1 f)^2 + \cdots + (\partial_{n-1}f)^2 + 1}\; dx_1 \ldots dx_{n-1}; \quad (A.19)$$

and the formula (A.18) is for $k = n$ verified for functions supported in V simply by the change of coordinates $x = (x', x_n) \mapsto (x', x_n - f(x'))$ that replaces $\partial\Omega \cap V$ with a subset of $\mathbb{R}^{n-1}$. From the Gauss formula one derives several other formulas, usually called *Green's formulas*, when u and v are suitably differentiable:

$$\int_\Omega \partial_k u \, \overline{v} \, dx = -\int_\Omega u \, \overline{\partial_k v} \, dx - \int_{\partial\Omega} \nu_k(x)u(x)\overline{v}(x)\,d\sigma,$$

$$\int_\Omega D_k u \, \overline{v} \, dx = \int_\Omega u \, \overline{D_k v} \, dx + i\int_{\partial\Omega} \nu_k(x)u(x)\overline{v}(x)\,d\sigma,$$

$$\int_\Omega (-\Delta u) \, \overline{v} \, dx = \sum_{k=1,\ldots,n} \int_\Omega \partial_k u \overline{\partial_k v} \, dx + \int_{\partial\Omega} \frac{\partial u}{\partial \nu}\overline{v}\,d\sigma, \qquad (A.20)$$

$$\int_\Omega (-\Delta u)\,\overline{v}\,dx - \int_\Omega u\overline{(-\Delta v)}\,dx = \int_{\partial\Omega} \frac{\partial u}{\partial\nu}\overline{v}\,d\sigma - \int_{\partial\Omega} u\frac{\overline{\partial v}}{\partial\nu}\,d\sigma;$$

$$\text{where} \quad \frac{\partial u}{\partial\nu} = \sum_{k=1}^n \nu_k\partial_k u,$$

the interior normal derivative. Here Δ is the Laplace operator $\partial_1^2 + \cdots + \partial_n^2$. The signs are chosen with applications in mind (it is the operator $-\Delta$ that is "positive").

Let $p \in [1,\infty]$. For a Lebesgue measurable subset M of $\mathbb{R}^n$, $L_p(M)$ denotes the vector space of equivalence classes of measurable functions $f : M \to \mathbb{C}$ with finite norm

$$\|f\|_{L_p(M)} = \Big(\int_M |f(x)|^p dx\Big)^{1/p} \text{ if } p < \infty,$$
$$\|f\|_{L_\infty(M)} = \operatorname*{ess\,sup}_M |f| \text{ if } p = \infty. \tag{A.21}$$

It is a Banach space with this norm. (The equivalence classes consist of functions that are equal almost everywhere (a.e.); we use the customary "abuse of notation" where one calls the equivalence class a function, denoting the class containing f by f again. If the class contains a continuous function — necessarily unique if M is an open set or the closure of an open set — we use the continuous function as representative. Note that the space $C^0(M)$ of continuous functions on M identifies with a subset of $L_1(M)$ when M is the closure of a bounded open set.) We recall that for a real measurable function u on M,

$$\operatorname*{ess\,sup}_M u = \inf\{\, a \mid u(x) \le a \text{ a.e. in } M \,\}. \tag{A.22}$$

When $p = 2$ we get a Hilbert space, where the norm is associated with the scalar product

$$(f,g)_{L_2(M)} = \int_M f(x)\overline{g}(x)dx. \tag{A.23}$$

Hölder's inequality

$$\Big|\int_M f(x)g(x)dx\Big| \le \|f\|_{L_p(M)}\|g\|_{L_{p'}(M)}, \quad \frac{1}{p} + \frac{1}{p'} = 1, \tag{A.24}$$

holds for $f \in L_p(M)$ and $g \in L_{p'}(M)$; it is the *Cauchy-Schwarz inequality* in the case $p = 2$. Note that $L_p(\Omega) = L_p(\overline{\Omega})$ when for example Ω has C^1 boundary.

When the measure of M is finite, we have an inclusion

$$L_p(M) \subset L_q(M) \quad \text{for } 1 \le q \le p \le \infty. \tag{A.25}$$

Recall that the proof for $p < \infty$ consists of observing that for $f \in L_p(M)$ one has, with $r = p/q$, $1/r + 1/r' = 1$, by the Hölder inequality:

$$\|f\|_{L_q(M)} = \left(\int_M |f(x)|^q dx\right)^{1/q} = \left(\int_M |f(x)|^{p/r} \cdot 1 dx\right)^{1/q}$$
$$\leq \left(\int_M |f(x)|^p\right)^{1/rq}\left(\int_M 1 dx\right)^{1/r'q} \quad \text{(A.26)}$$
$$= \|f\|_{L_p(M)} \operatorname{vol}(M)^{1/q-1/p},$$

where $\operatorname{vol}(M) = \int_M 1 dx$ is the volume (measure) of M.

When $M \subset V$ for some set V, we denote by 1_M (the indicator function) the function on V defined by

$$1_M(x) = \begin{cases} 1 & \text{for } x \in M, \\ 0 & \text{for } x \in V \setminus M. \end{cases} \quad \text{(A.27)}$$

When $f \in L_p(\mathbb{R}^n)$, $g \in L_q(\mathbb{R}^n)$, and

$$1 \leq p \leq \infty, \quad 1 \leq q \leq \infty, \quad \frac{1}{r} = \frac{1}{p} + \frac{1}{q} - 1 \geq 0, \quad \text{(A.28)}$$

then the convolution $(f*g)(x) = \int_{\mathbb{R}^n} f(y)g(x-y)\, dy$ defines a function $f*g$ in $L_r(\mathbb{R}^n)$, and

$$\|f*g\|_{L_r(\mathbb{R}^n)} \leq \|f\|_{L_p(\mathbb{R}^n)}\|g\|_{L_q(\mathbb{R}^n)}; \quad \text{(A.29)}$$

Young's inequality. In particular, if $f \in L_1(\mathbb{R}^n)$ and $g \in L_2(\mathbb{R}^n)$, then $f*g \in L_2(\mathbb{R}^n)$, and

$$\|f*g\|_{L_2(\mathbb{R}^n)} \leq \|f\|_{L_1(\mathbb{R}^n)}\|g\|_{L_2(\mathbb{R}^n)}. \quad \text{(A.30)}$$

When Ω is an open subset of $\mathbb{R}^n$, we denote by $L_{p,\mathrm{loc}}(\Omega)$ the set of functions on Ω whose restrictions to compact subsets $K \subset \Omega$ are in $L_p(K)$. In view of (A.25), one has that

$$L_{p,\mathrm{loc}}(\Omega) \subset L_{q,\mathrm{loc}}(\Omega) \quad \text{for } 1 \leq q \leq p \leq \infty. \quad \text{(A.31)}$$

In particular, $L_{1,\mathrm{loc}}(\Omega)$ is the space of locally integrable functions on Ω (containing all the other spaces $L_{p,\mathrm{loc}}(\Omega)$).

The lower index p on L_p-spaces (instead of an upper index) reflects the fact that p is placed in this way in the modern literature on function spaces, such as L_p-types of Sobolev spaces H_p^s, B_p^s (and their numerous generalizations), where the upper index s is reserved for the degree of differentiability.

When ϱ is a positive locally integrable function, we use the notation $L_p(\Omega, \varrho(x)dx)$ for the weighted L_p-space with norm

$$\|u\|_{L_p(\Omega,\varrho)} = \left(\int_\Omega |u(x)|^p \varrho(x)\, dx\right)^{\frac{1}{p}}. \quad \text{(A.32)}$$

Let us also mention the notation for ℓ_p-spaces. For $1 \leq p < \infty$, $\ell_p(\mathbb{Z})$ consists of the sequences $\underline{a} = (a_j)_{j \in \mathbb{Z}}$, $a_j \in \mathbb{C}$, such that

$$\|\underline{a}\|_{\ell_p} \equiv \left(\sum_{j \in \mathbb{Z}} |a_j|^p \right)^{\frac{1}{p}} < \infty; \qquad (A.33)$$

it is a Banach space with the norm $\|\underline{a}\|_{\ell_p}$. For $p = 2$ it is a Hilbert space, with scalar product

$$(\underline{a}, \underline{b})_{\ell_2} = \sum_{j \in \mathbb{Z}} a_j \bar{b}_j. \qquad (A.34)$$

The Banach space $\ell_\infty(\mathbb{Z})$ consists of the bounded sequences, with the sup-norm.

The index set $\mathbb{Z}$ can be replaced by many other choices, e.g. $\mathbb{N}$, $\mathbb{N}_0$, $\mathbb{Z}^n$, etc.

Exercises for Appendix A .

A.1.   Show the general Leibniz formulas (A.7).

A.2.   (a) Let $f \in C^1(\mathbb{R}^n)$. Show for any $x, y \in \mathbb{R}^n$ that the function $g(\theta) = f(x + \theta y)$ ($\theta \in \mathbb{R}$) satisfies

$$\frac{d}{d\theta} g(\theta) = \sum_{j=1}^n \partial_j f(x + \theta y) y_j,$$

and conclude from this that

$$f(x + y) = f(x) + \sum_{j=1}^n y_j \int_0^1 \partial_j f(x + \theta y)\, d\theta.$$

(b) Show Taylor's formula (A.8) for arbitrary N.

A.3.   Deduce the formulas in (A.20) from (A.18).
(*Hint.* Apply (A.18) to $\partial_k(u\bar{v})$.)

Appendix B
Topological vector spaces

B.1 Fréchet spaces

In this appendix we go through the definition of Fréchet spaces and their inductive limits, such as they are used for definitions of function spaces in Chapter 2. A reader who just wants an orientation about Fréchet spaces will not have to read every detail, but need only consider Definition B.4, Theorem B.5, Remark B.6, Lemma B.7, Remark B.8 and Theorem B.9, skipping the proofs of Theorems B.5 and B.9. For inductive limits of such spaces, Section B.2 gives an overview (the proofs are established in a succession of exercises), and all that is needed for the definition of the space $C_0^\infty(\Omega)$ is collected in Theorem 2.5.

We recall that a topological space S is a space provided with a collection τ of subsets (called the *open* sets), satisfying the rules: S is open, $\emptyset$ is open, the intersection of two open sets is open, the union of any collection of open sets is open. The closed sets are then the complements of the open sets. A neighborhood of $x \in S$ is a set containing an open set containing x. Recall also that when S and S_1 are topological spaces, and f is a mapping from S to S_1, then f is continuous at $x \in S$ when for any neighborhood V of $f(x)$ in S_1 there exists a neighborhood U of x in S such that $f(U) \subset V$.

Much of the following material is also found in the book of Rudin [R74], which was an inspiration for the formulations here.

Definition B.1. A topological vector space (t.v.s.) over the scalar field $\mathbb{L} = \mathbb{R}$ or $\mathbb{C}$ (we most often consider $\mathbb{C}$), is a vector space X provided with a topology τ having the following properties:

(i) A set consisting of one point $\{x\}$ is closed.
(ii) The maps
$$\begin{aligned} \{x, y\} &\mapsto x + y \quad \text{from } X \times X \text{ into } X \\ \{\lambda, x\} &\mapsto \lambda x \quad \text{ from } \mathbb{L} \times X \text{ into } X \end{aligned} \tag{B.1}$$
are continuous.

In this way, X is in particular a Hausdorff space, cf. Exercise B.4. (We here follow the terminology of [R74] and [P89] where (i) is included in the definition of a t.v.s.; all spaces that we shall meet have this property. In part of the literature, (i) is not included in the definition and one speaks of Hausdorff topological vector spaces when it holds.)

$\mathbb{L}$ is considered with the usual topology for $\mathbb{R}$ or $\mathbb{C}$.

Definition B.2. A set $Y \subset X$ is said to be

a) convex, when $y_1, y_2 \in Y$ and $t \in]0,1[$ imply

$$ty_1 + (1-t)y_2 \in Y ,$$

b) balanced, when $y \in Y$ and $|\lambda| \le 1$ imply $\lambda y \in Y$;
c) **bounded** (with respect to τ), when for every neighborhood U of 0 there exists $t > 0$ so that $Y \subset tU$.

Note that boundedness is defined without reference to "balls" or the like.

Lemma B.3. *Let X be a topological vector space.*
$1°$ *Let $a \in X$ and $\lambda \in \mathbb{L} \setminus \{0\}$. The maps from X to X*

$$\begin{aligned} T_a &: x \mapsto x - a \\ M_\lambda &: x \mapsto \lambda x \end{aligned} \tag{B.2}$$

are continuous with continuous inverses T_{-a} resp. $M_{1/\lambda}$.
$2°$ *For any neighborhood V of 0 there exists a balanced neighborhood W of 0 such that $W + W \subset V$.*
$3°$ *For any convex neighborhood V of 0 there exists a convex balanced neighborhood W of 0 so that $W \subset V$.*

Proof. $1°$ follows directly from the definition of a topological vector space.

For $2°$ we appeal to the continuity of the two maps in (B.2) as follows: Since $\{x,y\} \mapsto x + y$ is continuous at $\{0,0\}$, there exist neighborhoods W_1 and W_2 of 0 so that $W_1 + W_2 \subset V$. Since $\{\lambda, x\} \mapsto \lambda x$ is continuous at $\{0,0\}$ there exist balls $B(0,r_1)$ and $B(0,r_2)$ in $\mathbb{L}$ with $r_1, r_2 > 0$ and neighborhoods W_1' and W_2' of 0 in X, such that

$$B(0,r_1)W_1' \subset W_1 \text{ and } B(0,r_2)W_2' \subset W_2.$$

Let $r = \min\{r_1, r_2\}$ and let $W = B(0,r)(W_1' \cap W_2')$, then W is a balanced neighborhood of 0 with $W + W \subset V$.

For $3°$ we first choose W_1 as under $2°$, so that W_1 is a balanced neighborhood of 0 with $W_1 \subset V$. Let $W = \bigcap_{\alpha \in \mathbb{L}, |\alpha|=1} \alpha V$. Since W_1 is balanced, $\alpha^{-1}W_1 = W_1$ for all $|\alpha| = 1$, hence $W_1 \subset W$. Thus W is a neighborhood of 0. It is convex, as an intersection of convex sets. That W is balanced is seen as follows: For $\lambda = 0$, $\lambda W \subset W$. For $0 < |\lambda| \le 1$,

$$\lambda W = |\lambda| \frac{\lambda}{|\lambda|} \bigcap_{\alpha \in L, |\alpha|=1} \alpha V = |\lambda| \bigcap_{\beta \in L, |\beta|=1} \beta V \subset W$$

(with $\beta = \alpha\lambda/|\lambda|$); the last inclusion follows from the convexity since $0 \in W$.

$\square$

Note that in 3° the interior W° of W is an open convex balanced neighborhood of x.

The lemma implies that the topology of a t.v.s. X is translation invariant, i.e., $E \in \tau \iff a+E \in \tau$ for all $a \in X$. The topology is therefore determined from the system of neighborhoods of 0. Here it suffices to know a *local basis* for the neighborhood system at 0, i.e., a system $\mathcal{B}$ of neighborhoods of 0 such that every neighborhood of 0 contains a set $U \in \mathcal{B}$.

X is said to be *locally convex*, when it has a local basis of neighborhoods at 0 consisting of convex sets.

X is said to be *metrizable*, when it has a metric d such that the topology on X is identical with the topology defined by this metric; this happens exactly when the balls $B(x, \frac{1}{n})$, $n \in \mathbb{N}$, are a local basis for the system of neighborhoods at x for any $x \in X$. Here $B(x, r)$ denotes as usual the open ball

$$B(x, r) = \{y \in X \mid d(x, y) < r\},$$

and we shall also use the notation $\underline{B}(x, r) = \{y \in X \mid d(x, y) \le r\}$ for the closed ball. Such a metric need not be translation invariant, but it will usually be so in the cases we consider; translation invariance (also just called *invariance*) here means that

$$d(x + a, y + a) = d(x, y) \quad \text{for} \quad x, y, a \in X .$$

One can show, see e.g. [R74, Th. 1.24], that when a t.v.s. is metrizable, then the metric can be chosen to be translation invariant.

A *Cauchy sequence* in a t.v.s. X is a sequence $(x_n)_{n \in \mathbb{N}}$ with the property: For any neighborhood U of 0 there exists an $N \in \mathbb{N}$ so that $x_n - x_m \in U$ for n and $m \ge N$.

In a metric space (M, d), Cauchy sequences — let us here call them *metric Cauchy sequences* — are usually defined as sequences (x_n) for which $d(x_n, x_m) \to 0$ in $\mathbb{R}$ for n and $m \to \infty$. This property need not be preserved if the metric is replaced by another equivalent metric (defining the same topology). We have, however, for t.v.s. that if the topology in X is given by an *invariant* metric d, then the general concept of Cauchy sequences for X gives just the metric Cauchy sequences (Exercise B.2).

A metric space is called *complete*, when every metric Cauchy sequence is convergent. More generally we call a t.v.s. *sequentially complete*, when every Cauchy sequence is convergent.

Banach spaces and Hilbert spaces are of course complete metrizable topological vector spaces. The following more general type is also important:

Definition B.4. A topological vector space is called a Fréchet space, when X is metrizable with a translation invariant metric, is complete and is locally convex.

The local convexity is mentioned explicitly because the balls belonging to a given metric need not be convex, cf. Exercise B.1. (One has, however, that if X is metrizable *and* locally convex, then there exists a metric for X with convex balls, cf. [R74, Th. 1.24].) Note that the balls defined from a *norm* are convex.

It is also possible to define Fréchet space topologies (and other locally convex topologies) by use of seminorms; we shall now take a closer look at this method to define topologies.

Recall that a seminorm on a vector space X is a function $p : X \to \overline{\mathbb{R}}_+$ with the properties

(i) $p(x + y) \leq p(x) + p(y)$ for $x, y \in X$ (subadditivity) ,

(ii) $p(\lambda x) = |\lambda| p(x)$ for $\lambda \in \mathbb{L}$ and $x \in X$ (multiplicativity) .

$\qquad\qquad\qquad\qquad\qquad\qquad\qquad\qquad\qquad\qquad\qquad\qquad$ (B.3)

A family $\mathcal{P}$ of seminorms is called *separating*, when for every $x_0 \in X \setminus \{0\}$ there is a $p \in \mathcal{P}$ such that $p(x_0) > 0$.

Theorem B.5. *Let X be a vector space and let $\mathcal{P}$ be a separating family of seminorms on X. Define a topology on X by taking, as a local basis $\mathcal{B}$ for the system of neighborhoods at 0, the convex balanced sets*

$$V(p, \varepsilon) = \{x \mid p(x) < \varepsilon\} , \quad p \in \mathcal{P} \ \text{ and } \ \varepsilon > 0 , \qquad (B.4)$$

together with their finite intersections

$$W(p_1, \ldots, p_N; \varepsilon_1, \ldots, \varepsilon_N) = V(p_1, \varepsilon_1) \cap \cdots \cap V(p_N, \varepsilon_N) ; \qquad (B.5)$$

and letting a local basis for the system of neighborhoods at each $x \in X$ consist of the translated sets $x + W(p_1, \ldots, p_N; \varepsilon_1, \ldots, \varepsilon_N)$. (It suffices to let $\varepsilon_j = 1/n_j$, $n_j \in \mathbb{N}$.)

With this topology, X is a topological vector space. The seminorms $p \in \mathcal{P}$ are continuous maps of X into $\mathbb{R}$. A set $E \subset X$ is bounded if and only if $p(E)$ is bounded in $\mathbb{R}$ for all $p \in \mathcal{P}$.

Proof. That a topology on X is defined in this way, follows from the observation: When $x \in (x_1 + W_1) \cap (x_2 + W_2)$, then there is a basis-neighborhood W such that $x + W \subset (x_1 + W_1) \cap (x_2 + W_2)$. Moreover, it is clear that the topology is invariant under translation and under multiplication by a scalar $\neq 0$, and that the sets with $\varepsilon_j = 1/n_j$, $n_j \in \mathbb{N}$, form a local basis. The continuity of the p's follows from the subadditivity: For each $x_0 \in X$ and $\varepsilon > 0$, $V(p, x_0, \varepsilon) = x_0 + V(p, \varepsilon)$ is mapped into the neighborhood $B(p(x_0), \varepsilon)$ of $p(x_0)$ in $\mathbb{R}$, since

$$p(x) - p(x_0) \leq p(x - x_0) + p(x_0) - p(x_0) < \varepsilon,$$
$$p(x_0) - p(x) \leq p(x_0 - x) + p(x) - p(x) < \varepsilon,$$

for $x \in V(p, x_0, \varepsilon)$.

The point set $\{0\}$ (and then also any other point) is closed, since every $x_0 \neq 0$ has a neighborhood disjoint from $\{0\}$, namely, $V(p, x_0, \varepsilon)$, where p is chosen so that $p(x_0) \neq 0$ and $\varepsilon < p(x_0)$. We shall now show the continuity of addition and multiplication. For the continuity of addition we must show that for any neighborhood $W + x + y$ there exist neighborhoods $W_1 + x$ of x and $W_2 + y$ of y (recall the notation (A.13) and (A.14)) so that

$$W_1 + x + W_2 + y \subset W + x + y , \quad \text{i.e., so that } W_1 + W_2 \subset W. \qquad (B.6)$$

Here we simply use that when $W = W(p_1, \ldots, p_N; \varepsilon_1, \ldots, \varepsilon_n)$ is a basis-neighborhood at 0, and we set

$$W' = \tfrac{1}{2}W(p_1, \ldots, p_N; \varepsilon_1, \ldots, \varepsilon_N) = W(p_1, \ldots, p_N; \tfrac{1}{2}\varepsilon_1 \ldots, \tfrac{1}{2}\varepsilon_N) ,$$

then $W' + W' \subset W$ because of the subadditivity. For the continuity of multiplication we must show that for any neighborhood $W + \alpha x$ of αx there exist neighborhoods $B(0, r) + \alpha$ (of α in $\mathbb{L}$) and $W' + x$ (of x in X) such that

$$(B(0, r) + \alpha)(W' + x) \subset W + \alpha x . \qquad (B.7)$$

Here

$$(B(0, r) + \alpha)(W' + x) \subset B(0, r)W' + \alpha W' + B(0, r)x + \alpha x .$$

Let $W = W(p_1, \ldots, p_N; \varepsilon_1, \ldots, \varepsilon_N)$, and let $W' = \delta W$, $\delta > 0$. For c taken larger than $\varepsilon_j^{-1} p_j(x)$, $j = 1, \ldots, N$, we have that $x \in cW$ and hence

$$B(0, r)x \subset rcW .$$

Moreover,

$$B(0, r)W' \subset rW' = r\delta W$$

and

$$\alpha W' \subset |\alpha|W' = |\alpha|\delta W .$$

Now we first take δ so small that $|\alpha|\delta < \tfrac{1}{3}$; next we choose r so small that $r\delta < \tfrac{1}{3}$ and $rc < \tfrac{1}{3}$. Then

$$B(0, r)W' + \alpha W' + B(0, r)x \subset \tfrac{1}{3}W + \tfrac{1}{3}W + \tfrac{1}{3}W \subset W ,$$

whereby (B.7) is satisfied. All together we find that the topology on X makes X a topological vector space.

Finally we shall describe the bounded sets. Assume first that E is bounded. Let $p \in \mathcal{P}$, then there exists by assumption (see Definition B.2) a $t > 0$, so that $E \subset tV(p,1) = V(p,t)$. Then $p(x) \in [0,t[$ for $x \in E$, so that $p(E)$ is bounded. Conversely, let E be a set such that for every p there is a $t_p > 0$ with $p(x) < t_p$ for $x \in E$. When $W = W(p_1, \ldots, p_N; \varepsilon_1, \ldots, \varepsilon_N)$, we then have that $E \subset tW$ for $t \geq \max\{t_{p_1}/\varepsilon_1, \ldots, t_{p_N}/\varepsilon_N\}$.   □

Note that the sets (B.5) are open, convex and balanced in view of (B.3).

Note that x_k converges to x in X if and only if $p(x_k - x) \to 0$ for every seminorm $p \in \mathcal{P}$; and that a linear operator $T : Y \to X$, where Y is a t.v.s., is continuous if and only if all the maps $p \circ T$ are continuous.

Remark B.6. We remark that in the definition of the topology in Theorem B.5, the sets (B.4) constitute a *local subbasis* for the neighborhood system at 0, i.e., a system $\mathcal{B}'$ of neighborhoods of 0 such that the finite intersections of elements from $\mathcal{B}'$ constitute a local basis for the neighborhood system at 0. If the given family of seminorms $\mathcal{P}$ has the following property (which we shall call the *max-property*):

$$\forall p_1, p_2 \in \mathcal{P} \, \exists p \in \mathcal{P}, \, c > 0 : p \geq c \max\{p_1, p_2\} \,, \qquad (B.8)$$

then each of the basis sets (B.5) contains a set of the form (B.4), namely, one with $p \geq c \max\{p_1, \ldots, p_N\}$ and $\varepsilon \leq \min c\{\varepsilon_1, \ldots, \varepsilon_N\}$; here (B.4) is in itself a *local basis* for the topology. For a given family of seminorms $\mathcal{P}$ one can always supply the family to obtain one that has the max-property and *defines the same topology*. In fact, one can simply replace $\mathcal{P}$ by $\mathcal{P}'$, consisting of $\mathcal{P}$ and all seminorms of the form

$$p = \max\{p_1, \ldots, p_N\} \,, \, p_1, \ldots, p_N \in \mathcal{P} \,, \quad N \in \mathbb{N} \,. \qquad (B.9)$$

In the following we can often obtain that $\mathcal{P}$ is *ordered* (when p and $p' \in \mathcal{P}$ then either $p(x) \leq p'(x) \, \forall x \in X$ or $p'(x) \leq p(x) \, \forall x \in X$); then $\mathcal{P}$ has the max-property.

When the max-property holds, we have a simple reformulation of continuity properties:

Lemma B.7. $1°$ *When the topology on X is given by Theorem B.5 and $\mathcal{P}$ has the max-property, then a linear functional Λ on X is continuous if and only if there exists a $p \in \mathcal{P}$ and a constant $c > 0$ so that*

$$|\Lambda(x)| \leq cp(x) \text{ for all } x \in X \,. \qquad (B.10)$$

$2°$ *When X and Y are topological vector spaces with topologies given as in Theorem B.5 by separating families $\mathcal{P}$ resp. $\mathcal{Q}$ having the max-property, then a linear operator T from X to Y is continuous if and only if: For each $q \in \mathcal{Q}$ there exists a $p \in \mathcal{P}$ and a constant $c > 0$ so that*

$$|q(Tx)| \leq cp(x) \text{ for all } x \in X \,. \qquad (B.11)$$

Proof. 1°. Continuity of Λ holds when it holds at 0. Since the neighborhoods (B.4) form a basis, the continuity can be expressed as the property: For any $\varepsilon > 0$ there is a $p \in \mathcal{P}$ and a $\delta > 0$ such that $(*) \, p(x) < \delta \implies |\Lambda(x)| < \varepsilon$.

If (B.10) holds, then we can for a given ε obtain $(*)$ by taking $\delta = \varepsilon/c$. Conversely, assume that $(*)$ holds. We then claim that (B.10) holds with $c = \varepsilon/\delta$. In fact, if $p(x) = 0$, $\Lambda(x)$ must equal 0, for otherwise $|\Lambda(tx)| = t|\Lambda(x)| \to \infty$ for $t \to \infty$ whereas $p(tx) = 0$ for $t > 0$, contradicting $(*)$. If $p(x) > 0$, let $0 < \delta' < \delta$, then $p(\frac{\delta'}{p(x)}x) = \delta' < \delta$ so $|\Lambda(\frac{\delta'}{p(x)}x)| < \varepsilon$, hence $|\Lambda(x)| < \frac{\varepsilon}{\delta'}p(x)$. Letting $\delta' \to \delta$, we conclude that $|\Lambda(x)| \le \frac{\varepsilon}{\delta}p(x)$. This shows 1°.

2°. Continuity of T holds when it holds at 0; here it can be expressed as the property: For any $q \in \mathcal{Q}$ and $\varepsilon > 0$ there is a $p \in \mathcal{P}$ and a $\delta > 0$ such that $(**) \, p(x) < \delta \implies |q(Tx)| < \varepsilon$.

If (B.11) holds, then we can for a given ε obtain $(**)$ by taking $\delta = \varepsilon/c$. Conversely, when $(**)$ holds, we find that (B.11) holds with $c = \varepsilon/\delta$ in a similar way as in the proof of 1°. $\square$

Remark B.8. If the given family of seminorms is not known to have the max-property, the lemma holds when $\mathcal{P}$ and $\mathcal{Q}$ are replaced by $\mathcal{P}'$, $\mathcal{Q}'$, defined as indicated in Remark B.6. We can express this in another way: Without assumption of the max-property, the lemma is valid when p and q in (B.10) and (B.11) are replaced by expressions as in (B.9).

Note in particular that, whether the max-property holds or not, *the existence of a $p \in \mathcal{P}$, $c > 0$, such that (B.10) holds, is sufficient* to assure that the linear functional Λ is continuous.

We observe, as an outcome of the lemma, the general principle: *Continuity of linear mappings is shown by proving inequalities.* This is familiar for normed spaces, and we see that it holds also for spaces with topologies defined by seminorms.

The family of seminorms $\mathcal{P}$ may in particular be derived from a vector space Y of linear functionals on X, by taking $|\Lambda(x)|$ as a seminorm when $\Lambda \in Y$. An important example is where $Y = X^*$, the space of continuous linear functionals on X. Here the topology defined by the family of seminorms $x \mapsto |x^*(x)|$, $x^* \in X^*$, is called the weak* topology on X. Much more can be said about this important case (and the aspects of weak topology and weak* topology in connection with Banach spaces); for this we refer to textbooks on functional analysis.

Theorem B.9. *When X is a t.v.s. where the topology is defined by a countable separating family of seminorms $\mathcal{P} = (p_k)_{k \in \mathbb{N}}$, then X is locally convex and metrizable, and the topology on X is determined by the invariant metric*

$$d(x,y) = \sum_{k=1}^{\infty} \frac{1}{2^k} \frac{p_k(x-y)}{1 + p_k(x-y)} . \tag{B.12}$$

The balls $B(0,r)$ in this metric are balanced.

If X is complete in this metric, X is a Fréchet space.

Proof. The space is locally convex because of the definition of the topology by use of seminorms.

The series (B.12) is clearly convergent for all $x, y \in X$. Consider the function $f(a) = a/(1+a)$, $a \geq 0$. We want to show that it satisfies

$$f(a) \leq f(a+b) \leq f(a) + f(b). \tag{B.13}$$

Here the first inequality follows since $f(a) = 1 - 1/(1+a)$ with $1/(1+a)$ decreasing. For the second inequality we use this together with the formula $f(a)/a = 1/(1+a)$ for $a > 0$, which gives that

$$f(a)/a \geq f(a+b)/(a+b), \quad f(b)/b \geq f(a+b)/(a+b), \quad a, b > 0,$$

hence altogether $f(a) + f(b) \geq f(a+b)(a+b)/(a+b) = f(a+b)$. ((B.13) is immediately verified in cases where a or b is 0.) Define $d_0(x)$ by

$$d_0(x) = \sum_{k=1}^{\infty} \frac{1}{2^k} \frac{p_k(x)}{1 + p_k(x)}; \tag{B.14}$$

then $d(x, y) = d_0(x - y)$. We first show

$$d_0(x) \geq 0, \quad d_0(x) = 0 \iff x = 0;$$
$$d_0(x + y) \leq d_0(x) + d_0(y);$$
$$d_0(\lambda x) \leq d_0(x) \quad \text{for } |\lambda| \leq 1;$$

these properties assure that $d(x, y)$ is a metric with balanced balls. For the first line, we note that $d_0(x) > 0$ for $x \neq 0$ follows from the fact that $\mathcal{P}$ is separating. The second line is obtained by use of (B.13):

$$d_0(x+y) = \sum_{k=1}^{\infty} \frac{1}{2^k} \frac{p_k(x+y)}{1 + p_k(x+y)} \leq \sum_{k=1}^{\infty} \frac{1}{2^k} \frac{p_k(x) + p_k(y)}{1 + p_k(x) + p_k(y)}$$

$$\leq \sum_{k=1}^{\infty} \frac{1}{2^k} \frac{p_k(x)}{1 + p_k(x)} + \sum_{k=1}^{\infty} \frac{1}{2^k} \frac{p_k(y)}{1 + p_k(y)} = d_0(x) + d_0(y) \, .$$

The third line is obtained using that for $0 < |\lambda| \leq 1$,

$$d_0(\lambda x) = \sum_{k=1}^{\infty} \frac{1}{2^k} \frac{p_k(\lambda x)}{1 + p_k(\lambda x)} = \sum_{k=1}^{\infty} \frac{1}{2^k} \frac{|\lambda| p_k(x)}{1 + |\lambda| p_k(x)}$$

$$= \sum_{k=1}^{\infty} \frac{1}{2^k} \frac{p_k(x)}{1/|\lambda| + p_k(x)} \leq \sum_{k=1}^{\infty} \frac{1}{2^k} \frac{p_k(x)}{1 + p_k(x)} = d_0(x).$$

The metric is clearly translation invariant. We shall now show that the hereby defined topology is the same as the topology defined by the seminorms. Note that on one hand, one has for $\varepsilon \in {]}0,1{[}$ and $N \in \mathbb{N}$:

$$d_0(x) \le \varepsilon 2^{-N} \implies \frac{1}{2^k}\frac{p_k(x)}{1+p_k(x)} \le \varepsilon 2^{-N} \text{ for all } k$$

$$\implies \frac{p_k(x)}{1+p_k(x)} \le \varepsilon \text{ for } k \le N \implies p_k(x) \le \frac{\varepsilon}{1-\varepsilon} \text{ for } k \le N;$$

on the other hand, one has, if $2^{-N} < \varepsilon/2$,

$$p_k(x) \le \varepsilon/2 \text{ for } k \le N \implies$$

$$d_0(x) = \sum_{k=1}^{N} \frac{1}{2^k}\frac{p_k(x)}{1+p_k(x)} + \sum_{k=N+1}^{\infty} \frac{1}{2^k}\frac{p_k(x)}{1+p_k(x)}$$

$$\le \frac{\varepsilon}{2}\sum_{k=1}^{N}\frac{1}{2^k} + \frac{1}{2^N} \le \varepsilon.$$

Therefore we have the following inclusions between basis-neighborhoods $W(p_1,\ldots,p_N;\delta,\ldots,\delta)$ and balls $B(0,r)$ (with $\delta > 0$, $N \in \mathbb{N}$, $r > 0$):

$$B(0,r) \subset W(p_1,\ldots,p_N;\delta,\ldots,\delta),$$

when δ and N are given, $\varepsilon > 0$ is chosen so that $\varepsilon/(1-\varepsilon) < \delta$, and r is taken $= \varepsilon 2^{-N}$;

$$W(p_1,\ldots,p_N;\delta,\ldots,\delta) \subset B(0,r),$$

when r is given, N is chosen so that $2^{-N} \le r/2$, and δ is taken $= r/2$. This shows that $B(0,r)$ for $r > 0$ is a system of basis-neighborhoods at 0 for the topology defined by $\mathcal{P}$.

Finally, by Definition B.4, X is a Fréchet space if it is *complete* in this metric, i.e., if Cauchy sequences are convergent. □

The convergence of Cauchy sequences can be checked by the convergence criterion mentioned after Theorem B.5.

Note that the balls defined by the metric (B.12) need not be convex, cf. Exercise B.1. (However, there does exist, as mentioned after Definition B.4, another compatible metric with convex balls.) It will in general be most convenient to calculate on the basis of the seminorms rather than a metric, although its existence has useful consequences.

Remark B.10. One can ask conversely whether the topology of an arbitrary locally convex t.v.s. X can be defined by a separating family of seminorms. The answer is yes: To a neighborhood basis at 0 consisting of convex, balanced open sets V one can associate the Minkowski functionals μ_V defined by

$$\mu_V(x) = \inf\{t > 0 \mid \frac{x}{t} \in V\} \; ; \qquad\qquad\qquad \text{(B.15)}$$

then one can show that the μ_V's are a separating family of seminorms defining the topology on X. If X in addition is metrizable, the topology can be defined from a *countable* family of seminorms, since X then has a neighborhood basis at 0 consisting of a sequence of open, convex balanced sets. See e.g. [R74, Ch. 1] and [P89].

Like in Banach spaces there are some convenient rules characterizing the continuous linear operators on Fréchet spaces. By a bounded operator we mean an operator sending bounded sets into bounded sets.

Theorem B.11. *Let X be a Fréchet space and let Y be a topological vector space. When T is a linear map of X into Y, the following four properties are equivalent:*

(a) *T is continuous.*
(b) *T is bounded.*
(c) *$x_n \to 0$ in X for $n \to \infty \implies (Tx_n)_{n \in \mathbb{N}}$ is bounded in Y.*
(d) *$x_n \to 0$ in $X \implies Tx_n \to 0$ in Y.*

Proof. (a) $\implies$ (b). Let E be bounded in X, and let $F = T(E)$. Let V be a neighborhood of 0 in Y. Since T is continuous, there is a neighborhood U of 0 in X such that $T(U) \subset V$. Since E is bounded, there is a number $t > 0$ such that $E \subset tU$. Then $F \subset tV$; hence F is bounded.

(b) $\implies$ (c). When $x_n \to 0$, the set $\{x_n \mid n \in \mathbb{N}\}$ is bounded. Then $\{Tx_n \mid n \in \mathbb{N}\}$ is bounded according to (b).

(c) $\implies$ (d). Let d denote a translation invariant metric defining the topology. The triangle inequality then gives that

$$d(0, nx) \le d(0, x) + d(x, 2x) + \cdots + d((n-1)x, nx) = nd(0, x) \;, \qquad \text{(B.16)}$$

for $n \in \mathbb{N}$. When $x_n \to 0$, then $d(x_n, 0) \to 0$, so there exists a strictly increasing sequence of indices n_k so that $d(x_n, 0) < k^{-2}$ for $n \ge n_k$. Then $d(kx_n, 0) \le kd(x_n, 0) \le k^{-1}$ for $n \ge n_k$; hence also the sequence $t_n x_n$ goes to 0 for $n \to \infty$, where t_n is defined by

$$t_n = k \quad \text{for} \quad n_k \le n \le n_{k+1} - 1 \;.$$

According to (c), the sequence $t_n T x_n$ is bounded. It follows (cf. Exercise B.5) that $Tx_n \to 0$ for $n \to \infty$.

(d) $\implies$ (a). If T is not continuous, there exists a neighborhood V of 0 in Y and for any n an x_n with $d(x_n, 0) < \frac{1}{n}$ and $Tx_n \notin V$. Here x_n goes to 0, whereas Tx_n does not go to 0, i.e., (d) does not hold.   □

Remark B.12. It is seen from the proof that the assumption in Theorem B.11 that X is a Fréchet space can be replaced with the assumption that X is a metrizable topological vector space with an invariant metric.

We see in particular that a functional Λ on X is continuous if and only if $x_n \to 0$ in X implies $\Lambda x_n \to 0$ in $\mathbb{L}$.

Since "bounded" and "continuous" are synonymous for linear operators in the situation described in Theorem B.11, we may, like for operators in Banach spaces, denote the set of continuous linear operators from X to Y by $\mathbf{B}(X, Y)$.

One can also consider unbounded operators, and operators which are not everywhere defined, by conventions like those in Chapter 12.

We end this presentation by mentioning briefly that the well-known basic principles for Banach spaces which build on the theorem of Baire (on complete metric spaces) are easily generalized to Fréchet spaces.

Theorem B.13 (THE BANACH-STEINHAUS THEOREM). *Let X be a Fréchet space and Y a topological vector space, and consider a family $(T_\lambda)_{\lambda \in \Lambda}$ of continuous operators $T_\lambda \in \mathbf{B}(X, Y)$. If the set $\{T_\lambda x \mid \lambda \in \Lambda\}$ (called the "orbit" of x) is bounded in Y for each $x \in X$, then the family T_λ is equicontinuous, i.e., when V is a neighborhood of 0 in Y, then there exists a neighborhood W of 0 in X so that $\bigcup_{\lambda \in \Lambda} T_\lambda(W) \subset V$.*

Corollary B.14. *Let X be a Fréchet space and Y a topological vector space, and consider a sequence $(T_n)_{n \in \mathbb{N}}$ of continuous operators $T_n \in \mathbf{B}(X, Y)$. If $T_n x$ has a limit in Y for each $x \in X$, then the map $T : x \mapsto Tx = \lim_{n \to \infty} T_n x$ is a continuous map of X into Y. Moreover, the sequence $(T_n)_{n \in \mathbb{N}}$ is equicontinuous.*

We leave it to the reader to formulate a similar corollary for nets.

Theorem B.15 (THE OPEN MAPPING PRINCIPLE). *Let X and Y be Fréchet spaces. If $T \in \mathbf{B}(X, Y)$ is surjective, then T is open (i.e., sends open sets into open sets).*

Theorem B.16 (THE CLOSED GRAPH THEOREM). *Let X and Y be Fréchet spaces. If T is a linear map of X into Y and the graph of T is closed in $X \times Y$, then T is continuous.*

B.2 Inductive limits of Fréchet spaces

When defining the topology on $C_0^\infty(\Omega)$ for an open set $\Omega \subset \mathbb{R}^n$, we shall need the following generalization of Fréchet spaces:

Theorem B.17. *Let*

$$X_1 \subset X_2 \subset \cdots \subset X_j \subset \cdots \tag{B.17}$$

be a sequence of Fréchet spaces such that for every $j \in \mathbb{N}$, X_j is a subspace of X_{j+1} and the topology on X_j is the topology induced from X_{j+1}. Let

$$X = \bigcup_{j=1}^{\infty} X_j, \tag{B.18}$$

and consider the sets W that satisfy: $W \cap X_j$ is an open, convex, balanced neighborhood of 0 in X_j for all j. Then X has a unique locally convex vector space topology whose open, convex, balanced neighborhoods of 0 are precisely the sets W.

With this topology, all the injections $X_j \subset X$ are continuous.

The topology determined in this way is called the inductive limit topology, and spaces of this kind are called $\mathcal{LF}$ spaces (short for: inductive limits of Fréchet spaces). There is more information on this topology in Exercises B.15 and B.16, where it is shown how it can also be described in terms of seminorms.

Theorem B.18. Let $X = \bigcup_{j \in \mathbb{N}} X_j$ be an inductive limit of Fréchet spaces.

(a) A set E in X is bounded if and only if there exists a j_0 so that E lies in X_{j_0} and is bounded there.

(b) If a sequence $(u_k)_{k \in \mathbb{N}}$ is a Cauchy sequence in X, then there exists a j_0 such that $u_k \in X_{j_0}$ for all k, and $(u_k)_{k \in \mathbb{N}}$ is convergent in X_{j_0} and in X.

(c) Let Y be a locally convex topological vector space. A linear map T of X into Y is continuous if and only if $T : X_j \to Y$ is continuous for every $j \in \mathbb{N}$.

A variant of the theorem is (essentially) shown in [R74, Ch. 6]; see also Exercise B.16.

The fundamental property of the inductive limit topology that we use is that important concepts such as convergence and continuity in connection with the topology on X can be referred to the corresponding concepts for one of the simpler spaces X_j. This is seen from Theorem B.18, and we observe the following further consequences.

Corollary B.19. Hypotheses as in Theorem B.18.

(a) A linear map $T : X \to Y$ is continuous if and only if one has for each $j \in \mathbb{N}$ that when $(u_k)_{k \in \mathbb{N}}$ is a sequence in X_j with $u_k \to 0$ in X_j for $k \to \infty$, then $Tu_k \to 0$ in Y.

(b) Assume that the topology in each X_j is given by a family $\mathcal{P}_j$ of seminorms with the max-property (cf. Remark B.6). A linear functional $\Lambda : X \to \mathbb{L}$ is continuous if and only if there exists, for each j, a seminorm $p_j \in \mathcal{P}_j$ and a constant $c_j > 0$ so that

$$|\Lambda(x)| \le c_j p_j(x) \text{ for } x \in X_j. \tag{B.19}$$

Proof. (a) and (b) follow from Theorem B.18 combined with, respectively, Theorem B.11 and Lemma B.7. □

Exercises for Appendix B

B.1. Define d_1 and d_2 for $x, y \in \mathbb{R}$ by

$$d_1(x, y) = |x - y| \,, \quad d_2(x, y) = \left| \frac{x}{1 + |x|} - \frac{y}{1 + |y|} \right| \,.$$

Show that d_1 and d_2 are metrics on $\mathbb{R}$ which induce the same topology on $\mathbb{R}$, and that $(\mathbb{R}, d_1)$ is a complete metric space whereas $(\mathbb{R}, d_2)$ is not a complete metric space.

B.2. Let X be a t.v.s. and assume that the topology is defined by a translation invariant metric d. Show that (x_n) is a Cauchy sequence in X if and only if (x_n) is a metric Cauchy sequence in (X, d).

B.3. Consider the family of seminorms on $C^0(\mathbb{R})$ defined by

$$p_k(f) = \sup\{ |f(x)| \mid x \in [-k, k] \} \,, \quad \text{for } k \in \mathbb{N} \,,$$

and the metric defined by this as in Theorem B.9.
   Define

$$f(x) = \max\{0, 1 - |x|\} \,, \quad g(x) = 100 f(x - 2) \,, \quad h(x) = \tfrac{1}{2}(f(x) + g(x)) \,,$$

and show that

$$d(f, 0) = \frac{1}{2} \,, \quad d(g, 0) = \frac{50}{101} \,, \quad d(h, 0) = \frac{1}{6} + \frac{50}{102} \,.$$

Hence the ball $B(0, \tfrac{1}{2})$ is not convex. Is $B(0, r)$ convex for any $r < 1$?

B.4. Let X be a topological vector space.
(a) Show Lemma B.3 1°.
(b) Show that X is a Hausdorff space, i.e., for $x \neq y$ there exist neighborhoods V_1 of x and V_2 of y so that $V_1 \cap V_2 = \emptyset$.

B.5. Let X be a topological vector space.
(a) Show that $E \subset X$ is bounded if and only if, for any neighborhood V of 0 there exists a $t > 0$ so that $E \subset sV$ for $s \geq t$.
(b) Show that the only bounded subspace of X is $\{0\}$.

B.6. Consider $\mathbb{R}^2$, provided with the usual topology, and let $A \subset \mathbb{R}^2$, $B \subset \mathbb{R}^2$.
(a) Show that $2A \subset A + A$, and find an example where $2A \neq A + A$.
(b) Show that if A is closed and B is compact, then $A + B$ is closed.
(c) Show that $\overline{A} + \overline{B} \subset \overline{A + B}$ in general, and find an example where $\overline{A} + \overline{B} \neq \overline{A + B}$.

B.7.   Show Lemma B.7.

B.8.   Show that when X and Y are topological vector spaces, then the product space $X \times Y$ is a topological vector space.

B.9.   Give proofs of Theorems B.13–B.16.

The following exercises refer to the notation of Chapter 2.

B.10.   Show that $C^\infty([a,b])$ and $C_K^\infty(\Omega)$ are Fréchet spaces.

B.11.   Show that D^α is a continuous operator in $C^\infty(\Omega)$ and in $C_K^\infty(\Omega)$. Let $f \in C^\infty(\Omega)$ and show that the operator $M_f : u \mapsto f \cdot u$ is continuous in $C^\infty(\Omega)$ and in $C_K^\infty(\Omega)$.

B.12.   Let $f \in L_2(\Omega)$ and show that the functional

$$\Lambda : u \mapsto \int_\Omega u\, f\, dx$$

is continuous on $C_K^\infty(\Omega)$ for every compact $K \subset \Omega$.

B.13.   Show that $C_K^\infty(\Omega)$ is a closed subspace of $C^\infty(\Omega)$.

B.14.   Let A be a convex subset of a topological vector space E. Let x_0 be an interior point of A and x a point in the closure $\overline{A}$ of A. Show that all points $u \neq x$ on the segment

$$[x_0, x] = \{ \lambda x_0 + (1-\lambda)x \mid \lambda \in [0,1] \}$$

are interior in A. (*Hint.* Begin with the case where $x \in A$.)
Use this to show that when $A \subset E, A$ is convex, then

$$\overset{\circ}{A} \neq \emptyset \Rightarrow \overline{\overset{\circ}{A}} = \overline{A} \text{ and } \overset{\circ}{\overline{A}} = \overset{\circ}{A}.$$

Show by an example that $\overset{\circ}{A} \neq \emptyset$ is a necessary condition.

B.15.   Let E be a locally convex topological vector space and let $M \subset E$ be a subspace. Let U be a convex balanced neighborhood of 0 in M (with the topology induced from E).

(a) Show that there exists a convex balanced neighborhood V of 0 in E so that $V \cap M = U$.

(b) Assume that $x_0 \in E \setminus \overline{M}$. Show that V in (a) may be chosen so that $x_0 \notin V$.

(c) Let p_0 be a continuous seminorm on M. Show that there exists a continuous seminorm p on E so that $p|_M = p_0$.

B.16.   The inductive limit topology. Let X be a vector space and $E_1 \subsetneq E_2 \subsetneq E_3 \subsetneq \cdots \subsetneq E_j \subsetneq \ldots$ a strictly increasing sequence of subspaces of X so that $X = \bigcup_{j=1}^{\infty} E_j$. Assume that for all $j \in \mathbb{N}$, (E_j, τ_j) is a locally convex topological vector space; and assume that for $j < k$, the topology induced by τ_k on E_j is the same as τ_j (as in Theorem B.17).

(a) Define $\mathcal{P}$ by

$$\mathcal{P} = \{ \text{ seminorms } p \text{ on } X \mid p|_{E_j} \text{ is } \tau_j\text{-continuous } \forall j \in \mathbb{N} \}.$$

Show that $\mathcal{P}$ is separating. (One can use Exercise B.15 (c).)

(b) Thus $\mathcal{P}$ defines a locally convex vector space topology τ on X. Show that the topology induced by τ on E_j equals τ_j. (Cf. Exercise B.15 (a); show that τ has properties as in Theorem B.17.)

(c) Assume that Λ is a linear map of X into a locally convex topological vector space Y. Show that Λ is τ-continuous if and only if $\Lambda|_{E_j}$ is continuous $E_j \to Y$ for every $j \in \mathbb{N}$.

(d) Show that a set $B \subset X$ is bounded (w.r.t. τ) if and only if $\exists j \in \mathbb{N} : B \subset \overline{E}_j$ and B is bounded in $\overline{E}_j$. (Cf. Exercise B.15 (b).)

(e) The example: $X = C_0^0(\mathbb{R})$ (cf. (C.8)). Take $E_j = C_{[-j,j]}^0(\mathbb{R})$, with the sup-norm topology. Show that the topology τ on $C_0^0(\mathbb{R})$ determined from τ_j as in (b) is strictly finer than the sup-norm topology on $C_0^0(\mathbb{R})$.

B.17.   Show that when X is as in Theorem B.9, then

$$d_1(x,y) = \sum_{k=1}^{\infty} \min\{2^{-k}, p_k(x - y)\}$$

is likewise an invariant metric defining the topology.

Appendix C
Function spaces on sets with smooth boundary

We here define some additional function spaces, supplying those introduced in Chapter 2. The new types of spaces are associated with sets Ω where the boundary has a certain smoothness.

Definition C.1. An open set $\Omega \subset \mathbb{R}^n$ is said to be (of class) C^m for some $m \leq \infty$ when every boundary point $x \in \partial\Omega$ has an open neighborhood U with an associated C^m-diffeomorphism κ (a bijective C^m mapping with C^m inverse) such that κ maps U onto the unit ball $B(0,1) \subset \mathbb{R}^n$, and

$$
\begin{aligned}
\kappa(x) &= 0, \\
\kappa(U \cap \Omega) &= B(0,1) \cap \mathbb{R}^n_+, \\
\kappa(U \cap \partial\Omega) &= B(0,1) \cap \mathbb{R}^{n-1}.
\end{aligned}
\tag{C.1}
$$

We then also say that $\overline{\Omega}$ is C^m (in detail: closed and C^m).

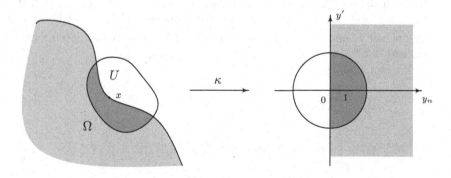

In the case $m = \infty$ one often says instead that Ω is *smooth*. In some texts, the above definition is taken to mean that $\partial\Omega$ is C^m. But in fact the boundary of Ω can be a C^m manifold also when Definition C.1 is not verified — think for example of the set $\Omega = \{x \in \mathbb{R}^n \mid |x| \neq 1\}$. The definition assures that Ω is locally "on one side of $\partial\Omega$"; this property involves Ω as well as $\partial\Omega$.

Examples: $\mathbb{R}^n_+$ is smooth. Balls in $\mathbb{R}^n$ are smooth. The set

$$\{x \in \mathbb{R}^2 \mid x_2 > 0,\ x_1 < \sqrt{x_2}\,\} \tag{C.2}$$

is C^1 but not C^2.

For a function u on $\overline{\mathbb{R}}^n_+$ it makes good sense to speak of the partial derivatives on $\overline{\mathbb{R}}^n_+$, and they exist up to order m when the partial derivatives of u on $\mathbb{R}^n_+$ up to order m extend to continuous functions on $\overline{\mathbb{R}}^n_+$. This carries over to smooth sets Ω, since one can define the derivatives on sets of the form $U \cap \overline{\Omega}$ in Definition C.1 by use of κ and the chain rule. [More precisely: κ maps x over to $y = (\kappa_1(x), \ldots, \kappa_n(x))$, so if we write $u(x) = u(\kappa^{-1}(y)) = \tilde{u}(y)$, we have that

$$(\partial_{x_j} u)(x) = \sum_{l=1}^{n} \frac{\partial \kappa_l(x)}{\partial x_j} \partial_{y_l} \tilde{u}(y)|_{y=\kappa(x)}\ . \tag{C.3}$$

In other words, the differential operator ∂_{x_j} on $U \cap \overline{\Omega}$ corresponds to the differential operator $\sum_{l=1}^{n} a_{jl} \partial_{y_l}$ on $B(0,1) \cap \overline{\mathbb{R}}^n_+$ with C^∞ coefficients $a_{jl}(y) = (\partial_{x_j} \kappa_l)(\kappa^{-1}(y))$.]

Regardless of the smoothness of Ω we denote by $C^k(\overline{\Omega})$ the vector space of functions on $\overline{\Omega}$ such that the partial derivatives up to order m defined on Ω are uniformly continuous on bounded subsets of Ω (hence extend to continuous functions on $\overline{\Omega}$). When $\overline{\Omega}$ is compact, this is a Banach space with the norm

$$\|u\|_{C^k(\overline{\Omega})} = \sup\{\,|\partial^\alpha u(x)| \mid x \in \overline{\Omega},\ |\alpha| \le k\,\}\ . \tag{C.4}$$

If $\overline{\Omega}$ is not assumed to be compact, we can define a Fréchet topology on $C^k(\overline{\Omega})$ by the family of seminorms (2.9) where k is fixed and K_j for example runs through the increasing sequence of compact sets $K_j = \overline{\Omega} \cap \underline{B}(0,j)$ satisfying $\bigcup_{j \in \mathbb{N}} K_j = \overline{\Omega}$. (In the compact case, the sequence $(K_j)_{j \in \mathbb{N}}$ can be replaced by one set $\overline{\Omega}$ and the seminorm is a norm.) For $k = 0$, one often writes C instead of C^0.

We can also define

$$C^\infty(\overline{\Omega}) = \bigcap_{k=0}^{\infty} C^k(\overline{\Omega})\,,$$

$$C^k_K(\overline{\Omega}) = \{\,u \in C^k(\overline{\Omega}) \mid \operatorname{supp} u \subset K\,\}, \quad C^\infty_K(\overline{\Omega}) = \bigcap_{k=0}^{\infty} C^k_K(\overline{\Omega})\,,$$

$$C^k_{(0)}(\overline{\Omega}) = \{\,u \in C^k(\overline{\Omega}) \mid \operatorname{supp} u \text{ compact } \subset \overline{\Omega}\,\}, \tag{C.5}$$

$$C^\infty_{(0)}(\overline{\Omega}) = \bigcap_{k=0}^{\infty} C^k_{(0)}(\overline{\Omega})\ .$$

The first space in this list is a Fréchet space with the topology determined by the family of seminorms (2.9) (k runs through all integers ≥ 0, the sequence (K_j) may be replaced by one set $\overline{\Omega}$ if this is compact). The space $C_K^k(\overline{\Omega})$ is a Banach space with norm $\sup\{\,|\partial^\alpha u(x)|\mid |\alpha| \leq k,\, x \in K\,\}$, while $C_K^\infty(\overline{\Omega})$ is a Fréchet space. Finally, $C_{(0)}^k(\overline{\Omega})$ equals the Banach space $C^k(\overline{\Omega})$ when $\overline{\Omega}$ is compact; otherwise it is an $\mathcal{LF}$-space (considered as $\bigcup_{K_j \subset \overline{\Omega}} C_{K_j}^k(\overline{\Omega})$ where $\bigcup_j K_j = \overline{\Omega}$ as above); $C_{(0)}^\infty(\overline{\Omega})$ is likewise $\mathcal{LF}$ when $\overline{\Omega}$ is not compact. (We use the index (0) to avoid confusion with $C_0^\infty(\Omega)$.)

One can show that

$$C_{(0)}^\infty(\overline{\mathbb{R}}_+^n) = \big\{\, u|_{\overline{\mathbb{R}}_+^n} \mid u \in C_0^\infty(\mathbb{R}^n) \,\big\}, \tag{C.6}$$

and that when Ω is smooth,

$$C_{(0)}^\infty(\overline{\Omega}) = \big\{\, u|_{\overline{\Omega}} \mid u \in C_0^\infty(\mathbb{R}^n) \,\big\}. \tag{C.7}$$

The proof of (C.6) is found in Seeley's paper [S64], and then (C.7) is deduced from this by use of diffeomorphisms as in Definition C.1. (When Ω is smooth and bounded, the localization arguments are similar to those used in Theorems 4.10 and 4.12.)

For an open set Ω and a compact subset K we define $C_K^k(\Omega)$ and $C_0^k(\Omega)$ by

$$\begin{aligned}
C_K^k(\Omega) &= \{\, u \in C^k(\Omega) \mid \operatorname{supp} u \subset K \,\}, \\
C_0^k(\Omega) &= \{\, u \in C^k(\Omega) \mid \operatorname{supp} u \text{ compact } \subset \Omega \,\},
\end{aligned} \tag{C.8}$$

the former is a Banach space (like $C_K^k(\overline{\Omega})$) and the latter is an $\mathcal{LF}$-space, namely, $\bigcup_{K_j \subset \Omega} C_{K_j}^k(\Omega)$.

Finally, we mention that one may need a notation for the following spaces (where M is open, or closed and C^k resp. C^∞):

$$\begin{aligned}
C_{L_p}^k(M) &= \{\, u \in C^k(M) \mid \partial^\alpha u \in L_p(M) \text{ for } |\alpha| \leq k \,\}, \\
C_{L_p}^\infty(M) &= \bigcap_{k=0}^\infty C_{L_p}^k(M).
\end{aligned} \tag{C.9}$$

They coincide with $C^k(M)$ resp. $C^\infty(M)$ when M is compact. The spaces $C_{L_p}^k(M)$ ($1 \leq p \leq \infty$) may be provided with norms

$$\Big(\sum_{|\alpha| \leq k} \|\partial^\alpha u\|_{L_p(M)}^p \Big)^{\frac{1}{p}} \text{ for } p < \infty, \quad \sup_{|\alpha| \leq k} \|\partial^\alpha u\|_{L_\infty(M)} \text{ for } p = \infty, \tag{C.10}$$

which for $p = 2$ take the form

$$\|u\|_k = \left(\sum_{|\alpha| \le k} \|\partial^\alpha u\|_{L_2(M)}^2 \right)^{\frac{1}{2}}. \tag{C.11}$$

This makes them normed but in general not complete spaces. Similarly, $C^\infty_{L_p}(M)$ may be topologized by a system of seminorms.

A special case is $C^k_{L_\infty}(\mathbb{R}^n)$; it is the space of C^k-functions on $\mathbb{R}^n$ with bounded derivatives up to order k. It is well-known that this is a Banach space with the norm

$$\|u\|_{C^k_{L_\infty}(\mathbb{R}^n)} = \sup\{\, |\partial^\alpha u(x)| \mid x \in \mathbb{R}^n, |\alpha| \le k \,\}. \tag{C.12}$$

Exercises for Appendix C

C.1. Let $M = \underline{B}(0,1)$ in $\mathbb{R}^n$. Show that $C^1(M)$ is complete with respect to the C^1-norm (cf. (C.4)), but not complete with respect to the norm (C.11) with $k = 1$.

Bibliography

[A05]   H. Abels, *Pseudodifferential boundary value problems with non-smooth coefficients*, Commun. Part. Diff. Eq. **30** (2005), 1463–1503.

[ADN64] S. Agmon, A. Douglis, and L. Nirenberg, *Estimates near the boundary for solutions of elliptic partial differential equations satisfying general boundary conditions, II*, Commun. Pure Appl. Math. **17** (1964), 35–92.

[A65]   S. Agmon, *Lectures on Elliptic Boundary Value Problems*, Van Nostrand Math. Studies, Van Nostrand, Princeton, NJ, 1965.

[AB64] M. F. Atiyah and R. Bott, *The index theorem for manifolds with boundary*, Proc. Bombay Symp. on Differential Analysis (Oxford), 1964, pp. 175–186.

[AS68] M. F. Atiyah and I. M. Singer, *The index of elliptic operators*, Ann. Math. **87** (1968), 484–530.

[ABP73] M. F. Atiyah, R. Bott, and V. K. Patodi, *On the heat equation and the index theorem*, Invent. Math. **19** (1973), 279–330.

[APS75] M. F. Atiyah, V. K. Patodi, and I. M. Singer, *Spectral asymmetry and Riemanninan geometry I*, Math. Proc. Camb. Phil. Soc. **77** (1975), 43–69.

[AE99] G. Avramidi and G. Esposito, *Gauge theories on manifolds with boundary*, Commun. Math. Phys. **200** (1999), 495–543.

[B56]   M. S. Birman, *On the theory of self-adjoint extensions of positive definite operators*, Mat. Sb. **38**: 90 (1956), 431–450, Russian.

[BW93] B. Booss-Bavnbek and K. P. Wojciechowski, *Elliptic Boundary Problems for Dirac Operators*, Birkhäuser Boston, Inc., Cambridge, MA, 1993.

[B66]   L. Boutet de Monvel, *Comportement d'un opérateur pseudo-différentiel sur une variété à bord, I-II*, J. Anal. Math. **17** (1966), 241–304.

[B71]   L. Boutet de Monvel, *Boundary problems for pseudo-differential operators*, Acta Math. **126** (1971), 11–51.

[BGW08] B. M. Brown, G. Grubb, and I. G. Wood, *M-functions for closed extensions of adjoint pairs of operators with applications to elliptic boundary problems*, arXiv:0803.3630 (to appear).

[BMNW08] B.M. Brown, M. Marletta, S. Naboko, and I.G. Wood, *Boundary triplets and M-functions for non-selfadjoint operators, with applications to elliptic PDEs and block operator matrices*, J. London Math. Soc. **77** (2008), 700–718.

[C63]   A. P. Calderón, *Boundary value problems for elliptic equations*, Proc. Joint Soviet-American Symp. on PDE, Novosibirsk, Acad. Sci. USSR Siberian Branch, 1963, pp. 303–304.

[C40]   J. W. Calkin, *Symmetric transformations in Hilbert space*, Duke Math. J. **7** (1940), 504–508.

[C90]   J. B. Conway, *A Course in Functional Analysis, Second Edition*, Springer-Verlag, New York, 1990.

[CH62]   R. Courant and D. Hilbert, *Methods of Mathematical Physics II*, Interscience Publishers, New York, 1962.

[DM91]   V. A. Derkach and M. M. Malamud, *Generalized resolvents and the boundary value problems for Hermitian operators with gaps*, J. Funct. Anal. **95** (1991), 1–95.

[DS58]   N. Dunford and J. Schwartz, *Linear Operators, Part I: General Theory*, Interscience Publishers, New York, 1958.

[EE87]   D. E. Edmunds and W. D. Evans, *Spectral Theory and Differential Operators*, Clarendon Press, Oxford, 1987.

[E81]   G. I. Eskin, *Boundary Value Problems for Elliptic Pseudodifferential Equations*, Transl. Math. Monogr. *52*, Amer. Math. Soc., Providence, RI, 1981.

[E98]   L. C. Evans, *Partial Differential Equations*, Amer. Math. Soc., Providence, RI, 1998.

[EM99]   W. N. Everitt and L. Markus, *Boundary Value Problems and Symplectic Algebra for Ordinary Differential and Quasi-differential Operators*, Amer. Math. Soc., Providence, RI, 1999.

[FGLS96] B. V. Fedosov, F. Golse, E. Leichtnam, and E. Schrohe, *The noncommutative residue for manifolds with boundary*, J. Funct. Anal. **142** (1996), 1–31.

[F69]   A. Friedman, *Partial Differential Equations*, Holt, Rinehart and Winston, New York, 1969.

[F34]   K. Friedrichs, *Spektraltheorie halbbeschränkter Operatoren und Anwendung auf die Spektralzerlegung von Differentialoperatoren*, Math. Ann. **109** (1934), 465–487.

[F68]   K. Friedrichs, *Pseudo-Differential Operators, an Introduction*, Courant Inst. Math. Sci., New York, 1968.

[FS70]   D. Fujiwara and N. Shimakura, *Sur les problèmes aux limites stablement variationnels*, J. Math. Pures Appl. **49** (1970), 1–28.

[G53] L. Gårding, *Dirichlet's problem for linear elliptic partial differential equations*, Math. Scand. **1** (1953), 55–72.

[GT77] D. Gilbarg and N. S. Trudinger, *Elliptic Partial Differential Equations of Second Order, Grundlehren Math. Wiss. vol. 224,* Springer-Verlag, Berlin, 1977.

[Gi74] P. B. Gilkey, *The Index Theorem and the Heat Equation,* Publish or Perish Press, Boston, MA, 1974.

[Gi85] P. B. Gilkey, *Invariance Theory, the Heat Equation, and the Atiyah-Singer Index Theorem,* Publish or Perish Press, Boston, MA, 1985, Second edition 1994, CRC Press.

[G60] I. C. Gohberg, *On the theory of multidimensional singular integral equations,* Dokl. Akad. Nauk SSSR **133** (1960), 1279–1282.

[GG91] V. I. Gorbachuk and M. L. Gorbachuk, *Boundary Value Problems for Operator Differential Equations,* Kluwer, Dordrecht, 1991.

[GH90] G. Grubb and L. Hörmander, *The transmission property,* Math. Scand. **67** (1990), 273–289.

[G68] G. Grubb, *A characterization of the non-local boundary value problems associated with an elliptic operator,* Ann. Scuola Norm. Sup. Pisa **22** (1968), 425–513.

[G70] G. Grubb, *Les problèmes aux limites généraux d'un opérateur elliptique, provenant de la théorie variationnelle,* Bull. Sc. Math. **94** (1970), 113–157.

[G71] G. Grubb, *On coerciveness and semiboundedness of general boundary problems,* Israel J. Math. **10** (1971), 32–95.

[G73] G. Grubb, *Weakly semibounded boundary problems and sesquilinear forms,* Ann. Inst. Fourier Grenoble **23** (1973), 145–194.

[G74] G. Grubb, *Properties of normal boundary problems for elliptic even-order systems,* Ann. Scuola Norm. Sup. Pisa Ser. IV **1** (1974), 1–61.

[G77] G. Grubb, *Boundary problems for systems of partial differential operators of mixed order,* J. Funct. Anal. **26** (1977), 131–165.

[G84] G. Grubb, *Singular Green operators and their spectral asymptotics,* Duke Math. J. **51** (1984), 477–528.

[G90] G. Grubb, *Pseudo-differential boundary problems in L_p spaces,* Commun. Part. Diff. Eq. **15** (1990), 289–340.

[G91] G. Grubb, *Parabolic pseudo-differential boundary problems and applications, Lecture Notes Math. vol. 1495,* "Microlocal Analysis and Applications", Montecatini Terme 1989, Springer-Verlag (eds. L. Cattabriga and L. Rodino), 1991, pp. 46–117.

[G95] G. Grubb, *Parameter-elliptic and parabolic pseudodifferential boundary problems in global L_p Sobolev spaces,* Math. Z. **218** (1995), 43–90.

[G96] G. Grubb, *Functional Calculus of Pseudodifferential Bound-ary Problems*, Prog. Math. vol. *65*, Second Edition, Birkhäuser, Boston, 1996, first edition issued 1986.

[G99] G. Grubb, *Trace expansions for pseudodifferential boundary problems for Dirac-type operators and more general systems*, Ark. Mat. **37** (1999), 45–86.

[G03] G. Grubb, *Spectral boundary conditions for generalizations of Laplace and Dirac operators*, Commun. Math. Phys. **242** (2003), 243–280.

[G08] G. Grubb, *The local and global parts of the basic zeta coef-ficient for pseudodifferential boundary operators*, Math. Ann. **341** (2008), 735–788.

[G57] N. M. Günther, *Die Potentialtheorie und ihre Anwendungen auf Grundaufgaben der mathematischen Physik*, Teubner, Leipzig, 1957.

[HP57] E. Hille and R. S. Phillips, *Functional Analysis and Semigroups*, vol. 31, Amer. Math. Soc. Colloq. Publ., 1957.

[H63] L. Hörmander, *Linear Partial Differential Operators, Grund-lehren Math. Wiss. vol. 116*, Springer-Verlag, Berlin, 1963.

[H65] L. Hörmander, *Pseudo-differential operators*, Commun. Pure Appl. Math. **13** (1965), 501–517.

[H66] L. Hörmander, *Pseudo-differential operators and non-elliptic boundary problems*, Ann. Math. **83** (1966), 129–209.

[H67] L. Hörmander, *Pseudo-differential operators and hypoelliptic equations*, Proc. Symp. Pure Math., vol. 10, 1967, pp. 138–183.

[H71] L. Hörmander, *Fourier integral operators I*, Acta Mat. **127** (1971), 79–183.

[H83] L. Hörmander, *The Analysis of Linear Partial Differen-tial Operators I, Distribution Theory and Fourier Analysis, Grundlehren Math. Wiss. vol. 256*, Springer-Verlag, Berlin, 1983.

[H85] L. Hörmander, *The Analysis of Linear Partial Differential Op-erators III, Pseudo-differential Operators, Grundlehren Math. Wiss. vol. 274*, Springer-Verlag, Berlin, 1985.

[H89] L. Hörmander, *Linear Functional Analysis*, Matematiska Insti-tutionen, Lunds Universitet, Lund, 1989.

[K66] T. Kato, *Perturbation Theory for Linear Operators, Grundleh-ren Math. Wiss. vol. 132*, Springer-Verlag, Berlin, 1966.

[K75] A. N. Kočubeĭ, *Extensions of symmetric operators and sym-metric binary relations*, Math. Notes **17** (1975), 25–28.

[KN65] J. J. Kohn and L. Nirenberg, *An algebra of pseudo-differential operators*, Commun. Pure Appl. Math. **18** (1965), 269–305.

[KV95] M. Kontsevich and S. Vishik, *Geometry of determinants of el-liptic operators*, Functional Analysis on the Eve of the 21'st

Century (Rutgers Conference in honor of I. M. Gelfand 1993), Vol. I, Birkhäuser (eds. S. Gindikin et al.), 1995, pp. 173–197.

[K47]   M. G. Kreĭn, *Theory of self-adjoint extensions of symmetric semi-bounded operators and applications I*, Mat. Sb. **20**: 62 (1947), 431–495, Russian.

[LSU68] O. A. Ladyzhenskaya, V. A. Solonnikov, and N. N. Uraltseva, *Linear and Quasilinear Equations of Parabolic Type*, Amer. Math. Soc., Providence, RI, 1968.

[LM54] P. D. Lax and N. Milgram, *Parabolic equations*, Ann. Math. Studies, Princeton **33** (1954), 167–190.

[LP67] P. D. Lax and R. S. Phillips, *Scattering Theory*, Academic Press, New York, 1967.

[LM63] J.-L. Lions and E. Magenes, *Problèmes aux limites non homogènes VI*, J. Anal. Math. **11** (1963), 165–188.

[LM68] J.-L. Lions and E. Magenes, *Problèmes aux limites non homogènes et applications. Vol. 1 et 2*, Editions Dunod, Paris, 1968, translated to English as *Grundlehren Math. Wiss. vol. 181–182*, Springer-Verlag, Berlin, 1972.

[L63]   J.-L. Lions, *Équations différentielles opérationelles et problèmes aux limites*, Grundlehren Math. Wiss. vol. *111*, Springer-Verlag, Berlin, 1963.

[LS83] V. E. Lyantze and O. G. Storozh, *Methods of the Theory of Unbounded Operators*, Naukova Dumka, Kiev, 1983, Russian.

[MM99] M. M. Malamud and V. I. Mogilevskii, *On Weyl functions and Q-function of dual pairs of linear relations*, Dopov. Nat. Akad. Nauk Ukr. **4** (1999), 32–37.

[MM02] M. M. Malamud and V. I. Mogilevskii, *Kreĭn type formula for canonical resolvents of dual pairs of linear relations*, Methods Funct. Anal. Topol. **8**: 4 (2002), 72–100.

[M93]   R. B. Melrose, *The Atiyah-Patodi-Singer Index Theorem*, A. K. Peters, Wellesley, MA, 1993.

[N68]   M. A. Naĭmark, *Linear Differential Operators. Part II: Linear Differential Operators in Hilbert Space*, Frederick Ungar Publishing Co., New York, 1968.

[N29]   J. v. Neumann, *Allgemeine Eigenwerttheorie Hermitescher Funktionaloperatoren*, Math. Ann. **102** (1929), 49–131.

[N55]   L. Nirenberg, *Remarks on strongly elliptic partial differential equations*, Commun. Pure Appl. Math. **8** (1955), 649–675.

[P89]   G. K. Pedersen, *Analysis Now*, Springer-Verlag, Berlin, 1989.

[P59]   R. S. Phillips, *Dissipative operators and hyperbolic systems of partial differential equations*, Trans. Amer. Math. Soc. **90** (1959), 193–254.

[RS82] S. Rempel and B.-W. Schulze, *Index Theory of Elliptic Boundary Problems*, Akademie-Verlag, Berlin, 1982.

⟩

[RN53] F. Riesz and B. Sz.-Nagy, *Leçons d'analyse fonctionnelle*, Akademiai Kradò, Budapest, 1953.

[R74] W. Rudin, *Functional Analysis*, Tata McGraw Hill, New Delhi, 1974.

[S91] X. Saint Raymond, *Elementary Introduction to the Theory of Pseudodifferential Operators*, CRC Press, Boca Raton, FL, 1991.

[S02] M. Schechter, *Principles of Functional Analysis*, Amer. Math. Soc., Providence, RI, 2002.

[S98] B.-W. Schulze, *Boundary Value Problems and Singular Pseudo-differential Operators*, Wiley, New York, 1998.

[S50] L. Schwartz, *Théorie des distributions I–II*, Hermann, Paris, 1950–51.

[S61] L. Schwartz, *Méthodes mathématiques pour les sciences physiques*, Hermann, Paris, 1961.

[S64] R. T. Seeley, *Extension of C^∞ functions*, Proc. Amer. Math. Soc. **15** (1964), 625–626.

[S66] R. T. Seeley, *Singular integrals and boundary value problems*, Amer. J. Math. **88** (1966), 781–809.

[S69] R. T. Seeley, *Topics in pseudo-differential operators*, CIME Conference on Pseudo-differential Operators (1969), 169–305.

[T81] M. Taylor, *Pseudodifferential Operators*, Princeton University Press, Princeton, NJ, 1981.

[T67] F. Treves, *Topological Vector Spaces, Distributions and Kernels*, Academic Press, New York, 1967.

[T75] F. Treves, *Basic Linear Partial Differential Equations*, Academic Press, New York, 1975.

[T80] F. Treves, *Introduction to Pseudodifferential and Fourier Integral Operators I–II*, Plenum Press, New York, 1980.

[V80] L. I. Vainerman, *On extensions of closed operators in Hilbert space*, Math. Notes **28** (1980), 871–875.

[VE67] M. I. Vishik and G. I. Eskin, *Elliptic equations in convolution in a bounded domain and their applications*, Usp. Mat. Nauk **22** (1967), 15–76, translated in Russ. Math. Surv. **22** (1967), 13–75.

[V52] M. I. Vishik, *On general boundary value problems for elliptic differential equations*, Tr. Moskv. Mat. Obšv. **1** (1952), 187–246, translated in AMS Transl. **24** (1963), 107–172.

[W84] M. Wodzicki, *Local invariants of spectral asymmetry*, Invent. Math. **75** (1984), 143–176.

[Y68] K. Yoshida, *Functional Analysis, Grundlehren Math. Wiss. vol. 123, 2nd Edition*, Springer-Verlag, Berlin, 1968.

Index

Graduate Texts in Mathematics

(*continued from page ii*)